Programming with Class

Programming with Class

A Practical Introduction to Object-Oriented Programming with C++

N. A. B. Gray
University of Wollongong, Australia

JOHN WILEY & SONS
Chichester · New York · Brisbane · Toronto · Singapore

Other Wiley Editorial Offices

John Wiley & Sons, Inc., 605 Third Avenue,
New York, NY 10158-0012, USA

Jacaranda Wiley Ltd, 33 Park Road, Milton,
Queensland 4064, Australia

John Wiley & Sons (Canada) Ltd, 22 Worcester Road,
Rexdale, Ontario M9W 1L1, Canada

John Wiley & Sons (SEA) Pte Ltd, 37 Jalan Pemimpin #05-04,
Block B, Union Industrial Building, Singapore 2057

British Library Cataloguing in Publication Data

A catalogue record for this book is available from the British Library

ISBN 0 471 94350 9 (pbk.);   0 471 94358 4 (disk)

Produced from camera-ready copy supplied by the author using
Microsoft Word.
Printed and bound in Great Britain by Bath Press Ltd, Bath, Avon

Contents

About this Book

This book has evolved from materials used in an undergraduate course on Object-oriented programming that has, in various forms, been taught since 1988. This course is intended for final year undergraduate students with at least one year's experience with the C programming language on a Unix system. Although intended mainly for students undertaking a structured course, the materials in this book should help those studying on their own.

The course has a strongly practical flavour. The earliest variant on this course was a practical programming elective on developing interactive applications for the Macintosh. Students first learnt to develop Macintosh applications using the procedural MPW-Pascal language and the Macintosh Toolbox, then they were shown the power of code reuse that came with Object Pascal and the MacApp class library. The programming language used in the current course is C++. This is the OO language that graduates are most likely to work with in their future employment. There is the further advantage that libraries of reusable classes, written in C++, are readily available for most computer systems. Several C++ compilers and class libraries are available in low cost student editions, making it practical for students to work on their own personal computers. The original emphasis on practical application has been retained. Object-oriented programming is presented as an enabling technology that allows complex programs to be constructed through the exploitation of reusable components taken from a library.

A practical focus

This book has four sections. The three chapters in section 1 introduce Object-oriented programming, providing some historical perspectives on the development of this style and a general explanation of how languages like C++ and Object Pascal actually work.

The four sections of this text,
1) Introduction

The second section focuses on C++. This section is not intended to be a comprehensive guide to C++. It is assumed that readers will already know C and so no consideration is given to those language constructs that are common to both C and C++. The three chapters in this section cover C++'s minor extensions to C, programmer defined types, and simple class hierarchies based on single inheritance. Examples used as illustrations include standard "abstract data types", such as linked lists, and simulations – like a "Space Invaders" game and a simulation of jobs in a computer's operating system.

2) Basic C++

3) Class libraries: exploiting reusable code

The aim of the third section is to enable students to build complex applications by combining and extending reusable components taken from a framework class library. The first chapter in this section gives a brief introduction to class browsers and other development tools, and provides some pointers to currently available libraries of reusable code. The next chapter illustrates analysis and design of a simple interactive application that can be built using a class library. The third chapter of this section covers some of the more advanced features of class libraries such as support for persistent objects. Finally, a design is presented for a specialized graphics-editor program that can be built with the frameworks.

4) Intermediate concepts

The final section covers a few more sophisticated topics. It is certainly not a treatment of advanced level ideas in Object-oriented programming – hence the section title "Intermediate OOP". Chapters in this section cover ideas on analysis and design techniques for developing OO software, some additional features of the C++ language, and a teasing nibble at the programming language Eiffel.

Normally, students taking this course are assessed solely on submitted practical assignments. The first assignment is an exercise involving the definition of an "abstract data type" (adt) and the implementation of some practical realization of this adt. Another early assignment would typically involve some form of simulation problem, because the necessary objects (and their classes) are usually easy to identify and characterize. Other small assignments provide experience with framework class libraries as students implement programs that are variations on standard examples. Students select to work in one environment – Macintosh, PC, or Unix – and utilize the frameworks and tools appropriate to that environment.

The major portion of the marks for the course is accorded to a project. For this project, students must identify their own application and justify why it will prove a good vehicle for demonstrating their understanding of OO concepts. Students provide analysis documents that illustrate how objects, and their initial classes, were identified for the chosen application. Design documents must be submitted that show how these classes were elaborated to develop a class hierarchy that integrates with an application framework. Finally, students must implement their applications. Examples of successful projects include an editor for musical scores, a graphics editor inspired by the commercially available KidzPaint™ program, and a Petri-net editor/executor.

Example code

Many example programs are discussed in the text. However, the complete code for these programs is not included. Most of the code of any program is uninteresting; only a few details will be needed to illustrate a particular concept being presented in the text. When complete programs are presented in textbooks, the interesting bits are often lost in pages of program listing. Of course, the complete programs have to be available to those who want to explore and extend the examples. The full text of programs, and accompanying make files and any "resource files", are available via "anonymous ftp" through the Internet from ftp.cs.uow.edu.au (/pub/oop directory). The files available on Internet include examples from Chapters 5, 6, and 12; these examples were implemented using ATT's C++ compiler on Unix. Implementations of the example programs from Chapters 8 and 10, along with additional examples, are also available. Several versions are provided for different common class libraries. The implementations include code for C++/MacApp on the Macintosh, Borland's OWL for PCs, ET++ (a public domain class library available from several sites on the Internet) on Unix, along with some others (e.g. the Symantec class libraries).

Introduction to Object-oriented Programming

Introduction to Object-oriented Programming

The three chapters in this section introduce the ideas behind object-oriented programming (OOP).

The first chapter is a justification for OOP. Why should programmers look at something other than the procedural style programs that have formed the main focus of their studies? The arguments used to justify OOP are derived from the presentations of Cox, Meyer, and other authors. They are based primarily on software engineering concerns. OO programming styles are seen as the best way of constructing reliable large scale software systems that are capable of evolving to meet the changing demands of their users.

Why OOP?

The second chapter is really an overview of the development of the OOP paradigm, starting with Simula and Smalltalk. These languages established OOP styles but, for various reasons, remained "niche" languages – outside the main stream of software development. The potential for using OOP in wider contexts was perceived in the early 1980s, and a variety of "hybrid" languages emerged. These hybrids apply a veneer of object orientation to a substrate of an existing procedural language such as Pascal or C. While new OOP languages continue to appear, it is the hybrids particularly those based on C that attract most current usage. Hybrid languages have the advantage of allowing organisations that have substantial investments in existing software, all written in procedural languages, to make an evolutionary change to an OO style.

A history of OOP

The final chapter in this section provides an overview of how the hybrid OO languages work. It treats both language/compile-time issues (how are objects defined) and run-time issues (what are objects like in a computer's memory, how do they "know" what functions they should execute). The examples used focus mainly on C++ with a few in Object Pascal. and some small fragments of Eiffel

How do OO languages and programs work?

The materials in these chapters should be supplemented with material from the following books:

Reference books

B. Cox, *Object-Oriented Programming: An Evolutionary Approach,* Addison-Wesley, 1986 (or second edition, 1993).

B. Meyer, *Object-oriented Software Construction*, Prentice-Hall,
   1988.
G.E. Peterson (Ed.), *Tutorial: Object-Oriented Computing*, IEEE, (1987).
B. Stroustrup, *The C++ Programming Language 2e*, Addison-Wesley, 1991.
A.L. Winblad, S.D. Edwards, and D.R. King, *Object-Oriented Software*, Addison-Wesley,
   1990.

Why Adopt Object-oriented Programming?

The programming techniques that are now known as "Object-oriented Programming" (OOP) filled a specialized niche for years, but they made little impact on the wider computing community. In the 1980s, this situation changed. Authors like Cox and Meyer argued the case for a wider use of OOP. Their arguments were primarily related to software engineering concerns. They identified problems common to most conventional software projects and presented reasons why OO programming techniques might be expected to ameliorate such problems.

By the end of the 1980s, these arguments had won general acceptance. OO programming has become a part of the main stream of software development. Of course, OOP is not the only way to go. Conventional procedural programming styles (as in C, Pascal, FORTRAN, and other languages), Lisp or other more strictly functional programming approaches, Prolog and more advanced logic-programming methodologies, all have their specific advantages and may prove more apposite for particular projects. But, in general, OOP seems to provide the best methodology for developing the user-oriented, interactive programs that are the typical products of modern software projects.

OOP is not the only way to go – but it is often the best

While each new program will embody some unique algorithm for manipulating specialized data, much of the functionality of a program will be similar to other user-oriented, interactive programs written for the same platform. A program that allows a physician to hunt for tumours in displays of three-dimensional body scan data, a program for graphing trend data that interrelate several variables in some model of the world economy, and a program for displaying nuclear flux in a reactor, must all provide similar options for their users to select data that are to be subject to further transforms, to "scroll" though large displays, and to save data to files. OOP offers a way of exploiting such similarities. A library of components can be created with these components embodying the standard behaviours of programs written for a particular platform. For example, there might be a component that looks after "scrolling" by attending to user actions involving scroll-

Enhanced programmer productivity

bars and then appropriately updating the coordinate offsets of views of the data. Rather than implementing their own version of scrolling behaviours, programmers writing a new application can utilise instances of the "scroller" abstraction provided in such a library. Often, such components can be used "off-the-shelf"; sometimes, they may require a small amount of adaptation. OOP techniques allow the construction of such libraries of reusable components that are open to further specialisation and adaptation. Systematic exploitation of reusable components can greatly enhance the productivity of a programming team.

Programs that can evolve

Another advantage of the OOP style is that it encourages the factorisation of a program into discrete, autonomous components (the "objects") that interact through well defined interfaces. This factorisation helps localise knowledge of details of the original program specification. If it becomes necessary to change the specification, the updates should affect only objects of one particular type. Localisation of change enhances a program's capacity to evolve to meet changing needs.

A more coherent development model

A third advantage of "going object-oriented" is that there tends to be a greater coherence among the various models of a program system. Similar "objects" will be present in the clients' domain model, the analysts' model, the designers' model, and even in the final code. Similarities between problem domain models and program solution models can make more practical the inclusion of a client representative, or of a domain expert, in the software development team. This is usually advantageous for it increases the chance that the clients will receive a program that satisfies their needs.

The increase in productivity through code reuse, programs that are easier to adapt to changed circumstances, and a more coherent model for software development are among the advantages that object orientation can bring. The rest of this chapter expands upon and illustrates these themes.

1.1 PROBLEMS WITH CONVENTIONAL PROGRAMMING STYLES

1.1.1 The low productivity of development programmers

In 1968, it was recognized that there was a "crisis" in software engineering and NATO's scientific branch started sponsoring conferences that focussed on how its condition might be cured. After 25 years the situation is no longer one of "crisis", the condition of software engineering processes is simply chronic.

A missing "components subindustry"?

McIlroy, a participant in the 1968 NATO Software Engineering Conference, suggested one reason why there might be a problem with software development:

> *"My thesis is that the software industry is weakly founded in part because of the absence of a software components subindustry ."*

Software "components"? They sound interesting. But how does work really get done: "*Ah, another program to write. We'll get the standard bits done with the usual components –*":

```
main(argc,arv)
int argc; char **argv;
{

}
```

"and now what?". From this point on, most programs are unique, individual creations. The software development process is more akin to bespoke tailoring than to any form of engineering.

In one of his presentations, Meyer asked an audience of professional programmers whether they had, in the previous year, written code for some variation on a symbol table that could store names and associated data and would provide a lookup function that retrieved the data for a given name. This is a fairly common programming task and a significant proportion of the audience had implemented such code. The programmers were then asked whether it was not essentially the same code as they had implemented the previous year in some slightly different context. Most of the programmers acknowledged that this was in the case. It was also the case that the code was essentially the same as that they had written on many other previous occasions and was pretty close to the code in books on data structures and algorithms.

Reinventing the symbol table

Yet each time a "symbol table" was needed, these professional programmers would start again, defining `structs` and then crafting routines complete with `"if (…) …; else …"` statements, `"switch(…) { case …: }"` statements etc. Their productivity was probably high enough in terms of lines of code per day; but what a waste. The real productivity of those expensive programmers was quite low as so much of their time was spent laboriously recreating things that should have been standard off-the-shelf components.

When analysing a similar problem, Cox came up with an illustrative analogy. He compared programmers building a program by combining `structs` and conditional statements etc., with electrical engineers building something like a digital signal processing system from individual transistors, capacitors, and resistors. He asked why there were no "Software Integrated Circuits" akin to the IC building blocks that engineers would really use.

Cox's Software ICs

Some reusable code is available in subroutine libraries such as those provided by the Numerical Algorithms Group or the Xlib libraries for X-Windows. The Macintosh Toolbox (a collection of ≈400 ROM routines implementing low level operations on Macs) represents another form of reusable code. Subroutine collections similar to these would have been familiar to McIlroy back in 1968 because the first versions of the big FORTRAN libraries for engineering and numerical algorithms had already been published. Such subroutines are reusable (many of those in the numerical and engineering libraries have been getting reused for 20 years or more, without ever requiring changes or extensions). But standard subroutines didn't really correspond to what McIlroy meant by a component.

But subroutine libraries are reusable!
Yeah, but only in a limited way.

A typical subroutine in one of the numerical libraries is defined something like the following example:

```
      SUBROUTINE ES001(ARRAY,N,EIGVALS,EIGVECS)
      DIMENSION ARRAY(100,100),EIGVALS(100),
    *     EIGVECS(100,100)
```

```
C   THIS SUBROUTINE USES JACOBI'S METHOD
C   TO FIND THE EIGENVALUES OF A REAL SYMMETRIC MATRIX
```

There is a very narrow interface between a subroutine and the rest of the program. Only simple data structures are used (normally just one- or two-dimensional arrays). The routine performs solely one well defined mathematical transform on the data given as its input.

A function, but not a component

There is no real sense of a "component". A reusable component would be more like a package of code that could be used to set up a symmetric matrix, read its elements from an input source, pretty-print the entire matrix, and provide a comprehensive set of transforms on the matrix including inversion, calculation of eigenvalues etc. While a numerical algorithms library might actually include subroutines to implement the various different data transforms, they would be disjoint.

Extending Cox's analogy, the subroutines in the libraries might correspond to small scale integrated circuits like a "flip-flop" (four or five transistors, a few resistors and capacitors in a single unit). The small scale integrated circuits provided a single function (e.g. the flip-flop circuit's function was to store one bit of data). They saved some work for the electronics engineer; but with LSI circuits, engineers have moved on to better things with a greater element of design and implementation reuse. What is wanted now is the software equivalent of a modern packaged LSI circuit (e.g. a 16 Megabit memory chip with address decoding, error detection and correction).

Using the symbol table as an example of a software storage structure, one wants a packaged version. The package should allow a running program to create empty symbol tables in properly initialized states, to add data, and then to lookup data. The package might need to define and support something equivalent to that memory chip's capabilities for error detection and correction. For example, if a program tries to store too much data in a symbol table, then some reasonably graceful failure handling mechanism should be invoked automatically.

It is the lack of packaging that leads to problems with the other subroutine libraries such as the Macintosh Toolbox or the Xlib routines. With these libraries, there are large numbers of routines that operate on global data structures. These data structures must be explicitly initialized (e.g. "Every Macintosh program must begin by initialising the Quickdraw globals with `InitGraf((Ptr) &qd.thePort);`"). There is no protection for these global data structures; data may be changed directly, or the routines that are supposed to be used for changing the data may be invoked in inappropriate circumstances.

The problem with development programmers is that they build everything from scratch. This takes time and therefore costs money. In the past, the programmers had no choice, there weren't any prefabricated, packaged components to work with.

1.1.2 The maintenance programmers

A tragic fate awaits most computer science graduates. Instead of becoming highly paid developers of exciting new software, their destiny is to become lowly paid "maintenance programmers". The proportion of resources expended on "maintenance" is surprising. Meyer suggests that approximately 70% of the cost of software is devoted to maintenance; Adele Goldberg of ParcPlace estimated that among their clients the figure is about 65%.

A need for "maintenance" seems odd because software cannot wear out. When the figures are examined in more detail, it becomes obvious that the term maintenance is inappropriate. Only a small proportion of maintenance costs go on bug fixes (for which the term maintenance might be appropriate). The largest proportion of these costs is incurred in adapting an existing program to meet changing user requirements.

Not maintenance, it is "evolution"

Programs are written to satisfy rigid specifications. Once written, the programs are given to the users. Inevitably, the users are not satisfied (even if the original specifications are). So maintenance programmers must start to change the programs.

A need for change in a program is not necessarily a sign of failure – in fact it is possibly a good sign of at least partial success. The original specification for a program will have described what the clients imagined would be useful, basing their choices on pre-existing work practices and knowledge. If a program proves useful, it changes the way in which its users work. They can identify additional related tasks that they would like the program to be extended to handle, or they may revise their ideas on the way some previously specified task is to be handled. (If a program is really pretty useless, it will be accepted by its intended "users" who will simply archive it. Such programs are not changed further, they merely get to be replaced.)

One might reasonably argue that the expression "software specification" is an oxymoron (a figure of speech with pointed conjunction of seemingly contradictory expressions: examples "faith unfaithful kept him falsely true", "honest politician", ...). "Specification" conveys the notion of a detailed description of a construction to be undertaken by an architect or an engineer. "Software" conveys a notion of an amorphous entity capable of infinite malleability and change.

Evolutionary change is an essential part of the life-cycle of a useful program; but most approaches to program design result in systems that are difficult to change. The difficulties associated with changes account for the high "maintenance" costs.

1.1.3 Disjoint processes for "analysis", "design" and "implementation"

Another problem with conventional approaches to program development is the disjoint nature of the analysis, design, and implementation processes. Just follow a typical problem through from the client's first request to the delivered system – and look at the number of different languages used.

First, there is the language of the problem domain (Figure 1.1). The client will specify requirements in terms of domain entities – such as the chemical structure, the table of properties, an information service bureau, and communications links of Figure 1.1.

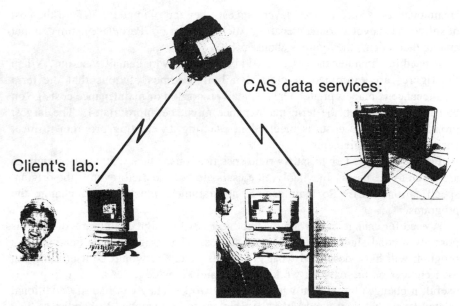

CAS data services:

Client's lab:

I require a system that allows me to enter a chemical structure graphically, converts it to a standard representation, transmits this to the Chem. Abstracts search (CAS) service in Colombus Ohio, retrieves details about its properties, files either or both structure and retrieved data and prints selections.

Figure 1.1   The client's initial request.

Translating the specification

As illustrated in Figure 1.2, the client's initial description is transformed through a whole series of reconceptualizations using a variety of representations. At each stage, there is a "translation" between different languages. At each translation step, there is the possibility of error and misinterpretation.

Also note how the client gets excluded early on. In a normal project, interactions with clients tend to be along the lines: "*Just tell us what you want. …. OK. It is 1] enter structure, 2] convert to standard form, 3] transmit, 4] receive response, and 5] display response. Now, go away! We will call you when the product is ready.*"

But, as illustrated in Figure 1.3, the original analysis may not have established exactly what was wanted. If the client is excluded early on, one may not find omissions until it is late and any changes will incur excessive costs. What is more, the client never understands much about the system and so cannot distinguish desirable extensions that are reasonable from those that are unreasonable.

It is all too much like an expensive version of the game of secrets. The client whispers a description of the desired system into the ear of the analyst; who whispers a rather different specification to the designers; they convey a slightly changed idea to the implementors. But no one can afford to laugh when there is a large discrepancy between the client's version and the implementors' version of a system.

The analyst:

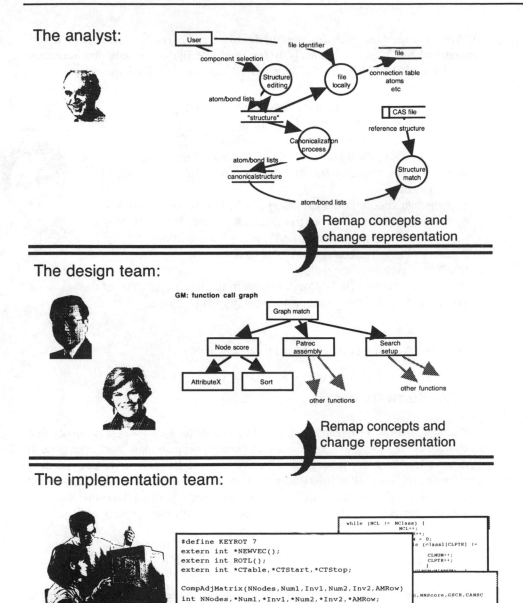

The design team:

The implementation team:

Figure 1.2   The software engineers' interpretations.

Even after final implementation and installation, all those "translation" steps are going to keep getting repeated. When "bugs" occur, the client will describe them in terms of the problem domain (e.g. *"When I used the atom drag tool to change a double bond from cis-configuration to trans-configuration, it didn't work because the program still retrieved data on the cis isomer"*). The poor maintenance programmer must translate such a bug report into something more meaningful

Translating the "bug" reports!

("*When you used the atom drag tool to do what? ...!*") The bug description has to be retranslated through those analysis/design and design/implementation interfaces in order to work out what part of the program might contain the error.

What I meant to say was that I need a system that:
1. allows structure editing as I described; but
2. processes several structures in one run;
3. performs structure searches as I've already described;
4. lets me enter tabular spectral data for three different kinds of
     spectra,;
5. transmits spectral data to Dialog or another search service,
6. retrieves references from the Chem.Abstracts literature files in
     addition to the structural info I mentioned before;
7. stores these various data in local files,
8. allow searches of local files;
and while you are at it, I'd like the system to plan a synthesis of the compound, make my coffee, and write my Ph.D. thesis.

Figure 1.3   The client's intended request.

1.1.4 Software costs

Of course, the real problem as perceived by commercial developers is simply that software costs a lot. Software costs a lot because programmers start from scratch (more or less) on every new project. Software costs a lot because we keep needing to change the code after it has been written, and such changes are difficult and error-prone. Software costs a lot – at least in part because of all those disjoint steps in its development and all those translations between different representations.

1.2   THE SOURCE OF THE PROBLEMS

These problems of high maintenance costs and lack of reuse are related. They stem in part from the way that software is usually designed: *Start with the functional specification of what the program is going to do, and use a top-down decomposition approach to re-express this function in terms of small procedural components that will be executed in sequence.*

Top-down decomposition works fine – for stable programs.
Long ago, this approach was helpful. But the programs then being developed tended to be relatively stable. They were things like payroll programs, or rocket trajectory calculations. The laws of physics don't ever change. The processing of a payroll is pretty much the same as when it was achieved by clerks moving decks of cards among various mechanical tabulators, sorters, and printers.

But, as illustrated in the next two subsections, top-down functional decomposition can have a couple of problems. Firstly, there is a strong tendency for the original specification of a program to become embodied in every part of the code that is written, making the code difficult to adapt and change. Secondly, as Meyer has argued, "*a program doesn't necessarily have a top*".

1.2.1 The original specification pervades the code

The following example, based on an idea presented by Cox, illustrates how the top-down decomposition approach fares less well with modern programs that must frequently adapt to changing needs. As illustrated in Figure 1.4, the example program is to handle electronic mail for users in some large research organisation. It is to deal with email from an external wide area network, memos generated on personal computers in the administration department, pictorial data received on a computerized facsimile machine, and short notes entered by a receptionist.

The electronic office example

These various types of mail item are to be queued up for users who can review their mail item queues, and select mail items from the queue. Once a mail item has been selected, its summary and/or content may be displayed, it may be requeued, or it may be copied to a file.

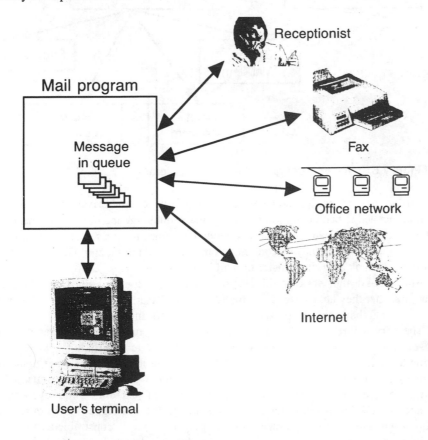

Figure 1.4   The "mailer" program.

A part of a design for this program is shown in Figure 1.5. This particular design was developed using a top-down decomposition approach. The mail program will require some initialisation and termination routines and a main loop that responds to user commands and checks whether any of the various sources of mail items have additional items that should be appended to the queue.

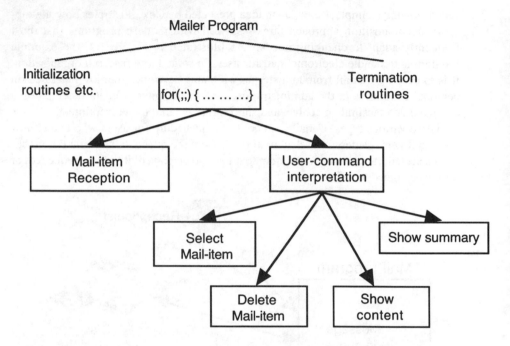

Figure 1.5   A partial design for the "mailer" program.

Further decomposition steps would lead to the characterisation of various routines needed to handle mail item reception and queuing, and of routines that will handle the user commands. These will include routines for the selection of an item from the queue, the display of content or summary of an item, and others – such as routines that look after the transfer of mail items to files.

Additional design work would flesh out the detail of these routines, and the data structures that they are to use. Eventually, the specifications would be sufficiently complete that they could be passed to programmers for implementation.

Over-specification discourages code reuse   The very nature of the focussed top-down design approach discourages code reuse. Code reuse comes more readily after a little "lateral thinking". Instead of trying to establish the unique features of the current problem, the designer must try to recognise similarities with other previously solved problems. The reuse of a design has greater economic benefit than the reuse of just some code; even code cannot be reused if its relevance is not recognized. If the designer works purely in a top-down fashion, the implementation is likely to be overly constrained.

By the time the programmers start coding, they will be working in an extremely tightly specified context. The form of mail items will have been refined and resolved down to some complicated `union`. The specific queuing operations as required in the context of this program will have been determined and used to design the queue data structures. Inevitably, the programmers implement a new, special purpose solution to the specific queuing problem at hand.. They don't modify old code (*"Joe wrote that code to deal with some strange special case, and rewriting is easier than changing Joe's obscure old code"*). They don't use a general purpose set of queuing subroutines (*"Those routines are too general. They don't take advantage of the special features of this problem and, anyway, we have to have some extra features in our queue"*).

Further, the top-down development process will have had the effect of distributing details of the original problem specification throughout the code that must be implemented. All those ShowSummary and ShowContent routines "know" about the different types of mail item that are supported. All will have code of the form –

The original specification pervades the code

```
void ShowContent(MailItem* mailitem)
{
    ...
    switch(mailitem->type)
    {
case EMAIL:        ShowContentofEmail(mailitem); break;
case FAX:          DrawPictureofFax(mailitem);  break;
    ...
    }
    ...
}
```

Knowledge of the structures chosen for queues will pervade another group of routines. These problems are illustrated in Figure 1.6.

The top-down approach to design and implementation creates *brittleware* rather than software. After all, the specification is bound to get changed. Because the original specification has been spread throughout the design and implementation, a later change causes major disruption, breaking the brittle code.

Brittleware?

Shortly after the programmers who implemented the mailer system have left the company, management will inevitably request extensions – "*Just add another type of mail item; we can now get digitized sound message from our voice-mail device and we would like to be able to output them via the speakers on the users' computer terminals*".

The result of the management's request is that every one of those switch(...) {...} constructs is going to have to be changed. Many of the files comprising the system are going to have to be re-edited, recompiled and relinked. This process provides lots of opportunities for errors!

1.2.2 Meyer's topless program

After I start a word processor program and open a document, I may today:

• add a new paragraph

• cut-and-paste some of the existing text

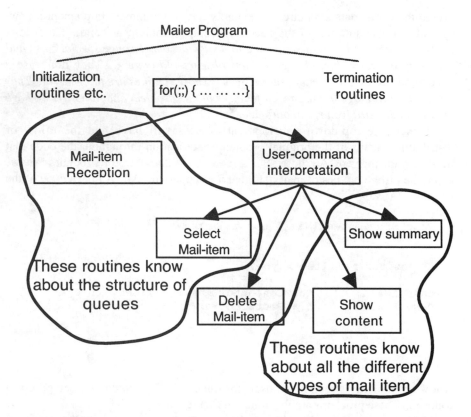

Figure 1.6   Top-down approaches to design and implementation often pervade
an entire program with details from its original specification.

- run the spelling checker

- italicize a phrase and change a header to use a different font face and size

- paste in a diagram from a drawing program

- copy out a table of numbers that I want to paste into a spreadsheet

It does not matter to me how this word processor program works, provided that it
does all that I require. The workings of the program do matter enormously to the
maintenance programmers employed at the software company that sells the word
processor program. The design of the program determines how many problems
they will face.

A simple "functional   The first version of the word processor was probably specified as being a
design" for a text   program that could: accept character input (in a single font with fixed face, style,
processor   and size), allow cursor-based cut-and-paste editing, display text in a scrollable
window, and read and write text files. With this specification, the program might
well have been designed around the supposed primary role of character input and
storage. If so, it could have used a single large array for a character buffer and a
supplementary array integers that identify the positions where lines start.

A simple organisation, with a collection of routines for adding and removing text from those two global array structures, would allow the program to be developed rapidly. File input and output would be trivial, just transfer a block of bytes.

But the problems would start to come when management asked for extra features to make the program competitive with rivals. The spelling checker would not be too much of a problem – it is after all just a function that munges through the character array pulling out groups of letters separated by white space or punctuation. The use of different fonts would have presented more of a challenge; but could probably have been handled by adding another couple of array structures (one to record the format details for runs of successive characters, the other to record the size of the largest character in each line of text). The real difficulties would come when the program had to start to handle non-textual data such as embedded pictures and tables.

Such additions would have to be patched in, maybe with a system of "magic lines" (line index entries that flagged the presence of a separate picture or table data block). Such patches would then have to be worked through all the code for cutting and pasting data and saving data to disk. After a few such cycles of development, the program would have lost any coherent structure. It would be an almost unmaintainable maze of switch statements and special case cut outs.

An alternative starting point for the design of the word processor could have focussed on the role of a document that owned a list of "paragraphs". Each paragraph would be represented as a separately allocated block of bytes in the heap that could be copied into a separate, larger buffer structure when selected for editing. Although it would involve lists and dynamic (heap) storage this design might well lead to faster code because the cutting and pasting of paragraphs would not necessitate as much copying of characters as would the design based on the use of a single large character buffer.

A design focussing on component data structures

Data transfers to and from disk would involve some function that iterated along the document's list of paragraphs, transferring in turn each paragraph. Code for text display would involve another controlling function that iterated along the list of paragraphs and called a display routine for each paragraph in turn.

In this design, adding the spelling checker would involve a little more work than in the approach that used the block text buffer. The routine that munges around through a block of bytes extracting words would be the same, but there would have to be a driver routine that iterated through paragraphs. Support for multiple fonts would involve about the same amount of coding, but rather than a couple of large arrays holding all formatting data there would be rather similar format records associated with each individual paragraph.

This design would have started to show its advantages when management asked for support for pictures. Only relatively small changes would have been required. The paragraph list associated with a document would have been generalized to a list of data items – an item being either a picture or a text paragraph. It would probably have been done using `unions` with a tag field that distinguished pictures from text. The processing routines would have used `switch` statements that use the tag fields to select specialized routines for the different data types. The changes made to accommodate pictures would suffice, more or less, when tables were also required.

Of course, a program built in this fashion with `unions` and `switch` statements would still have the various maintenance problems outlined in section 1.2.1. But at least the program would have retained a coherent structure based upon that original design which focussed on the data.

Data as a better basis for design than function

Often, data provide a more stable base for design than does some presumed function. The function of a program tends to change as the program evolves to meet new needs; much of the data will remain the same.

Meyer takes these arguments further and tries to show that an attempt to focus on the "function" of a program can lead to a variety of design problems. His illustrations show how slightly different concepts as to the function of a program can lead to quite different designs. A top-down approach leads to unnecessarily early commitments, and an excessive focus on temporal sequence of function calls. Often, a program will not really have a single well defined *function* that can serve as the basis of its design. For example, it is not clear what the starting point for a top-down analysis should be in the case of the evolved word processor that supports a number of different styles of use. Meyer summarises his arguments with the slogan:

"Real systems have no top."

1.2.3 The problems of top-down design

The main problems that result from a top-down design approach are:

- The approach discourages the reuse of standard code. The strong focus on the specific task of the program makes it more difficult to recognise parts that resemble other programs.

- The approach permeates the code with details of original specification – so making change more difficult.

- the function of a program (i.e. the "top" for the top-down design process) tends to be less clearly defined in the context of modern, highly interactive programs that are expected to evolve and adapt to changing needs.

1.3 SAVED BY "MODULAR PROGRAMMING"?

If software is to be made a little more "soft" and "pliable" (so as to facilitate evolutionary change), a way to build systems must be found that does not permeate the *entire* program with details from the original specification. Instead, knowledge should be more localized. There has to be a way to group together data and the functions that operate on data so that everything concerning a particular data type is collected together.

The idea of packaging data and functions together is not new. Parnas was suggesting something along these lines back in the early 1970s when the procedural programming languages C and Pascal were just coming into use. During the 1970s, a variety of experimental programming languages provided support for abstract data types that package data and related functions. By the end of the 1970s, languages such as Modula2 and Ada had emerged. These languages provide considerable support for a "modular" style of programming that is based on packaging of data and functions.

1.3.1 Programming with modules

A design for the "mailer" program that focussed on the data would identify a variety of abstract data types that could combine data and associated functions. For example, the mailer program requires a queue. An abstract data type can be defined to represent a queue; it will define services such as the ability to report its length, to append items, and to return either the item at the head of the queue or, possibly, an item selected from a specific position in the queue.

"Mail items" as abstract data types

"Mail items" could be another abstraction. There are a variety of different types of mail item. But they all have similar characteristics. Each mail item will have an arrival time, a summary, a sender's address of some form, a content field, and other analogous data. They are subject to similar processing:

- they are created by particular "mail item sources",

- they are put in queues,

- their summaries and contents are displayed,

- they get filed on disk,

and

- they get destroyed.

If a "module" is built that defines the structures of mail items and provides procedures for using them, then all detailed knowledge about these mail items can be localized in this one part of the code. If, subsequently, new varieties of mail item are required, changes to the code are also localized to this module.

"Modules" as an implementation of abstract data types

Modula2 modules and Ada packages provide an approach to building software whose design focussed on the data. As illustrated in Figure 1.7, a module defines an abstract data type. Each module will have a public interface and a private implementation part. The public interface will contain details of the functions and procedures that are "exported" by that module. These exported functions define the "services" that the module provides to the rest of the program. A MailItems module would presumably provide services for creating and destroying mail items, getting their summaries or content displayed, and writing them to a file created by the calling program.

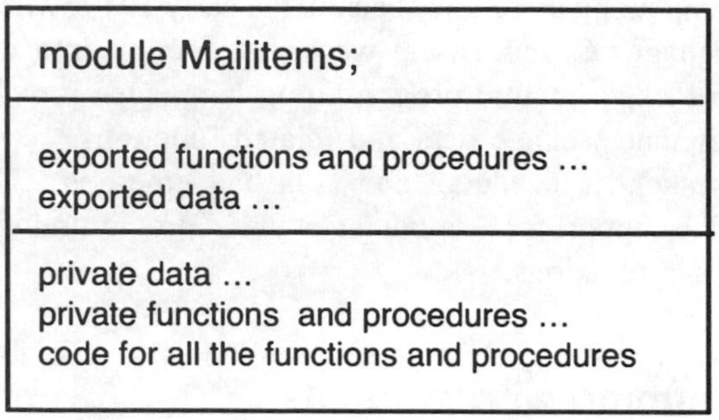

```
┌─────────────────────────────────────────────┐
│                                               │
│  module Mailitems;                            │
│                                               │
├─────────────────────────────────────────────┤
│                                               │
│  exported functions and procedures ...        │
│  exported data ...                            │
│                                               │
├─────────────────────────────────────────────┤
│  private data ...                             │
│  private functions  and procedures ...        │
│  code for all the functions and procedures    │
│                                               │
└─────────────────────────────────────────────┘
```

Figure 1.7   Going "modular".

"Opaque" data types   A module may export details of the types of data structure that it uses. More typically, the public interface will export only an "opaque" data type. Essentially, all details of the data structures used by the module are kept private to its own implementation. Instances of these data structures, in this case individual mail items, can be created on the heap by creation routines that form part of the module's public interface. A calling program can hold addresses of mail items in pointers, but cannot deal directly with their structures. When a mail item must be manipulated, its address must be passed in a call to one of the other functions exported by the `MailItems` module.

A well written version of a `MailItems` module can conceal the fact that there are different types of mail item. A main program can be written using just the knowledge that mail items can be created, can have summaries and content displayed, and can be saved. As far as the main program is concerned, the same routines, as exported by the `MailItems` module, are used in all cases.

The fact that mail items are `unions` with a tag field distinguishing their types is known only within the implementation code of the `MailItems` module. Code of the form:

```
        switch(mailitem->type) {
case FAX: ...
    ...
    }
```

will only occur in this implementation part.

Because other parts of the program can never see inside an opaque data type, there can be no dependencies on the actual structures of these types. If the structures need to be changed, the change will not propagate into the rest of the code. Thus, when management requests a "VoiceMail" variant, it is only the implementation code of the `MailItems` module that need be changed.

Programming in this style is possible using a conventional language such as C – provided that the programmers are well disciplined. But a C compiler can do little to help enforce correct usage of data structures and routines. If a language is to "support" a programming style (rather than simply permit that style), then its compiler should provide some specialized constructs – such as Modula2's export and import lists for modules and its facilities for opaque types.

1.3.2 Exploiting "generic" packages

Ada supports another advanced feature that facilitates a modular programming style. In Ada it is possible to define a package (module) of routines for manipulating queues. Queues of what? Well, queues of an unspecified data type Thingy:

```
module QUEUESTUFF[Thingy] is
begin …
    x : Thingy;
    xt : array[…] of Thingy
    …
```

Such a "generic "package is not real code. It is a kind of template that a compiler can use to create packages that manipulate queues for specific types of data element.

In an application program, one can have a queue of integers (i.e. instantiate a copy of the QUEUESTUFF module with Thingy=integer), and a queue of pointers to mail items (instantiate a copy of the module with Thingy= mailitemptr).

1.4   LIMITATIONS OF MODULAR PROGRAMMING?

Modules and generic packages can certainly help. But they do *not* solve all the problems that arise when trying to make software more "pliable".

1.4.1 Changing the software to meet a change in specification

Even with modules, it will still be necessary to change existing code when new variants of data types are defined. For example, in the mailer program, many of the functions and procedures in the MailItems module are going to have to be edited in order to accommodate the additional VoiceMail variant on a mail item.

It does help a lot to have the changes localized. But, it is necessary to edit files that contained code that had previously been completed and tested. Whenever code is changed, there is a possibility of error. Consequently, the changed modules are going to have to be extensively re-tested to verify that the changes have not disrupted previously correct processing behaviours.

1.4.2 "The standard library modules is never what I need"

It is all very well to have a generic QUEUESTUFF module in some library, but it is a standard queue handler. Often, it turns out that one requires a queue that is basically similar to the standard one in the library but differs because it has at least one slightly aberrant behaviour.

The Customer Queue in a Norwegian Post Offices   In their book on Simula, Birtwistle *et al.* describe a simulation exercise that involved modelling many activities in a Norwegian post office. One aspect of the model entailed simulating the behaviour of customers entering the post office and joining a queue to wait for service. So the implementation of the simulation needed a "Norwegian Post Office Customer Queue" (NPOCQ) data type. Now, a NPOCQ is much like an ordinary queue; it has functions to:

- check its length (e.g. potential customers check the length – and maybe decide to come back another day, counter staff look at the length – and choose when to schedule their breaks),

- "append" an additional element (if the queue is not too long, a potential customer becomes a queuing customer by appending themselves to the end of the queue),

- remove leading element (one of the counter staff starts serving the customer who had been at front of queue).

An NPOCQ is thus similar to an ordinary queue, and one might be tempted to use the standard generic QUEUESTUFF package.

A non-standard "append" behaviour   However, an NPOCQ has some non-standard behaviours with regard to the "appending" of additional elements (customers). According to Birtwistle, if a little old lady enters a Norwegian post office, she does not join the queue at the end, instead she goes to the front of the queue and joins there or, if there are already other little old ladies at the front of the queue, she inserts herself in the queue in front of the first queuing customer who is not a little old lady.

The generic QUEUESTUFF module is inadequate – it has to be changed. The normal situation with modules, or packages, is *"The generic XXX module is inadequate for our application – it has to be changed!"*

A particular application will almost invariably require either some additional data fields, or some additional function associated with the data type provided by the module, or some change to one of the functions already defined in the module.

The result tends to be duplication of the original QUEUESTUFF module, with each copy having its own distinct "patches". In a big software system, one may end up with several distinct variants of the QUEUESTUFF module. Although basically similar, each incorporates some unique modifications. Once again, maintenance problems arise.

1.4.3 Meyer's modules that are simultaneously "Closed" and "Open"

Meyer provided a neat characterisation of the problems that exist with modules. Although modules and generic packages seem to offer a basis for developing better software, one requires something more. One requires modules that are, in Meyer's terms, simultaneously "Closed" and "Open".

Closed modules   Modules should be "Closed" in the sense that their code is executable, it defines the behaviour of some standard component in a software system, it has been checked (possibly proven) to be correct in doing whatever it is supposed to do, and *it does not get edited again*!

Modules must be "Open" in the sense that the code is extendable; it must be possible to define additional or changed functionality when necessary. *Open modules*

The standard modular languages, such as Modula2 and Ada, do not satisfy this Open-and-Closed requirement.

1.5 OBJECT-ORIENTED PROGRAMMING: INHERITANCE AND DYNAMIC BINDING

However, modules that are both "Open" and "Closed" can be realized using Object-oriented Programming languages. The two features of OO languages that distinguish them most from the modular programming languages are the use of "inheritance" and mechanisms for the "dynamic dispatch" of function calls. Combined, these features provide the basis of extendable library modules and encourage a much more modular style of code.

1.5.1 Inheritance

As in the modular languages or in conventional procedural languages, in an OO programming language one can define new data types or "classes". A simple class definition is similar to the definition of an abstract data type in a modular language. The class definition will, in some manner, incorporate a public interface (i.e. a description of what instances of the class can do), and a private implementation part (i.e. a definition of how things are achieved). For example, one might have a class Queue that specifies how Queue objects (i.e. instances of class Queue) can be created and how they may then be asked to state their length, append items, or return a queued item: *Defining a simple class*

```
class Queue:
public interface :
    make a new Queue object, …
    Append(Item),  Length returns integer,
    First returns Item,
    …
private implementation :
    …
end Queue;
```

OO languages vary greatly in their syntax. But all will be able to express something equivalent to the ideas expressed in the summary shown above. A class declaration, e.g. class Queue, introduces the new data type. The "public interface" will usually provide a function that can be used to create instances of the class (it is possible for a language to define such a function implicitly). The rest of the public interface will specify the things that Queue objects can be asked to do — they can be asked their Length (they will return an integer value), they can be asked to `Append` a data item, they can asked to return the `First` queued item, and so forth.

The "private implementation" part will define the code required to perform such operations and identify the types of data that must be owned by each individual `Queue` object in order for it to work. For example, a simple implementation for `class Queue` might use a circular buffer. Then, each `Queue` object would have to own an array in which items could be stored – the buffer – and a couple of integer index values.

Using inheritance

The "inheritance" mechanism in an OO programming language permits the definition of specialized forms of simple classes. Such specialisations are defined by specifying how they differ from a standard class. So, one could for instance define a "Norwegian Post Office Customer Queue" as a specialized kind of queue with a different mechanism for appending items:

```
class NPOCQ isa Queue:
public interface :
    replace Append(Item)
    make a new NPOCQ
private implementation :
    ...
end NPOCQ;
```

An `NPOCQ` object is a kind of `Queue`. Just like any ordinary `Queue`, an `NPOCQ` can be asked its `Length` or asked to return the `First` queued item; when performing these operations it will use exactly the same mechanisms as any ordinary `Queue`. An `NPOCQ` can be also asked to `Append` an item; but in this case, it will use its own version of `Append` in preference to the `Append` mechanism defined in `class Queue`.

Reuse is encouraged when library classes are open for specialisation

The creation of a new more specialized class, such as `class NPOCQ`, does not require any changes to the simpler class (`class Queue`) on which it is based; nor does it require any duplication of code. It is this "openness" of existing classes to extension that makes reuse of library code much more common in an OOP environment. The existing library class may not define exactly the version of some data type that is required in a particular situation; but, the standard version is probably $\approx 90\%$ correct. Rather than start again building a new version of some data type, it is simpler for programmers to replace the remaining 10% of the standard library version with alternative functions that fit the particular requirements of an unusual application.

As well as encouraging the use of libraries of standard classes, such as queues and lists, the use of classes and inheritance can provide a basis for a more effective organisation of program specific code. For example, a revised version of the "mailer" program (section 1.2.1) may well make use of standard queues and lists from some library, but it must also deal with a variety of specialized mail items and mail item sources.

A hierarchical approach can also help here. A possible class hierarchy for mail items is illustrated in Figure 1.8. Class `MailItem` is a very general abstraction. It defines mail items as they are to be seen by the rest of the program. Mail items are things that can be created, can be asked to show a summary, or show their content, or write themselves to disk. All mail items will have such abilities.

While the main parts of the program can be written using solely this general abstract class MailItem, the programmers concerned with providing mail items will see the class hierarchy.

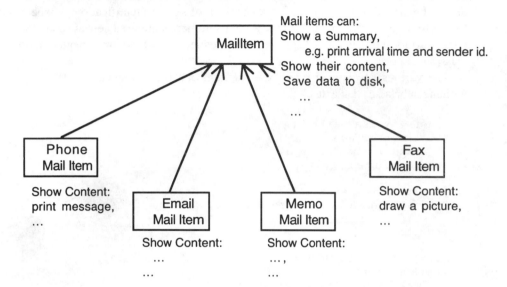

Figure 1.8   "Inheritance" – defining specialized variations on the theme of "mail item".

Class MailItem may not be able to specify very much about the behaviour of mail items or about the types of data that they require. After all a Fax item would have to own some form of compressed bit-map picture and have some way of unpacking and drawing this information when asked to show its content; in contrast, a Phone item might require just a simple character array that is used to record a short message, and its ShowContent() function might involve simply a printf statement.

So, class MailItem would probably leave a lot of details to be defined in the more specialized subclasses:

```
class MailItem:
public interface :
    ShowSummary,
    ShowContent: defined in subclasses,
    Save: defined in subclasses,
    …
private implementation :
    every mail item has: an arrival time, a sender id, …
    ShowSummary = print arrival time, sender identifier,
                  header line, …
    …
end MailItem;
```

Although in this sense incomplete, the public interface part of class MailItem will specify everything that a mail item can do. The private implementation part will define how to handle those aspects that are common to all kinds of mail item. Thus, as in the example above, all mail items can be defined as having data that define the arrival time of a message, some form of sender identification, maybe a header line etc. If most mail items show their summaries in the same way, code for a standard form of summary display can be provided as part the implementation of class MailItem.

The various specialized types of mail item can then be accommodated by defining subclasses that extend class MailItem:

```
class Fax isa MailItem
public interface :
    make a new Fax object,
    ShowContent, Save, …
    …
private implementation :
    each Fax has a compressed picture, and …
    …
end Fax;

class Phone isa MailItem ·
public interface :
    make a new Phone object,
    ShowContent, Save, …
    …
private implementation :
    each Phone has a small text buffer, and …
end Phone;
```

Programming by differences

Only those things that make a subclass different are specified; the programming style for class definition is one of "programming by differences". Each Fax and each Phone object has data fields for arrival time, sender identification etc. These data fields are not specified. They are present implicitly because Fax and Phone objects are MailItems. Faxs are special in that they have pictures and a specialized ShowContent() function to draw a picture; Phones are special because they have character arrays and a ShowContent() function that prints the array. These differences are specified.

Polymorphism

The code for a main program that works with mail items will be written entirely in terms of references to objects that are instances of class MailItem. There never will be any such objects in the running system. All the mail items that are actually created in a running program will be instances of one or other of the more specific subclasses; they will be Faxs, Memos, or EMail messages. The main program does not need to make any provision for these multiple forms of mail item; it works provided each form supports the standard functionality defined by class MailItem. In OO jargon, the mail item references in the main program are "*polymorphic*" (many formed).

1.5.2 Dynamic dispatch of function calls

If mail items were being handled in a modular programming language, the MailItem module would export a ShowContent(MailItem*) function whose implementation would select the specific function required to show the contents of any given mail item that was passed as an argument. Obviously, the selection of the required function has to happen at run-time (i.e. *dynamically*), because it depends on the type of mail item whose content is to be displayed.

```
void ShowContent(MailItem* mailitem)
{
    ...
    switch(mailitem->type)
    {
case EMAIL:      ShowContentofEmail(mailitem); break;
case FAX:        DrawPictureofFax(mailitem);   break;
    ...
    }
    ...
}
```

Of course, in a modular style programming language, the programmer must enumerate all possible choices of data type and associated function (using case labels in a switch (...) ... statement, or explicit tests in some if(...) ... else if(...) ... else ... construct). It is this explicit enumeration of alternatives that makes it impossible for a conventional module to be both "open" (allowing the addition of a new VoiceMail type of mail item), and "closed" (being immune to later changes and therefore not editable).

Asking an object to perform a task

In an OO language, each different type of mail item provides its own ShowContent() function. The code in the calling program that wants a mail item (themailitem) to show itself will be something like:

```
themailitem->ShowContent();
```

Nowhere in the code is there an explicit list of all the different mail item types and associated functions. Somehow, everything is just left implicit. The correct functions just get auto-magically invoked.

Of course, there is no magic involved. There is still a need at run-time to select from among a number of different functions. But the mechanism has been changed to one that can be handled entirely by the compiler.

A simplified view of the mechanism used in an OO language is given in Figure 1.9. One should imagine that each individual mail item has a pointer to a table of functions that are to be used when that item is manipulated. All the items of given type (class) share a single table of functions; so at run-time, there could be several Fax items each with a pointer to the Fax function table, a Phone item pointing to the Phone table, and other Memo, EMail etc. items linked their tables.

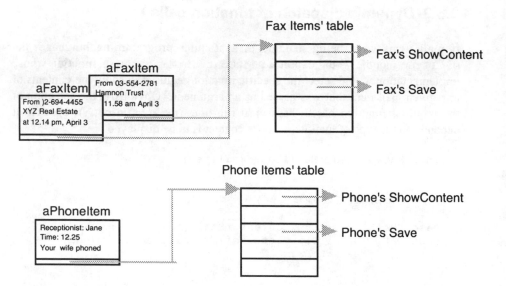

Figure 1.9   Simplified view of the kind of mechanism used in an OO language
to achieve "dynamic dispatch" of function calls.

The code compiled for a statement such as:

```
themailitem->ShowContent();
```

Dynamic binding   involves following the link from the mail item to the associated function table and
using the ShowContent() function specified there. This indirect function call
mechanism is called "*dynamic binding*". The process is *dynamic* because it
happens at run-time. The "*binding*" bit relates to the idea of the specific function
address that gets attached or bound to a particular subroutine call instruction.

Even when other kinds of mail item are incorporated into the system, this code
compiled for the statement "themailitem->ShowContent();" remains correct. In
fact, it does not even need to be recompiled to accommodate additions. If
maintenance programmers must add support for voice mail, their work is limited to
the definition a new class:

```
class VoiceMail isa MailItem
public interface :
    make a new VoiceMail object,
    ShowContent, Save, …
        …
private implementation :
    each VoiceMail has a buffer for a digitized signal,
            and …
        …
end VoiceMail;
```

and the implementation of a ShowContent() function that sends the digitized
signal to the loud-speakers on the user's workstation. Once this code is complete, it
can be compiled and linked to the code of the existing mail handling program.

Subsequently, when `VoiceMail` objects get created, they will function correctly. Each `VoiceMail` object will be created with a pointer to a `VoiceMail` function table (generated by the compiler when it processed the class definition and added to the program code by the linker). This table will contain a reference to the `ShowContent()` routine that does output digitized sound. So, when a `VoiceMail` item is asked to show its content, it will speak.

Although the interplay of inheritance and dynamic binding is possibly a bit subtle and difficult to grasp initially, it is this feature of OO programming that can greatly simplify code. Using these techniques, programs can become easier to write in the first place, and much easier to modify and extend.

1.6   AN OO EXAMPLE

The exploitation of features such as inheritance and dynamic binding necessitates a distinct approach to program design and implementation. Naturally, there is some similarity to design and implementation approaches based on the use of abstract data types and modular languages. The difference is that the OO languages provide greater flexibility.

If the advantages of an OO programming language are to be exploited in the implementation of a new program, then:

- The initial design should focus on the data processed, and not on some presumed "function" of the program (just as with modular languages).

- Many of the data types required in a program should be recognized as variants of standard abstract data types, as implemented in terms of classes in some class library. Existing code that handles the standard behaviours of such abstract data types should be reused. In fact, these data elements should simply be created as instances of library classes.

- Where library classes don't provide exactly the behaviours required of some data type used in the program, it may be possible to create new specialized subclasses that adapt or extend one of the library classes to meet the new requirements.

- New abstract data types (classes) should be defined for any data elements that are unique to the program. Often, though not invariably, these program specific classes can be formed into hierarchies during the process of design refinement.

Quite frequently the most effective way to pick useful data elements, around which a program may be designed, is to regard that program as involving some form of *simulation* of the real-world environment where the programming problem arises. One should first try to identify "*objects*" that exist in the real-world problem: these will include such concrete things as the mail items in the "mailer" program or customers in the post office, and more abstract objects such as queues. The behaviours of these objects must then be determined: mail items can display their summaries, display their content, write themselves to disk, etc. Classes can be defined that describe the behaviour of the various types of object present. Next, the overall computational process has to be described in terms of interactions among the identified objects.

Objects present in the running mailer program

If the mailer program is reconsidered in terms of objects, one can quickly identify a number of different types. Obviously, there are the mail items themselves. As described in section 1.5.1, it is possible to define a simple class hierarchy involving a general class `MailItem` and specific subclasses for the various different kinds of mail item that can actually occur. Some of the other objects that might be present in the program are shown in Figure 1.10.

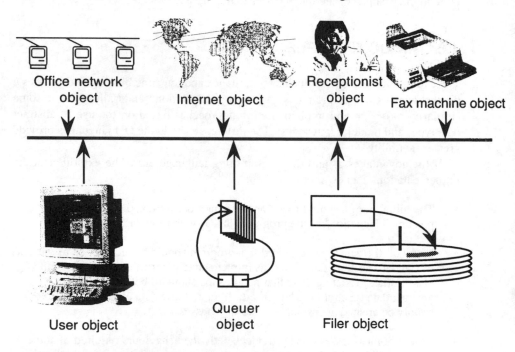

Figure 1.10 An object based model for the "mailer" program.

"Source" objects

There are several "source" objects that act as sources of incoming mail items. One looks after the Internet connection and creates `EMail` objects, another attends to the local network and creates `Memo` objects, still another monitors the facsimile machine and creates `Fax` objects, and the last enables a receptionist to enter details of calls into `Phone` objects. In a simulation, these source objects would first be asked to initialise themselves (they have to open "sockets" and other connections to separate processes running on the computer). Subsequently, at regular intervals they would be given a chance to "run", at which times they would check their input channels and, if data were available, would create new mail items.

Class hierarchy for "Source objects"

These source objects would most probably be implemented as instances of classes from a class hierarchy that parallels the `MailItem` hierarchy shown in Figure 1.8:

```
class MailSource
public interface :
```

```
        Initialise (deferred for implementation in subclasses)
        Run (deferred for implementation in subclasses)
        ...
    private implementation :
        ...
    end MailSource;

    class InterNetSource isa  MailSource
    public interface :
        make new InterNetSource object
        Initialise
        Run
        ...
    private implementation :
        each InternetSource object has a Unix socket, and ...
        ...
    end InterNetSource;
```

The `Run()` function of each of these `MailSource` classes would be something along the following lines:

```
XXX MailSource Run()
{
    Listen to XXX Input Channel;
    while XXX Input Channel has data {
            create a new XXX mail item
            fill it with data from XXX Input Channel
            send new XXX mail item to Queuer for
                appending to Queue
            Listen to XXX Input Channel
            }
}
```

There would be several rather similar implementations; the `Run()` function of class `InterNetSource` would create `EMail` items; the `LocalNetSource`'s `Run()` function would create `Memos` and so forth. There would be some minor differences in implementation related to how they listen to their input channels and how they copy data into a newly created mail item.

The `Queuer` object would probably be an instance of some standard `class` *A Queuer object*
`Queue` taken from some class library. It is possible that the class library would provide something akin to Ada's parameterized types (e.g. "`class Queue of Thingys`"); if this were the case, the `Queuer` object would have been created as an instance of `class Queue of MailItem`. The `Queuer` object would handle requests such a `First()` (it would return a reference, i.e. pointer, to the mail item that had been at the front of the queue, after first removing it from the queue), `Append(reference to mail item)`, and `Length()`.

The `Filer` object would assist in the process of saving mail items in disk files, *A Filer object*
and possibly retrieving mail items from disk files. It wouldn't write the mail items to the file; the actual data transfer is the responsibility of each individual mail item because only the mail item knows what information it owns and can save. The role of the `Filer` object would be to create files, open files, and set up input/output streams etc.

The User Object   The "user" object would supply the interface to the person using the mailer program. Like the "source" objects, a "user" object would have a Run function that would accept and act on the actual user's commands. A very simple version would be something like:

```
void Run()
{
    char commandkey;
    if(!User_Input()) return;
    scanf("%c",&commandkey);
    switch(commandkey) {
case 'd': /* "Delete" command. */
            delete currentmailitem;
            break;
case 'g': /* "Get" command. Dispose of any  existing mail
            item and get the first from the queue. */
            if(currentmailitem != NULL)
                    delete currentmailitem;
            currentmailitem = gQueuer->First();
            break;
case 'f': /* "File" command.  Ask 'filer' object to arrange
          for the current mail item to save itself to disk */
            gFiler->Save(currentmailitem);
            break;
case 's': /* "Summary" command. Get mail item to show its
                    summary */
            if(currentmailitem != NULL)
                    currentmailitem->ShowSummary();
            break;
case 'c': /* "Contents" command. Get mail item to show its
                    content */
            if(currentmailitem != NULL)
                    currentmailitem->ShowContent();
            break;
            .
            .
            .
    }
}
```

The user's input would be translated into "commands" or "messages" that get sent to other objects in the running system. Thus, an input 'c' results in a request to the current mail item that it show its content. If the item is a Fax it will draw a picture, a VoiceMail item will send sounds to the speakers, and so forth. This is achieved using the *dynamic binding* mechanism as explained in section 1.5.2.

"Sending messages to objects"   This style of "flow control" is very typical of object-oriented programming. In fact, in some languages (notably Smalltalk) programs are described in terms of "objects" and "messages". Work is done by "sending a message" to an object asking it to do something (e.g. "*hey filer, please arrange a disk file and get this mail item to save itself*", "*hey queuer, please give me the first one of your mail items*"). The "sending a message" terminology is slightly misleading because it can be interpreted to imply concurrency (e.g. it is possibly to imagine the "user" object getting on with further work while the "filer" is busy setting up files and i/o streams). While there are experimental languages that support some measure of

concurrency, in most existing OO languages these interactions among objects are implemented using normal function call/return sequences and do not involve concurrent operations.

The code for the individual functions that define the various behaviours of a particular kind (class) of object is typically simpler than the code found in normal procedural programs. Thus, the code in the user object's `Run()` function that gets the current mail item to show its content is simply:

Simpler code!

```
currentmailitem->ShowContent();
```

while in a conventional procedural style code it would involve some complex switch on a tag field in a `union` data structure:

```
    switch(currentmailitem->type) {
case EMAIL:
        ShowContentofEmail(mailitem);
        break;
case FAX:
        DrawPictureofFax(mailitem);
        break;
    …
    }
```

The elimination of all these switches helps reduce the complexity of the code, making it easier for maintenance programmers and others to understand. Further, because work is partitioned out amongst many cooperating objects each of which handles some restricted subtask, one tends to have many small functions to accomplish a particular task rather than a single complex procedure.

The complete system would have

Overall program organisation:

* 1 "user" object – that handles commands entered by the person using the mailer program,
* 1…*n* "phone receptionist", "email handler", "localnet", and similar source objects that create mail items,
* 1 "Queuer" object,
* 1 "Filer" object,
* 1 "current mail item" that may be displayed, filed …,
* a large number of other mail item objects in the queue.

This structure is typical of a large class of OO programs. A few more or less permanent objects get created at initialization time and remain in existence throughout the execution of the program; large numbers of other objects get created, asked to carry out various tasks, and then get destroyed.

The main part of the mailer program would consist of an initialization routine that creates the principal players:

```
/* Pointer to a 'user' object. */
User*    gUser;
/* A pointer to 'Filer' object. */
Filer*   gFiler;
/* A pointer to a 'Queue manager' */
```

```
Queue*      gQueuer;
/* A pointer to the current mail item */
MailItem* currentmailitem;
/* A list to store the "source objects" */
List        gSourceList;

void Initialise()
{
    /* create and initialise principal players */
    currentmailitem = NULL;
    gUser = new User;
    gFiler = new Filer;
    ...
    {
    /* Create the 'source objects and add them to a list */
        Source*        aSource;
        aSource = new EmailListener;
        gSourceList.Append(aSource);
        aSource = new Receptionist;
        gSourceList.Append(aSource);
        ...
    }

    ...

}
```

and a `Run()` routine that gives each of the active participants a chance to check for inputs from external channels etc.:

```
void Run()
{
    for(;;) {
        Source*        aSource;
        /* Iterate along the list of source objects */
        /* Giving each in turn a chance to "Run" */
        /* and check their input channels */
        for(    aSource=gSourceList.First();
                aSource != NULL;
                aSource= gSourceList.Next())
                    aSource->Run();
        /* now give the 'user' a chance to run. */
        gUser->Run();
        }
}

int main(int argc,char* argv)
{
    Initialise();
    Run();
}
```

That is all. All the other control flow is handled through the mechanisms where one object asks another to perform a task.

• *"queuer, put the first mail item from your queue in 'currentmailitem'"*;

* *"filer, open a disk file and help the currentmailitem save itself"*;

* *"currentmailitem, show your summary"*;

Note that there are *no* details of the types of mail item in the main program, nor in the filer-code, nor in the user-code. There is simply a `currentmailitem` pointer to some kind of mail item. The "calling routine" tells the item identified by this pointer to perform some task using whatever code is appropriate – so if it is a `Phone` mail item then the code for phone mail items is used, if it is a `Memo` mail item then the code for memos is used etc. *We no longer have those messy problems associated with details of the original specification permeating the entire code.*

Details of the original specification do NOT permeate the code

If a new mail item type must be added, e.g. for voice mail, this will *not* necessitate changes to all the code segments. In order to support a new variant mail item type, e.g. a voice mail type, the "maintenance" programmers must simply:

Relatively easy to extend for a new type of mail item

* define a new class for the additional mail item type, and implement the code that specifies how this new type differs from the standard mail item type;

* define a new class of source objects to deal with the new input channel;

* add a couple of lines to the initialisation routine that create an instance of the new class of source objects and append it to the list of sources that must get a chance to run.

Thus, this design focussing on the objects results in a structure that allows for much easier extension.

1.7 CONSISTENCY OF MODEL: ANALYSIS-DESIGN-IMPLEMENTATION

"Going object-oriented" is not simply a matter of using C++, or Object Pascal, or Smalltalk, or Eiffel or any other OO programming language. To get the benefit of an OO language – with all the opportunities for improved code reuse, simpler code, and ease of extension – one must design a program that embodies an object-oriented philosophy.

In fact one should go OO all the way from analysis, through various design levels, into implementation. When this is done in a project, it usually results in a more consistent model in all of the analysis, design, and implementation stages.

The old ideal project development cycle was the "waterfall" model. The model suggested an analogy between the progress of a software project through various development stages and the experience of a cork floating down a stream with small pools separated by waterfalls. The first pool corresponds to "requirements analysis and definition". The cork gets battered and tossed around in this pool for a little while before being thrown over a cliff to fall into the "system and software design" pool. There is no going back up the cliff. Similarly, the analysts work over a problem statement of a software project, and when satisfied with how they've transformed it, they document their version, sign it off, and drop it on the designers.

The old "waterfall" model for software development

In similar fashion, the project (cork?) is tossed around by designers who eventually drop it over the next cliff onto the implementors.

The various stages of project are far less distinct when an object-oriented approach is taken. The analysis processes merge more naturally into design and then into implementation. There are no longer the big barriers introduced by radically different representations.

In an OO approach, one focuses on objects, their responsibilities and interactions. This focus helps to partition the overall problem. An initial analysis can usually factor the original problem into a system that involves several separate clusters of objects of different types or classes. Figure 1.11 provides an illustration. At the problem domain level, it shows a few of the principal clusters such as might be identified for the chemistry program introduced in Figure 1.1. The running program will involve a number of different kinds of objects that get used to represent chemical structures, another group that handle communications, a cluster of objects that deal with user interaction, and possibly a final group that deal with local files.

Each of these clusters will involve objects of several interrelated classes and there will be many interactions amongst the objects in a cluster. However, there will be relatively fewer interactions between objects in different clusters. It is quite practical to dive down into design, and even prototype implementation, for those classes needed in any one cluster. Often such prototyping forays are needed for the development group to clarify issues related to the original requirements specification. Information gained through prototyping can be used to refine an initial analysis model.

OO analysis

As suggested by the illustration in Figure 1.11, the analysis phase of an OO project involves identifying the major categories of object in the system and their principal interactions. The analysts need to try to identify the "resources" (data) managed by such objects and the services that they provide. In the example of Figure 1.11, a `Structure` object will own any lists of atoms and bonds that may be used to define the form of a chemical. An editing tool object will hold details of any new atoms or bonds that are to be added to the structure. Editing tool objects must interact with `Structure` objects; they will need to be able to ask the `Structure` object to perform tasks such as adding and removing atoms and bonds from its lists. Such relationships are described in terms of client-server interactions among objects; the editing tool is a "client" that requires the `Structure` object to provide services such as `AddAtom`. The analysts will also need to identify how principal objects might get to be created and destroyed. Apart from any hierarchical relationships that arise naturally in the problem domain, the analysts will not be particularly concerned with the discovery of hierarchies.

OO high level design

As the project moves from the analysis into design, the initial rather fuzzy categories of objects identified by the analysts start to become more firmly defined classes. There are often differences between the analysts' initial categories and rough classes and the actual classes as developed by the designers.

The analysts may simply see class `Structure`, but the designers take a more detailed view and see that this class `Structure` is built from instances of other classes such as class `Atom` and class `Bond` stored in some kind of dynamic collection data structure such as a list or a dynamic array. The analysts may work with a category of "editing tools"; the designers may find that this maps into a group of quite distinct classes – "dragger tool", "atom naming tool", "component adding tool" etc. – that have relatively little in common except for the fact that they are used for editing. So there are changes between levels, but there is still a much greater commonality than there would be between a conventional analysis specification and design document.

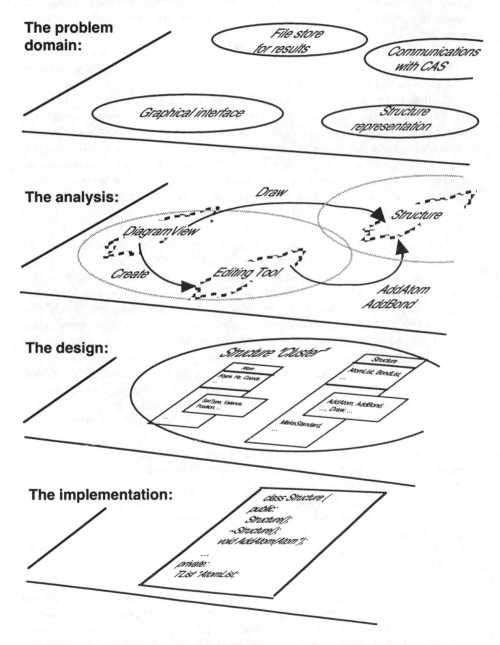

Figure 1.11   OO Analysis, Design and Implementation enhances consistency.

As the design develops, the fuzzy classes acquire well defined boundaries. The data that will be needed by a `Structure`, or by a `DraggerTool`, or any other type of object, will become clearer. The public interface for the classes will emerge as the designers follow through little scenarios that show how the objects in a running program would work together to handle some request from the user. When one object (a client) requires action by another (a server), the kind of action requested must get to be included in the public interface of the class that is being developed to define that type of server object. In addition to the functions specified in the public interface, many other private housekeeping and implementation functions will get to be identified for the different classes.

OO detailed design

As the design becomes more detailed, the use of standard components becomes an issue. *"So, class `Structure` requires some form of dynamic storage structure to hold its atoms and its bonds – how about an instance of class `LinkedList` from the class library; and that 'diagram view' – hadn't it better be made a specialisation of class `View` provided in the graphics user interface class library."* The interactions among instances of different classes as previously identified get to be reconsidered in more detail; the data passed as arguments in requests and the information returned are both characterized. Once this is done, the details of the communication protocols can be finalized.

Iterating through the analysis/design process

Sometimes commonalities emerge when the designers look at the protocols for classes that were originally considered to be quite distinct. For example, in the chemistry program one would eventually find that `Structures` were being asked to draw themselves and save themselves to disk, as were `SpectrumTables`, and `LiteratureReferences`, and a few other classes of data element. It might be worth exploiting such a commonality in behaviour. A new abstraction, class `DataItem`, could be introduced to capture such common behaviours and serve as a superclass for the various specialized forms. Introduction of such a new abstraction would result in a further cycle through the analysis and design phases. The analysts might then realise that rather than the having distinct "pointer to `Structure`, pointer to `SpectrumTable`, ..." data elements in some main data structure, they could use a list of pointers to `DataItems` and that this change would simplify many other parts of the system.

OO: Implementation:

Eventually, the analysis and design develops far enough for at least prototype implementation. The designers' classes are directly mapped into the classes of the OO programming language.

Essentially, the same model is used throughout the development process. Although working at different levels of detail, analysts, designers, and implementors are all trying to characterise objects that own resources and have specific behaviours. All are trying to describe the system in terms of interactions among such objects.

Typically, the client who commissioned the program can be kept much more involved. The anthropomorphic descriptions of objects with particular behaviours and responsibilities can usually be understood by domain experts. So, the domain expert can become part of the development team, right though analysis to at least the end of the initial design process. This increases the chance that the product will be the system desired by the client.

1.8   WHY ADOPT OOP?

Because an object-oriented approach gives you:

- Greater resilience in the face of changing specifications;

- Reusability of code;

- Simpler code;

and

- A consistent model throughout the software development cycle.

REFERENCES

Most books on object-oriented programming are primarily language manuals. They seem to assume that the reader already knows why one should be interested in OOP ("The reader wouldn't have bought an OO language book if he/she didn't know that OOP was useful and already understand how it could be used!" ?) and so these texts focus almost entirely on language details. Most OO analysis and design texts are presentations of an individual's proposed methodology and/or notation. So, finding out "Why adopt OOP?" is not that easy. But, you can get more ideas from the following references.

B. Cox, *Object-Oriented Programming: An Evolutionary Approach,* Addison-Wesley, 1986.   *Reference books*
   One of the first OO books aimed at convincing practicing programmers of the advantage
   of OO style programming. Still worth reading, it has now reappeared in a second
   edition.
B. Meyer, *Object-oriented Software Construction*, Prentice-Hall, 1988.
   OK, Meyer is trying to sell you Eiffel – his own language. It might be a good idea for
   you to buy Eiffel (it is a very attractive high level language). In the meantime, his
   arguments for OO style are still among the best. (There is a revised edition due in 1994.)
A.L. Winblad, S.D. Edwards, and D.R. King, *Object-oriented Software*, Addison-Wesley,
   1990.
   An overview and justification. Not for programmers, more for managers, still useful as
   an initial overview.

$\mathcal{A}$ History of OO Programming

2

Object-oriented programming has a long history; beginning perhaps with:

```
element Pat;
activity secretary(redhaired, thumbs);
    Boolean redhaired; integer thumbs;
begin - - - end;
...
Pat := new secretary(true,10);
```

This fragment of Simula code appeared in a paper in the Communications of the ACM back in 1966. It creates an instance of class `secretary` and sets her working (despite the handicap of having ten thumbs).

Classes and objects for simulation

During the next few years, most "objects" were parts of programs that performed simulations. These simulations were of things like air traffic control systems, in which the objects were planes. Other simulations were of biological systems wherein objects represented infected individuals spreading a disease in some population. Many of the studies were part of standard "operations research" tasks, like simulations of customers arriving at a car-wash. In these simulations, numerous objects would be created as instances of particular classes; each would follow some form of script that defined its actions; the simulation would follow their quasi-random interactions. Object-oriented programming was then synonymous with simulation.

Classes and objects for human-computer interaction

In the 1970s, classes and objects were adopted as the basis of an approach to humanizing computer systems. For the first time, a computer system was designed working from the premise that the needs of the end user should be the centre of attention. The system was "programmed" by the end users, who could build data manipulation tools that matched their specific needs. The programming involved users taking existing components, combining and extending them, and then getting them to obey commands. By 1977, objects were performing live on bit-mapped computer screens (see Figure 2.1).

Message Interaction   ## Pictorial Effect

```
box new named "joe"
```

```
joe turn 30
```

```
joe grow -15
```

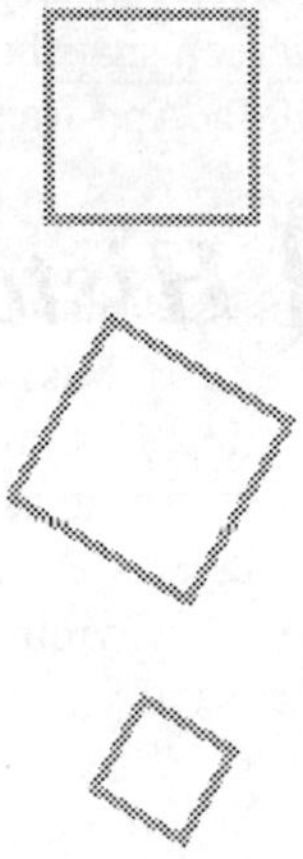

Figure 2.1   Interacting with an early version of Smalltalk.

The record of object manipulations, shown in Figure 2.1, comes from one of the early versions of Smalltalk, as developed by researchers at Xerox in the 1970s. Smalltalk intrigued many, but it was outside the existing streams of software development. The Smalltalk project resulted in a different, but again overly narrow perception of object-oriented programming. For a while, OOP became to be viewed by the uninitiated as a matter of pretty graphics, expensive workstations, and speculation about futuristic, idealized computer systems. A programming style that had objects interacting by the exchange of messages was seen as a cute form of simulation.

Classes and objects for software engineers?

In the early '80s, Cox and others argued the case for OOP to a wider audience of software engineers. The case was twofold. First, classes, objects and messages were presented as a means of making software more malleable – more easily changed, and enhanced. Second, it was argued that existing languages could be extended with object-oriented features, so allowing an evolutionary change in software. "Hybrid languages", in which OO features were grafted onto an existing procedural language base, were to facilitate transfer to an OO programming style. There were to be component libraries associated with these hybrid languages. In these libraries, the graphic components used for end-user programming in the Smalltalk environment would be replaced by "Software Integrated Circuits" for use by real programmers. These Software ICs were standard data structures packaged with associated code as attractive, reusable components.

By the mid-1980s, practical implementations of such ideas were available on at least an experimental basis. Hybrid languages were emerging, along with collections of "reusable" components.

Classes and objects for practical programmers

One of the more painful tasks facing programmers at that time was the construction of programs that utilized the graphics user interfaces that came with the new breed of user-friendly computers. These computers may have been user-friendly, but they were decidedly programmer-hostile. Each application required the careful coding of hundreds of lines of detailed, fairly low-level code, replete with obscure system calls for manipulating windows etc. It seemed that every new application required all the work to be done over again. The emergence of the first large application frameworks reduced this requirement.

The screen snapshot shown in Figure 2.2 is from a demonstration program built with the MacApp framework in 1986. Frameworks like MacApp provide classes whose methods expressed all the standard behaviours expected of an application, a document, or a window. A programmer creating a new application had to define new subclasses that extended the standard classes by adding application specific behaviours. Thus, in the example, the class `TQuadDocument` describes a specialization of class `TListDocument` – the specialization being that `TQuad-Document` needs to store a list of quadrilaterals and a pointer to a view where these were displayed. An instance of class `TQuadDocument` is similar in most of its behaviours to any other document; but it does have some special behaviours, such as its own way of making a window. Initially little more than a curiosity, these framework libraries, and related development support tools, gradually entered the general programming culture.

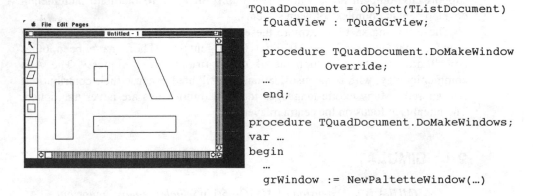

```
TQuadDocument = Object(TListDocument)
    fQuadView : TQuadGrView;
    ...
    procedure TQuadDocument.DoMakeWindow
            Override;
    ...
    end;

procedure TQuadDocument.DoMakeWindows;
var ...
begin
    ...
    grWindow := NewPaltetteWindow(...)
```

Figure 2.2   MacApp, an early framework class library, circa 1986.

In the late 1980s, the steam-roller started moving. C++ emerged from the AT&T research laboratories and rapidly crushed the other hybrid OO languages and experimental forms. C++ is a multi-paradigm language. It is "a better C". C++ can simply be used in the traditional style of C; but it provides far more in terms of compile-time checks. These checks catch many programming errors that would otherwise have remained in code to manifest themselves later as rare and obscure run-time bugs. C++ is particularly strong in the area of support for abstract data types, as realized through its *class* construct. C programmers can make one evolutionary step by abandoning C's structs and adopting C++'s classes. Finally, C++ supports the employment of an OO programming style, so allowing C programmers to progress with a second evolutionary step. The existence of C++ validates Cox's earlier thesis that there is an evolutionary approach to the adoption of OO (though it is possibly not quite the approach that he had planned).

Unlike other experimental forms and hybrid languages, C++ did not come with libraries of useful components or support tools. This omission somewhat slowed

C++ a.k.a. "the steam-roller"

initial adoption, but also provided a niche for small software companies. By about 1990, C++ compilers were available for most platforms. A few class libraries had appeared, some as commercial products and others in the public domain. Within another year, development environments for C++ programming appeared with some support for OO design and class library usage. C++ is now well on the way to supplanting C as the most widely applied programming language. The language is continuing to evolve with new versions every two or three years and regular updates to commercial implementations.

Classes and objects for software engineers!

Experimental languages continue to emerge. One of the more interesting is Eiffel, an OO language designed by Meyer. Eiffel appeared around 1985/86, but its most widespread variant is Eiffel 2.x (x=1,2,3) which was released about 1989. An Eiffel 3 has undergone a very prolonged period of gestation and is just becoming available. Eiffel has consistently been an innovative language. When released, the compiler for Eiffel was only part of a comprehensive package of software engineering tools. Eiffel was first in delivering features such as multiple inheritance, genericity, and exception handling. Promised for Eiffel 3 and its support environment are a whole new family of tools to facilitate incremental development of complex software systems.

The following sections explore these topics in slightly more detail, and also touch on the work of the Artificial Intelligentsia. There have been other contributors to the development of object-oriented techniques. The Ada community may, with some justification, feel slighted because their contributions are ignored. Many exotic languages that had minor roles are never mentioned. The intent is to focus on the central players.

2.1   SIMULA

"SIMULA is an extension of ALGOL 60 in which the most important new concept is that of quasi-parallel processing."

Those words introduced, and possibly biased the reception of the Simula language.

The use of computer models to simulate complex systems (e.g. nerve networks, traffic flow, social systems etc.) became practical with the second generation and early third generation machines. The processing involved in simulations requires list structures, complex control flows and other features that made it difficult to write programs in available high level languages like Fortran-II or Algol-60. In the early 1960s, a variety of specialized simulation languages appeared. The first dialect of Simula was implemented around 1962.

Generalize upon Algol's "own" variables

Simula was the work of a Norwegian group, with Nygaard and Dahl the main contributors. They built on top of Algol. Algol is the ancestor of most modern languages including Pascal, Modula2, C and others. Algol was primarily a stack-based language that allowed the nesting of procedures (as in Pascal but not C); a nested, inner procedure could access the variables of its enclosing procedure(s). Space for Algol variables was normally allocated on the stack on entry to a procedure and freed on exit. Algol had one feature that disappeared from most of its successors – the concept of "own" variables. A procedure could declare an "own" variable that would remain in existence from one invocation to the next (much like a "static" variable declared inside a C function). Simula first replaced "own" variables with new constructs that packaged data with code that represented the activity of one element in a simulated system. An activity was physically a block of data in memory that contained the data local to some modelled concurrent activity and some record of a locus control (something like a local program counter); an activity was referenced by an "element" (a kind of pointer).

A coroutine structure allowed each activity to progress. Figure 2.3 provides a simple illustration where there are two activities in a simulated office (secretary activity and boss activity). Each has a loop defining its behaviour. In the body of the loop a RESUME( ) statement allows the executing coroutine to suspend its processing and restart the processing in another activity. Eventually, the activity that suspended itself will be restarted; it will continue execution at the statement following the RESUME( ) statement.

Add coroutine capabilities

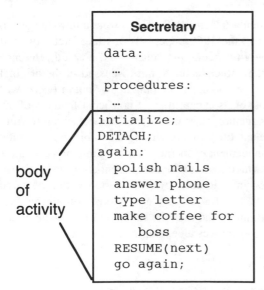

Figure 2.3   A simple illustration of two activities using coroutines to simulate concurrent execution.

In a simulated system, there may be many instances of a particular activity (e.g. a boss may have several secretaries). The "activity" declaration in Simula defined a class of processes that could be created. Particular instances of the class were created using a "new" operator (as in Pat := new secretary(…);). When an instance was created, it would execute the code defined in the body of its activity (class).

Allow multiple instances of the same "class" of activity

If the object created was essentially a data structure (e.g. something representing an individual link in a linked-list, or a complete linked-list), the code in the body would simply initialize the data, run to completion, and return control to the point of call. If the object was to represent an active process (like the secretary or the

"Data structures" and active processes

boss in an office simulation), the body would start with initialization code and then a DETACH statement. DETACH created all the necessary accoutrements of a separate process – a small segment of stack space for internal procedure calls, a pseudo program counter etc. Once these had been created, control was returned to the caller. The newly created object would restart at the point after the DETACH statement when it was first "resumed".

Remote access

The attributes (data fields and procedures) of an object could be "remotely accessed". Thus, a linked-list object could be accessed to get its first or last elements, or boss could query data kept by a secretary. A process was associated with a data structure that could be accessed via a reference. "*It is worth noting the similarity between a process whose activity body is a dummy statement and the record concept proposed by Hoare and Wirth*". It seems that the Simula construct was already more general than Hoare/Wirth records (the records of AlgolW and then Pascal); as well as local variables, local procedures of Simula objects existed and could be "remotely accessed" e.g. a list object could be asked to append an extra item.

Simula-67

Simula was standardized, as "Simula-67", in 1967. There were a few changes to the syntax and nomenclature used in the 1966 paper. Terms like "activity" and "element" were replaced, with "activity" becoming "class" and a typed reference replaced the earlier "element". Many other features, not discussed in the original CACM paper, were also refined. A textbook on the language, *Simula begin*, was published in 1973 and covered most of its capabilities. Significant features of the language included the basic class structure, class inheritance, a mechanism for run-time identification of objects, "virtual" functions providing a combination of static and dynamic typing, automatic management of memory for dynamic objects, and a library of useful base classes from which to build the components of a simulation.

Classes and references

Simula classes formed a single syntactic unit. A class started with a declared name and argument list, then had a begin … end block containing first declarations for local data and procedures, and then the statements that made up the body of the class. The text *Simula begin* gives an example of class Line:

```
class Line(a,b,c); real a,b,c;
begin
    Boolean procedure parallelto(l); Ref(line) l;
            if l=/=None then
                    parallelto:=abs(a*l.b-b*l.a) < 0.0001;
    Ref(Point) procedure meets(l); Ref(line) l;
    begin
        …
    end;

    real d;
    d := sqrt(a**2+b**2);
    if d = 0.0 then error else
            begin
                    d := 1/d; a := a*d; b:= b*d;  c := c*d;
            end;
end ***line***
```

An instance of class `Line` is essentially a data structure with three reals defining the line equation ($ax + by + c = 0$). When a line is created, suitable arguments must be provided (e.g. `ref(line) line1; …; line1 := new Line(2.5, -1.0, 4.0)`). The "body" of class `Line` starts after the declaration `"real d;"`. The code of the body consists of a few statements that "normalize" the line coefficients. Since there is no `DETACH` statement in the body, the code is run to completion; and an initialized instance of line is returned. (Because there was no `DETACH` statement, none of the extra processing needed to establish a pseudo program counter etc. was performed. It would be a mistake to try to `RESUME` a line. It is an inactive object.)

A line can be asked (true, false) whether it is parallel to some other line or where (if at all) it intersects another line. In each case it needs a "reference to a line" as an argument. "References" are essentially typed pointers – in the example there are references to lines (`Ref(line)`) and references to points (`Ref(Point)`). In the code for the procedures `meets` and `parallelto` the coefficients defining the second line are "remotely accessed" (e.g. `l.a` – remote access to `"a"` data value of line `"l"`).

Simula provided a single inheritance mechanism that allowed new classes to be defined as extensions of existing classes. So, one might have a class `aircraft` that defined the general abilities of aircraft in some simulation, and specializations – jets, propeller planes, and helicopters. The program would contain the definition of the base class `aircraft`:

Inheritance in Simula

```
class aircraft;
begin
    procedure accelerate(…)
    begin
    …
    end
    procedure turn(…)
    begin
    …
    end
    …
end ***aircraft***;
```

and the other classes that would each be declared as specializations of class `aircraft`:

```
aircraft class jet;
begin
    …
end ***jet***;
aircraft class helicopter;
begin
    procedure hover
    begin
    …
    end;
    …
end ***helicopter***;
```

The declarations of these subclasses are qualified by the name of their parent class.

Naturally, a ref(aircraft) variable could be made to point to a jet, propeller plane, or helicopter:

```
ref(aircraft) x;
...
if inputval = 0 then x := new helicopter
else
if inputval < 0 then x := new jet
else x := new propellerplane;
...
```

This is quite acceptable because helicopters, propeller planes, and jets are all aircraft.

Run-time class
identification

In an air-traffic simulation, one might have an array of aircraft:

```
ref(aircraft) array     targets[1:100]
```

that contained a mix of jets, propeller aircraft and helicopters. In the program it might be necessary to find the helicopters, and tell them to hover. Simula had a kind of case-statement (inspect) and a test-operator (is) either of which could be used to check whether a particular target was indeed a helicopter.

However, code like:

```
if target[x] is helicopter then
    target[x].hover;
```

wouldn't get past the compiler. Variable target[x] is defined as a reference to aircraft, and aircraft cannot hover. The variable has to be "type-cast" to a reference to helicopter before the code is acceptable. Simula provided a kind of type cast operator (qua ≈ "consider as"):

```
if target[x] is helicopter then
    target[x] qua helicopter.hover;
```

Run-time class identification and related features have been omitted from some later languages. If a language does not directly support run-time class identification, some similar facility is often provided in library classes.

Static and dynamic
binding

If one had code like:

```
target[z].land;
```

one would expect different types of aircraft to land in quite different ways. Helicopters could plonk themselves down in car-parks; propeller planes could bounce down onto grass fields; but jets would have to find airports with thousand metre runways. The ability to "land" would have to be specified in the class hierarchy at the level of "class aircraft". But each subclass would want to replace this behaviour:

```
class aircraft;
    XXXXXX
```

```
      procedure land;
      begin
      ...
      end;
end;

aircraft class helicopter;
begin
      procedure land;
      begin
      ...
      end;
end;
```

It wasn't sufficient to simply define different "land" procedures in each specialized class. By default, a Simula compiler uses the type of a reference variable to determine the routine that was to be called. In the following code:

```
      ref(helicopter)  h1;
      ref(jet) j1;
      ...
      h1.land;              invoke helicopter.land
      ...
      if userinput = 1 then target[x] := h1
      else target[x] := j1;
      ...
      target[x].land;       invoke plane.land
      ...
      j1.land;              invoke jet.land
      ...
```

the variables h1, j1, and target[x] have types "reference to helicopter", "reference to jet", and "reference to plane" and so the routines that would be invoked are helicopter.land, jet.land, and plane.land. Choosing the routines according to the type of the reference is called "static binding".

Static binding

Of course, the programmer would prefer that target[x] land in an appropriate manner. If it were a helicopter it should use helicopter.land but if it were a jet it should use the function jet.land. This requires "dynamic binding" where the function "bound" to the call statement is determined dynamically at run-time according to the nature of the object being referenced by the pointer.

Dynamic binding

Simula requires that the programmer identify those functions where dynamic binding is required. They are identified when first declared. A function that is to be called using a dynamic binding scheme has to be labelled as virtual in the class where it is first declared.

In order to get the desired effect of target aircraft landing in individually appropriate manners, the declaration for class aircraft would have to be:

```
class aircraft;
      virtual: procedure land;
      ...
end;
```

C++ makes very similar use of both static and dynamic binding, and it too requires that the programmer identify functions as being "virtual" if they are to be called using dynamic binding. These issues are treated in more detail in Chapter 6.

Automatic management of dynamic memory

Objects were explicitly created using Simula's "new" operator, but they did not have to be explicitly freed. *"Mechanisms for explicit destruction of objects exist in some languages. Their purpose is to release storage space for used by new objects. Such mechanisms are deliberately excluded from SIMULA since they are likely to lead to errors in model operation, by oversight or by mistake."* The run-time environment provided as part of the system automatically reclaimed the space occupied by an object when it ceased to be referenced by any of the reference variables.

Access control

The early versions of Simula did not have any controls on remote access to data fields and procedures of an object. Later, a three level scheme was introduced with unprotected items, "protected" items (not available to clients but accessible in descendant classes), and "hidden" items (inaccessible by both clients and descendants).

Limitations on the use of Simula

Though decidedly innovative, Simula was not a great success. It filled a particular niche in the area of discrete event simulation but wider use, taking advantage of its OO features, was never achieved. Poor marketing was one factor limiting its use. Simula had been developed by a government agency; such organizations don't actually market products but, unlike academic institutes, they do charge for them. Simula was costly; too costly to be purchased on the whim of a researcher who simply wanted to investigate an unusual language.

In the 1980s, Simula-67 began to feel a little coy about its age and changed its name to Simula. The language is still in use with implementations available on many platforms. (A version that runs within the MPW shell on Macintosh computers is in the public domain having been contributed by Lund University. This version can be obtained using "ftp" file transfer from any of a number of archive sites on the Internet.) Simula is not a serious contender for OO development because many aspects of the language are dated (e.g. the i/o facilities bring back memories of decks of punched cards) and it lacks the development environments and class libraries that are available for other languages.

Model for "active objects"

The problems of "quasi-parallel programming" are being investigated in many current research projects. In a Simula program, one would typically have many passive objects (which had completed their scripts in the "body" of their class definitions) and some active objects (one of which would be running, the others would be waiting to resume their activities). This model with active and passive objects is one that is still being explored as a means of handling concurrency. The new studies are investigating true concurrency achieved with shared memory multiprocessors or, even more problematically, with distributed network computing resources. True parallelism with multiple processing threads introduces extra problems. In a coroutine model, as there is only one processing thread it is not necessary to bother with "critical regions" where data must be updated. But, if there are multiple processing threads, access to data by one thread may disrupt an updating process being performed by another thread.

2.2 SMALLTALK

"Ideally, the personal computer will be designed in such a way that people of all ages and walks of life can mold and channel its power to their own needs."

The quote is from Alan Kay. It expresses his desire for humanized computer, a desire that was partially satisfied through the development of the Smalltalk language and environment.

Kay was dropped into object-oriented programming by his professor at Utah. His research work was to be part of a project on the creation of a single user, graphics workstation with the starting point for ideas on graphics interaction being the earlier work of Sutherland. Sutherland had developed a graphics system, Sketchpad, in the early 1960s. Sketchpad allowed users to manipulate graphic elements, e.g. parts of circuit diagrams, with a light pen. In some of the publications on the work, Sutherland described the system's graphic displays in terms of "masters and instances" – e.g. a master graphic element that specified the representation of a resistor and the instances for each resistor in a particular circuit. After having a week or two to read Sutherland's description of Sketchpad, Kay, as the most junior team member, was given the additional task of getting a strange new compiler to work. It was a Simula compiler.

Sketchpad graphics meets Simula

Kay recognized a coherence between the graphic world of Sketchpad and the classes and objects of Simula. Rather than try to build a complex graphics software system composed of subroutines that manipulated common data structures, Kay realized that there was a more elegant approach. The various graphic elements could be thought of as communicating virtual computing devices. Each owned information such as a position, a scale, a rotation, a label and specialized data (e.g. resistors had Ohm values). Each could be asked to perform actions that depended on their data. Such ideas were utilized in the FLEX system that Kay helped develop at Utah.

Kay joined the Xerox Palo Alto Research Centre (Xerox Parc) at the start of the 1970s. Xerox had a uniquely long-term view of computing. At a time when most other companies were still obsessed with the creation of elaborate multiprocess batch and time-shared operating systems that could maximize CPU usage, Xerox recognized that it was "people usage" that really mattered. Long before other companies, Xerox realized that hardware costs would move rapidly downwards and that it would be possible to dedicate CPU power to individual users. The research group was charged with planning products for the offices of the 1980s and 1990s.

Xerox Parc: planning the future

Some aspects of this long term planning required intelligent but conventional engineering design. Users employing computers to solve problems, as opposed to those solving computer problems, want simple rapid interaction invoking small amounts of processing. A moderate power CPU devoted to an individual user is more appropriate than a large, complex, shared processor. High bandwidth communications require things like bit-map displays. Such features could be foreseen by any group prepared to consider changes to the computing styles that existed then.

Future hardware

*Future human-
computer interaction*
But the Xerox plan also required creativity. The computers were to be humanized. It is wrong to require that people wishing to apply computing power must first learn some arcane set of incantations (accorded the name of a command "language"). There had to be some simpler, more direct way for users to select data and then choose how those data were to be transformed. A major achievement of the Xerox research group was the creation of a new style of user interaction. Crudely summarized, the style is to use a graphical display that allows the user to "pick an object and give it a command".

Children as users
The work of the Xerox group started with an investigation of ways to build the "Dynabook" – a kind of "personal digital assistant" that was to combine information storage with active processing capabilities allowing it to respond to queries and experiments. Various experimental computers and software simulators were built in order to identify the problems that might arise with such personal computers. The researchers were influenced by the work of Papert on the Logo language that had been developed to enable children to explore computational processes. It was felt that problems of human-computer interaction could be brought strongly into focus when children were considered as users.

The Smalltalk language was designed with the intent of making simple tasks easy to program, so that children might access computer power. But the language was not to be limited, it had to be possible to perform complex tasks. Many of the early studies with Smalltalk did focus on use by children (including the development of an OO programming curriculum for high school students). But, particularly after about 1978, the Smalltalk system became more a development environment for programmers.

Smalltalk-7x
The language is partially interpreted; a compiler converts Smalltalk code into byte-codes that are interpreted by a "Smalltalk Virtual Machine". Most of the early interpreters were developed by Ingalls, with the overall project being lead by Goldberg. Smalltalk went through several successive incarnations.

The first Smalltalk, Smalltalk-72, started life as a Basic program, mutated into Nova assembly language, and finally ended up running on Xerox's own Alto computers. It had objects and classes, but no inheritance. Whereas Simula distinguished built in data types like integers from user defined classes, in Smalltalk everything was an object – an instance of some class. This enhanced the uniformity of the language making it easier for users, particularly children, to learn how it worked. Everything worked the same way. Smalltalk-72 provided classes for numbers, lists and similar structures, and of course Logo style "turtles" for graphics. Many of the applications illustrated in Kay's paper in *Scientific American* were developed in this early Smalltalk dialect.

By 1976, Smalltalk had been redesigned. Class hierarchies could be defined. The definitions of classes were themselves objects and could be manipulated by programs, or could have behaviours that represented properties of the class as a whole rather than properties of individual instance objects. Further refinements led to Smalltalk-80 – the basis for current dialects.

As it evolved, Smalltalk became an increasingly complete environment. It subsumed the role of editor, compiler/interpreter, file-system, and to a large extent that of the operating system itself. The language had been designed to facilitate the construction of software systems where users could pick an object and give it a command. Naturally, this model for a user interface became the model underlying the development environment wherein Smalltalk programs were created.

Smalltalk programs were primarily the creation of individuals, not of large software teams. A programmer would start with a standard "Smalltalk image". This "image" contained the code of the Smalltalk byte-code interpreter (virtual machine) together with the source code and compiled forms of a large number of predefined classes that, using a single inheritance scheme, formed an extensive tree structured hierarchy. Smalltalk programs were created by changing this image and then saving the modified image. The changes could be effected through the addition of new classes or modifications of the standard classes. By changing the standard Smalltalk image, a programmer could produce an individualized system that had been extended to solve a particular type of problem.

Smalltalk images

The classes supplied in the standard image included "container classes", "compiler classes", "process management classes", "file classes", "display classes", and others. The container classes specified the forms and behaviours of various abstract data types for storing information. These varied from simple fixed size arrays, through trees, to complex dictionary structures. The "compiler classes", whose behaviours defined how Smalltalk source code was converted into byte-codes, naturally made use of these container classes. For example, the "parser" components of the compiler built syntax-trees (which are special kinds of trees), utilized a system dictionary (an instance of a special kind of dictionary), and produced byte-codes in "compiled methods" (instances of a special kind of array).

All of this standard code was accessible to the Smalltalk programmer. Since new classes were defined as extensions of existing classes (even if just the base "Object" class), a Smalltalk programmer needed to become familiar with the classes that made up the standard Smalltalk image. As there were a couple of hundred classes, each having many different behaviours, this was a formidable task. Conventional documentation techniques were of limited efficacy.

Lisp interpreters are examples of a simpler situation where conventional documentation techniques suffice. Many Lisp systems are delivered as an "image". The Lisp image would contain a few hard coded functions ("eval", "apply", "cons", "car", "cdr", etc.) that defined the basic Lisp interpreter and anything up to a thousand or more additional functions that would be represented in source code form. Like the Smalltalk programmer, the Lisp programmer could modify the standard image by adding extra functions. The Lisp programmer too had to become familiar with those functions already provided as part of the standard image. But in the case of Lisp, these functions could be documented individually. Each function took arguments and delivered a result. Each was independent and could be discussed in isolation. Lisp programmers might use existing functions in the implementation of a new function, but they didn't have to modify existing functions.

In Smalltalk, the classes describe objects which have behaviours and state data and interact with objects of other classes; these classes can be modified through the creation of subclasses. Documentation that simply lists classes, and describes each class in terms of a set of behaviours, is of limited utility. One constantly requires cross-referencing among the class descriptions.

Smalltalk "class browser"   The development environment provided by the Smalltalk group incorporated new tools that facilitate the exploration of interrelations among members of complex class hierarchies. The environment used the multi-windowing graphics capabilities of the Smalltalk language. Code defining new classes, or additional features for existing classes, could be entered in one window. Other windows could be used to display lists of existing classes (or a diagram of the tree-structured class hierarchy) or details of any selected class. Information was retrieved by "picking an object and giving a command". Thus, details of a class could be obtained by using a mouse to select that class in the hierarchy diagram and giving a "display details" command. Fragments of code could be selected, again by using the mouse, and tested by entering a "do it now" command.

One version of the standard Smalltalk environment is illustrated in Figure 2.4. This screen image was obtained using Digitalk's Smalltalk/V(Mac). The main window contains three subwindows. The top-left subwindow contains a list of the classes that are defined in the current Smalltalk image. One of the process management classes has been selected and is highlighted. The top-right subwindow is here used to display a list of some of the things that a Process-Scheduler can do; a ProcessScheduler can create a copy of a stack, it can "fork" etc. The third subwindow displays the code that specifies how a Process-Scheduler actually does fork. This subwindow permits editing. A Smalltalk programmer might want to change the fork code, possibly getting extra trace information displayed when a process does fork. The change could be entered and the Smalltalk image saved; subsequently, the system would exhibit the modified behaviour.

Smalltalk development environment is widely copied   As well as the "browser" tool for facilitating the inspection and editing of code, the Smalltalk environment included tools that allowed a programmer to inspect objects at run-time. Development environments that provide standard classes in a library, a "browser" for viewing these classes, and an "inspector" for looking at specific objects are now becoming standard aids for all object-oriented languages. Almost all these browsers, inspectors etc. are recognizably derivatives of the early Smalltalk prototypes.

Smalltalk language principles   Although the language evolved through many versions, three principles characterized all the variants: 1) data are stored as objects with automatic storage management, 2) objects are all instances of classes, with the class containing a definition of the object's behaviours, and 3) processing is accomplished by "sending messages" to objects.

The fragment of code from Kay's 1976 paper, that was shown in Figure 2.1, typifies the Smalltalk style (although its syntax is a little dated):

```
box new named "joe"
```

This statement involves two messages sent to different objects. First, a "new" message is sent to an object that knows how to create "box" objects (the recipient here is a "meta-object" that represents the box class). This message results in the creation of a new instance of the box class. This new box is then sent the "named" message with an argument "joe" (making variable "joe" a reference to the box).

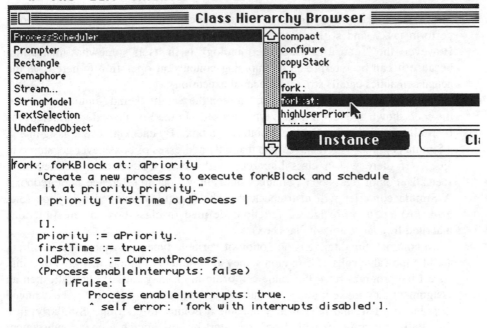

```
fork: forkBlock at: aPriority
    "Create a new process to execute forkBlock and schedule
     it at priority priority."
    | priority firstTime oldProcess |

    [].
    priority := aPriority.
    firstTime := true.
    oldProcess := CurrentProcess.
    (Process enableInterrupts: false)
        ifFalse: [
            Process enableInterrupts: true.
            ^ self error: 'fork with interrupts disabled'].
```

Figure 2.4   A version of the Smalltalk class Browser.

```
joe turn 30
```

The box object is then sent another message, this one asking it to turn itself by 30°. A "turn" function is executed that recomputes the positions of the corners and then redraws the box on a terminal screen.

Smalltalk introduced some novel terminology for OO systems, e.g. "objects are sent messages". A Smalltalk class definition specifies those messages that are accepted by the class meta-object and those that are accepted by individual instances of that class. The code that processes a message is a "method" of the class.

Smalltalk terminology

The term "method" for a function of a class has acquired much wider usage. In general discussions of OO programming languages, the term method is often used in the context of C++ (where the correct term would be "member function") or in relation to Eiffel (where "feature" should be used).

The term "method" is commonly used with other OO languages

Actually, "sending a message to an object" is simply a procedure call; a Smalltalk statement like:

```
joe turn 30
```

is a variation on a more conventional expression like the Simula equivalent:

```
ref(box) joe;
joe := new box;
...
joe.turn(30);
```

Smalltalk's syntax may help to reinforce the concept of objects that can be asked to perform tasks, and so facilitate an appreciation of an OO programming style. However, the "send a message" terminology is in itself somewhat misleading because it can be interpreted as implying concurrent operation (whereas as the actual semantics entails strictly sequential execution).

Typed and typeless languages

There is one important difference between the Smalltalk and Simula versions of the code shown above. Simula is type checked. In the Simula code, variable "joe" is specified as being a reference (pointer) to a box. By checking every assignment, a Simula compiler will make certain that only addresses of boxes ever get stored in "joe". (If there was a class hierarchy with "class coloredbox" defined as a specialization of "class box", Simula would allow "joe" to reference a coloredbox.) A Simula compiler will also check that the functions invoked using joe (e.g. joe.turn(30)) are indeed functions defined in class box (or are functions inherited from ancestors of class box).

Smalltalk is typeless

In contrast, Smalltalk has no notion of variable types. Variables in Smalltalk are like the value cells of Lisp atoms, they are pointers to things, any kind of thing. As a Lisp program runs, the value cell of an atom may initially be null, then an assignment may make it point to a numeric value, subsequently it may be changed to point to a list, and later it may become a pointer to a string. Similarly, in a Smalltalk program a variable "joe" may start by referencing a box; a subsequent assignment may change it to reference a numeric data item.

Method calls are always dynamic

Because there are no types, the Smalltalk compiler never tries to restrict the messages sent to an object referenced by a particular variable, and never tries to use anything like Simula's "static binding". Methods are always invoked dynamically. The code produced for joe turn 30 is always of the form "*hey object referenced by joe, find your* turn *method and use it with argument 30*".

Of course, if the "turn" message is sent to joe after joe has been reassigned to point to a numeric datum, there will be a run-time error (numbers won't have a turn method, because numbers don't know how to turn). "Method not found" errors are part of programming life when working with untyped languages.

There are situations where untyped languages are more flexible. But, compiler checks made in the typed languages do have the advantage of reducing the scope for errors in a program.

Control structures as objects

In Smalltalk, everything is supposed to be an object. "Everything" includes program control structures. Programmers trained in more conventional languages often take some time to acclimatize themselves to this feature. The following fragment of Smalltalk code is a method of class OrderedCollection (in the Digitalk Smalltalk/V implementation):

```
include:  anObject
   | index |
   index := startPostion - 1;
   [ (index := index + 1) > endPosition ]
         whileFalse:  [
                 anObject = (contents at: index)
                          ifTrue:  [ ^true]].
   ^false
```

Basically, this is a routine that takes one argument (anObject) and returns a Boolean value. The code for the routine starts with the declaration of a local variable index. The variables startPostion, endPosition, and contents are instance variables of the "ordered-collection object" that is being searched; Pascal and C programmers can imagine them as being defined in some enclosing block. The main part of the body of the routine involves a loop in which successive data elements are checked. If an element is found to be equal to the argument anObject, the routine returns the value true; otherwise, it returns false (Smalltalk's ^ is approximately synonymous with C's return).

The novelty lies in the loop structure:

```
[ code block1 ]  whileFalse:  [ code block2 ].
```

The *code blocks* are objects, and therefore can be sent messages, such as the whileFalse: message. In this type of loop, *code block1* should be something that creates a Boolean valued object. If this object has value false, *code block2* will be executed; then, *code block1* will be re-evaluated. This process continues until *code block1* returns a true value.

Around 1979, Xerox decided to disseminate its latest Smalltalk. This was intended to help achieve a number of objectives including 1) a desire to influence hardware designers to produce machines on which Smalltalk might perform better, and 2) the intent of establishing Smalltalk as a standard OO programming language and program development environment. Collaborations were established with a number of companies, such as Apple and Tektronix, who implemented Smalltalk on their machines. A series of books was published. Widespread coverage was achieved in August 1981, with an issue of *Byte* magazine devoted to the language

Out of the lab. and into the world

Although greatly increasing awareness of Smalltalk and interesting many in its potential, the articles in *Byte* did tend to emphasise its distinctness. This is particularly true of the paper by Ingalls that summarizes the design principles behind Smalltalk. Some of the principles are just statements of a particular view of programming and programming languages:

A system to serve the creative spirit

• A language should be designed around a powerful metaphor that can be uniformly applied in all areas.

• A computer language should support the concept of "object" and provide a uniform means for referring to objects in its universe.

• Computing should be viewed as an intrinsic capability of objects that can be uniformly invoked by sending messages.

• To be truly "object-oriented", a computer system must provide automatic storage management.

Other principles were:

• An operating system is a collection of things that don't fit into a language. There shouldn't be one.

- The purpose of the Smalltalk project is to provide computer support for the creative spirit in everyone.

Now, a system to serve the creative spirit may be fine for folks at a "non-profit" organization like Xerox, but it does seem remote from the needs of the average programmer. Joe Blow, programmer at Acme Computer Systems, didn't need a system for his creative spirit; he needed a faster C compiler to help him finish Acme's new IBM-PC spreadsheet program. As for "computing being an intrinsic …; intrinsic what of what?".

Applications emerge   The dissemination of Smalltalk into companies like Tektronix and Hewlett Packard did eventually lead to broader use. Its value as a prototyping language began to be appreciated. Love has described some early scientific applications developed for Schlumberger Research where geologists had to develop many different, short-lived, specialized programs for the display and interpretation of various geophysical data. A Smalltalk system, with various specialized display classes, was found to be a much more productive environment for developing such programs (which had previously been implemented in Fortran or Pascal).

Commercializing   In the mid-1980s, Xerox corporation finally started more seriously on the
Smalltalk: ParcPlace   commercialization of the work of its Palo Alto Research Centre. A separate company, ParcPlace, was established to develop and market Smalltalk. This company subsequently broadened its product range to include development environments and libraries for C++. ParcPlace, and other companies, can now supply Smalltalk environments that are substantially platform independent. Consequently, a software house can develop a Smalltalk application and have it run on Unix, IBM-PC, and Macintosh machines.

Smalltalk is used for the development of deliverable applications. These applications tend to be highly interactive "editors". For example, there are Smalltalk based graphic CASE tools, pretty interfaces to electronic mail systems, systems for modelling and displaying traffic in computer networks, and systems for displaying and manipulating the images produced by CAT-scanners and X-ray machines. On current machines, Smalltalk applications execute rather sluggishly. Enhanced hardware, such as 66 MHz Intel-486 and Pentium CPUs, will make performance more acceptable. The use of Smalltalk for product development will increase as high performance hardware gets cheaper; but, at the moment, the language remains primarily a tool for prototyping and for more long-term experimentation and research.

2.3 THE ARTIFICIAL INTELLIGENTSIA: ACTORS, FRAMES, AND OO EXTENSIONS TO LISP

Hewitt's actors   Many ideas related to objects, classes, hierarchies and so forth have moved in and out of fashion in the AI world over the past 30 years. Actually, one strand has been running steadily (though very slowly) for about 25 years. At about the time that Kay was imagining graphics computer systems being built from a horde of little virtual computers, Hewitt at MIT was inventing a somewhat similar model as the basis of an intelligent machine.

Hewitt's "actors" model arose out of one of those academic squabbles. This particular squabble was between the AI "declaritivists" and the AI "proceduralists" and took place in the late 1960s to early 1970s. The declaritivists wanted an intelligent machine's knowledge represented as a uniform collection of facts operated on by some generalized inference system. (Modern Prolog programs constitute a somewhat simplified version of the declaritivists' ideal.) Many of the proceduralists were associated with research groups that had some robotics experience. They felt that it was non-intuitive to require a robot to perform a logical proof that it could actually move its hand to pick up a block, and so sought other ways of representing and using knowledge. The first "actors" model was one suggestion. An intelligent computer system was to be composed of a host of little virtual machines, or "actors", each of which could solve one specific kind of subproblem. A task given to an intelligent machine would come to some initial actor that would know how to break the task into subproblems. This initial actor would then communicate with other actors; each would either solve a subproblem and return a result, or invoke help from yet other actors.

Actors are somewhat like objects; they provide specified services based on private knowledge. They are not necessarily organized in class hierarchies. Any individual actor can serve as a prototype from which another actor, with modified or extended behaviour, can be created. Since the system is all embedded in Lisp, everything is highly dynamic and can be changed as a program executes.

The actors project soon came to focus on problems of organizing massively parallel systems. It has been an active research project for over 20 years. (Somewhat misleadingly, an OO language called Actor was created in the late 1980s. Crudely summarized, Actor is a Smalltalk dialect for the IBM-PC that uses conventional procedural code for methods. It has nothing to do with the actors project.)

Other research into knowledge representation for AI programs lead to concepts similar to classes and instances (objects). Minsky's "frames" are in some ways similar to class definitions. A frame is a bit like a declaration of a record structure in Pascal or C; it specifies the form of a structure with various "slots", just as Pascal record declarations define the form of a structure with various data fields. Programs could create instances of different frames and fill in the slots with appropriate individualized values. However, "slots" are more complex than mere data fields. Like a data field, a slot can hold a value; but in addition, a slot may have numerous attached Lisp functions. There can be functions that will be automatically invoked if the value in a slot is read or is changed. Other slots in a frame might have functions that can compute a value, or provide some default value if a specific value has never been assigned to the slot.

Minsky's frames

Relations among different frame definitions could be described. Where appropriate, such relations could be used to form a tree-structured classification hierarchy for frames similar to the class hierarchies in Simula or Smalltalk. But more general structures could also be created. For example, a "person" frame might be declared as being related to a "legal entity" frame and to a "mammal" frame. A program manipulating data about some specific person could use these relations as needed to determine either the legal responsibilities or the biological characteristics of that person.

"Frame" like constructs were used in a number of the experimental knowledge representation languages of the late 1970s. In the 1980s, some of the "expert system shells" also utilized frames as a way of organizing domain knowledge.

Lisp meets Smalltalk

In the late 1970s, there was some cross fertilization between the Smalltalk project at Xerox Parc and Lisp users from Stanford. This lead to a variety of extensions to Lisp that allowed for some form of object-oriented programming. Other rather similar OO extensions to Lisp were being explored simultaneously at MIT.

Some of the earliest of these extensions were very obviously Smalltalk based. A class declaration template was provided:

```
(setq aircraft (class (Object)
    (ivs  (vx) (vy) (vz) (callsign) …)
    (methods      (method1 (arglist))
                  (method2 (arglist))
                  …
                  (accelerate (a1 a2 a3))
                  (turn (d))
                  …)
    (cvs (counter))
    (metamethods  (meta1 (arglist))
                  …)))
```

Code something like this would define `aircraft` as a new class (derived from some base `Object` class provided in the system). Each individual aircraft (instance of this class) has a number of instance variables (`ivs`) that specify x, y, and z components of its velocity, its call sign etc. Aircraft have methods like `accelerate` and `turn`; these were listed along with their argument lists in the class declaration. Like Smalltalk, these Lisp dialects had "class variables" (`cvs`) and "meta methods" (class methods). Class variables hold data characteristic of all members of the class.

Code implementing the methods could be given in the class definitions or could be specified separately. Once the class had been defined, and the methods implemented, instances could be created and manipulated:

```
(setq a1 (send aircraft 'New))
…
(send a1 'turn 15)
```

The coding style emphasized Smalltalk's idea of sending a message to an object (or to a class). Thus, the first statement creates an individual aircraft by sending a "New" message to the aircraft class. Later, the aircraft `a1` is sent a "turn" message.

Multiple inheritance

These early OO Lisp dialects provided the first implementations of multiple inheritance. Concepts analogous to multiple inheritance were commonplace in AI programs. The semantic networks of the late 1960s had multiple links among nodes. Variations on "frames" had defined relations among classes some of which could be represented as multiple inheritance schemes.

CLOS

The Smalltalk-like dialects of Lisp had the disadvantage of resulting in programs containing two distinct styles of code. Most Lisp code has the standard functional style with statements having the form (`<functionname>`

<argument1> <argument2> ...); the "OO" bits had the quite different style (send <object> <message ...>). On the whole, the Lisp community felt that this mixture was undesirable, and as the OO dialects of Lisp evolved they were reworked so that the invocation of methods of objects could be expressed in Lisp's normal functional style. The early OO dialects of Lisp were eventually replaced by the "Common Lisp Object System" (CLOS).

CLOS uses a simple macro to define the data structure aspect of objects. Macro defclass is similar to the defstruct macro that previously existed in Lisp for defining simple structures.

CLOS's defclass macro

```
(defclass aircraft ()
    (vx     ; x-component of velocity
     vy
     ...
     callsign))
(defclass helicopter (aircraft)
    (numrotors    ; number of rotor blades
     ...))
(defclass jet (aircraft)
    ...))
```

In the simplest use, the macro specifies any inheritance relations and names additional data fields ("slots"). Here, class aircraft is defined; as no inheritance is specified it is implicitly derived from a "standard-object" class provided by the system. An aircraft has a number of slots – components of velocity etc. Class helicopter is a specialization of class aircraft; helicopters have extra slots, such as a count of rotor blades. Usually, more complex versions of the defclass macro would be used so as to specify default initial values for data elements, and to provide access functions for setting or reading fields. CLOS follows Smalltalk, and earlier OO-Lisps, in allowing the definition of class variables ("shared slots") as well as instance variables ("local slots"). A CLOS program creates instances of a class using the make-instance function:

Make-instance function creates objects

```
(setf aircraft1 (make-instance 'helicopter))
```

Normally, a call to make-instance would include a series of key-value argument pairs that initialize slots in the new object.

"Type" information is kept in objects. There are functions provided to ask an object whether it is a member of a particular class.

Run-time object type identification

As explained earlier, CLOS wanted to use Lisp's normal function-call syntax for method calls. Lisp's function calls take the form:

```
(<function-name>  <arg1>  <arg2>  ...)
e.g.
(cons 4 '(3 2 1))
(turn x 35)
(land second_aircraft)
...
```

The Lisp interpreter looks up the function name (`cons`, `turn`, or whatever) in its symbol tables and checks that the name is associated with a built-in function or with a user-defined (`defun`) function. If there is no such function, Lisp reports an error. If a function is found, Lisp checks that it has been given the correct number of arguments; again, if there are incompatibilities, an error gets reported. Otherwise, Lisp proceeds by evaluating the arguments, binding the values to the formal parameters of the function, and then proceeding to interpret the body of that function.

Defgeneric and defmethod

CLOS adds the ability to "overload" function names. So, one can have several different `land` functions. The function dispatch mechanism involves additional processing steps that select the appropriate version. Different overloaded functions can be defined in CLOS using the `defgeneric` and `defmethod` macros. The `defgeneric` macro identifies a function as one that can have multiple versions, specifies the number of arguments, and possibly provides some default implementation. Specialized variants appropriate to different classes are created using the `defmethod` macro (all variants must have the same number of arguments):

```
(defgeneric
    land (thing)
    (:documentation "code to get some aircraft to land"))

(defmethod
    land ((thing jet))
       (let* …    ; code to land a jet
        ))

(defmethod
    land ((thing helicopter))
       (let* …    ; code to land a helicopter
        ))
```

The different specialized methods refine the function's argument list by specifying classes for the argument(s) and provide distinct function bodies. When calling such a generic `land` function, as in (`land second_aircraft`), CLOS evaluates the argument and checks whether it references an object. If an object is referenced, CLOS uses its class to help select the variant of the generic `land` function that should be used. So if `second_aircraft` references a jet, the specialized jet-landing method will be used; but if it had referenced a helicopter, then the helicopter-landing function would have been invoked. Thus, CLOS can provide a dynamic call mechanism equivalent to Smalltalk's scheme.

Multi-methods

Actually, CLOS allows much more complex forms of function dispatch. The checking process is not limited to the first argument. One can have something like:

```
(defgeneric
    rear_end_collision (stationary moving)
       )
(defmethod
      rear_end_collision ((stationary  truck)  (moving
                                          car))
       …)
```

```
(defmethod
        rear_end_collision ((stationary  bicycle)
                                        (moving truck))
          ...)
(defmethod
        rear_end_collision ((stationary  truck)
                                        (moving bicycle))
            ...)
```

with the code for the different methods describing the various outcomes for collisions between differing kinds of vehicles. A call such as:

```
(rear_end_collision vehicle1 vehicle2)
```

will entail a consideration of the types of both vehicles during the process of choosing the appropriate function to invoke.

Complexities such as multi-methods, class and method precedence rules, and multiple inheritance all add to the challenge of programming in CLOS.

2.4 OBJECTIVE C

Objective C was the hybrid language that Cox devised to provide an evolutionary pathway from conventional procedural programming to an object-oriented style. Cox saw object-oriented programming as a way of dealing with the problem of change:

> "(A way of building) systems malleable enough to keep up with organizations propensity for creating, transporting, and manipulating a tremendous variety of data types.
>
> This is done by adding a thin layer of object-oriented structure on top of conventional hardware, languages, and operating systems. ... It allows programmers to define new data types and install them in working systems, often without changing the rest of the system. ..."

In his papers, Cox contrasted the operator/operand model (function/data structure) of conventional programs with the message/object model. He pointed out that "operators" are not independent of their operands; as in the example from previous sections, the "operator" that lands an aircraft depends on the type of the aircraft. In the conventional programming style, such differences must be tracked in the calling environment. Hence, it is inevitable that the calling environment must be cognizant of all the different variant data structures and the distinct functions that are appropriate for each variant. If new types are added (e.g. hovercraft get added to the range of aircraft managed), changes are necessary throughout the programs that use aircraft (the environment). In contrast, the message/object model encapsulates knowledge of the different implementations of a function with the data; the environment need only know of the existence of the function and is not affected when other variant types are added. Using such arguments, Cox helped convince many software engineers of the potential advantages of an object-oriented approach.

A veneer of Smalltalk on a C substrate

Cox's Objective C language was very strongly influenced by Smalltalk but uses standard C as its basic substrate. An Objective C program will contain some conventional C programming constructs interleaved with statements utilizing objects. Objective C is handled by a preprocessor that translates its class declarations and implementation code into conventional C structures and code.

Objective C's objects

The programming style for the object-oriented aspects is the same as in Smalltalk: messages are sent to objects. Even the syntax is similar; messaging is encoded in constructs of the form [<object-id> <message-id> <arguments>]. (Such statements get translated into C code in which a message dispatcher function is called with the <object-id> and <message-id> as arguments. This message dispatcher is part of the run-time support environment. It uses the object's class, and if necessary data on the inheritance hierarchy, to identify the C function that corresponds to the method being invoked.) In normal usage, objects are really typeless; all variables used with objects are of type id (a C pointer type defined in a typedef statement).

Objective C's classes

The language supports single inheritance class hierarchies. Classes can have shared class data and class methods ("factory methods" in Objective C). Instances are explicitly created (by sending a "new" message to the factory) and of course are really just heap-based structures. Smalltalk's automatic memory management system for reclaiming unused objects was abandoned as being too costly, so programmers became responsible for memory management.

Class libraries as an essential part of an OO environment

As well as presenting the case for an OO programming style, Cox also argued for the use of reusable code. Using analogies to hardware design, he recommended the use of "Software Integrated Circuits" – pre-built components that could be purchased from a supplier. Thus, the provision of a class library was an intrinsic part of Cox's model for evolutionary change from conventional to object-oriented programming. Consequently, Objective C came with class libraries, such as its IC-Pak library. IC-Pak provides facilities similar to those of Smalltalk's collection classes (i.e. useful data storage structures such as trees, and lists.)

Use of Objective C

Much of the software for Next computers was developed in Objective C. There are also low cost compilers available from Stepstone for Unix, IBM-PC and Macintosh machines. However, Objective C has not made much impact in the market place.

2.5   APPLE'S PASCAL DIALECTS: CLASSCAL AND OBJECT PASCAL

According to folklore, Steve Jobs then head of Apple Corporation visited the Xerox Parc research laboratories in the late 1970s. He came away convinced that he had seen the future of computing in the Smalltalk systems running on prototype Xerox Alto and Dorado computers. He believed that if Xerox couldn't commercialise the concepts, then Apple would.

Apple Corporation had two projects developing machines to succeed the Apple II and the failed Apple III. One project lead to the short-lived Lisa machine that became available in 1983; the other project resulted in the Macintosh in 1984.

The Lisa was considerably more powerful, and costly than the Macintosh. The Lisa system didn't actually use the Smalltalk language, and it abandoned the Smalltalk model of an integrated operating system, language compiler/interpreter and user interface. But, the software and operating system were basically object-oriented. Applications were built from reusable components – the Lisa Toolkit. (The Lisa Toolkit was defined by a set of classes, quite unlike the more familiar Macintosh Toolbox which is simply a horde of subroutines.) These Toolkit classes, and much of the operating system, were written in the language "Classcal" (≈classes in Pascal) which was an OO hybrid based on Pascal and Simula.

Lisa and the Classcal language

The Lisa was not a commercial success. It delivered a Smalltalk-like environment for a fraction of the cost of a genuine Smalltalk machine; but the world wasn't ready for Smalltalk-like systems. The cost, ≈$10,000 in 1983, was too high for a personal computer system. Further, software product development for the Lisa was slow. At that time, there were very few programmers with experience in object-oriented techniques; consequently, developers had difficulties exploiting the Classcal language and Lisa Toolkit classes. Within a couple of years, this experiment died.

Constraints on the design of the Macintosh, particularly limitations on memory, precluded a wholehearted adoption of an object-oriented style. Many of the components in the Macintosh operating system are clearly "object inspired" (e.g. windows, menus, and other components are defined as data structures with a pointer to a table of associated functions – the WDEF or MDEF functions). But overall, the operating system lacks the intelligibility and consistency that might have been obtained from a carefully thought out OO design and implementation.

Macintosh developments

Initially, the main language for the Macintosh was a Pascal dialect derived from UCSD Pascal. In many ways, this dialect resembled Modula2 more than the old standard for Pascal. The language had "Units", equivalent to Modula2 modules, that supported modular program construction and that could be separately compiled. In cooperation with Niklaus Wirth, Larry Tessler at Apple made a minimal set of further extensions to this Pascal dialect to allow for object-oriented programming.

Wirth and Tessler define Object Pascal

Object Pascal extends Pascal's standard "record" type declaration by allowing methods to be specified along with the data fields:

```
type
    aircraft     = object
          vx, vy, vz   : integer;
          callsign     : string;
          ...
          procedure turn(angle : integer);
          procedure land;
          ...
          end;

    helicopter   = object(aircraft)
          numrotors    : integer;
```

```
...
procedure hover;
procedure land; override;
...
end;

jet      =      object(aircraft)
...
procedure land; override;
end;

propellerplane -      object(aircraft)
...
end;
```

Such type declarations define the forms of the data structures that represent individual objects (instances of a class) and provide forward references to the associated methods.

These Tessler-Wirth extensions permit the creation of tree-structured class hierarchies, as in the example where helicopter, jet, and propellerplane are defined as specializations of aircraft. Object Pascal is more flexible than Simula in respect to the redefinition of inherited methods. A method can always be redefined, as is `procedure land` in these examples. (The compiler requires the programmer to explicitly confirm, by use of the keyword "override", that redefinition is intended and that there hasn't just been an accidental name clash.) All invocations of methods are, in principle, dynamic; a statement `x.land` is compiled into code of the form *object x, lookup your land method and execute it*. (However, in many situations the compiler-linker system can optimize the code to employ direct subroutine calls.)

In Object Pascal (OP), objects are only created on the heap. A variable defined as being of an object type is actually a reference (usually, a four-byte pointer), like a Simula `ref(X)` variable. (In fact, because of a quirk in the Macintosh operating system's approach to memory management, these variables are actually double indirection pointers. The Macintosh OS needs to be able to move data in its heap. When a data structure is created on the heap, the Macintosh OS records the address of the structure in a "master pointer", that is located at some fixed address, and then returns the address of this "master pointer" rather than the address of the data. Each heap-based data structure has a link to its own master pointer; if the OS moves a structure, it can find and update the master pointer. Accesses to the data from program code use two indirection cycles: the first goes from pointer to master pointer, the second from master pointer to data.)

As in Simula, instances of objects must be created explicitly; this is achieved using an extended version of Pascal's `new()` dynamic memory allocator:

```
var
    x        : aircraft;
    ...
    if inputval = 0 then x := new(helicopter);
    else
    if inputval < 0 then x := new(jet);
    else x := new(propellerplane);
```

...

The invocation of methods of objects is again in the style of Simula:

```
x.turn(30);
...
```

(The syntax of the OP language hides the fact that multiple indirection operations are being performed on the reference variable.)

Object Pascal did not attempt to provide sophisticated automatic memory management; the programmer became responsible for reclaiming heap-memory from objects that had been discarded. Access controls were not defined; the data fields of an object can be modified by code other than the nominated methods. The language did not provide anything equivalent to Smalltalk's "class methods" or "class data". The extensions to the base Pascal dialect were deliberately minimal; providing just enough to permit the adoption of an OO programming style.

Consequently, Object Pascal is not in itself particularly interesting as an OO language. But, in 1986, Apple released the MacApp framework class library written in Object Pascal. Here was the first realistic attempt to get commercial software developers to switch from conventional procedural programming styles to the use of object-oriented techniques.

MacApp

The MacApp library was devised to tempt developers who had been creating Macintosh applications using Apple's Pascal language. There was no need for their programmers to switch languages, and the syntax and basic use of the new object-oriented language features were easy to learn. Moreover, the classes in the MacApp library offered major advantages to developers.

Programmers can benefit by reusing list classes, tree classes and so forth; but the benefits of such reuse are relatively minor. Handling of data is rarely a major issue, "collection class" libraries are useful but they cannot influence development costs in any dramatic way. A framework library is quite different. Rather than assorted individual components, a framework provides an integrated set of classes targeted at a particular type of usage. There will be many collaborations among instances of these classes, many interconnections. The advantage of a framework comes from having such inter-connections sorted out within the framework itself.

Framework class libraries

Macintosh programs are expected to support multiple documents, scrollable and selectable displays in windows, cut-and-paste editing and are supposed to be consistent with certain semi-standardized user interface conventions. The implementation of such features requires a lot of code – code that is more or less orthogonal to the application specific issues on which a developer would really prefer to focus.

The designers of the MacApp library abstracted out the essence of a Macintosh program and devised a set of classes that represented its various principal components. These classes included i) an Application class, ii) a Document class, iii) Window and Frame classes, iv) View classes, and v) Command classes. An Application object organized interaction with the operating system and translated low-level system events into messages for other objects. A Document object organized data in memory and data transfers between files and memory. View objects displayed data and handled some aspects of user-interaction such as mouse-

MacApp classes

based selection and drawing. `Frame` and `Window` objects were created to group views. `Command` objects provided support for reversible changes – situations where a program is expected to offer Edit/Undo and Edit/Redo options. (The MacApp model was influenced by Smalltalk's *model-view-control* paradigm for program organization. The "model" part contains problem specific objects; "views" are for visualization; a "control" part handles user interaction. But in MacApp the control responsibilities are a little more diffuse, being present in several classes.)

Code provided to handle typical patterns of collaborations amongst objects

The code defined as methods of these library classes provided a form of skeletal program. An `Application` object would respond to a user's File/New and File/Open menu-selections by creating a new `Document` object and, if appropriate, telling the `Document` to read data from an existing file; then, the `Application` object would ask the `Document` to create its windows and views. `Document` objects could respond to a user's File/Save or File/Revert menu-requests. Keyboard and mouse events were picked up by the `Application` and translated into requests for actions by the appropriate views, windows, scroll-bars or other objects present in the running system. All the standard kinds of collaborations among instances of different classes had been worked out and provided for in the code of the library classes.

Specializing the library classes

Of course, the `Document` class in the library didn't contain any data; the windows that it opened contained blank views; when, as a result of a File/Save menu selection, it was asked to write data to a file, it did nothing. A developer using MacApp would create a new specialized subclass of class `Document` that did hold data, that created views that displayed its data, and which could transfer its data to files. Only a few methods had to be written. The developer could rely on the framework calling these methods at appropriate times and did not have to get involved in all the low-level event decoding and dispatching.

The power of reuse

One could obtain a real appreciation of the power of code reuse by implementing a small Macintosh program using conventional coding techniques, and then exploit the MacApp library in a reimplementation of the same system. Using conventional coding, a minimal cut-and-past simple text editor required some 2500 lines of code. A MacApp variant, with greater functionality and greater consistency with user interface standards, could be implemented in ≈100 lines of code. All that was required was the definition of specialized `Application`, `Document` , and `View` classes and the implementation of three or four short methods in each of these classes.

Evolution of MacApp

The MacApp framework library has been put to practical use by commercial developers for several years and has evolved through two major revisions. The second version reworked the view classes used for display of information and provided many additional classes; it also supported multiple languages with interfaces to C++ and to a dialect of Modula2 that was available on the Macintosh. The final version was reimplemented in C++.

Apple fails to exploit MacApp

But, to a large extent, Apple Corporation failed to exploit MacApp fully. The company claimed that it was planning for OO programming to be basis of future software systems and asserted that it was creating MacApp to facilitate future product development on the Macintosh. However, the development of its own MacApp product was not what one would expect from a major computer company, being more in the style of an underfunded university research project. The development team was so small that support tools such as class browsers could not be provided with the first release of MacApp. Documentation lagged far behind new software releases. The various versions of MacApp (releases 1, 2, and 3 along with intermediate versions) had significant incompatibilities and no aids for transition; this level of support must have delighted those companies that were using MacApp for product development. Other development groups within Apple failed to coordinate with the MacApp group so that new OS features, e.g. QuickTime, did not integrate well with MacApp.

With MacApp, Apple had had a brief opportunity to take a lead and provide a general multi-platform framework library, but either it did not perceive this chance, or it did not accord any value to such developments. However, MacApp has served as a model for several other framework class libraries that have been developed for other platforms. These other frameworks have been implemented using C++ on Unix and IBM-PC platforms and, in some cases, now offer much more functionality.

MacApp as a model for other framework libraries

Implementations of Pascal dialects with OO extensions appeared for the IBM-PC around 1988-89. The dialect implemented by Microsoft is quite similar to Apple's Object Pascal. Borland's TurboPascal dialect is significantly different; it represents an attempt to map some of the features of C++ onto a Pascal substrate. The existence of such radically different dialects has limited the growth of object-oriented Pascal systems. Apple Corporation essentially abandoned further development of its Object Pascal around 1990.

QuickPascal, TurboPascal

Nevertheless, the Pascal language continues to evolve. The ANSI standards committee for Pascal has already redefined the standard to provide for separately compiled modules and has almost completed a specification of a standardized set of OO extensions for Pascal. This new standard Pascal will allow the definition of "classes" all implicitly derived from a standard ROOT class. Multiple inheritance is permitted. The language distinguishes concrete classes (classes that can be instantiated), abstract classes that can only be used by the derivation of concrete subclass, and "property classes" (these essentially define a set of behaviours, e.g. the abilities to read and write a disk version of an object; property classes are useful when multiple inheritance schemes are used to create concrete classes with a mix of standard behaviours).

ANSI X3J9 Pascal standard

Although the existence of a standard OO Pascal will benefit users, it is unlikely that many major software packages will be developed in this language. Most developers will work in C++; the more innovative may try Eiffel.

2.6   C++

Stroustrup began the development of C++ at AT&T laboratories around 1980. The first widely available description was a paper describing a 1981 version of the language. This paper was published in *Software – Practice and Experience* in 1983. The stated aim was to provide the C language "with facilities for expressing the structure of a medium-sized multi-source-file program". The type of facility desired was described in the following terms:

"For example, it should be possible to specify that a data structure can be accessed only by a specific set of functions, and that it must be initialized in a specified way before use. This should be done in such a way that no 'hidden costs' would preclude its wide distribution within the C community."

It was further explained that "It was not the aim to turn C into a 'very high level language'." Noting the existence of languages like Smalltalk, Simula67, Ada, CLU and others that already provided various abstraction facilities, Stroustrup felt it necessary to explain the need to develop "classes" for C:

"Why bother with a relatively old and imperfect language like C? First, the author does not know of a language, or a set of ideas that could be made into a language, that has all the advantages of C plus the advantages of a data abstraction facility without significant new disadvantages. Second, the effort needed to introduce a new language, however ideal, into the thousands of systems and organizations that now use C is unimaginable"

Those sentences were written in 1981, when C was still a relatively little used language. The 1980s was the decade of C programming. A task that was unimaginable in 1981 has been compounded about one hundred fold. C++ is required as a bridge that allows existing programmers and existing systems to transfer over to the use of object-oriented methodologies. Most organizations can accommodate evolutionary change such as is allowed by the use of C++ whereas revolutionary change, like a switch to a totally new language, is generally impractical.

C with Classes

The 1981 version of the C++ language already provided strong support for abstract data types (classes) that had a public interface specifying the services that they provided, and a private implementation part that defined how those services were implemented. Initialization routines could be defined for classes. Such initialization routines were automatically invoked whenever an instance of a class was created, and so guaranteed that objects were always in well-defined states. The language also changed the syntax of function declarations; argument lists were required so that a compiler could check for correct usage.

Objects as instances of "first class types"

Support for "first class types" was fairly extensive even in this earliest version. Class objects could occur as data members of other classes, could be returned as results of functions, and passed as arguments. A programmer could define the meanings of operators, such as the assignment operator =, to provide a behaviour tailored to the requirements of a particular class.

Support for class hierarchies

The support for class hierarchies was much more limited. Single inheritance class hierarchies could be defined; but the mechanics used to support function redefinition were quite apparent to the programmer. Memory management for dynamically created instances of classes, that had been automatic in Simula, became an explicit responsibility of the programmer.

Efficient code

In keeping with the general tenor of C, there was a focus on efficiency – "any facility that appears to be inefficient will be unused". "Inline functions" could be defined; these provide the semantics of a function call while avoiding the overheads associated with building stack frames and executing call/return instructions. Another construct to support efficiency allowed a cluster of inter-dependent classes to access each other's private data without function call overheads (the "friend" construct).

Stroustrup considered providing parameterized (generic) classes along the lines of Ada packages. Generic classes allow the specification of something like class vector(THING), i.e. a general definition of vector properties that could be used in a program to create vector(int) (an instance of type "vector of integers") or vector(MailItem) (a "vector of MailItems"). This language feature was *not* incorporated because experienced programmers could achieve the desired effect using C macros.

Generic classes considered, but macros thought to be adequate

"C with Classes" was released from the AT&T laboratories in 1983 and began to experience increasing use in universities and research institutes. The name C++ ("increment the existing state of the C language?") is ascribed to R. Mascitti and became standard about 1983. In the period around 1983-1985, there was considerable interplay between the C++ development group and the committees that were starting to define a standard dialect for C. C++'s function signatures were adopted by the ANSI C committee. The C++ language was also tidied up with, for example, improvements being made to the mechanism for defining and redefining member functions in class hierarchies.

C++ enters general use

The language was extensively reworked based on practical experience in the period 1981-1985. Support for classes as first class types was further extended, with programmers being able to redefine the meanings of all operators. So, it became possible for a programmer to declare class Matrix, define the meaning of operators '+', '-', and '*' for matrices, and write code like "Matrix a,b,c; …; c = a + b;". Stroustrup published a book describing the C++ language in 1986. OO programming styles still formed only a minor part being covered in just one chapter of the book, and even this chapter was more concerned with the mechanics of declaring class hierarchies. By 1988, university course texts had begun to appear (e.g. the text by Wiener and Pinson), with the first popular text being that of Lippman in 1989.

C++ Version 1

AT&T laboratories were not in the business of building and distributing class libraries. C++ came with only two small class libraries, and both of these were atypical of the "reusable component classes" that programmers might wish to use or employ as models for new classes. The libraries provided were a "class complex" (complex numbers) and a group of related classes in the "task" library that provided facilities closely similar to those that came with Simula for discrete event simulations. The lack of class libraries hindered the adoption of C++.

The case of the missing class libraries

Gorlen and co-workers at NIH provided one of the first C++ class libraries. This class library, the first version of which was released in 1986, was inspired by the Smalltalk libraries. It contained a number of simple concrete classes that merely repackaged Unix data structures (e.g. a Date class) or provided some special type such as a class for rational numbers. In addition, the library contained a set of "collection classes" (e.g. lists, trees, dynamic arrays) and a single inheritance hierarchy that provided a root Object class (with abilities to read/write disk version, print itself, perform comparisons, etc.) which was elaborated in a

Gorlen's NIH class library; ET++ framework

series of subclasses. As well as being widely used in itself, Gorlen's class library served as a model for many later developments. The ET++ framework library was released in 1989. ET++ was a reimplementation and extension of the early MacApp framework to suit Unix systems that had greater power and more graphics capabilities than a small Macintosh system.

C++ evolves: 85-92

"C++ evolved to cope with problems encountered by users, and through discussions between the author and his friends and colleagues." The original access controls – public interface and private implementation – are fine for a programmer defined data type but are overly restrictive in the context of class inheritance. New access control mechanisms were devised. C++ acquired "static member functions" that provided functionality more or less equivalent to "class methods" in Smalltalk. Multiple inheritance offers power (at the cost of complexity and potential for abuse); extensions permitting multiple inheritance were added to the language. The use of macros for generic class was only ever an option for real C-hackers. The revised language provides templates for creating generic classes. A system for handling run-time exceptions, as pioneered by Ada and CLU, was added to the language.

C++ standard

And, around 1989, the international standards committees got involved. A new standard for C++ has been outlined in the book by Ellis and Stroustrup. The new features are demonstrated in the second editions of Stroustrup's C++ language text, and of Lippman's primer. One illustration of the difference in the language is that several chapters of Stroustrup's second edition are devoted to a discussion of how to design object-oriented programs and class libraries. Some of the features described in these newer texts are only just becoming available on standard compilers. International standards committees prefer lengthy deliberations and several compiler manufacturers have waited for the standard to be approved before implementing newer features such as parameterized types and exceptions.

C++ and the outside world

Interest in object-oriented databases was growing slowly in the period 1985-1990. Early OO databases typically worked with proprietary OO extensions to C, or with CLOS, or Smalltalk. From around 1990, C++ began to emerge as the standard programming language for use with such products. Commercial environments for C++ development on Unix workstations started to come onto the market around 1991 from software houses such as ParcPlace. Software development for personal computers, both IBM-PC and Macintosh, switched substantially to C++ in the period 1990-1992. C++ framework libraries for the IBM-PC were released by Borland (1991) and Microsoft (1992) while on the Macintosh the MacApp framework was also switched to C++.

2.7   EIFFEL

Meyer, the inventor of Eiffel, had been chairman of the Association of Simula users for a number of years and was firmly convinced that object orientation offered the best approach to software development. In the mid-1980s, he established a company, Interactive Software Engineering (ISE), that was originally intended to develop CASE tools. The development environment was problematic as Simula, his preferred choice, was not available on the machines that were to be used.

Meyer decided to implement a modernised Simula. His extensions to Simula include multiple inheritance, a facility for handling exceptions, generic types, and mechanisms for combining assertions with class constructs (preconditions and postconditions for methods, invariants for classes). Automatic memory management was retained from Simula (it had been dropped in Objective C, C++, and the Object Pascal dialects). Nevertheless, Eiffel was kept simple, much simpler than Ada – comparable in complexity with Pascal. It was a one paradigm language; go OO, or don't go at all. A major design objective was to combine flexibility with safety:

> *"From a programming viewpoint, what is really fascinating is the combination of flexibility and safety. Thanks to multiple inheritance and dynamic binding you can get software structures that are so decentralized that reversals of design decisions cease to be a nightmare; they become a normal part of the design process. For all this flexibility, the static typing mechanism, the assertions, and the disciplined exceptions bring a degree of reliability which is unheard of in traditional software developments."*

In 1988, Meyer published a book, *Object-oriented Software Construction*, in which he presented his views on how an OO approach could resolve many software engineering problems, and he described his Eiffel language. The book became an instant classic. Even C++ aficionados admit to having benefited from reading the first 60 pages of Meyer's book. (A later book by Meyer focuses solely on the Eiffel language; the version described is that proposed for the new Eiffel 3 environment just now becoming available.)

"Object-oriented Software Construction"

Rather than sell a CASE tool, ISE sold Eiffel. The implementation of Eiffel from ISE came with a comprehensive set of class libraries and two development environments (both running on Unix systems, one working with X-Window displays and the other with ordinary terminals). The Eiffel compiler converts Eiffel to C code. The idea was that as C is available on almost every machine it could act as a general purpose "assembly language" and facilitate migration of Eiffel programs. Such a view was overoptimistic because the Eiffel development and graphics user interface classes depend on a host windowing environment. Eiffel has been largely confined to Unix platforms with X-Window support. A version of Eiffel for the IBM-PC is available from SIG Computer Gmbh (Germany). The first release of this version received somewhat mediocre reviews, but it has been improved and is so low cost that it might be worth getting for experimentation even if it is not yet suitable for serious development work.

Availability of Eiffel

Adoption of Eiffel has been relatively limited. Some of the problems are technical. Eiffel is resource hungry. Compile times are lengthy; a multitude of temporary files are created during compilations (Eiffel-3 provides a combination of a compiler and an interpreter that can reduce compilation times for classes that are being experimented with). Executable programs are large and their execution speed is a little leisurely. Most of the limitations on adoption are managerial. It is

Why is Eiffel such a rarity?

difficult for a commercial organization to risk a project by adopting a new development methodology and language. Even if use of a higher level language like Eiffel could bring long term benefits, most managerial decisions have to give higher weight to short term considerations. Long term planning requires investment (e.g. training of development teams) and resources for investment are limited in recessionary economic cycles.

REFERENCES

Simula
G.M. Birtwistle, O.J. Dahl, B. Myhrhaug, and K. Nygaard *Simula begin*, Petrocelli/Charter 1973.

O.J. Dahl, and K. Nygaard, *SIMULA – an ALGOL-Based Simulation Language*, Communications of the ACM, **9**, 671-678 (1966).

B. Kirkerud, *Object-Oriented Programming with Simula*, Addison Wesley, 1989.

Smalltalk
D.H.H. Ingalls, *Design Principles Behind Smalltalk*, Byte, **6**, 286-298 (1981).

A.C. Kay, *Microelectronics and the Personal Computer*, Scientific American, September 1977, 231–244.

G. Krasner, *Smalltalk-80: Bits of History, Words of Advice*, Addison Wesley, 1983.

T. Love, *Experiences with Smalltalk-80 for Application Development*, in Tutorial: Object-oriented Computing, G.E. Peterson, (Editor), IEEE, pp. 52-56, 1987.

AI
G. Agha, *ACTORS: A Model of Concurrent Computation in Distributed Systems*, MIT Press, 1988.

C. Hewitt, P. Bishop, and R. Steiger, *A Universal Modular ACTOR formalism for Artificial Intelligence*, in Proceedings of 3rd International Joint Conference on Artificial Intelligence, pp. 235-243, 1973.

J.A. Lawless, and M.M. Miller, *Understanding CLOS*, Digital Press, 1991.

M. Minsky, *A Framework for Representing Knowledge*, in "The Psychology of Computer Vision", P.H. Winston (Editor), McGraw-Hill, 1975.

M. Stefik, and B.G. Bobrow, *Object-oriented Programming: Themes and Variations*, AI Magazine, **6**, Winter 1986, 40-62; (reprinted in Tutorial: Object-oriented Computing, G.E. Peterson, (Editor), IEEE, pp. 182-204, 1987).

Objective C
B. Cox, *Message/Object Programming: An Evolutionary Change in Programming Technique*, IEEE Software, January 1984, 50-61.

B.J. Cox, *Object-oriented Programming: An Evolutionary Approach*, Addison-Wesley, 1986.

Apple's class library
K.J. Schmucker, *Object-oriented Programming for the Macintosh*, Hayden, (1986).

D. A. Wilson, L.S. Rosenstein, and D. Shafer, *C++ Programming with MacApp*, Addison-Wesley, 1990.

C++

M.A. Ellis and B. Stroustrup, *The Annotated C++ Reference Manual*, Addison-Wesley, 1990.

K.E. Gorlen, S.M. Orlow, and P.S. Plexico, *Data Abstraction and OO Programming in C++*, John Wiley, 1990.

S.B. Lippman, *C++ Primer*, Addison Wesley, 1989.

S.B. Lippman, *C++ Primer, 2nd edition*, Addison Wesley, 1991.

B. Stroustrup, *Adding Classes to the C Language: An Exercise in Language Evolution*, Software–Practice and Experience, **13**, 139-161 (1983).

B. Stroustrup, *The C++ Programming Language*, Addison Wesley, 1986.

B. Stroustrup, *The C++ Programming Language, 2nd edition*, Addison Wesley, 1991.

A. Weinand, E. Gamma, and R. Marty, *Design and Implementation of ET++, a Seamless Object-Oriented Application Framework*, Structured Programming, **10**, 1989.

R.S. Wiener, and L.J. Pinson, *An Introduction to Object-oriented Programming and C++*, Addison Wesley, 1988.

Eiffel

B. Meyer, *Object-oriented Software Construction*, Prentice Hall, (1988).

B. Meyer, *Eiffel: the Language* , Prentice Hall, (1992).

Miscellaneous

G.E. Peterson (Editor), *Tutorial: Object-Oriented Computing*, IEEE, (1987).

EXERCISES

1. If you have a Macintosh computer with the MPW development environment and have access to the Internet, you can obtain and experiment with Simula. Find "lund-simula" (or simula-lund) using "archie" or a similar utility program. A number of archive sites have Lund University's public domain version of Simula and its associated Hypercard documentation. Import the Simula compiler and related files and try the example programs provided.

2. Find out about some of the OO languages not mentioned in this short overview: Self, Trellis, Sather, Neon, Actor, Beta.

$\mathcal{H}$ow OO Languages Work

3

This chapter provides an overview of how object-oriented languages work. The focus is mainly on the "hybrid languages", primarily extensions to Pascal and C, where a conventional procedural language has been extended with at least some support for an object-oriented programming style.

As illustrated in Figure 3.1, there are two main concerns. First, there are issues of how the forms of objects are defined. This is a language issue, a compile time issue. The language must have new programming constructs that let a programmer define a new "type" or "class" of objects – e.g. the MailItems, PhoneMessages and Facsimiles of the example Mailer program in Chapter 1. These "type specifications" or "class declarations" will define the data resources owned by each object of a particular type (e.g. all MailItems have an arrival time, Facsimiles also have picture data) and will identify the "services" such objects can provide (e.g. ShowContent and ShowSummary routines). The language must also provide some way of specifying how different classes are related e.g. Facsimile objects need to be defined as specializations of MailItems. Other language extensions may be needed so that objects can be created by a running program.

Compile-time definition of objects,

Secondly, there are all the run-time issues. As suggested in Figure 3.1, most objects will be dynamic and created in the heap. The run-time form of these objects needs to be defined as must the mechanisms for realizing features such as dynamic binding. If there are Facsimile (fax), PhoneMessage, and other objects, one wants the fax-objects to use the Facsimile version of the code for a function like ShowContent while the phone-message objects should use the PhoneMessage version of this code.

Run-time representation of objects

The definition of objects is examined in sections 3.2 to 3.7. The run-time representation and behaviour of objects is covered in sections 3.8 to 3.13. This chapter concludes with consideration of some other issues such as support for automatic garbage collection (a form of memory management).

Small code fragments are used when illustrating many of the concepts. Most code fragments are in C++; some use Object Pascal (Apple's MPW dialect), and a few are in Eiffel. Object Pascal examples are mainly used to introduce a simplified version of a concept. C++ versions of the same examples are then used to show more refined forms. For instance, Object Pascal does not provide controls on access to an object's data, while in C++ access can be controlled. In Object Pascal,

Illustrative code

a type definition describing the form of an object simply describes the data it owns and the behaviours that it exhibits (i.e. its "methods"). Naturally, the C++ declaration of an equivalent class of objects is a little more complex because it must also describe the appropriate access controls on data and behaviours. Eiffel examples are used where they illustrate an interesting alternative to the C++ approach.

These code fragments are so simple that they can be read directly. A more complete introduction to C++ is given in Chapters 4-6.

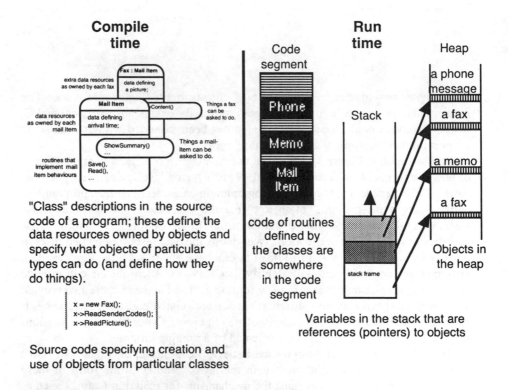

Figure 3.1   Compile-time and run-time aspects of objects.

3.1 "SPACE INVADERS": AN EXAMPLE PROGRAM THAT USES OBJECTS

One main example will be used through most of this chapter; this example is also used in Chapter 6. Features of this program will be used to show how class hierarchies can be employed, and how polymorphism can be exploited. Other concepts, such as multiple inheritance, will be illustrated in terms of possible extensions to the basic program.

The example is a hybrid program that incorporates procedural code for a main program, along with the use of objects. It is a simulation game, originally inspired by the Strategic Defence Initiative of the 1980s.

"Space Invaders" is a simple, arcade style video game. The game pits a human player against waves of program-generated "invaders". The game will employ some form of graphic display and a sound device for output and a mouse, trackball or joystick incorporating a button for user input. As indicated schematically in Figure 3.2, the basic version of the "Space Invaders" program incorporates the following elements:

- *a "Gun"*. This is controlled by the user. The gun is displayed at the bottom of the screen. The lateral position of the gun can be changed using the trackball (or equivalent input device). Firing of the gun is controlled by the button, but is subject to a maximum rate of fire. When the gun is fired, a "laser flash" (vertical coloured line emanating from the gun barrel) will be briefly displayed on the screen. All space invaders intersected by a laser flash will be "destroyed". The score of the player will be incremented by the number of invaders destroyed.

- *Space Invaders*. Space invaders are generated by the program. In the initial version of the program, a constant number of invaders will be in play. As invaders are destroyed by gun fire, they will be replaced automatically. Space Invaders are created with an initial vertical position slightly above the top of the displayed area and a random lateral position within the display bounds. They move continually downwards towards the base of the display.

    Space Invaders may also move horizontally, but are limited to the bounds of the display area. A Space Invader that reaches the left or right bound of the display area will change its horizontal motion so as to move back toward the centre of the playing area.

    If a Space Invader reaches ground level (as represented by a horizontal line positioned above the gun), the game terminates. Termination is to be accompanied by suitable visual and sound effects. A final screen display will then show the score obtained by the player.

- *Saucer, Rock, Bomb*. The initial version of the game will provide three distinct types of Space Invader, namely Saucer, Rock, and Bomb. The behaviours of these different types of invader are generally similar but each type has a distinct visual form and a different pattern of motion. "Rocks" tumble slowly downwards with small random fluctuations in their horizontal location. "Bombs" travel directly downwards with no horizontal motion; their vertical speed is initially small, but they accelerate as they descend. "Saucers" can move across the entire width of the screen; saucers will attempt to avoid gunfire by changing their direction of horizontal motion in attempts to avoid becoming directly above the gun.

- *Sound effects*. The choice of sound effects is left to the imagination and discretion of the programmer.

- *Purpose of game*. The game has no purpose – save possibly that of leaving the players with the satisfaction that they obtained a high score by shooting many invaders before their own inevitable destruction.

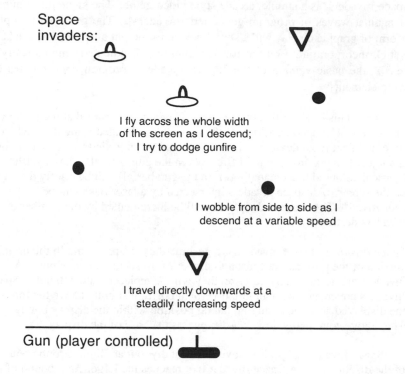

Space
invaders:

I fly across the whole width
of the screen as I descend;
I try to dodge gunfire

I wobble from side to side as I
descend at a variable speed

I travel directly downwards at a
steadily increasing speed

Gun (player controlled)

Figure 3.2   "Space Invaders V1.0" – a simulation program devised for the
training of commanders of Strategic Defence Initiative installations.
Space Invaders can all move – but they all move in distinct ways.

Program structure   The program is to be implemented using objects to represent the gun and the
various invaders. Procedural code will be used to set up the game structure and to
handle aspects such as mouse-tracking, recording of button presses, maintenance of
scores, and keeping a game clock running. The procedural code of the game will
have calls requesting actions by various objects. At appropriate intervals, the gun
will be instructed to adjust its lateral position toward the current mouse-cursor
position ($\approx$1/30th second) and, if there has been a button press, to fire (checked at
$\approx$1/10th second intervals). If the gun is fired, all invaders will be instructed to
check whether they were destroyed. Destroyed invaders should be removed and
new invaders generated. At regular intervals, all invaders will be instructed to
"Run". The Run behaviour of an invader causes it to erase its current display
image, compute a new position, and display an image at the new position. The
Run() method of an invader should return a Boolean value indicating whether that
invader has successfully destroyed the gun with its last move. The control loop
terminates the game when an invader reports success.

Polymorphic   The game code works with references (pointers) to Invaders. Invaders are
references   obviously polymorphic (polymorphic = "many shaped" – just look at Figure 3.2).
Their overall characteristics are similar. They all can Run(). They can check
whether they would have been destroyed by a particular burst of gunfire. Although
the general behaviours are common, the different types of invader handle specifics
in different ways.

The program should take advantage of the commonalities that exist amongst different types of Space Invader. There are two aspects to this use of commonality. One aspect is the sharing of code. If all different types of Space Invader perform some task in exactly the same way (e.g. the check for destruction by gunfire), the code for this task should appear just once and be shared by each different special invader type.

Code reuse among similar types of object

A higher level objective is the separation of the program into a component that handles the game and knows merely of the existence of "Space Invaders", and a component that handles the different types of invader. If a complete separation can be made, the system can subsequently be extended by the definition of a new type of invader – without any change to the code that manages the general structure of the game. This aspect will illustrate how OO features such as polymorphism and dynamic binding can improve the overall structure of an application and facilitate extension.

Exploiting polymorphism and dynamic binding to simplify program structure

The OO components of the Space Invaders application require a language that can define types (classes) of object and specify simple, single inheritance class hierarchies. The type of inheritance hierarchy appropriate for this application is illustrated in Figure 3.3.

Defining class hierarchies

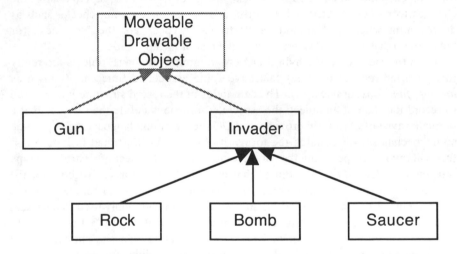

Figure 3.3   A possible class hierarchy for objects in the Space Invaders application.

The application requires the existence of Gun objects, and of Saucers, Rocks, and Bombs. In the source text of the program, it must be possible to define the forms of such objects and specify their behaviours. The Saucers, Rocks and Bombs are all variants of some basic type Invader. An implementation language must provide some means of specifying this relationship in the type (class) definitions. It might be appropriate to add a further level to the class hierarchy. Both Guns and Invaders are "Moveable-Drawable Objects" – possibly this feature could be exploited and some saving of code achieved. (In the actual program implementation, it was felt that the gain obtained from the introduction of a class of

Moveable_Drawable objects was marginal. The program uses a class Gun, and the small hierarchy Invader/Bomb,Invader/Rock, Invader/Saucer.)

3.2 CLASSES: DEFINING THE STRUCTURE AND BEHAVIOUR OF OBJECTS

What are objects?

Object "state" and object behaviours

From the viewpoint of the overall program, an object is something that owns "resources" and provides some services based on these resources. Usually, these "resources" are just simple data elements but they can be things like input-output ports, files, "graphics contexts" that characterise a display device, or other more complex forms of data. In the Space Invaders example, each Invader is an object that owns at least two integers that hold information about its position, along with other data such as horizontal and vertical velocities. (The particular values in these integer coordinates and other variables define the "state" of an Invader. Many texts use the terms "object state" and "object behaviour".) The application structure relies on each individual Invader to take charge of its own display, to control the changes to its coordinates and velocity, and to perform special checks such as determining whether it should be destroyed. These are the "services" (or behaviours) that an Invader provides to the rest of the application.

Object ≈ data structures + associated functions

From the viewpoint of a hybrid OO programming language, objects are really just glorified record structures that have some associated functions (known as *member functions*, or *methods*). The data fields of the record structure will be used to record the state of an object; the associated functions define the services that a particular type of object can provide. An OO programming language must provide some mechanism for defining the forms of objects. As illustrated in Figure 3.4, this will involve i) specifying the data resources that each object of a particular type will own, ii) identifying the things that an object can be asked to do, and iii) providing some implementation routines that define how an object handles tasks.

Design diagram for a class

Diagrams, like the simple version in Figure 3.4, are commonly used to represent classes as conceived by a designer. The notation used in Figure 3.4 is similar to that of Coad and Yourdon, or of Henderson-Sellers. A class is shown to own resources and have behaviours. Such diagrams can distinguish between "public" and "private" behaviours. The behaviours shown as extending outside of the main box, such as Run(), represent the services that an instance of the class will provide. The other behaviours (e.g. Draw()), shown as totally enclosed within the class outline, are support routines required to implement the services that are provided to clients. Every individual Invader object, i.e. each instance of this class, will have its own data resources – its own set of integer variables defining a position etc. They will have some way of sharing the code that defines how they Run() and Move() etc.

Access control on data fields

A hybrid OO programming language may provide some forms of control on access to the data fields of an object. If a language enforces access controls, then each object really does own its resources (because these data can only be modified by the object executing one of its associated routines). This idea is known as *encapsulation* . Each object is a little walled-off capsule of data, with its data fields only accessible through the service functions that it offers.

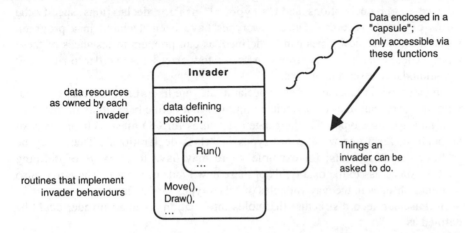

Figure 3.4   An OO language must provide a means for defining the forms of "objects". What data do they own? What can they do? How do they do things?

In Eiffel, Smalltalk, and Simula a "class" is a syntactic unit. In contrast, as illustrated in Figure 3.5, hybrid languages such as Object Pascal, Common Lisp Object System, and C++ have a weaker model. A "class" is represented as a form of "type declaration" and set of separately defined routines.

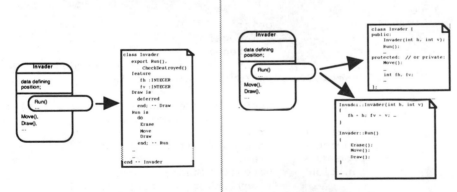

Figure 3.5   Contrast between OO languages with "classes" as syntactic units and hybrid languages with classes based on existing facilities for defining record structures.

Base language
support for "record"
/ "struct"
declarations
Languages like Pascal and C already have mechanisms for specifying the form of new record types and for creating instances of these types. Pascal's `Type record … end` declarations, and C's `typedef struct` declarations, specify the forms of simple records. Once new types have been declared in a program, variables of these new types can be defined, as can pointers to instances of these types. These existing mechanisms just need a few extensions in order to be able to accommodate classes and objects.

Hybrid OO
extensions associate
functions with struct
declarations
In these hybrid languages, "class declarations" are like `struct` declarations but with the extra feature of the associated routines that define behaviours of instances of a new `struct` type. The language extensions for OO must include a way of associating functions with record types in "class declarations". There may be additional requirements; for example, there may have to be ways of defining controls on access to the data fields of these new kinds of record. There may also be minor changes in the way variables of "class types" are defined.

Pascal style object
declaration
In Pascal, a record structure that holds information about an invader could be defined as:

```
Type
    Invader        =        record
            h        : integer;    { horizontal coordinate}
            v        : integer;    { vertical coordinate }
            …                      { any other data fields needed }
            end;
```

With object extensions (as in Apple's MPW Object Pascal), this declaration can become:

```
Type
    Invader        =        object
            h                : integer;
            v                : integer;
            …                { any other data fields needed }
            function         Run(GunPosition : Point) :
                                        Boolean;
            function         CheckDestroyed(GunPosition : Point) :
                                        Boolean;
            procedure        Erase;
            procedure        Move(GunPosition : Point);
            procedure        Draw;
            function         HaveHitGround() : Boolean;
            …                { any other "methods" needed }
        end;
```

The keyword `object` replaces the keyword `record` at the start of a declaration ("object" is really not appropriate, they should have used "class"). In Pascal style, the data fields are followed by the list of functions that an `Invader` can perform. Here it is specified that an `Invader` can `Erase` its display image, it can `Move`, it can `Draw` itself. An `Invader` can also be asked to `Run`; the `Run` function will entail calls to `Erase`, `Move`, and `Draw` and returns a `Boolean`. There would be additional "methods" declared, such as functions for initializing a newly created `Invader` with appropriate coordinates and velocities. These method declarations are like standard "forward declarations" in Pascal. They specify only the function signature (name, argument types, return type). The function definition will appear somewhere later in the program source text.

The method definition will appear like an ordinary function declaration save for the fact that the function name will be qualified by the type (class) name:

Pascal style method definition

```
procedure Invader.Erase;
var
    aRect  : Rect;
begin
    { compute bounding Rectangle }
    SetRect(aRect, h-kHalfWidth, v-kHalfHeight,
            h+kHalfWidth, v+kHalfHeight);
    EraseRect(aRect);
end;

function Invader.Run(GunPosition : Point) : Boolean;
begin
    Erase;
    Move(GunPosition);
    Draw;
    Run := HaveHitGround
end;
```

Because the declaration of the function in the object's type-declaration is a Pascal "forward declaration", it is not actually necessary to repeat the argument list and return type of functions at their point of definition. However, programs are more readable when the argument list is repeated.

The function (or procedure) name is qualified with the type name at the point of definition because there could be several functions with the same name, each belonging to a different type, and the compiler must be informed as to which is currently being defined. Thus, in the Space Invaders example, the Gun object type might also have Run, Move, Erase methods. When a Run function is being defined, the compiler must be told whether it is Gun.Run(...) or Invader.Run(...).

The general form of a C++ class declaration is similar. It is possible to specify something almost exactly equivalent to the Object Pascal version:

C++ style object declaration

C++ structs have data fields and associated functions

```
struct Invader   {
    int     h, v
    ...              /* any other data fields needed */
    Boolean     Run(Point GunPosition);
    Boolean     CheckDestroyed(Point GunPosition);
    void        Erase();
    void        Move(Point GunPosition);
    ...              /* any other "methods" needed */
};
```

But such a declaration fails to take advantage of the additional features of C++ that support access control. The C++ declaration that specifies the form of an Invader would be something like the following:

C++ classes add data encapsulation and access controls

```
class Invader {
public:
    Boolean        Run(Point GunPosition);
    Boolean        CheckDestroyed(Point GunPosition);
    …              /* Other services as required */
private:
    void           Erase();
    virtual void   Move(Point GunPosition);
    virtual void   Draw();
    Boolean        HaveHitGround();
    …        /* Other "housekeeping" functions as required */
    int            h, v;
    …              /* Other data as required */
};
```

In C++, `class` and `struct` declare typenames as well as start a definition of a type. (This version of class `Invader` will need a little refinement before it can be used in the proposed `Invader/Rock`, `Invader/Bomb`, `Invader/Saucer` hierarchy. The necessary refinements are covered in section 3.4 where the interplay of access controls and inheritance is explored.) The meaning of the keyword `virtual` is explained in section 3.3.

Public interface

There are two parts to such a C++ class declaration: a public interface part and a private implementation part. The public interface specifies the services that an `Invader` object will perform for the rest of the program. As far as the main program is concerned, `Invaders` can be asked to `Run(…)` and they can be asked to check whether they have been destroyed by gunfire. (`Invaders` can also be asked to initialize themselves when created, so there will be other functions in the public interface part. The "constructor" functions used to create and initialize objects in C++ are explained in Chapter 5.)

Private implementation

The other details of class `Invaders` are only of concern to the programmer who writes the code for `Invaders`. `Invaders` can execute a `Move()` function, but it is not intended that the `Move()` function be called from the main program. The `Invader::Move()` function is purely a housekeeping/support routine required for the implementation of `Run()`. `Invaders` shouldn't be simply told to move. If the main program asked an `Invader` to move, the subsequent visual displays would be incorrect (an image of the `Invader` would remain at the position where it was before being told to move). In C++, housekeeping functions are placed in the "private" part of the declaration. A C++ compiler can then act to prevent inappropriate calls to housekeeping functions of `Invaders` from code written by the "client" programmer who implements the main game routines.

Member functions

In C++ terminology, the functions associated with a class are *member functions* of that class. The member functions include both the services as defined in the public interface and the housekeeping functions defined in the private part of a class declaration.

Private data fields or data members

The data fields, or in C++ terminology *data members*, would also appear in the private part of the class declaration. The C++ compiler will enforce access control. Data members can only be accessed by the code of the member functions.

C++ method (member function) definition

In most cases, a C++ class declaration contains simply the declarations of the signatures (name, arguments, and return type) of member functions (and so is very much like the Object Pascal type-declaration with its forward declaration of methods). As illustrated in Chapter 5, C++ does allow member functions to be defined in a class declaration. But, usually, a class will simply declare the member functions and their definitions will appear in some other file.

The compiler converts a class declaration into a symbol table entry that contains a summary of the type and accessibility of data members and member functions (Figure 3.6). This table would be referenced by the compiler during subsequent processing of the code defining member functions, and of the code of any application using the class. The information in the symbol table entry will allow the compiler to verify usage of data members; performing checks such as confirming that private data are only accessed from the code of those functions identified in the class's symbol table entry.

Compiler's symbol table entry for a class

```
class Invader {
public:
   Boolean Run(Point);
   Boolean
      CheckDestroyed
         (Point);
   ...
private:
   void Erase();
   ...
   int h,v;
};
```

Name	Type	Access
Invader::h	int	private
Invader::v	int	private
Invader::Erase	void (void)	private
...		
Invader::Run	Boolean (Point)	public

Figure 3.6   Compiler converts class declaration into a symbol table entry that will be referenced when checking subsequent source code and when generating final executable code.

As with Object Pascal, the member function definitions must appear with the function names qualified by the class name. There are minor syntactic differences, but the style is generally similar:

Member function definitions

```
void Invader::Erase()
{   // Use Macintosh Toolbox routines to erase image:
    Rect   aRect;/* compute bounding Rectangle */
    SetRect(&aRect, h-kHalfWidth,v-kHalfHeight,
        h+kHalfWidth,v+kHalfHeight);
    EraseRect(&aRect);
}

Boolean Invader::Run(Point GunPosition)
{
    Erase();
    Move(GunPosition);
    Draw();
    return HaveHitGround();
}
```

*Eiffel class
declaration*
Eiffel provides a final variation on these declarations. In Eiffel, the declaration of the form of a class and the definition of its behaviours are packaged as a single entity. The `Invader` class could be defined in Eiffel (version 2) along the following lines:

```
class  INVADER export
        Run, CheckDestroyed
    feature
        h, v   : INTEGER -- coordinates
        ...
        Run (GunPosition : POINT) :BOOLEAN is
        do
                Erase;
                Move(GunPosition);
                Draw;
        end -- Run

        Erase is
        do
                ...
        end -- Erase
        ...
    end -- INVADER
```

Eiffel uses the term feature where C++ would use member. Here the class has features that are simple data fields, like h and v, and features that are functions, such as `Run` and `Erase`.

Eiffel controls access by "exporting" features that are to be visible to clients. (A data attribute can be included in the export list; clients can see it as a read-only value. Clients can't tell whether they have read access to a data attribute or are obtaining a result from a call to a member function that requires no arguments.) Here, the functions `Run` and `CheckDestroyed` are exported – for these are the services that an `Invader` provides to the rest of the program. In Eiffel version 2, a class starts with an `export` clause listing all exported features. In version 3 of the language, controls on export are on a feature by feature basis. Exports can be selective; an Eiffel class can nominate the classes of favoured clients that have access to particular features.

3.3   DEFINING CLASS HIERARCHIES

One of the key features of Object-Oriented Programming is the ability to define classes that are extensions of, or specializations of existing classes. In the Space Invaders example, there is a need to define the classes `Saucer`, `Rock`, and `Bomb` as specializations of class `Invader`.

A simple way of representing designs for class hierarchies is shown in Figure 3.7; arrows point from derived classes to their base class(es). Derived classes must know their base class. A base class does not know about derived classes.

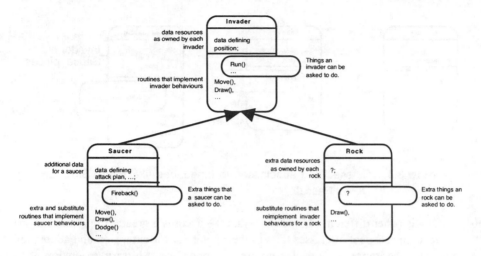

Figure 3.7   The design of a simple class hierarchy

The definition of specialized subclasses involves: i) the identification of a "base" class that is to be extended, ii) the specification of those additional data resources that are required in instances of the specialized subclass, iii) the definition of additional member functions, and usually iv) the definition of functions that *replace* some of those defined in the base class. Sometimes, the additional functions present in a specialized subclass will be part of that class's "public interface", as when instances of a subclass can be asked to do extra things that are not expected of all instances of the more general base class. Often, additional functions will be "private"; a specialized class may have to do extra work when fulfilling standard request.

All the languages provide rather similar ways of defining hierarchical relations among classes. But there are differences, particularly in relation to the ability to override inherited behaviours and in the access allowed to data members declared by base classes.

In most OO languages (e.g. Eiffel, Object Pascal, Smalltalk), derived subclasses are really just expansions of their base classes. As illustrated in Figure 3.8, they combine the data members and member functions declared in both base class and subclass. Figure 3.8 shows a class "Real Saucer" as combining the features separately declared in the Invader base class and the Saucer extension subclass. As far as the compiler is concerned, the class declarations are combined to give something having the form suggested by the "Real Saucer" class design. Data members defined in the base class, Invader, can be accessed quite freely in the member functions defined by the Saucer subclass. A subclass can freely *override* any inherited functions, replacing them with new versions with altered or additional capabilities. If a child class dislikes some characteristic that it inherited from a parent, grandparent or more distant ancestor, then it simply "overrides" the inherited behaviour with something of its own choosing. Such languages are, in Meyer's terminology, "open". A programmer can always extend and adapt an existing class by the definition of a new subclass.

Languages with classes that are open for extension

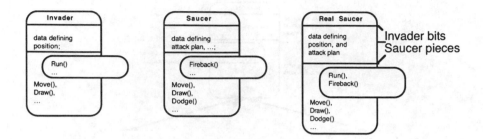

Figure 3.8   Classes and subclasses in languages like Eiffel, Smalltalk and Object Pascal.

Restricted, preplanned inheritance in C++

C++ is rather different. In many ways, C++ accords a greater importance to the encapsulation aspect of classes than to their potential for extension and adaptation through inheritance. C++ classes are not open for arbitrary extension. As illustrated in Figure 3.9, a base class (like `Invader`) does to some degree retain its identity even within a larger entity such as "Real Saucer" (the entity obtained by combining the `Invader` base class with the `Saucer` extensions). By default, the base class components would still be encapsulated; therefore, data members declared in the base class would not be accessible in the code of `Saucer` member functions, and the author of the `Saucer` subclass would not be able to override arbitrarily the inherited `Invader` behaviours. The only extensions allowed will be those that have been preplanned by the author of class `Invader` who must specify which `Invader` functions are open for adaptation in derived subclasses. The author of a C++ base class may also have to make special provision to allow subclasses to access a class's data members.

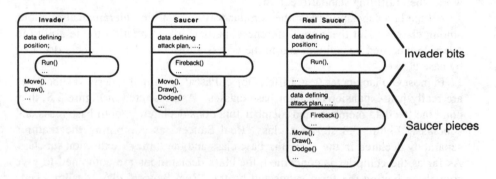

Figure 3.9   Classes and subclasses in C++.

Permitting changes the "virtual" keyword

In class `Invader`, member functions are defined for `Draw()` and `Move()`. But each different specialized type of `Invader` draws itself differently and may have a different movement pattern. The author of class `Invader` knows that these member functions must be left open so as to allow subclasses, like `Saucer` and `Rock`, to specify specialized variants. Just as in Simula, C++ requires the author of the base `Invader` class to declare the member functions `Draw()` and `Move()` as being *virtual*. In general, the author of a C++ class must nominate those member functions that can be modified or extended in subclasses. The keyword `virtual` is used to qualify the declaration of these functions.

`Invader::Draw()` and `Invader::Move()` typify two distinct situations. The *"Deferred functions" or "pure virtual functions"* author of class `Invader` may be able to define some reasonable default `Move()` function that only needs to be overridden in subclasses with non-standard movement patterns. But, in the case of something like `Draw()` there is no meaningful default (how do you draw an arbitrary shaped invader?). The author of class `Invader` must declare a `Draw()` member function, but cannot provide any implementation. The implementations must be provided by subclasses. Both Eiffel and C++ have linguistic features that allow the author of class `Invader` to specify that `Invader::Draw()` *must* be overridden in derived subclasses. In Eiffel, this is achieved by declaring such functions as "deferred"; C++ permits member functions to be specified as being "pure virtual functions".

Although there are these variations in the scope allowed for the adaptation of a *Specifying class inheritance* base class, the actual specification of class hierarchies is done in much the same way in all OO languages. The declaration of a specialized subclass names the "base" class from which it is derived. For the Space Invaders example, the declarations are:

Pascal style hierarchy

```
Type
    Invader        =        object
              h              : integer;
              ...            { As declared in section 3.2 }
              end;

    Saucer =         object(Invader)
              colour         :  integer;
              ...
              procedure      FireBack(GunPosition : Point);
              procedure      Draw; Override;
              procedure      Move(GunPosition : Point); Override;
              ...
              end;
```

C++ hierarchy

```
class Invader {
    ...        /* NOT quite as defined above, */
               /* The base class programmer must allow */
               /* subclasses to access Invader's data members */
               /* See section 3.4 */
};

class Saucer : public Invader {
public:
    ...             /* Any additional services */
private:
    /*      Implementation of new features */
    ...
    void   FireBack(Point GunPosition); // Or public?
    /*      Overrides of inherited behaviours */
    virtual void Move(Point GunPosition);
    virtual void Draw();
    ...
```

```
/*        Additional data members */
int            colour;
...
};
```

These new declarations of derived classes specify only those features that differ from the base class. `Saucers` may come in different colours; if so, there had better be an additional data member that records the colour for a saucer (this data member will be set in an initialization procedure and used in `Draw()`). `Saucers` may possess additional functionality – possibly they are allowed to fire back at the gun (a purely visual effect having no relation to the outcome of the game). If so, this additional functionality has to be declared – as in the illustrated procedure `FireBack()`. If it were intended that the `FireBack()` method could be invoked from client code (the main game routines) then it would be declared in the public interface part of class `Saucer`. It would be more likely that `FireBack()` was a private implementation feature of class `Saucer`, only called as part of the handling of `Move()` and so it would appear, as shown, in the private implementation part. (Of course, Object Pascal does not support these public *v.* private distinctions.)

Only the extra features are declared. A `Saucer` already has data members to hold its horizontal and vertical coordinates because it is a kind of `Invader`. A `Saucer` can erase its image from the screen because this is one of its existing capabilities as a kind of `Invader`. These declarations are all implicit once a `Saucer` has been declared as being a kind of `Invader` (using Pascal's `object(Invader)` construct, or C++'s `: public Invader` construct).

As well as extending a base class by adding new behaviours and new resources (data members), a derived class may need to change an inherited behaviour. For example, the default `Move()` behaviour of an `Invader` might be to move steadily downwards with small random horizontal fluctuations (the behaviour as actually exhibited by `Rocks`). `Saucers` need to do something different, they need to "override" the inherited behaviour that was specified in class Invader.

Such overrides will appear in the declarations of derived classes. In both C++ and Object Pascal the replacement, "overriding" function declaration *must* have the same signature as the original function that is to be replaced. Pascal requires the keyword `Override` in any such overriding declaration. This is redundant, but provides a little extra security when working with complex class hierarchies. C++ does not require any explicit indication when a derived classes overrides an inherited method. The keyword `virtual` does not even need to be repeated with the member function declaration in the derived class (although it is legal to do so). Of course, as explained above, the author of a C++ base class must explicitly give permission for a function to be overridden (by specifying it as `virtual` in the base class declaration). Usually, it is convenient to simply "copy and paste" a method declaration from a base class to a derived class and to include the `virtual` keyword.

*Pascal's override
keyword as a cross-
checking device*

A compiler can easily recognize an overriding declaration because of the match of signatures of function declarations in a base class and some derived descendant class. Pascal requires the `Override` keyword to get the programmer to confirm that a replacement of behaviour is intended. In complex class hierarchies, one can have deep inheritance paths with many functions being declared at each level. It is quite easy for a programmer to define a new specialized class (e.g. `MyPopperButton = object(TRadioButton)`) and pick a name for some supposed extra functionality (e.g. `procedure Reset;`) that happens to have the same name as an obscure method defined three or four levels up through the class hierarchy. If the `Reset` method in `MyPopperButton` is not declared as an `Override`, the Pascal compiler catches the declaration and identifies it as an error. (If `Reset` is declared as an `Override` then the programmers presumably know what they are doing and are fully aware of the default inheritable behaviour.) Pascal does not allow overloaded names for methods (i.e. functions with the same scope and having the same name but distinguishable by different sets of arguments) and so will reject an attempt to define a method `procedure Reset;` if there is an inherited method such as `procedure Reset(Status : Boolean)`.

Caution on C++ function overloading and overriding

Unlike Pascal, C++ permits overloaded function names and does not have any check mechanism that a programmer of a derived class could use to confirm that they intend to override an inherited behaviour. This opens a window for errors. If a derived class defines a function with the same name but different arguments from an inherited virtual function then this is *not* an override but a replacement that, in some circumstances, will hide the inherited behaviour. (Compilers generate warnings because this code pattern is almost always an error.) A more subtle problem arises when the programmer implementing a derived class tries to override an inherited method that was not specified as `virtual` in its original declaration. Such a redefinition of a member function is permitted, but it would not have the intended meaning. These issues are examined in more detail in Chapter 6.

Choosing the virtual functions

If a member function has not been initially declared as `virtual`, it cannot be properly overridden in a derived class. How does a C++ programmer identify the member functions that should be made `virtual` functions? One approach is to make all member functions `virtual`; but this is not generally necessary. The programmer who implements a class that is to form the base of some hierarchy must determine whether a task can be accomplished in one unique way, or can conceivably involve different actions. If there is only one way to erase an object from the screen, the `Erase()` function should be implemented in the base class and *not* be made `virtual`. If there is a default behaviour, such as the invaders' movement pattern, that could be changed, then the function should be declared as `virtual` in the base class and a default implementation defined. If a behaviour is known to exist but is inherently unique to each specialized subclass (e.g. the `Draw()` method of invaders), the function must be declared as `virtual`. As shown in section 3.12 there is a mechanism whereby the base class programmer can require that definitions of a method be provided in subclasses.

Again, Eiffel deserves a brief illustration. Class Saucer would be defined along the following lines:

```
class SAUCER export
        Run, Destroyed
    inherit INVADER
        redefine Move, Draw
    ...
end -- SAUCER
```

"Inherit" clauses in the class declaration identify base class(es). Eiffel classes are open with respect to inheritance. Just as in the simple Object Pascal language, the programmer implementing a derived class can replace any undesired inherited behaviour. Methods that need to be redefined are identified, in a "redefine" statement, that forms an extension to the inherit clause.

3.4   ACCESS CONTROLS AND INHERITANCE

Languages differ significantly in the ways in which they control access. Object Pascal is totally open. Code of methods of derived classes have free access to the data fields and methods defined in their base classes; unfortunately, so do the client programmers who use, and can misuse, objects from these unprotected classes. Eiffel is more restrictive. In Eiffel, a class can hide information from its clients, but leave the same information accessible to derived classes. C++ is the most restrictive of all. When C++ says private, it means private!

Essentially, a data member or member function declared as private can only be accessed by the code of other functions specifically nominated in the same class declaration. (These will include the member functions of that class together with any "friend" functions that are also identified in the class declaration. C++'s friend functions are covered in Chapter 5.)

Consequently, if one does have class Invader and class Saucer defined as:

```
class Invader {
public:
    Boolean      Run(Point GunPosition);
    ...                /* Other services as required */
private:
    void         Erase();
    virtual void Move(Point GunPosition);
    ...      /* Other "housekeeping" functions as required */

    int          h, v;
    ...                /* Other data as required */
};


class Saucer : public Invader {
public:
    ...                /* Any additional services */
private:
    /*      Implementation of new features */
    void         FireBack(Point GunPosition);
    ...
    /*      Overrides of inherited behaviours */
    virtual void Move(Point GunPosition);
    virtual void Draw();
    ...
    /*      Additional data members */
    int          colour;
    ...
};
```

then the programmer implementing class Saucer would find that a C++ compiler would reject any code, such as function Saucer::Draw() and Saucer::Move(), that referenced the h and v coordinates that define the Saucer's position. A Saucer has h and v coordinates because it is an Invader – but they can't be used in Saucer code! These variables are private to the base class Invader and can only be used in the methods defined for class Invader. Similarly, member Erase() is also private to class Invader and couldn't be called from Saucer code.

C++'s public and private access qualifiers are sufficient for the definition of simple programmer defined types ("abstract data types") such as those illustrated in Chapter 5. A wider range of controls is required to build class hierarchies. A class may introduce some member data and member functions that should be concealed from clients but which should be accessible in the code of derived classes. To allow this, C++ has an additional access control – protected. A protected member can not be accessed by clients but is available in derived classes.

Public, private and protected

Protected data fields are common, but there are advantages in keeping the data members introduced in a class private to that class. If they are private, then they can only be changed by member functions of their class. If derived classes will need access to the data, protected access functions should be provided to read and to set these data values. This simplifies debugging. If the value of a data member is found to be invalid it is usually fairly easy to trace the call that sets it to an incorrect value. If the data member were directly accessible in derived classes then it would be necessary to search through the code of the class, and the code of all of derived classes, to find an incorrect assignment. Other housekeeping member functions should also be made protected, so that they can be used by descendants, rather than private.

Keep data private, provide protected access mechanisms

Using the additional protected access controls, the class declarations become:

```
class Invader {
public:
    Boolean       Run(Point GunPosition);
    ...           /* Other services as required */
protected:
    void          Erase();
    virtual void  Move(Point GunPosition);
    ...      /* Other "housekeeping" functions as required */

    /* Protected access functions for private data fields */
    int           H();   // Read h
    int           V();   // Read v
    void          SetH(int newh);
    void          SetV(int newv);
private:
    /* private data fields */
    int           h, v;
    ...           /* Other data as required */
};

class Saucer : public Invader {
public:
    ...           /* Any additional services */
protected:
```

```
     /*       Implementation of new features */
     void              FireBack(Point GunPosition);
     ...
     /*       Overrides of inherited behaviours */
     virtual void Move(Point GunPosition);
     virtual void Draw();
     ...
     int          Colour();
private:
     /*       Additional data members */
     int          colour;
     ...
};
```

In this example, the access functions defined by Invader are H() and V() to read the h and v coordinates; and modifier functions SetH(int), SetV(int) are provided so that Saucer code can change the coordinates. Class Saucer has also been reorganized so that it too could act as a base class for some more specialized subclass

If h and v were declared by class Invader as being protected rather than private, they would be directly accessible in the code of class Saucer and the access functions would not be needed. In this simple situation, this would be reasonable. However, in more complicated examples it is appropriate to make the data members of a class private to that class and provide the protected access functions.

The general form of a compiler symbol table entry for Saucer is shown in Figure 3.10. The entry would show the complete structure, distinguishing inherited members from the additions and replacements defined in the class. As shown in 3.10, a Saucer should use its own Saucer::Move() function but relies on Invader::Erase() for the implementation of an erase operation. Such tables exist only at compile time; they are used to check source code and guide the generation of final executable code.

```
class Saucer : public
        Invader {
public:
    ...
protected:
    virtual void
       Draw();
    virtual void
       Move(Point);
    ...
private:
    int color;
};
```

Name	Type	Access
Invader::h	int	private
Invader::v	int	private
Saucer::color	int	private
Invader::Erase	void (Point)	protected
...		
Invader::Run	Boolean (Point)	protected
Invader::H	int (void)	protected
...		
Saucer::Draw	void (void)	protected
Saucer::Move	void (Point)	protected

Figure 3.10 Schematic form of a compiler's symbol table entry for class Saucer.

3.5 FORMS FOR CLASS HIERARCHIES THAT USE SINGLE INHERITANCE

Smalltalk uses a single inheritance mechanism to build class hierarchies. It also provides a root class, `Object`, from which all other classes are derived. This architecture leads to tree-structured class hierarchies, as illustrated in the first part of Figure 3.11.

Neither C++ nor the object extensions of Pascal require derivation of classes from a specific initial root class. Consequently, one can have different class hierarchies – "forests" of small trees (bushes, and stubs). This style is illustrated in the second part of Figure 3.11 using the classes from Space Invaders.

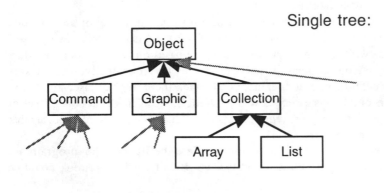

Single tree:

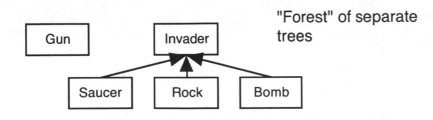

"Forest" of separate trees

Figure 3.11   Forms of class hierarchy using single inheritance.

Many current class libraries for C++ and Object Pascal are organized as single inheritance hierarchies. There is a root class defined (usually named `Object`, or `TObject`) from which all the classes in the library will have been derived. If the library provides for "collection classes" (useful data structures like linked-lists and dynamic arrays), it will probably be a requirement that any objects stored in such collections also be members of classes that are also derived from the library's `Object` base class. (In future, C++ class libraries are likely to take advantage of multiple inheritance and templates and will exhibit more complex graph structures).

Applications using the current single inheritance class libraries most typically have the majority of their classes defined as parts of some large tree structure rooted in the library's `Object` base class. Usually, there will be a few classes that are separate from this main hierarchy. Most of these separate classes would not be part of any class hierarchy, instead they will simply be programmer defined types like those discussed in Chapter 5.

3.6   MULTIPLE INHERITANCE

Although some languages, such as Smalltalk and Object Pascal, provide only single inheritance, other languages such Eiffel and C++ (in versions $\geq$ 2.0) support multiple inheritance. Multiple inheritance provides some increase in flexibility, but at a significant cost of increased complexity.

One use for multiple inheritance can be illustrated in terms of an extended version of Space Invaders. The new version of the game is to involve two phases. The first phase will be long-range defence where bases on Earth's moon, on Mars and elsewhere attempt to fend off interstellar vehicles. Eventually, the attackers will break through these outer defences and attack the terrestrial laser-guns as in the earlier version. In this extended version, `Saucers` can appear at both phases of the game. Sometimes they are required to act as `InterStellarVehicles`, at other stages they are `Invaders`.

If the language provides only for single inheritance, the programmer implementing the `Saucer` class would face a difficulty. Class `Saucer` could be declared as:

```
class Saucer : public Invader {
public:
    /*
    Here I've got to repeat the specification of the
    services provided in class InterStellarVehicle.
    */
    ...
private:
    /*
    Here I've got to repeat the specification of the
    implementation of class InterStellarVehicle.
    */
    ...
};
```

or as:

```
class Saucer : public InterStellarVehicle {
public:
    /*
    Here I've got to repeat the specification of the
    services provided in class Invader.
    */
    ...
private:
    /*
    Here I've got to repeat the specification of the
```

```
        implementation of class Invader.
        */
        ...
    };
```

Either construction necessitates the duplication of code. If class `Saucer` can only inherit features from one base class, all those features that represent the second aspect of its behaviour must be replicated.

If the language supports multiple inheritance, then the declaration of a derived class can simply identify all classes from which it wishes to inherit features: *Declaring multiple ancestor classes*

```
    class Saucer : public Invader, public InterStellarVehicle {
        ...
    };
```

3.6.1 Class hierarchies as Directed Acyclic Graphs (DAGs)

The first complication introduced by multiple inheritance is just the very form of the class hierarchies themselves. Instead of simple trees, one has directed graphs as in Figure 3.12.

Standard multiple inheritance:

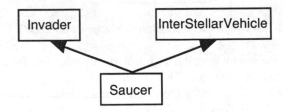

Repeated Inheritance:

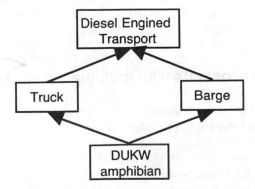

Figure 3.12 Examples of multiple inheritance including a case with repeated inheritance.

There is one obvious restriction on multiple inheritance, there must not be any cycles. It is impossible to have a class A (derived from something not yet met called class E), that along with class B is parent to class C which is, along with class D, a parent to class E (see Figure 3.13). The compiler would obviously get confused with such a class structure. (Class A has an integer data member in addition to whatever it is going to inherit from class E; C adds a char* to the float from B and the int from A; class E gets everything C has, adds an int[10] from D and a char[256] of its own, and passes everything on to A. How big is an instance of class A? Are you sure?) Class hierarchies created using multiple inheritance are not just directed graphs, they must be directed *acyclic* graphs (hence the acronym "DAGs" that sometimes appears).

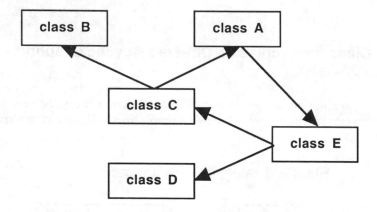

Figure 3.13   An impossible situation of "cyclic inheritance".

Repeated inheritance   Although one cannot have cycles it is perfectly plausible for a class hierarchy to exhibit repeated inheritance. This is illustrated in Figure 3.12 which shows a possible class hierarchy with a root class "Diesel Engined Transport", the derived classes "Barge" and "Truck", and the class "DUKW amphibian". A DUKW (a World War II military amphibious supply vehicle) combines features of both Truck and (motorized) Barge and so it reasonably inherits from both classes. Since both Truck and Barge inherit from "Diesel Engined Transport", DUKW is repeatedly inherited from this class.

3.6.2 Name conflicts with multiple inheritance

Once multiple inheritance is permitted, the language must provide a mechanism for resolving name conflicts. In an extended version of a Space Invaders program, a method DetectTarget() might be defined in both class Invader and class InterStellarVehicle. The implementations would be distinctly different in the two classes. This would be an example of a name conflict. In code written for class Saucer, calls to DetectTarget() would be ambiguous – should the saucer use its radar (like an Invader) or its other device (whatever is provided for an InterStellarVehicle).

There are basically two approaches for resolving such an ambiguity:

- name conflicts can be resolved by systematically renaming inherited members when a class is declared;

- name conflicts can be resolved by qualifying the names, to obtain an unambiguous reference, at the point of usage.

Eiffel uses the renaming strategy. One (or both) of the inherited `Detect-Target()` functions would have to be renamed (e.g. to `LongRangeDetect()`, and to `BattleDetect()`) when class `Saucer` was declared. When writing code for class `Saucer`, the programmer would know from context which function to invoke.

```
class SAUCER inherit
    INVADER
            rename
                    DetectTarget as BattleDetect
            end;
    INTERSTELLARVEHICLE
            rename
                    DetectTarget as LongRangeDetect
            end;
    ...
```

Eiffel style declaration with renaming

(Only one version of `DetectTarget` has to be renamed.)

C++ uses the alternative name qualification scheme. Again, when writing code for class `Saucer`, the programmer would know from context whether to invoke `Invader::DetectTarget()` or `InterStellarVehicle::DetectTarget()`.

Both strategies are workable, and neither causes particular problems when the methods are part of the private implementation of a class. The programmer implementing a class that has methods inherited from multiple ancestors will know from context which method to invoke.

It is a bit more difficult if the functions with name conflicts form part of the public interface of a class, because then client programmers have to understand the naming schemes. It is disconcerting to client programmers if one class offers services, similar to those provided by related classes, under a name that is quite distinct from that used by those related classes. It is also disconcerting to client programmers if services have to be invoked using some complex call sequence that involves describing an object's ancestry. Either way, you lose; programming becomes more complex.

3.6.3 Ambiguities with repeated inheritance

Multiple inheritance introduces no end of problems. Another contorted side effect, that has to be addressed by the language designer, occurs with repeated inheritance. Consider the repeated inheritance hierarchy of Figure 3.12:

```
class DieselEnginedTransport {
public:
```

```
      ...
   protected:
      ...
   private:
      DieselEngine theMotor;
      ...
   };

   class Truck : public DieselEnginedTranport {
   public:
      ...
   protected:
      ...
   private:
      Wheels        theWheels[10];
      ...
   };

   class Barge: public DieselEnginedTransport {
   public:
      ...
   protected:
      ...
   private:
      Propeller     thePropeller;
      Rudder        theRudder;
      ...
   };

   class DUKW : public Barge, public Truck {
   public:
      ...
   private:
      ...
   };
```

How many diesel engines should a DUKW possess?

Logically, it is the designer of the DUKW who should decide how many engines it gets. The designer could go for one engine with a strange gear box (neutral, two reverse, six forward gears and if you go left-up-left-down you engage the propeller). Alternatively, the designer could choose to have two engines – one for driving wheels like a truck and one driving the propeller like a motorized barge (a bit wasteful, but it does help a lot just as the DUKW hits the beach).

Since there is a choice, the programming language needs to be sufficiently rich to allow the programmer to express their choice. Here, Eiffel's approach seems quite appropriate. Eiffel uses renaming (see Chapter 13 for some limitations):

Eiffel: choosing one engine

```
class DUKW inherit
   TRUCK;
   BARGE;
   ...
```

Eiffel: choosing two engines

```
class DUKW inherit
   TRUCK
```

```
           rename
                   theMotor as truckMotor;
           end;
      BARGE;
      …
```

The approach used in C++ is more complex, and in some ways less satisfactory. It is not the designer of class DUKW who determines the number of engines – it is the designers of the trucks and the barges (who probably never considered the possibility of their classes being reused in any way and certainly weren't considering the requirements of DUKWs). (As noted earlier, in C++ inheritance and reuse have to be planned.)

In C++, the replication/non-replication of data members is specified by defining inheritance structures to use "virtual" or "non-virtual" inheritance. As illustrated in Figure 3.14, a class that repeatedly inherits from a base class will get one copy of the base class's data members for all the virtual inheritance paths, and one additional copy of the data members for each non virtual inheritance path.

C++ virtual inheritance

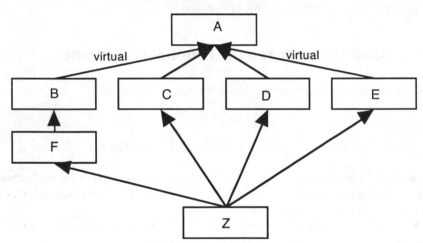

Z will get 3 copies of A's data members; one for two virtual inheritance paths (via E and B/F) and one for each of the other inheritance paths (via C and D).

Figure 3.14 C++ has a complex mechanism for determining the number of copies of replicated data members obtained under repeated inheritance.

It is the classes that directly derive from the base class that get to specify whether inheritance is virtual (non-replicating) or non-virtual (replicating). If the class declarations were not using virtual inheritance:

```
class DieselEnginedTransport { … };
class Truck : public DieselEnginedTransport { … };
class Barge : public DieselEnginedTransport { … };
```

```
class DUKW : public Truck, public Barge { … };
```

or, virtual was specified on only one path:

```
class DieselEnginedTransport { … };
class Truck : virtual public DieselEnginedTransport { … };
class Barge : public DieselEnginedTransport { … };
class DUKW : public Truck, public Barge { … };
```

then a DUKW gets two engines. If the declarations were:

```
class DieselEnginedTransport { … };
class Truck : virtual public DieselEnginedTransport { … };
class Barge : virtual public DieselEnginedTransport { … };
class DUKW : public Truck, public Barge { … };
```

then a DUKW gets one engine. The designer of DUKW can't choose what he wants to do, he is constrained by his inheritance.

3.6.4 Realistic applications of multiple inheritance

Multiple inheritance is commonly introduced with examples like:

```
class Boat { … };
class Plane { … };
class FlyingBoat : public Boat, public Plane { … };
```

Situations like this rarely arise in practice. One just doesn't get "concrete" classes (classes that get instantiated in a program) like Boat and Plane that can also be usefully combined to get new classes.

Multiple inheritance as type combination

Most practical uses of multiple inheritance are quite different. Multiple inheritance is used more as type combination device.

Consider a program that has objects of many different classes and which needs to store these objects to file. For example, an extended version of Space Invaders with a "game save" option would need to write to disk instances of classes Saucer, Rock, Gun, Bomb, Player (name, current score etc.) and possibly some others. If the classes are being developed by separate programmers, the program is likely to end up with:

```
class Saucer {
public:
    void   Read(FILE*);
    void   Write(FILE*);
    …
};

class Rock {
public:
    void   ReadFrom(FILE*);
```

```
    void    WriteOn(FILE*);
    ...
};

class Gun {
public:
    void    Get(FILE*);
    void    Put(FILE*);
    ...
};

class Player {
public:
    void    ReadFrom(FILE*);
    void    PrintOn(FILE*);
    ...
};
```

Each programmer has implemented some minimal version of the concept of a data type that can save itself to a disk file. Each programmer has implemented the concept differently. The code that saves the game state to disk becomes contorted.

It would be far more effective to define the abstract concept "storable":

*Abstract class
storable*

```
class Storable {
public:
    virtual long Size();
    virtual void ReadFrom(FILE*);
    virtual void PrintOn(FILE*);
    ...
};
```

Class Storable would provide an interface that captures the idea of a "storable" data type. This interface would be more than the minimal Read/Write functionality. For example, it is often useful to know how much space an item will require before an attempt is made to write it to disk, so class Storable would provide a Size() method that computes the space requirement.

No instances of class Storable are ever going to be created in a program. Class Storable does not have any data members. Class Storable probably has no default definitions for its methods. It is simply an expression of an abstract concept.

However, many classes can inherit from class Storable:

```
class Saucer : public Invader, public Storable { ... };
class Rock : public Invader, public Storable { ... };
class Gun : public Storable { ... };
```

Here, Saucer is really being defined as a data type that combines all the features of the classes Invader and Storable. Class Saucer will have to provide the implementations of Size(), PrintOn() etc:

```
long Saucer::Size()
{
```

```
        /* two ints for coordinates, one for colour */
        long   s;
        s = 3 * sizeof(int);
        s += …;           /* plus … for other data fields */
        return s;
    }

    void Saucer::PrintOn(FILE* f)
    {
        fprintf(f, "%d %d %d",h,v,color);
        …
    }
```

No code saving, but Similar implementations will be needed in the other instantiable classes like
greater consistency Rock. The use of class Storable doesn't provide any code savings, but it does
improve the consistency of different parts of the same program. Within the
program, there is now one consistent concept of a storable type.

Most of the common applications of multiple inheritance seem to be variants
upon this idea of composing concrete types from abstract types. One variant is the
implementation of container classes (i.e. useful data structures for storing things
in). This use is well illustrated using examples from the Eiffel version 2 class
library (1989 edition). There, class Stack is defined as an abstract concept – it is a
type that provides services such as storing data (its "put" function), reporting its
state ("full" and "empty" functions), return the top item and so forth. The class
uses Eiffel's support for generic types (now also available in C++ as class
templates); a STACK is a stack of "T"s – and a T is whatever the programmer wants
to have in a stack.

```
    deferred class STACK[T] export
            count, empty, full, put, item, …
        feature
            empty : BOOLEAN
                    deferred
            item : T -- access item at top of stack
                    deferred
                …
    end -- STACK
```

Class Stack is a nice definition of the abstraction of "stackiness" – it says what it is
to be a stack. But in itself, it is useless. Programs cannot create instances of class
Stack because nothing in Stack is actually defined. In Eiffel terms, everything is
"deferred" i.e. left to be specified in specialized subclasses.

The class library also contains many classes that provide storage. There are for
instance:

```
    class ARRAY[T] export
        …
    end -- ARRAY
```

and

```
class LINKED_LIST[T] …
    …
end -- LINKED_LIST
```

These are both instantiable classes; a program can create and use instances of ARRAY[POINT] or LINKED_LIST(INVADER). They can also be used to turn Stack from an abstraction into a practical component. The Eiffel library includes some practical version of class Stack that combine the basic abstraction with these different storage structures. So, one can use:

```
class FIXED_STACK[T] export
    …
    inherit STACK[T]
    inherit ARRAY[T]
    …
end -- FIXED_STACK
```

or

```
class LINKED_STACK[T] export
    …
    inherit STACK[T]
    inherit LINKED_LIST[T]
    …
end -- LINKED_STACK
```

Problems of name conflicts and repeated data fields just don't typically arise when multiple inheritance is used to create some required type through the combination of several abstract types. The abstract types being combined usually have no data fields. They should represent orthogonal, non-overlapping concepts and consequently they should have quite different behaviours and so methods of the same name should be avoidable. *Combining abstract types rarely leads to name conflicts*

Don't expect to use multiple inheritance to build "FlyingBoats" from "Boats" and "Planes". If multiple inheritance is used at all in a program, its likely use will be as in the example of class Storable. As with class Storable, a group of related behaviours is captured in a single abstraction and a type or class is defined to represent this abstraction. This defined type is then used to construct, in a consistent manner, all of those real classes required in an application that happen to share a set of behaviours. *Using multiple inheritance*

3.7 SUMMARY ON DECLARING THE STRUCTURES OF OBJECTS

• The syntax used, in a hybrid language, for declaring the structure of objects of a particular type (class) will be based on the language's pre-existing syntax for declaring record structures.

- Derived classes (that represent specialized subclasses) will specify their base class in their declaration. (Often, the same "base" class will be used by several derived classes and so constitutes their "superclass".)

- A language may provide a base class from which all other classes are derived. This leads to a class hierarchy with a unique root; if only single inheritance is used, the class hierarchy will then be tree structured.

- The declarations of classes specify the names and types of the data fields that will be present in each object that is created as an instance of the class. The class declaration will also identify associated methods (member functions). These function declarations are usually "forward declarations"; the code defining the methods will be located later in the same source file or in a separate text file.

- A derived class need only identify its additional data fields and methods; implicitly, it inherits all the data fields and methods of its base class.

- A derived class may be able specify that it wants to change some inherited behaviour. In "open" languages (Smalltalk, Object Pascal, Eiffel) such changes are at the discretion of the programmer implementing a derived class. Normally, the declaration of the derived class would then include some explicit statement to the effect that an inherited method is being redefined.

  In more "closed" languages (i.e. C++), the provision for changes must be preplanned. The author of the base class, that introduces a particular behaviour, must specify whether this behaviour is fixed or can be modified in derived classes. The base class declaration includes the specification as to whether a function can be adapted (a "virtual" function) or is fixed (no "virtual" specification).

- There may be a scheme for defining multiple inheritance. If such a scheme exists, then the language will provide mechanisms for resolving any ambiguities related to the inheritance of identically named functions that occur in the different base classes. Some other mechanism will be provided to allow the choice of replicated or non-replicated data members in the situation of repeated inheritance.

- There should be controls on access to data and behaviours. In C++, three distinct levels of access are provided: 1) there is a *public* interface that defines the "services" provided by objects that are instances of a particular class, 2) there are "*protected*" functions that typically perform housekeeping tasks needed by the class and by any derived classes, and 3) there are *private* data (and, possibly, some private functions).

3.8   CREATING INSTANCES OF CLASSES

Typically objects are dynamic

In an OO program, most objects are going to be "dynamic". They will be created explicitly, and will remain in existence until they are freed. The space needed for these objects will be taken from the heap. This is the mechanism used in Object

Pascal. Unlike Pascal records, which can also be created in the stack, objects in Object Pascal can only be created in the heap (using an extension to Pascal's new(...) procedure e.g. var aSaucer : Saucer; new(aSaucer);) and must be explicitly freed (aSaucer.Free()). Smalltalk, Eiffel, and Simula are a little more subtle. Objects are explicitly created; but they don't necessarily have to be explicitly freed. These languages can use "garbage collection" methods, outlined in section 3.15, to reclaim the heap-space occupied by objects that are no longer being used.

C++ is more complex. In C++, objects (instances of particular classes) can be created in the "static data segment", on the stack, or in the heap. C++ uses the static data segment for all variables that are declared outside of the body of a function and for those variables that are local to functions but which have been declared, explicitly, as being statics. Space is permanently allocated for statics. The stack is used for automatic variables, those that are declared as local to C++ functions and those needed for call by value arguments in calls to C++ functions. The C language has always supported dynamic heap-based structures; space for these was managed through calls to the run-time library routines malloc() and free(). C++ has replaced C's heap-management routines with more flexible facilities for creating objects in the heap. In C++, dynamic instances of objects are created using the operator new and explicitly freed using the operator delete

Static and automatic objects

Students who have completed Pascal or C assignments that use dynamic heap-based structures will be familiar with the use of pointer declarations and pointer-dereferencing. A heap-based Pascal record or C struct has to be accessed via a variable that is of a pointer-type; the value of this variable is an address. The value in the pointer is set when a heap-based structure is actually created using the free-storage management system. Every access to the heap-based record involves explicit pointer dereferencing; such explicit pointer manipulations tend to make code more complex and, possibly, obscure. Typical code involving manipulation of heap based records in Pascal and C is:

```
Type                             struct employee {
    employee = record                   int idnum;
            idnum : integer;            ...
            ...                  };
            end;                 typedef struct employee
    eptr = ^employee;                        employee;
...                              ...
var                              {
    anemployee : eptr;           employee* anemployee;
begin                            anemployee = (employee*)
    new(anemployee );                malloc(
    anemployee^.idnum := 1;              sizeof(employee));
    ...                          anemployee->idnum = 1;
    ...                          ...
```

Of course, the use of pointers and pointer dereferencing is inherent in the use of heap-based structures. But a language can hide such detail. Simula chose to conceal the details, and many languages have followed.

In Simula, Smalltalk, Eiffel and Object Pascal, all objects are heap based and therefore the variables that are used for objects should logically be pointers. But

Object reference variables

these languages all use "object reference variables" that conceal their "pointer" nature. For example, in an Object Pascal version of some variant on Space Invaders, one might have the following:

```
Type
    Invader = object
        …
        ;
    Saucer = object(Invader)
        …
        ;
var
    CommandShip  : Saucer;
    InvasionFleet: array[1..100] of Saucer;
    currentTarget: Invader;
    …
```

CommandShip is an object reference variable (i.e. just an uninitialized pointer); InvasionFleet is an array of 100 such pointers. These variable definitions wouldn't result in the creation of any actual Saucer objects.

The objects would be explicitly created using new(…):

```
        new(CommandShip);
        if(CommandShip=NIL) then begin
            writeln('Sorry. Out of memory');
            halt;
            end;
        CommandShip.InitSaucer(50,-10);
        …
        CommandShip.Run();
        …
        …
        currentTarget := CommandShip;
```

Where methods are invoked, there is no explicit pointer dereferencing; so, the code is CommandShip.Run() (and not CommandShip^.Run()).

Assignment of a derived class to a base class entity

The assignment, currentTarget := CommandShip, is really setting a second pointer to reference a single structure on the heap. The variable currentTarget is of type (reference to) Invader while CommandShip is of type (reference to) Saucer, but the assignment is legal because a Saucer is an Invader. A Saucer can do anything that one might ask of an Invader.

Reference semantics

The "object reference variables" of Object Pascal have "reference semantics". Variables declared of object types are not really instances of those types, they are really just "pointers". Assignments of such variables don't involve copying of data structures, instead an address is copied from one "pointer" to another. The same reference semantics are used for variables of object types in Simula, Smalltalk, and Eiffel.

C++ class variables and pointers

In a C++ function, a local variable definition "Saucer CommandShip;" results in an instance of class Saucer being constructed in the stack frame of the function. If an application requires dynamically allocated instances of classes, it must declare pointers. The C++ code equivalent to the Pascal listed earlier would be:

```
class Invader {
    ...
};

class Saucer : public Invader {
    ...
};

...

    Saucer*     CommandShip;
    Saucer*     InvasionFleet[100];
    Invader*    currentTarget;
    /* extra variables used to illustrate */
    /* value based manipulations */
    Saucer*     EscapeShip;
    Saucer      X;
    Saucer      Y;
    Invader     Z;
    ...
    CommandShip = new Saucer;
    EscapeShip = new Saucer
    if(CommandShip == 0) {
            printf("Out of memory\n");
            exit(1);
            }
    CommandShip->InitSaucer(10,-5);
    ...
    CommandShip->Run();
    ...
    currentTarget = CommandShip;
    ...
    *EscapeShip = *CommandShip;
    X = *CommandShip;   /* extra code manipulating */
    Y = X;              /* Saucers and Invaders */
    Z = Y;
    if(Z.CheckDestroyed(...)) ...;
```

Here, the pointers CommandShip, InvasionFleet[] and currentTarget correspond to the object reference variables in the Pascal code fragment. Some additional variables, instances of classes Saucer and Invader, are also defined. These are actual stack-based instances of their respective classes.

As illustrated in the code fragment, the invocation of methods of an object identified by a pointer requires explicit pointer dereferencing, as in CommandShip->Run(). If the actual object is being manipulated, the "." operator is used to invoke a method, as in Z.CheckDestroyed().

In hybrid languages like Object Pascal and C++, the mechanisms for referring to members (data fields and method functions) of an object are, naturally, based on the language's pre-existing mechanisms for referring to the fields of a structure. The C language provided the "." selector operator for accessing the fields of a struct and the "->" selector for accessing fields from a pointer to a struct. C++ uses these selector operators when invoking member functions of an object. (If any data members are included in a class's public interface, they too can be accessed using these operators.) As shown above, Object Pascal uses its "." selector for

Mechanisms for accessing members

accessing methods of objects as an extension to its previous use for accessing fields of records.

In the C++ code fragment, the assignment "currentTarget = CommandShip;" involves an assignment of one pointer to another. After the assignment, both pointers reference the same structure in the heap. There is just the one object but it is now referenced from two places. This is the same situation as in the Object Pascal fragment. Again, the assignment is legal because a Saucer is an Invader and can perform any action expected for an Invader referenced by the pointer currentTarget.

Value semantics   The assignment "*EscapeShip = *CommandShip" involves copying the values from the data members of the Saucer pointed to by CommandShip into the data members of the other heap-based Saucer named EscapeShip. Similarly, the assignment "X = *CommandShip" involves copying the values from the data members of the Saucer pointed to by CommandShip into the data members of the stack-based Saucer named X. After each of these assignments, there are then two distinct objects that, at least temporarily, have the same values in their data members. The assignment "Y = X;" again involves copying of data from one Saucer object to another. The final assignment "Z = Y;" also involves the copying of values. Z is an instance of class Invader, it has only those data members defined in class Invader (the h, v coordinates etc). Y is a Saucer with additional data members (such as a colour field). When a value assignment is performed from a Saucer to an Invader, only the values in the class Invader data members get copied. This time one gets two objects that have a subset of equivalent data members having the same values.

The consistent use of reference semantics adopted in the other languages is inappropriate for C++. C++ allows more flexibility in the use of classes. Actually, one does not often have a class which gets used extensively both as a dynamic and as an automatic. In a C++ program, one typically has a few classes whose instances are mainly heap based, and some other classes that are most frequently created as local stack-based variables. Classes that represent simple programmer defined types, such as the Points and Rectangles used in Chapter 5, are most often instantiated as local variables (and as data members within more complex classes). With these data, value assignments (e.g. "Point pt1,pt2; … pt2 = pt1;") that involve copying of data are common. More elaborate types, such as "Views" and "Documents" in an applications framework, will typically be instantiated only as dynamically created instances in the heap. It is very unusual for a complex object like a View to be actually duplicated by a value assignment. But duplicate references to such an object are frequently created by the copying of pointers.

Space allocated for objects   Usually, the space allocated for an object in the heap (or, in C++, in the stack, or static data segment) would be equal to the sum of the size of its data members (rounded up by any byte-alignment requirements) plus the space needed for one additional pointer. This extra pointer field will always be there in Object Pascal, Eiffel and similar languages. In C++, space for the pointer is only reserved if the class of the object uses virtual functions. The pointer contains the address of a class "method table" as described in the following section. The pointer is automatically initialized with the address of the method table. Code to initialize these method table pointers is included in the `new` operator used when creating dynamic objects. Similar instructions to set links to method tables might be incorporated in the preamble code of any function that has to create automatic instances of classes in the stack.

Although any link to a method table will always be automatically initialized, the data members of a newly created object will either contain garbage bits or will all have been set to zero. Such initial values, even if all zero, probably don't represent a valid state for an object; so, immediately after creation a call will typically be made to an initialization routine. The *create, then initialize* sequence became so typical that the processes were folded together. C++ introduced the idea of automatically invoked "constructor" routines that would initialize the data members of newly created objects. This feature has been copied in later versions of Eiffel and in the proposed ANSI standard for an object-oriented Pascal dialect. C++ constructors are illustrated in Chapter 5.

Initialization of
objects

3.9   REPRESENTATION OF OBJECTS, AND CODE FOR CLASSES, IN MEMORY

The following account presents a simplified view of how objects from various classes and the code for those classes might be represented in memory at run-time.

When a program like Space Invaders has been compiled, the code segments will contain the code for the various member functions; so there will be code for `Invader::Erase()` (the general erase procedure used by all types of `Invader`), `Invader::Move()` (the default movement behaviour which happens to suit `Rocks`), `Saucer::Move()` and `Bomb::Move()` (the replacement move functions for these specialized subclasses) and a whole series of other routines such as the various `Draw()` methods. Additionally, somewhere in memory there will be class-specific "method tables"; these will be either in the code segment or the static data segment. These method tables will contain the addresses of the member functions for a class.

When objects are created at run-time, each is linked to the "method table" of its corresponding class. Figure 3.15 illustrates the situation in memory once the "Space Invaders" program has started. There are a number of different objects representing instances of each different type of `Invader`. Each has its own individual space in the heap. The space allocated in the heap is that necessary to store all the data members of particular type, plus one additional pointer. This extra pointer holds the address of a method table. So each `Rock` has a pointer to the `Rock` method table; each instance of class Saucer has a pointer to the `Saucer` method table, and each `Bomb` references the `Bomb` method table.

The method tables contain the addresses of appropriate functions for each class. These tables are constructed from the data that the compiler had in its symbol table defining each class. So, for class `Saucer`, the entries in the table will specify that an "erase" operation uses `Invader::Erase`, a "move" operation uses `Saucer::Move` and so forth. Other entries in these method tables might be as follows:

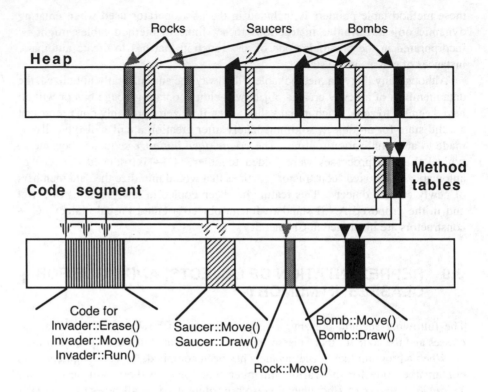

Figure 3.15   Run time organization of objects and code.

```
Method     Saucer              Rock                Bomb
Erase      Invader::Erase()    Invader::Erase()    Invader::Erase()
Run        Invader::Run()      Invader::Run()      Invader::Run()
Move       Saucer::Move()      Invader::Move()     Bomb::Move()
Draw       Saucer::Draw()      Rock::Draw()        Bomb::Draw()
...        ...                 ...                 ...
```

(Actually, some optimizations are possible. If there is only one implementation of
a method that gets used by all classes, no entry is required in the method tables; a
method lookup is not necessary because a simple fixed subroutine call will suffice.
In general, this kind of optimization requires some global analysis of a program at
compile-time or at link-time. Further, C++ needs to consider only those functions
that have been declared as virtual – so, in a C++ program, the tables wouldn't
actually need an entry for `Erase()` or `Run()`.)

The compiler can convert code invoking methods for objects into an indexed
jump through a method table. For example:

```
     Source                      Approximate translation
Saucer  CommandShip;        ...
...                             ...
CommandShip.Draw();         pick up the address of class
```

```
                              Saucer's method table
                         (Draw is entry 3), extract address
                             from 3rd entry in table
                         push arguments
                         make indirect call to subroutine at
                             address taken from table
```

The compiler will have associated an index with each method when building the symbol table entry for a class. When a call is made, the compiler determines the index of the required method and uses this index in the code as illustrated.

If the code involves an automatic or static instance of a class, a C++ compiler can, at compile time, identify the required routine and encode a simple subroutine call. When an object is a dynamic instance of a class addressed via a pointer (or a C++ "reference variable") e.g. :

```
Invader*  target;
…
if(…) target = new Saucer;
else target = new Bomb;
…
target->Draw();
…
```

the code generated for `target->Draw()` has to pick up the base address of the correct method table. After all the pointer `target` might reference either a `Bomb` or a `Saucer`. The object referenced by pointer `target` will have its own internal pointer to the method table for its class. So the code for `target-> Draw()` will involve first accessing the structure on the heap, then extracting the address of the method table, and finally indexing into the method table to get the address of the required routine.

```
        Source          Approximate translation
    …                           …
    target->Draw();   load address register with "target"
                             (i.e. address of struct in heap)
                      add in offset corresponding to extra
                             data field that holds the address
                             of the method table
                      fetch address of method table
                             (Draw is entry 3), extract address
                             from 3rd entry in method table
                      push arguments
                      make indirect call to the subroutine
                             at the address taken from table
```

This scheme will get the address for `Saucer::Draw()` when the target is a `Saucer`, and the address for `Bomb::Draw()` when the target is a `Bomb`.

A compiler can arrange that any pointer to a method table is at some known offset from the start of an object's record structure in the heap. It is simplest to think of the pointer as occupying the first few bytes of the record structure. In practice, more complex layouts have to be used for objects in order to maintain compatibility with C `struct`s.

Virtual tables   The scheme outlined in this section is a somewhat simplified version of that used in C++. In C++, the tables are called "virtual tables" (because they are used for virtual functions) rather than method tables. Multiple inheritance does add quite a number of complications to this basic scheme. The major advantage of the scheme is that there is very little overhead associated with calls to virtual functions.

Alternative mechanisms for invoking methods   Many other languages use less efficient schemes for invoking methods. These alternative mechanisms typically involve some form of run-time search for the correct method. The invocation of a method for an object will be translated into a call to a "dispatch function" that takes the name of the required method and a reference to the object. The dispatch function first retrieves the method table for the object (using the object's method table pointer). This method table will contain the names and addresses of those methods that were added or redefined in that object's own class. For example, the `Saucer` table would only contain entries for `Draw()` and `Move()`. The table is searched by comparing the names (or compiler tokens corresponding to the names) of available methods with that requested; if a match is found, the address of the function is taken from the table and used. If a required method, e.g. `Erase()`, is not in the table, the dispatch function follows a link back to the method table of the base class (in the example, this would be a link back to the method table for class `Invader`). There will then be another loop through the names searching for the required method. This "dispatch function" approach does make the method tables smaller. But, the run-time costs are significant.

3.10 METHODS HAVE AN IMPLICIT OBJECT REFERENCE PARAMETER

Code for a typical method in Object Pascal is:

```
procedure Invader.Move(GunPosition : Point)
begin
    { default movement is 2 pixels down screen }
    { and a random movement to left or to right }
    v := v + 2; { v coord increases downwards! }
    h := h + irandom(-4,4);
    { check still within horizontal bounds }
    ...
end;
```

This code quite obviously manipulates the h and v fields of some record structure. What isn't obvious is the identity of the record structure that is being manipulated.

When methods are invoked they are being performed on behalf of a particular instance of a class. The address of that instance must be available in the method code.

Implicit object reference parameter for methods   The compiler arranges this internally by redefining all methods to take an extra argument. This argument will be a reference to (or pointer to) the actual object for which the method is invoked:

Source code

```
Invader = object
    h, v   : integer;
    procedure Move(
          gp : Point)
    function Run(
      GunPosition : Point)
      : Boolean;
    ...
end;
```

Code as interpreted by
the compiler

```
Invader = object
    h, v       : integer;
    procedure Move(Self:
        Invader; gp Point);
    function Run(Self:
        Invader;
        GunPosition : Point)
                    : Boolean;
        ...
    end;
```

Pascal

Similarly, in C++:

```
class Invader {
public:
    ...
    Boolean Run(Point);


protected:
    void Move(Point);


private:
    int    h, v;
    ...
};
```

```
class Invader {
public:
    ...
    Boolean Run(const
        Invader* this,
        Point);
    ...
protected:
    void Move(const
        Invader* this,
        Point);

private:
    int    h, v;
    ...
};
```

C++

The compiler automatically adds the extra parameter to all function signatures. The extra implicit parameter is named "Self" in Smalltalk and Object Pascal, "current" in Eiffel, and "this" in C++ and Simula.

Similarly, the compiler interprets the definitions of procedures as if they had been written with an appropriate reference to the object in all references to data members and member functions:

C++'s "this"

References to members are implicitly qualified in the code of methods

Source code

```
procedure Invader.Move(
        gp : Point);
begin
    ...
    v := v + 2;
    ...
end;
```

Code as interpreted by
the compiler

```
procedure Invader.Move(Self:
        Invader; gp : Point);
begin
        with Self do begin
            ...
            v := v + 2;
            ...
            end
end;
```

Pascal code with implicit object references made explicit

```
function Invader.Run(
    gp : Point) : Boolean;
```

```
function Invader.Run(
        Self : Invader;
        gp : Point) : Boolean;
```

```
begin                          begin
    Erase;                         Self.Erase;
    Move(gp);                      Self.Move(gp);
    Draw;                          Self.Draw;
    Run := HaveHitGround           Run = Self.
                                          HaveHitGround
end;                           end;
```

C++ code with implicit object references made explicit

```
void Invader::Move(Point)      void Invader::Move(const
                                   Invader* this,
                                   Point)


{                              {
    ...                            ...
    v += 2;                        this->v += 2;
    ...                            ...
}                              }


Boolean Invader::Run(Point gp)Boolean Invader::Run(const
                                   Invader* this,
                                   Point gp)

{                              {
    Erase();                       this->Erase();
    Move(gp);                      this->Move(gp);
    Draw();                        this->Draw();
    return HaveHitGround();         return this->
                                          HaveHitGround();

}                              }
```

There are two ways that Pascal can do this. It can bracket the entire body of a method with a "with Self" clause or it can qualify the individual references with "Self."; the effect is equivalent. In C++, the compiler's interpretation is that all data members and member function calls are qualified with "this->".

Explicit qualification using "this->" or "Self." is legal

Although the extra object reference parameter is never explicitly declared in a function signature, it is legal to make explicit use of "Self." or "this->" in the code of methods. Explicit usage of the qualifier is redundant. However, some programmers like an explicit style, and so code in the form:

```
procedure Invader.Move(GunPosition : Point)
begin
    { default movement is 2 pixels down screen }
    { and a random movement to left or to right }
    Self.v := Self.v + 2; { v coord increases downwards! }
    Self.h := Self.h + irandom(-4,4);
    { check still within horizontal bounds }
    ...
end;
```

or

```
void Invader::Move(Point)
{
    this->v += 2;
```

```
    this->h += irandom(-4,4);
    ...
}
```

can be found in some class libraries. The main argument for this style is that it distinguishes references to members from references to globals. The programmers at Microsoft liked the style so much that when they implemented a dialect of Object Pascal they wrote the compiler so that it *required* such explicit use of "Self.".

On the whole, programmers do not use explicit qualification of members with "this->" (or "Self."). Naming conventions are relied on to distinguish between global variables and data members. C++ has an alternative mechanism for highlighting references to global functions (using the "scope qualifier operator" that is described in Chapters 4 and 5).

The most common use for the implicit this (Self) parameter is in the construction of networks of objects that require mutual references for intercommunication. For example, in an application framework, a "Document" object might have to create a "View" object for display of its data. The Document naturally keeps a pointer to the View that it created and will use this pointer when requesting that the View invalidate itself or perform other display actions. But the View may also require a pointer to the Document. For example, the View may need to query the Document as to the number of data elements that must be displayed so that it can set appropriate scaling factors. This is a typical situation; the Document needs, and has, a reference to a View, the View must be given a reference to its Document. Code of the following general form is common:

Passing "this" as a parameter

```
class     MyDocument : public Document {
public:
    ...
    void    OrganizeViews();
    ...
private:
    MyView*        fmyView;
    ...
};

class MyView : public View {
public:
    ...
    void   LinkToDocument(MyDocument* adoc);
    ...
private:
    MyDocument*   fmyDoc;
}

void MyView::LinkToDocument(MyDocument* adoc)
{
    fmyDoc = adoc;
}

void MyDocument::OrganizeViews()
{
```

```
...
/* create the main view */
fmyView = new View;
...
fmyView->LinkToDocument(this); /* <=== "this" as arg. */
...
}
```

When a document is executing `OrganizeViews()`, the implicit `this` parameter
holds its address. The call to the `LinkToDocument()` method of `View` passes `this`
as an argument so that the `View` object can save it. Subsequently, the `View` object
can use its `fmyDoc` pointer to communicate with its document.

3.11 EXPLOITING POLYMORPHISM IN A PROGRAM'S STRUCTURE

If one has code that involves creation and manipulation of objects of known types,
there really isn't much point in having method tables and method dispatch
mechanisms. To illustrate this, imagine a version of Space Invaders with a few
more of the methods being in the public interface. Then, in principle, one could get
C++ code like:

```
Saucer      CommandShip;
Saucer      Escorts[3];
Bomb        theBomb;
Invader     something;
...
CommandShip.Draw();
for(i=0;i<3;i++) Escorts[i].Draw();
...
theBomb.Draw();
...
CommandShip.Move();
...
theBomb.Run();
...
something = CommandShip;
something.Draw();
something.Move();
```

"Static binding"   A compiler can actually sort out this code and determine, unambiguously, which
routine should be called at each point. `CommandShip` is a `Saucer`, `Escorts` are
`Saucers`, `theBomb` is a `Bomb`, `something` is an `Invader`; therefore, the compiler can
perform the translations:

```
Escorts[i].Draw()        push address of ith Escort onto stack
                         jump to subroutine Saucer::Draw

theBomb.Draw()           push address of theBomb onto stack
                         jump to subroutine Bomb::Draw
```

```
CommandShip.Move()    push address of CommandShip onto stack
                      jump to subroutine Saucer::Move

theBomb.Run()         push address of theBomb onto stack
                      jump to subroutine Invader::Run

something.Draw()      push address of something onto stack
                      jump to subroutine Invader::Draw

something.Move()      push address of something onto stack
                      jump to subroutine Invader::Move
```

Since this is all sorted out at compile time, it is done "statically" – it doesn't involve any consideration of what is happening at run-time. (The `Invader` "something" is, of course, an oddity. Its coordinates are set by copying from the `CommandShip`, then it is drawn and told to move. Because it is an `Invader`, it naturally uses the `Draw()` and `Move()` methods of `Invader`; at least it has a default movement behaviour, but it won't have any real drawing mechanism.)

Of course, that code fragment shown above is quite different from the code one would really have in a Space Invaders program. A Space Invaders program would not be manipulating individual instances of specific classes. The intent of inventing an "Invader" abstraction was to separate concerns. The programmer implementing the game code needs to know only that invaders exist and that they have certain abilities (which for these examples include `Draw`, `Run`, `Move`). The programmer implementing invaders is concerned with the fact that Invaders do come in different types that have similar but distinct behaviours.

So, the code in the game would be more like the following:

Code exploiting polymorphism of Invaders

```
Invader*   gAttackers[10];    // global array
void Play()
{    int i;
     // get gAttackers array to hold some randomly
     // chosen invaders
     GetInvaders();
     ...
     for(i=0;i<10;i++)
           gAttackers[i]->Erase();
     ...
     for(i=0;i<10;i++)
           gAttackers[i]->Move();
     ...
     for(i=0;i<10;i++)
           gAttacker[i]->Draw();
```

The code would manipulate a list or an array of `Invader*` items. Each entry in this list would be a pointer to a `Saucer`, or to a `Rock`, or to a `Bomb`; but this is of no concern to the implementor of the game.

The compiler will process the method calls differently according to whether they are calls of virtual or non-virtual functions. If it is a non-virtual function, the compiler can generate a simple subroutine call:

```
        code                          Approximate translation
gAttackers[i]->Erase();
                                push value in gAttackers[i] onto stack
                                subroutine call to Invader::Erase
```

"Dynamic binding" But, if it is a call to a virtual function, the compiler generates the code involving a run-time, i.e. *dynamic*, lookup of the correct function:

```
        code                          Approximate translation
gAttackers[i]->Move();
gAttackers[i]->Draw();
                                pick up address in gAttackers[i]
                                add offset to get method table field
                                extract address of method table
                                index into method table (Draw => 3,
                                    Move => 4) to get address of routine
                                push address in gAttackers[i] onto stack
                                call routine determined by lookup
```

As illustrated in Figure 3.16, this code gives the required behaviour.

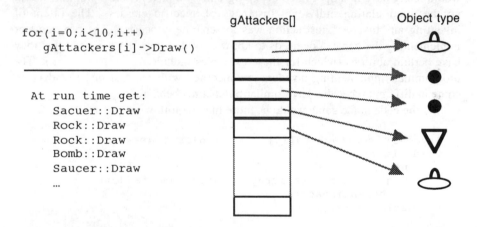

```
for(i=0;i<10;i++)
    gAttackers[i]->Draw()

At run time get:
    Sacuer::Draw
    Rock::Draw
    Rock::Draw
    Bomb::Draw
    Saucer::Draw
    ...
```

Figure 3.16 Dynamic binding, with method lookup at run-time, results in the correct calls to different member functions for the polymorphic invader references in the array gAttackers[].

It is possible to set up class `Invader` in such a way that the programmers implementing the game never need know how many different specialized types of `Invader` actually exist (code in Chapter 6). A function will provide a randomly chosen `Invader` and this can be used to fill `gAttackers[]`. The programmers implementing the Invader code can add a new type, e.g. a "`GuidedMissile`" that homes in on the gun, and this need only become known to the game implementors when it appears on their test screens.

Dynamic calls in   In the code of class member function, the compiler will encode a call to another
member functions member function as a dynamic lookup if it is a virtual function and as a simple

subroutine call if its a non-virtual function. For example, in the code given earlier for the `Invader::Run()` method:

```
Boolean Invader::Run(Point gp)
{
    Erase();
    Move(gp);
    Draw();
    return HaveHitGround();
}
```

The calls to `Move()` and `Draw()` will use the dynamic lookup method while the calls to the non-virtual member functions `Erase()` and `HaveHitGround()` will use simple subroutine calls to the `Invader::Erase()` and `Invader::HaveHit-Ground()` routines.

This might seem odd on first reading. After all, if it is the code of `Invader::Run` why doesn't it simply call the default implementation of `Invader::Move`? But of course, the method code is really:

```
Boolean Invader::Run(
        /* compiler fakes the argument const Invader* this */
        Point gp)
{
    this->Erase();
    this->Move(gp);
    this->Draw();
    return this->HaveHitGround();
}
```

The functions `Erase()`, `Move()`, `Draw()` etc. are being invoked for some kind of invader, whatever kind is referenced through the pointer "`this`". The `Move()` or `Draw()` routine that gets called had better be the one that is appropriate for whatever that object actually is. Consequently, it is essential that these calls use dynamic binding.

3.12 EXPLICIT SUPPORT FOR ABSTRACT CLASSES

In real object-oriented programming using inheritance, "abstract" classes occur very frequently. Abstract classes express general concepts, such as the concept of an invader, or of a "storable item", or of a Stack of "T"s. The objects actually created at run-time are instances of "concrete" classes that are specialized subclasses of the general abstract classes.

It is appropriate to make this distinction in the class declarations so that a client programmer can understand how to use classes, and so that a compiler can check that use is appropriate. Simple languages such as Object Pascal don't bother with such refinements, but both Eiffel and C++ have mechanisms for specifying that a class is abstract.

Eiffel is quite explicit. A class declaration will begin with the keyword "deferred" if it is an abstract class. Examples might be:

```
deferred class Storable
    export Size, ReadFrom(FILE), PrintOn(FILE)
    feature
            PrintOn(FILE) is
                    deferred
            ReadFrom(FILE) is
                    deferred
            Size : INTEGER is
                    deferred
end -- Storable

deferred class Invader
    export Run, CheckDestroyed
    feature
            h, v   : INTEGER;
            Run(p : POINT) : BOOLEAN is
                    do
                            Erase;
                            Move(p);
                            Draw;
                            Result := HaveHitGround;
                    end -- Run
            Draw is
                    deferred
            Move(p : Point) is
                external irandom(INTEGER, INTEGER) : INTEGER ...;
                    do -- default movement is down and random l/r
                    v := v + 2;
                    ...
                    end -- Move
            ...
    end -- Invader
```

Abstract class, partially implemented abstract class

Class Storable is purely abstract, no implementation is provided. It is simply a description of a "type" – a set of behaviours. The behaviours are listed; but each is "deferred" (i.e. left to be implemented in some concrete subclass). Class Invader is a "partially implemented abstract type". The class definition describes the data that all invaders must own (h, v coordinates etc.) and the behaviours that all must exhibit. Class Invader provides the implementation of standard behaviours, e.g. Run, and the default implementation of some other behaviours, e.g. Move, while specifying others as "deferred" (e.g. Draw). Because Storable and Invader are both deferred classes, the Eiffel compiler will prevent a programmer from creating an instance of either of these classes. Any Invaders created must be created as instances of other, more specialized derived classes. A class Storable object will only occur if an instance is created of a concrete class that includes class Storable among its ancestors.

C++ abstract classes and "pure virtual functions"

C++ provides the same capabilities, but it is less direct in its expression of the idea of an "abstract" or "deferred" class. C++ eschewed the introduction of an extra keyword such as "abstract" or "deferred"; extra keywords make code too verbose (and readable). (Introducing a new reserved word does break all programs that had used that word to name a variable or a type; C++ has tried to avoid such incompatibilities.) Instead, a class is made abstract by defining one (or more) of its

member functions as being a "pure virtual function"; and this definition is achieved
by declaring a virtual function to have a 0 implementation:

```
class Storable {
public:
    virtual long Size() = 0;
    virtual void PrintOn(FILE*) = 0;
    virtual void ReadFrom(FILE*) = 0;
    ...
};

class Invader {
public:
    Boolean           Run(Point);
    ...
protected:
    virtual void      Draw() = 0;
    virtual void      Move(Point);
    void              Erase();
    ...
private:
    int           h, v;
};
```

In languages like Object Pascal that don't explicitly support the concept of
abstract classes, it is necessary to provide "implementations" for all the methods of
all classes. So, an Object Pascal version of Invader would have to provide an
implementation for Draw. This would be done by providing an empty procedure
body, or a procedure that reported an error.

*Empty methods as a
substitute for proper
abstract functions*

```
Type
    Invader = object
            h, v   : integer;
            procedure Draw;
            ...
            end;

procedure Invader.Draw;
begin
{$IFC  defensive }
    write('Something is amiss. An invader can not draw itself');
    halt;
{$ENDC }
end;
```

Actually, empty methods occur quite often in classes that form parts of
"application framework" class libraries. In a C++ library, these will always be
virtual functions but *not* pure virtual functions. For example, a "Document"
class might have an empty method AboutToSave():

*Other uses of empty
methods*

```
class Document : public ... {
public:
    ...
```

```
            virtual void AboutToSave();
            ...
      };

      void Document::AboutToSave()
      {
      }
```

Such functions are provided by the class designer to assist in the implementation of derived classes with unusual behaviours. Very few applications need to do anything special before saving a document – but it is plausible that an application might need to perform some special action. So, the designer of a good class library provides a virtual AboutToSave() method with a default do-nothing implementation. Every time a document is saved, it first executes its AboutToSave() method. In the vast majority of applications using the class library, this means that two or three instruction cycles are wasted calling and returning from the empty routine. In the odd case where special behaviour is required, a derived subclass of Document can provide an effective definition for its AboutToSave() method.

Arguments for a virtual function

The signature (arguments and return type) of a virtual function, as defined in an abstract class, has to take into account the requirements of its intended subclasses. (This seems like one of those bumper-sticker jokes – "madness is inherited, you get it from your children".) The reason for this is best illustrated by a simple example.

In the Space Invaders program as defined in section 3.1, three distinct movement behaviours were specified. Bombs moved directly down; Rocks tumbled downwards with small lateral motions; Saucers could try to avoid the gun. Of course, Saucers need to know the position of the gun if they are to avoid it. So, the Move() method of Saucer must take as an argument either a reference to the Gun (so that it can be asked where it is) or, as actually chosen, a Point value holding the Gun 's position:

```
      void Saucer::Move(Point gp)
      {
            /* Move generally downwards but avoid that gun */
            int    hnew, vnew;
            vnew = V() + 3;
            hnew = H() + hvelocity;
            /* if less than 10 points horizontally from gun */
            /* start to take evasive action */
            if(10 < abs(gp.h - hnew)) ...
            ...
      }
```

If Saucer's Move() member function is to replace the default Move() defined by class Invader, these functions *must* have the same signature. So, the need for a Point argument to the Move() member function sort of bubbles-up to the level of class Invader. Both the Rock and Bomb Move() functions must then take Point arguments even though these arguments are never used.

Features flow upward through class hierarchies!

This effect is common. Requirements of specific subclasses may force the reworking of the interface planned for a general abstract class. Member functions may finally have quite complex argument lists (i.e. three or four arguments), but with many of the arguments, required only by sibling subclasses, passing unused in the body of any specific variant. (If methods have more than four arguments, too much data is being transferred among objects. The design should be reviewed.)

Many class libraries have been designed using predominantly single inheritance starting from some library provided `Object` class as a root. These libraries often exhibit results of a tendency to move "useful behaviours" up through the hierarchy. For example, the "storable" behaviours might have been found useful at many points in the class library. If multiple inheritance is not available, or is not used, "storablility" can only be obtained as a general behaviour by floating it up through the hierarchy and making it a property of the root `Object` class. This has the disadvantage that the root class may become overburdened with pure virtual functions that must be implemented in every class created for use with that library.

Other features flow to the root of a tree structured hierarchy

3.13 SUMMARY ON THE DEFINITION AND USE OF OBJECTS

- C++ allows the definition of static, automatic and dynamic instances of classes; variables of class types can be defined as automatics and statics, typed-pointers will be defined to reference dynamically created objects. In other languages, objects are essentially all dynamic, heap-based entities; variables defined of class types are actually "object reference variables" (i.e. pointers).

- In hybrid languages like C++ and Object Pascal, dynamic instances of classes are created using an extension to the languages' existing facilities for creating structures in the heap. Object Pascal extends the functionality of the "`new(...)`" procedure provided by its run-time environment. C++ replaces C's `malloc(...)` function with its `new` operator.

- Objects have an extra data member in addition to those declared; this is a pointer to their class's "method table". (In C++, the pointer is only included in objects that are instances of classes that have virtual functions. Simple programmer defined types such as `Points` and `Rectangles`, illustrated in Chapter 5, would not include such a pointer field.) This pointer is set automatically when an object is created.

- Most languages provide a mechanism whereby an initialization routine will be automatically invoked to set the data fields of a newly created object into an acceptable state. In C++ terminology, these special initialization routines are "constructors".

- Code that uses an object can reference its member functions (and any visible data members) using a syntax that is based on the language's existing mechanisms for referencing the data fields of a record structure.

- The invocation of a member function for an object can be described in terms of a request for action by that object, or even in terms of "sending a message asking an object to do something". So, "`Saucer CommandShip; ...;`

`Command-Ship.Run(p)"` is a request to a particular `Saucer` to execute its `Run` member function; while the code `"Invader* firstInvader; …; firstInvader = gAttackers[i]; …; firstInvader->Draw()"` involves a draw-yourself request to the `Invader` referenced by the pointer `firstInvader`.

- The member functions of a class have an extra argument, a pointer to an object, inserted by the compiler. Within the body of the code of a member function, all references to data members or to member functions are implicitly qualified by this pointer. When a call to a member function is made, this pointer is initialized with the address of the object for which the call is made; so, very approximately:

```
CommandShip.Run(p)   ==> compiler ==>   Run(&CommandShip,p)
```

This pointer argument contains the address of a data structure which represents the state of the object that is executing the function.

- In C++, a compiler may have sufficient information to allow a call to a member function to be encoded as a simple subroutine call ("static binding"). This will be the case whenever the object is a static or automatic variable; because then its actual type and, therefore, the appropriate version of the function are known at compile time. The compiler can also generate simple subroutine calls whenever there is only one implementation of a member function (as in a class hierarchy where the function is defined as a non-virtual function in the base class).

Code involving static (compile time) binding typically involves the use of instances of programmer defined classes that are not parts of inheritance hierarchies. Examples include classes such as the Points and Rectangles illustrated in Chapter 5. The coding style may be "object based":

code   "translation"
```
    Point    pt1, pt2;
    Rectangle  r1;
    …
    pt1.Move(4,5);
                    "pt1, move yourself across and down"
    …
    if(pt1.Equal(pt2)) …
                    "pt1, are you the same as pt2?
                            if so …"
    …
    r1.Insct(3,3);
                    "r1, shrink by … "
```

but such code is not object-oriented.

In class hierarchies, calls to non-virtual functions do get statically bound at compile time. Essentially, this constitutes a programmer designed optimization of the code.

- "Dynamic binding" entails some run-time action to identify a specific implementation of a member function that is to be executed next. In the code:

  ```
  Invader*        target;
  ...
  target->Draw();
  ...
  ```

  the code generated for the statement target->Draw() will entail i)accessing the record structure referenced by the pointer target to get the address of the appropriate Saucer, orRock, orBomb method table, ii)"looking up" method Draw() in this table to find the address of the appropriate version – Saucer::Draw(),Rock::Draw() or Bomb::Draw(), and iii) calling this dynamically identified routine.

  The code that actually "looks up" the member function in the method table may use a simple, efficient indexing scheme, or some much less efficient but potentially more flexible iterative search through the table(s).

- In Object Pascal, Eiffel, Simula, Smalltalk etc. all calls to methods of objects are "dynamically bound". (Actually, a global analysis of programs at compile-link time is sometimes used to eliminate a few dynamically bound calls by identifying those situations where there is only one possible version of a function. Although the effect is somewhat similar to a C++ programmer specifying a non-virtual function, it does not involve planning by the programmer. It is an implementation optimization; not a programmer planned optimization.)

- Inheritance and dynamic binding are necessary for an object-oriented programming style. In the OOP style, client code using objects works in terms of (pointers to) variables of a generalized abstract class e.g. Invader* pointer variables in C++.

  The abstract class specifies the existence of certain behaviours providing default implementations for some and deferring others for definition in subclasses. Subclasses provide definitions for deferred methods and may redefine inherited methods.

  Client code uses the general capabilities defined by the abstraction ("all Invaders can 'run' and can check whether they should be destroyed") and can be made completely independent of the existence of specialized subclasses.

  In C++, an abstract base class (e.g. class Invader) must specify functions that may be changed in derived subclasses as "virtual" and provide the default implementation (e.g. virtual void Move(Point) { v += 2; h += irandom(-4,4); }). The abstract class must specify deferred functions as pure virtual functions (e.g. virtual void Draw() = 0;)

- A major advantage of the OO programming style is the almost complete separation of concerns. Game implementors know there are "Invaders" but need not know how many variants there are, or in what ways their particular behaviours may differ. Implementors of Invaders must be consistent with the narrow published public interface of class `Invader`; apart from this restriction, they are largely free to make their own implementation choices.

3.14 EXTENDED OO FEATURES: "GENERIC" OR "PARAMETERIZED" CLASSES

Among academics, there are many arguments as to what exactly are the features that make a language "object-oriented". Most agree that OO languages should have i) some mechanism, like C++ classes, for representing abstract types which can be instantiated as run-time objects (*abstraction*), ii) some packaging mechanisms that guarantee that the data resources that belong to an object can only be accessed by requesting that object to execute one of its member functions (≈"*encapsulation*"), iii) *inheritance* is needed to allow specialized derived classes to be defined as specializations of existing base classes, and iv) it should be possible to design a program to exploit *polymorphism* and *dynamic binding*. In this section, and the following section, a couple of additional features are presented. Not all OO languages exhibit these features, but if they are available they can greatly facilitate programming.

The first feature is support for "generic" or "parameterized" classes. Eiffel introduced "generic classes" at an early stage as Meyer was interested in combining the "Ada package" style of code reuse with an object-oriented approach. Equivalent functionality appeared much later in C++ with the "template classes" of C++ V. 3. Many of the current C++ class libraries were developed before templates became available; but some of the later products, such as Borland's class libraries for the IBM-PC, do take advantage of templates..

Container classes   One major use for generic types is with "container classes". Smalltalk provided the original inspiration for container classes. When it first became available, it already had a collection of every useful variant of boxes for putting things in. Every conceivable "stack", "queue", "dynamic array", "linked-list", "useful-other" was already implemented. Smalltalk is untyped (apart from the fact that everything is an object). An object could be put into one of these containers; and an object could be retrieved.

Similar libraries of container classes were developed for the typed languages, Object Pascal, C++, Eiffel etc. The issue arose of what type of thing could be put into a container, and what could be retrieved.

If the containers are to be useful, they must not be overly restrictive on the types of permitted contents. They must allow general use. But, the only really general "thing" that can be put into a container is a pointer – something equivalent to a C "`void*`". The container object doesn't store instances of other classes, it stores pointers to them. (In class libraries that provide some root class, e.g. `TObject`, the pointers stored in containers are of type `TObject*` rather than `void*`. Then, only objects from classes derived from class `TObject` may be stored in containers.)

There aren't any problems about putting things into a container. The problems come when taking items out of the container.

For example, one might have some form of queue of pointers to "things". Anything can be "saved in the queue" by an `Append()` operation that takes a pointer argument. But the queue can only give back a `void*`:

```
class Q {
public:
    void Append(void*);
    void* Remove();
    …
};
```

Such an implementation loses type information concerning the datum that was queued. When the datum is retrieved as a `void*`, an explicit type cast operation must be used to convert the pointer back to a pointer to an object of known type.

Every time a compiler encounters an explicit type cast, it is being told "Trust me. I'm a programmer" and must just accept it. If the programmer does the right thing, all goes well. But the compiler does not get an opportunity to perform any safety checks. All too often, the result is a program that aborts at run-time.

Don't trust type casters

Compile time checks are possible if, when instantiating an instance of the queue class, the programmer is required to specify the type of datum that will be queued – "This queue will be used for 'mail-items'". Subsequently, the compiler can check that anything added to the queue is of class `MAILITEM` (or some more specialized subclass) and items retrieved from the queue are of type `MAILITEM` .

Specify the type of datum that a container will contain

This is achieved with "generic classes" ("parameterized classes"). Eiffel examples were illustrated earlier:

```
deferred class STACK[T] export
          count, empty, full, put, item, …
     feature
          empty : BOOLEAN
                deferred end;
          item : T -- access item at top of stack
                deferred end;
          …
end -- STACK

class FIXED_STACK[T] export
          …
     inherit
          STACK[T]
          ARRAY[T]
     …
end --
```

Class `Stack` is simply an abstraction; there are a number of specialized, concrete implementations such as class `FIXED_STACK` which relies on a limited array for storage. The declaration specifies that "`T`"s can be stacked – where a `T` is any kind of Eiffel object.

A program can instantiate class `FIXED_STACK`. But it must specify what the stack will contain; so, one might have a declaration like:

Defining an instance of a generic class

```
letterstack        : FIXED_STACK[MAILITEM];
```

which defines `letterstack` to be a stack onto which mail items can be pushed, and from which `MAILITEM`s can be popped. The compiler will verify that `letterstack` is used appropriately.

Nothing happens to a datum that gets pushed on a stack until it is popped off again. A stack is totally unconcerned with the nature of the objects that it stores. One does not always have this degree of independence.

For example, a program might need a "priority queue" that keeps objects in ascending order. When an item is added to a priority queue, the queue object has got to find the right place for the item. The queue isn't responsible for actually performing the comparison operation. The new item can be asked to compare itself with a stored datum and report its ranking with respect to that stored item. The queue object can have a method with a loop in which the new item to be stored is compared with successive queued items until the appropriate insertion point is found. Although not responsible for actually doing the comparison, the queue object does depend on the stored items having the capability of comparing themselves. This is a requirement for use of a priority queue. The requirement forms part of the declaration of the class. For example, the Eiffel "structures library" defines a particular version of a priority queue:

```
class SLP_QUEUE[T->PART_COMPAR]
    export
         ...
    inherit
         ...
end --
```

A program can create such queues to store objects of class `T` – provided that `T` is derived from class `PART_COMPAR`. (`PART_COMPAR` is an abstract class that Eiffel uses to capture the concept of objects that can compare themselves $>, \geq, =, \leq, <$ with other objects of the same class. `PART_COMPAR` would be used in some form of multiple inheritance scheme rather like `STORABLE` as described in section 3.6.)

Some examples of C++ parameterized types are given later in Chapter 12. As parameterized types are a recent addition to the language, many class libraries use the older less reliable style of having containers of `void*` or `TObject*` pointers and using type coercions. The Borland class libraries are one of the first to start to exploit parameterized types.

3.15 EXTENDED OO FEATURES: AUTOMATIC GARBAGE COLLECTION

OO languages differ significantly in their support for "automatic garbage collection". If a language and its run-time environment provide automatic garbage collection, programming can be considerably simplified. Programmers can focus more on the problem that is to be solved rather than on details of memory allocation; overall, this leads to higher programmer productivity. However, automatic garbage collection can entail significant run-time costs.

The types of problem largely solved by automatic garbage collection are illustrated here in terms of another extension to the Space Invaders program. The idea behind this extended version is illustrated in Figure 3.17. A new class of Space Invader has been added – a transporter for an invading army:

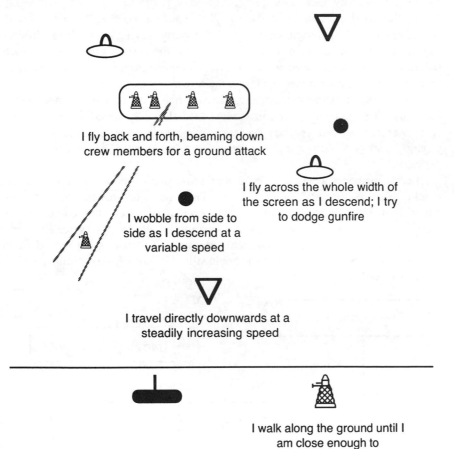

Figure 3.17 Class Transport added to the Space Invaders. Transports attack by beaming down Dalek crew-members that then advance on the ground.

```
class Transport : public Invader {
public:
    ...
private:
    Dalek*  crew[6];
...
```

The Run() behaviour of Transports has them moving across the screen occasionally "beaming down" a Dalek that then attacks on the ground. (The Gun class had better be extended to work with a three-button mouse that allows it to fire left or right as well as straight up.) These Daleks are still part of the crew of the Transport, as recorded in the crew[] data field. If their attack is successful, they will be picked up by their Transport.

Multiple references to objects

As illustrated in Figure 3.18 there might be several Daleks forming a ground attack party, with different members of the party having come from different Transports. The program would use a list for the current ground attack party. Individual attacking Daleks would then be referenced from two places – the list of ground attackers, and the crew record in their own Transport.

Problems with memory management arise when a Transport is shot down by the gun. The Transport and its crew should all be destroyed, except for those crew members who have been successfully beamed down and are a part of a ground attack party. It is simple to delete the Transport. The problems relate to the crew.

Dead space in the heap, or dangling pointers

If crew members were not deleted, then those that were on board the transport would remain as "ghosts" in the system. The data structures for those Daleks would remain in the heap, uselessly occupying "dead space" that could never be reclaimed. But it would also be wrong to simply iterate through the crew[] array deleting Daleks. For that would leave the ground-attackers list with dangling pointers – references to Daleks that should still exist but whose defining data have been destroyed.

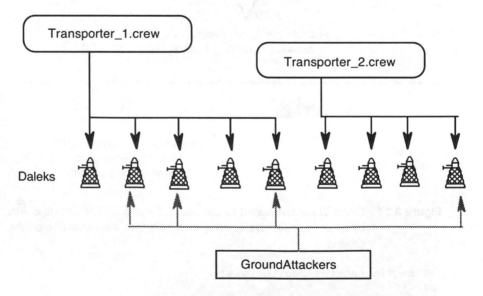

Figure 3.18 Daleks in ground attack party will each be referenced in two different collections – their Transport's crew record, and the list of ground attackers.

In situations such as this, the programmer is faced with doing a lot of "housekeeping" to keep things consistent. Possibly, extra data fields would be added to the Daleks (Boolean on_board_a_transport?) that would be checked before deletion. The trouble with this approach is that it leads to a proliferation of extra data fields and special processing checks. An on_board_a_transport data field in a Dalek would handle the current problem, but other special cases would arise as the program was further extended.

Ad hoc solutions to memory management problems

A more systematic alternative is for each object to have an associated count that records the number of places from which it is referenced. When created, this count is set to 1. When a further reference is created by assignment, the count is incremented. When a reference is cancelled, the count is decremented. The object is finally deleted when its count gets to zero. (Self referencing and circular data structures can cause problems in reference counting schemes; the internal references may mean that reference counts never reach zero even though all external references to the structure have gone.)

Reference counts

The straightforward implementation of reference counts involves the introduction of an extra class of objects that keep pointers to those real objects whose storage is to be controlled. ("All problems in computing science can be solved by adding an extra level of indirection.") In the Space Invaders example, the Transporter crew[] array and the GroundAttackers list wouldn't hold simple Dalek* pointers, they would contain instances of some DalekPtr class. When a Dalek gets beamed down and joins the GroundAttackers a new DalekPtr object would be created and added to the list; as part of the process of its creation, a DalekPtr object would tell its associated Dalek to increment an internal reference count. The code handling the shooting down of a Transporter would have a loop in which each DalekPtr object in the crew[] array was destroyed. When a DalekPtr was destroyed, it would send a message to its associated Dalek notifying it that its reference count was to be decremented. Dalek objects would be coded so as to self destruct when their internal reference count reached zero.

Such coding is pretty standard; there are examples in all the text books. But it is convoluted code, and it is tiresome to have to organize the same sort of code for every class that may need this kind of storage management. The straightforward implementation also involves an extra level of indirection every time the code needs to access a controlled object. (For example, instead of iterating along the GroundAttackers list with code like "*for each Dalek* in GroundAttackersList do Dalek->Exterminate()*", one needs "*for each DalekPtr in GroundAttackersList do DalekPtr.Dalek-> Exterminate()*".)

Disadvantage of reference counts

The handling of these memory management problems involves tiresome, repetitive work hacking out the code. All tiresome, repetitive work is better done by computers.

In an automated approach, the programmer never need be concerned with returning dead space to the heap. In the Space Invaders example, the Transport object would have been originally allocated on the heap and a pointer returned; this pointer would have been part of an array, or list, of pointers to current attacking invaders. When a Transport was shot down, the pointer that referenced it would simply be set to NULL. There would be no code to "free" or "delete" the Transport; and its crew would never be considered. Nevertheless, the dead space in the heap occupied by the `Transport` object, and those unfortunate crew members who were on board, would eventually be reclaimed by a "garbage collector". (It is a "garbage collector" because it clears the heap of the "garbage" of discarded and forgotten data structures.)

Garbage collectors   In primitive implementations of garbage collection, a program is allowed to run until heap space has been exhausted. Then, a garbage collector routine is given the task of reclaiming all dead structures so freeing up space in the heap. In more practical systems, the garbage collection process takes place at regular intervals (this prevents the heap from growing unnecessarily large and avoids the effect of a program stopping dead for several seconds as can occur with single shot garbage collection). The "garbage collector" may be implemented as a subroutine that gets invoked from some point in the memory allocation code of the run-time support library. In more sophisticated systems, a garbage collector will be implemented as a coroutine, or maybe a separate process thread, that can run in parallel with the main program.

Garbage collection systems were originally used mainly in Lisp environments and the way they work is most easily illustrated in terms of a simplified version of a Lisp system. A simple Lisp system will pre-allocate all available memory into four zones. One zone will hold strings for atom (variable) names, and prompts. Another will hold numeric values. The third holds the Lisp atoms. In a simple implementation, a Lisp atom comprises a pointer to its name and a pointer "value" field. An atom's value field can hold the address of another atom, or the address of a numeric value, or the address of the start of some list structure. The fourth and largest zone is used for "cons-cells"; these are the small data structures (consisting really of just two pointer fields and a couple of bits for flags) that are used to construct lists and whose allocation and reclamation has to be managed. The two pointer fields (the "car" field and "cdr" field) in a cons-cell can hold addresses of atoms, or addresses of numeric values, or addresses of other cons-cells. When memory is initially allocated, all the cons-cells are joined together in a long list – the "free list" – (the "cdr" field, tail pointer, of each cons-cell holds the address of the next cons-cell).

If a user types in:

```
input>(setq a '(b c))
(b c)
```

the Lisp interpreter would build a list from two cons-cells, filling the head ("car") fields of these cells with the pointers to the atoms 'b' and 'c' and linking the first to the second. The address of the first cell would be stored in the value field of atom 'a'. After construction, the list would be printed as shown in the code fragment. If the user then entered:

```
input>(reverse a)
(c b)
```

the Lisp interpreter would build a second list and print it. The situation in memory would then be as shown in Figure 3.19; there would be two lists represented in the cons-cells in addition to the free list containing all remaining cons cells (not shown). The first list is still "live", because it can be accessed via the value cell of atom 'a'. The second list is "dead"; there are no pointers referencing its first element.

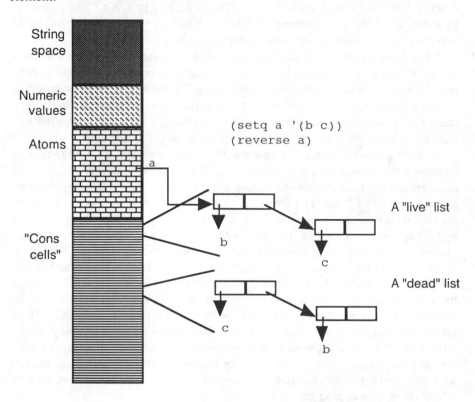

Figure 3.19 Memory organization in a simple Lisp interpreter.

If Lisp's garbage collector were invoked at this point, it would start by *Mark* identifying all those cons-cells that were in live lists. These are easy to find. Live lists must be referenced in the value cells of atoms; because different address ranges are used, it is easy to distinguish references to lists from pointers to numeric values or other atoms. Once the start of a list has been found, a mark can be placed in the first cons-cell (in the tag field provided). Then, the "car" and "cdr" pointers are followed. If a pointer leads to another cons-cell, this too is marked and used as the starting point for a recursive search.

When all live chains of cons-cells have been marked, the garbage collector *Sweep* enters its second phase. It runs through the entire cons-cell space checking each cell in turn. If a cons-cell has been marked, it is part of a live chain and should be left alone (once the mark has been cleared). If a cell is not marked, it is part of a dead structure and can be reclaimed. Each dead cell is threaded back into the free-list.

Garbage collection is relatively easy in Lisp. The regularity of the structures helps enormously. Only cons-cells need to be reclaimed. These are known to consist of two pointer fields that have to be followed, and a tag field that can be used by the garbage collector when marking live structures. The starting points for a search are also known. Live lists must be referenced from the value cells of atoms, so the garbage collector need only iterate through the array of atoms to find all possible lists. (The recursive mark phase of the mark-sweep algorithm described may use a lot of stack space, but that is another problem.) The process is so standardized that manufacturers of Lisp workstations have found it worthwhile to provide special hardware to assist garbage collection.

Garbage collection in a hybrid environment

It is a *lot* harder to do garbage collection when there are arbitrary data structures in the stack and on the heap. Unlike Lisp's atoms, there are no obvious starting points for a search for live structures in the heap. It is true that the remaining live structures must be those referenced by pointers that form parts of structures in the stack (or in the static data segment). But in ordinary compiled code, there is nothing to distinguish a pointer from a character sequence or a number.

In order to provide garbage collection, the run-time system is likely to require additional data from the compiler, such as data describing the layout of objects of different classes (so that pointers can be more readily identified). The compiler may have to insert "housekeeping" code that records extra information needed by the collection process. (Alternatively, a compiler may fake everything and arrange that all accesses to heap-based objects proceed through extra intermediaries that perform reference counting. This still leaves problems with circular structures.)

Timing problems

In addition to any overheads in storage and time involved in setting up the data needed by a garbage collector, there is an additional problem – the CPU time that the collector uses. Collection will occur at times determined by the load on memory and not at times determined by the logic of the application. If an application has any time-critical steps, it cannot permit these to be delayed by a switch of the CPU to a garbage collector coroutine. A global flag can be set – `no_garbage_collection = true;` – prior to entry to time critical areas; but, this approach is fraught with risks (the flag may not be set when it should have been, or it may not be cleared after having been set).

Languages providing automated garbage collection

Both Smalltalk and Eiffel provide automatic garbage collection. (In fact, the authors of both these systems identify automated garbage collection as an essential feature of an OO language.) For Smalltalk, the choice of automated garbage collection is obvious. The language is primarily for prototyping – for exploring computational solutions to a given problem. Smalltalk programmers are "problem solvers" not "coders"; they need to focus on problem specific issues and not on machine-related details. The Smalltalk environment has all the information describing the forms of objects (the class definitions are themselves objects present at run-time); consequently, the information needed by a garbage collector is available. Smalltalk systems are interactive; their responses are on human time scale ($> \approx 0.1$ seconds). There is no need to be concerned with time-critical constraints.

Eiffel has provided automated garbage collection as part of its effort to raise the level of programming languages. Languages influence programmers; if a language seems to focus on low level issues, such as pointer arithmetic and hints to a compiler regarding register usage, then programmers frequently focus on the same low level issues. In the long run, such a focus is not productive. Hardware costs go down; software costs go up. Software packages may need to last for twenty years (e.g. IBM's transaction system software has survived two complete replacements of its original 360-hardware). It is better for a module in such a big software system to be designed so that it is clearly a solution to a particular subpart of the problem rather than appearing as a contorted effort to coerce desired behaviour out of a recalcitrant compiler and machine. It is very easy for memory management hacks to obscure the structure of a software component. As Eiffel is intended to be a high-level, design and implementation language, automated support for garbage collection is necessary.

The various object-oriented extensions to Pascal obviously do not attempt to provide for automated garbage collection. These Pascal dialects include only the minimal extensions necessary to permit an OO programming style.

Languages without support for automated garbage collection

C++ has also chosen not to support automated garbage collection. This choice reflects the broader aims of C++. It is not "just an OO language". It is intended to replace C in areas where C has dominated – such as time-critical communications components in operating systems, utility programs and so forth. The stronger type checking in C++, and the "first class programmer defined types", allow programmers to write much more reliable and clearer C-style programs. A tenet of the C++ programming language is "you only pay for what you use". If a program does not require a feature, it should not pay a cost at run-time that support for such a feature would entail. Support for garbage collection could entail the compiler generating extra tables providing a run-time system with information concerning every different type of data structure used; every structure defined in a program might have to have additional tag fields appended for use by a run-time garbage collector. If a program does not leave garbage in the heap, these provisions for garbage collection are pure overhead – paying for things that aren't used. Further, it has been argued that a programmer should be able to adopt storage management methods that exploit details specific to the application and so are far more efficient than a general purpose automated garbage collector. Consequently, C++ as a language does not support garbage collection. There have been numerous papers in C++ conferences describing various implementations that can be used as models if a garbage collection system is required.

C++: "you only pay for what you use", so garbage collection is an extra

3.16 EXTENDED OO FEATURES? EXCEPTION HANDLING

"Exception handling" is not an object-oriented feature. It is a software engineering feature. But it is a feature that has concerned the designers of OO languages. Eiffel incorporates some exception handling features. C++ will include exception handling when the appropriate international committees have agreed on the features that are to be supported; several compilers already support an exception mechanism based on the proposed standard.

When something in the run-time environment fails, it is usual for a program to have to first perform a considerable amount of tidying up, and then provide a user with some choices for alternative actions. For example, consider what might happen in a program that is trying to read data from a file when the disk decides to be unreadable.

Run-time exceptions

The read operation will have been initiated by the user picking a File/Open option from a menu and selecting a file from the list shown in a dialog. The "application object" would then start the read sequence by creating a "document" object and asking that Document to execute some OpenOld() method. The document might then create one or more View objects, and some other objects in memory that are to hold the data that are to be read from the disk file. These objects will then be told to read themselves from the file. A few might succeed but one, while attempting to execute its ReadFrom() method will get hit with a run-time exception.

The exception will have started as a "fail" interrupt from the disk. This interrupt will be handled by the operating system and repackaged as an "exception" that is to be handled by the application.

It is logical for the exception to be handled by the chain of currently active objects. Each object in the chain gets a chance to deal with the exception. The object may be able to resolve the problem. If not, it should tidy up, and then pass the exception back to the next object in sequence.

An individual data object is most unlikely to be able to deal with a disk error. Its response will be to abandon processing of its ReadFrom() method and pass the buck back to the Document object.

A Document object should have code to deal with a read error on disk. It must be a hard error (the operating system will have tried re-reading the disk). Basically, the Document should abort the OpenOld method that its executing, but first it must tidy up. It must delete all those objects that represent data elements that were read in whole or in part. Then it must delete any views that it created. Only when this tidying up is complete can the Document object pass the buck to the Application object.

Finally, the application must delete the Document object and present an alert to the user warning of an unreadable file.

As suggested by this simple example, knowledge about what must be done when an error occurs can be localized in objects and their methods. Organizing the error handling mechanisms on the objects is therefore likely to be a sensible strategy.

The "exception handling" facilities that are being added to OO languages are intended to facilitate the description of error handling processes such as that just described.

Exception handling is a little beyond the scope of initial programming exercises. The example programs in this text will not make provision for recovery from run-time errors (other than that already provided in the class libraries that are used). The exception handling features of C++ and of Eiffel will be illustrated very briefly in examples in Chapters 12 and 13.

3.17 EXTENDED OO FEATURES: ASSERTIONS

Eiffel allows a programmer to specify "pre-" and "post-" conditions for individual methods and "class invariants". These constructs allow the definition of tests that verify correct usage of an object of a given class. For example, consider the put (i.e. push) method of a FIXED_STACK object (a stack that uses a fixed sized array for storage). Now, a FIXED_STACK has finite storage capacity and, consequently, it is possible that a put operation would cause a stack to exceed its capacity. If allowed to occur when the stack was full, a `put` would probably overwrite some memory locations that belong to the memory manager and so cause chaos. Therefore, a `put` operation had better be checked:

```
class FIXED_STACK[T] export
        ...
    inherit
            STACK[T]
            ARRAY[T]
    feature
    ...
    put(v:T)
            require
                    stack_full : not full
            ...
            end -- put
    ...
end --
```

Method "put" with a require clause (pre-condition test)

The code for the `put` method starts with a `require` clause. This consists of one (or more) labelled checks. If a stack is full when a `put` is attempted, the error will be caught by the `require` code which will cause termination of the program with an error message (in this case, "stack_full"; the label on the checking statement forms a part of the error message). A few additional examples of `require`, `ensure` and `invariant` will be illustrated in the chapter on Eiffel (`ensure` statements express postconditions, `invariant` statements describe class invariants.)

Similar checks can be included in C++ programs. There is no specific support for these checks in C++. However, the `assert` macros available with most C implementations can be utilized.

It is, however, somewhat misleading to regard Eiffel pre-/postcondition statements as just run-time checks akin to Pascal assertions or C `assert` macros. These features really are of greater importance as a mechanism that allows a class designer to characterize class behaviour to a client and to explain requirements to implementation programmers.

Not a run-time check, rather a design tool

Eiffel can be used as a design tool as well as an implementation language. When used as a design tool, the designer constructs the class definition with each method defined by a precondition (`require` statement) and a post-condition (`ensure` statement), and the class as a whole described by an invariant. Sometimes, these various conditions will be given as Eiffel statements; sometimes, they will be more in the form of a comment. While building these class descriptions, the designer can use the tools of the Eiffel environment to obtain diagrams of class hierarchies, listings of relations among classes (e.g. Which classes does this class make use of? Which other classes use instances of this class?).

The designer doesn't provide any code for the classes, that is left for the implementors; but the designer's descriptions of the methods are sufficiently detailed that coding should be easy. For example, part of the design description for the `remove` method of class FIXED_STACK is:

Require and ensure statements to guide implementors

```
class FIXED_STACK[T]
    ...
    remove
            -- remove top item
            require
                    not_empty:   not empty
            ensure
                    not full
                    count = old count - 1
    ...
    end
```

The `ensure` clause reminds the implementor to update a count of data items and to clear any Boolean flag that might have been used to indicate a full stack. The `require` clause can be included in the actual program as a run-time check; it also makes it clear to the programmer that there will be no need to include other checking code in the actual implementation of the `remove` method.

Require and ensure statements to inform clients

Eiffel comes with tools that can be used to print indexes of its class libraries. These indexes identify the services provided by the various available classes so that an applications programmer may more easily find existing reusable components. Any `require` or `ensure` clauses associated with a method will appear in the class description given to the prospective client. They provide a brief form of "contract". When applications programmers see that a `remove` method has

```
require
        not_empty:   not empty
```

they know that it is *their* responsibility to ensure that they do not attempt to remove things from an empty stack and that their code must therefore be along the following lines:

```
s    : FIXED_STACK[LETTER];
l    : LETTER;
...
shuffle is
    ...
    do
        ...
        if(s.empty) then ...
        else
                l = s.item;
                s.remove;
                ...
                end
```

If they don't check for an empty stack but simply go ahead and attempt to access or remove an item, they know that the library class will cause a run-time exception.

REFERENCES

The creators of Eiffel and C++ have both published explanations of why their languages have their chosen forms. The languages do have rather different aims and this is reflected in their forms. Meyer presents the case for a pure object-oriented, very high level language in his text *Object-oriented Software Construction*. The *Annotated Reference Manual* by Ellis and Stroustrup explains the choices made for the "hybrid" C++ language.

There aren't many simple explanations of how objects actually work in the memory of the machine. The Ellis/Stroustrup text does include some explanations of how method tables ("virtual tables") can be organized in the context of multiple inheritance.

B. Meyer, *Object-oriented Software Construction*, Prentice-Hall, 1988.

M.A. Ellis and B. Stroustrup, *The Annotated C++ Reference Manual*, Addison-Wesley, 1990.

In section 3.2, design notations for describing classes were briefly foreshadowed. For example, Figure 3.4 illustrates the form of an Invader class using a notation similar to those developed by Coad and Yourdon and by Henderson-Sellers. Chapter 11 elaborates on such design notations.

Introduction to the use of C++ as an Object-oriented Implementation Language

Introduction to C++

The next three chapters introduce C++. Cover C++ in only three chapters? Impossible! The coverage is limited. It is expected that the materials here will be supplemented by one of the standard C++ texts, e.g. Lippman's *C++ Primer*, and by the implementation notes that would be provided with the compiler that is used. Previous experience with C, and some acquaintance with Modula2 or Pascal, is assumed.

C++ retains all the standard procedural programming features of C. It uses a basically similar style for variable and function declarations; has selection statements (`if(…)` `…;else` `…;` and `switch(…)` `{ … };`); iteration statements (`for(…;…;…)`,`while(…)` `…;` and `do … while (…);`); and flow control statements (`break;` `continue;` `return …;` and even `goto`). The "block structure" is similar to C, as are the scope rules for all variables that are not parts of classes. None of these standard aspects of coding are covered in the next three chapters; all are assumed to be familiar to readers.

The first rather short chapter covers some extensions that C++ provides that are relevant to normal, procedural-style code. These include minor differences, such as new styles for comments, as well as more significant changes, such as "overloaded function names" and the new type-secure "stream i/o" that should be used in preference to the ISO/ANSI C i/o libraries.

Minor extensions to C

The next chapter adds a touch of class. C++ classes are introduced through examples. Rather than present a definition of C++'s class with all its features, a series of examples is given that successively add more capabilities.

Classes as "abstract data types", "reusable components", and "object-based" programming styles

The first examples, `Point` and `Rectangle`, are archetypical "abstract data types". The class declarations define the form of new structures (like a `typedef struct` in C, or a Pascal `Type x = record … end` declaration). But, as well as specifying the data fields ("data members") in these new record types, the C++ declarations specify controls on access to these fields and nominate the procedures ("member functions and friends") that should be used with instances of the new record type. (In the rest of this text, the term "programmer defined type" will be used in preference to "abstract data type"; it is a more apt description of the entities like Points, Rectangles, and Lists that are being defined and used.) The Point and Rectangle examples illustrate how the author of a class can, by defining the semantics of operators such as = (assignment) and '+', integrate a new type so completely into the language that it appears to be a "1st class type" having the same status as built-in types like integer (int) or character (char). It seems a fairly safe prediction to assert that all C programmers, no matter how hidebound and reactionary, will eventually adopt C++ if only to obtain these capabilities for defining "1st class" data types.

Another example, class Player, introduces some of the complexities in classes, such as the need for the class author to manage any separately allocated storage, or other resources, that must be controlled by instances of a class. In the next section, class LinkedList and its associates Link and LinkedListIterator illustrate how a group of classes can be designed to work together; in this example, an instance of class LinkedList uses many instances of class Link when building up its data store. While simpler classes like Point and Player may be conceived and used in isolation, most of the more complex classes exist as parts of small "clusters" with strong ties existing among instances of the different classes in a cluster.

Class Assoc, that implements a form of association list, is used to illustrate another standard object-based programming style – the delegation of work to a "subcontractor". Class Assoc doesn't handle any storage management, preferring to delegate this work to an instance of class LinkedList. Class LinkedList was invented for one application, and ends up being reused by many other classes as a component or subcontractor.

Finally, an example class ChemicalStructure is used to explain the role of "static" members in C++ classes.

Inheritance and OO programming

The third chapter in the section illustrates classes with inheritance and support for an object-oriented style of programming. Here, the Space Invaders program is revisited and a simple implementation is given that illustrates single inheritance, polymorphism, dynamic binding – and all the other fun features of an OO programming style.

Advanced features of C++

Some more advanced features of C++, such as support for multiple inheritance, are taken up Chapter 12.

Reference books

The materials in these chapters should be supplemented with material from one or more of the following books:

S.B. Lippman, *C++ Primer 2e*, Addison-Wesley, 1991.
Pretty much the standard C++ text. It covers the whole language. The first 150 pages is really C revisited, with data types, expressions and statements, functions etc. all being reviewed. Overloaded functions are detailed in 50 pages. Classes in all their complexity, including templates, are covered in the next 170 pages. There are 100 pages on inheritance in C++ and 50 pages on OO design. The book has appendices, including a quite useful one giving many details of the stream libraries.

B. Stroustrup, *The C++ Programming Language 2e*, Addison-Wesley, 1991.
Well, it is his fault, lets hear his explanation. This book gives the creator's view of his language and some advice on its use. The text is a little more concise than Lippman on the standard C components of the language. It has a detailed coverage of classes, derived classes and template classes and a lengthy chapter on operator overloading. Facilities for

exception handling are covered; these are only just becoming available with latest releases of commercial C++ compilers. The last quarter of the book is advice and suggestions on the design of C++ programs and class libraries. This section of the book should be used as a reference even if an alternative text is used for initial study of the language.

S. Meyers, *Effective C++*, Addison-Wesley, 1992.
The C++ gotcha book. Explains why things are not quite the way you thought they were. Essential at every stage of learning (surviving?) C++.

T. Cargill, *C++ Programming Style*, Addison-Wesley, 1992.
Nine cautionary tales from C++ land. Each chapter examines code that abuses a particular feature of C++; all examples are said to be derived from illustrations of C++ in other text books. Those thinking of using "operator overloading" or "multiple inheritance" should first read the appropriate cautionary tales.

M.Andrews, *Visual C++ Object Oriented Programming*, SAMS, 1993.
K. Chriatian, *Microsoft Guide to C++ Programming*, Microsoft, 1992.
T. Faison, *Borland C++ 3 Object Oriented Programming*, SAMS, 1992.
S. Holzner, *Miccrosoft Foundation Class Library Programming*, Brady, 1993.
S. Holzner, *Borland C++ Windows Programming*, 3e, Brady, 1994.
N. Rhodes and J. McKechan, *Symantec C++ Programming for the Macintosh*, Brady, 1993.
N.C. Shammas, *What Every Borland C++ 4 Programmer Should Know*, SAMS, 1994.
D.A. Wilson, L.S. Rosenstein, and D. Shafer, *C++ Programming with MacApp*, Addison Wesley 1991.
M.J. Young, *Mastering Microsoft Visual C++ Programming,*, Sybex, 1993.
These are typical of the books that describe specific implementations of C++, and associated class libraries, that are available on various platforms.

C++ : Minor Extensions to C

4

C++ is multi-faceted. It is different from the other OO languages considered here; it is inherently more complex. Apple's original "Object Pascal" was developed to be the "smallest" extension to Pascal that would permit at least some limited style of object-oriented programming. Smalltalk and Eiffel each represent an "ideal" OO programming language as conceived by particular individuals or small groups. Objective C simply provides a way of layering some Smalltalk constructs on a C substrate, and thus providing some support for OO programming. C++ has wider aims. C++ was to be a "better C". It was to provide full support for programmer defined data types. It was to extend C with Simula-like class constructs, and so provide support for OO programming styles. While accomplishing all these tasks, it was to maintain compatibility with existing C code and have the same level of run-time efficiency as C.

A better C? Primarily, this means a C with much more compile-time checking, particularly type checking. C++ was so successful at being a better C, circa 1983, that it had a substantial influence on the development of the C programming language. Current ISO C incorporates a number of features, such as function prototypes, that were originally pioneered in C++. The old style C function declaration did not permit the compile-time checking of argument lists. Many errors in C++ programs resulted from mismatches between the required arguments and the values passed in a call. Function prototypes were introduced to eliminate such errors. *A better C?*

Many other features in C++ were inspired by observations of C language features that were error prone or subject to misuse. Alternative constructs were provided in C++ that reduce or eliminate the possibility of error. Examples of these extensions include the replacement of most #define macros by either const declarations for constants, or inline declarations for functions. *Avoid C's more error prone features*

Some features of C have survived into C++ for reasons of compatibility despite their having proven to be unsatisfactory. For example, C provides very little support for the concept of an array. C's array concept is very low level and much too tightly bound to the autoincrement-pointers that existed on primitive computer architectures. It would have been possible for C++ to have defined arrays properly, providing optional support for array bound checking and other commonly provided *Unimproved C features, arrays* features. However, such a change would have made C++ incompatible with all existing C code. In language design, compromises may have to be made.

Another retention from C is the syntax of variable declarations. These are moderately complex because of the options for modifiers on basic types (e.g. `short`, `long`, `register` etc.). The basic declarations are simple enough:

```
<type>                    <variablename>;
int                       count;
register int              i;
static unsigned short int c;
```

C allows the definition of derived types using postfix and prefix forms. The postfix forms, `()` for function declaration and `[]` for array declaration, rarely cause trouble. A form like:

```
short int c, f(), m[10];
```

is easily interpreted as defining a single short integer variable, declaring the existence of a function that takes no arguments and returns a short integer, and defining an array of ten short integers. The problems seem to come with the prefix forms, particularly the `*` pointer forms. The prefix `*` belongs with the variable that is being declared, but there are no constraints on layout and the following are all acceptable:

```
long int       *ptr;
long int*      ptr;
long int   *   ptr;
```

The most common mistake is a declaration such as:

```
long int* ptr1, ptr2;
```

The type of `ptr1` is indeed "pointer to long integer"; but, `ptr2` is a "long integer". The declaration style seems to be a hangover from the earliest days of C on a PDP-7 where programmers thought in terms of "variables" and "pointers". The declaration "`long int *ptr;`" meant that `ptr` was a pointer to a "long integer variable". With a proper typed language it is better to think of `ptr` as a variable whose type is "pointer to long integer"; the difference being that the `*` modifier would then really be part of the type specification rather than a prefix part of a variable name. Although C's declarations do cause some problems, the requirement for compatibility meant that C++ could not make changes. In this text, the declaration style for pointers will be:

```
<type>*        <variable>;
int*           ptr1;
char*          bufptr;
Link*          nextentry;
```

The $*$, pointer derived type, will be written as part of the type specifier. Consequently, each pointer will be individually defined.

The differences from C noted here are:

Changes covered in this chapter

- changed comment styles;
- function prototypes;
- overloaded functions;
- default arguments for functions;
- automatic type coercions;
- `const` declarations;
- `inline` function definitions;
- types `void` and `void*`;
- tidier syntax for the declaration of enumerated types and `structs`;
- declarations as statements;
- references;
- new dynamic storage manipulations;
- "scope qualifier operator";

and

- stream input/output.

Some of these features (function prototypes, type `void`, etc.) are in ISO C but are covered here because there will be readers whose experience has been with earlier C compilers.

4.1 COMMENTS

A noticeable, but really quite minor change from C is the reintroduction of BCPL style comments. While C inherited much from the earlier BCPL language it did drop some features such as `resultof` blocks and the BCPL line-comment facility. Line-comments have made a comeback. (Line-comments are described first, not because they are important, but because they will appear in most subsequent code fragments.)

In C++, the sequence '`//`' introduces a comment block that will be terminated by the next newline. These line comments provide a convenient means of annotating a procedure call or documenting the role of a variable that is being defined; examples of usage are:

```
Delay(60,t);            // wait for ≈1 second
…
…

Link*      temp;        // pointer to previous link in chain
char*      astr;        // pointer to text buffer
…
```

C style comments, started by the opening sequence '`/*`' and terminated by the closing sequence '`*/`', are also valid in C++. Commenting styles in major programs and packages are usually determined by "house-rules". Typically, the standard C style comments, with begin-end bracketing, are used for any lengthy explanations in the code (and for inserting comments in the middle of a statement) while the line-comment style is mainly for documenting variable definitions.

4.2   FUNCTION PROTOTYPES

A function declaration should give:

- the name of the function;
- the type of value returned;

and

- details of the arguments required.

It is the final requirement, details of the arguments, that differentiates C++ (and now ISO C) function prototypes from the older C function declarations.

Unchecked function declarations and calls can lead to errors

C allows the following declarations and function calls:

```
extern char* strchr();
extern int strncmp();
extern double lgamma();
...
char*     ptr;
char*     id1;
char*     id2;
double    x;
...
    ptr = strchr(id1,'@');
    ...
    if(strncmp(id1, id2)) ...;              // ?
    ...
    x = lgamma(2);                          // ?
    ...
    printf("%s", id1, x);                   // ?
```

The trouble with this style is that there cannot be any compile-time checking to verify that a function is called with appropriate arguments. Many errors in C programs were caused by the omission of arguments (as in the illustrated call to `strncmp()`), or by arguments of the wrong types (as in the call to `lgamma()`). The 'lint' program could detect such errors, but its reports did not highlight them well. Significant problems, like incorrect arguments, could easily be lost in chaff such as lint's report that the return value from `printf()` was unused.

Declaring arguments for functions

C++ introduced function prototypes where a function declaration specifies the argument types.

```
extern char*    strchr (const char *s, int c);
extern int      strncmp (const char *s1, const char *s2,
                            size_t n);
extern double   lgamma(double);
```

Given prototypes, a compiler can check that all function calls use an appropriate number of arguments, and that these arguments have the correct types.

A declaration does not have to name the arguments, it must simply give their types. Consequently, the declaration double lgamma(double) is correct. Obviously, the definition of the function will name arguments, so the definition of lgamma() would read something like:

```
double lgamma(double x)
{
    … x …
    return …
}
```

Sometimes, there are circumstances where arguments will be unused. For example one might have a function that is to draw some graphic symbol with colour highlighting, but during development the colour features might not be used:

Unused arguments

```
void doodle(int h, int v, RGBcolor hue)
{
    MoveTo(h,v);
    PlotIcon(263); // When colour stuff works, add colour!
}
```

This code would result in a warning message "hue not used". An excess of "trivial" warnings can obscure a more serious warning (e.g. "x used but not set") and the compilation process may abort if too many warnings are generated. In situations such as this, the function definition may have the argument name commented out, or the argument may simply be omitted with only its type being specified:

```
void doodle(int h, int v, RGBcolor)
{
    …
}
```

In C++, a function that takes no arguments is declared in the form:

Functions with no arguments

```
extern int rand();
```

this differs slightly from the ISO C declaration of a function with no arguments:

```
extern int rand(void);
```

The ISO C style of declaration is also valid C++, and many programmers prefer this style.

There are many functions in the standard C libraries, e.g. printf(), that take a variable number of arguments. Since C++ has to maintain backwards compatibility with existing C code, a mechanism is required to notify the compiler that a particular function does have a variable number of arguments and that calls to the function should not be checked. The notation uses three dots (like an ellipsis, but the actual '...' ellipsis character is not used). For example, the C++ version of <stdio.h> declares printf() as:

Functions with an variable number of arguments

```
int printf(const char *format, ...);
```

so specifying that in every call to `printf()` there must be a `char*` format string, and that this string may be followed by an arbitrary number of additional arguments of unchecked types.

4.3 OVERLOADED FUNCTIONS

Quite commonly, an application will require a whole series of very similar functions. Consider for instance the series of functions provided in the input/output libraries of a Modula2 system; there one will find:

```
WriteReal(x);
WriteInt(i,10);
WriteString(s);
WriteChar(c);
```

In many C libraries, there are rather similar series of functions; for example, the functions for computing the absolute value of a numeric argument:

```
int       abs(int);     // int is short on this machine!
long      labs(long);
double    fabs(double);
```

In these situations, there is really one concept – the concept "write a value" or the concept "get absolute value" – but the library-author must create, and the library-user must remember, a whole series of different names that distinguish functions by argument type. This kind of tiresome work is best left to a computer.

In C++, function names can be "overloaded", i.e. there can be many different functions with the same name (provided that the compiler can distinguish among them on the basis of argument values). So, in C++ it is possible to define the functions:

```
void      Write(short,short);
void      Write(long, short);
void      Write(double);
void      Write(char*);
void      Write(char);
short     abs(short);
long      abs(long);
double    abs(double);
```

The compiler can sort out which function to invoke based on the types of the arguments used in the call. (There are some complexities when a function is called with an argument that could be interpreted as having a type that is one of a number of different possible types, for each of which there is a different overloaded function definition. These complexities are explored in texts such as the *C++ Annotated Reference Manual*, (ARM), by Ellis and Stroustrup.)

The way C++ deals with function overloading is to rename *all* functions.   *Name mangling*
Details of a possible renaming scheme are given in ARM. Using the scheme
described there, the example Write functions would be renamed along the lines:

```
void        Write(short);        => Write__Fs()
void        Write(long);         => Write__Fl()
void        Write(double);       => Write__Fd()
void        Write(char*);        => Write__FPc()
void        Write(char);         => Write__Fc()
```

The linker will see calls to, and code for these distinct functions. These renamed
functions are also the names that would appear in debugging outputs. The new
names for the example `Write()` functions are easy to understand. The renaming
schemes come up with much more complex forms when there are multiple
arguments and arguments of programmer defined types. It has been suggested that
the function name gets "mangled" by the C++ compiler. (On most systems, there is
an "unmangle" utility that will take a name created by C++ and translate it back to
the original form.)

As functions are normally defined in different modules, and referenced via   *Type safe linkage*
`extern` statements, a compiler cannot do all the checking on function definition
and usage (because the compiler never gets to see the complete program text).
Some checking must be left to the linker. The checking done by the linker will
detect errors such as where one module defines `void foo(short)` while another
module specifies `extern void foo(long)`. The compiler will have renamed the
defined function, e.g. as `foo__Fs` while the calls compiled in the second module
would request `foo__Fl`. The linker would report function `foo_Fl` as being missing
and terminate program construction.

Name-mangling can result in problems in mixed language programs. Consider   *'extern "C" {...}'*
a C++ program that needs to use the random number functions from the standard C   *specifier*
libraries. The declarations:

```
extern void        srand(int);
extern int         rand();
```

would be processed and the calls all handled, but the functions would have been
renamed `srand__Fi` and `rand__Fv`. At link-time, the linker would not be able to
find these functions. Once again, the programmer must be able to inform the
compiler of particular circumstances that require different treatment. In this case,
the compiler must not "mangle" names of functions in the C library. This is
accomplished by a special form of the `extern` specifier:

```
extern "C" {
void        srand(int);
int         rand();
}
```

The C++ compiler will not mangle names so declared as being C functions. (Some
environments may have further variants that permit specification of external
Fortran or Pascal functions. Usually, this entails more work by the compiler.

Quite apart from omitting the renaming process, the C++/C compiler system must then arrange to handle arguments and return values in different ways.)

On most systems, the standard C library header files will have been updated so that they can be used in either C or C++ compilations. Compiler directives have been inserted so that when a header file is being read by a C++ compiler all the C functions will be properly declared as being extern "C". For example, the <math.h> header files for the Macintosh MPW system read in part:

```
#ifdef __cplusplus
extern "C" {
#endif
extended sin(extended x);
…
…
…
extended hypot(extended x, extended y);

#ifdef __cplusplus
}
#endif
```

The macro __cplusplus is defined by every C++ compiler.

4.4 DEFAULT ARGUMENTS FOR FUNCTIONS

Often, libraries provide functions with many arguments, one or two or which are unique in every call but most of the others having the same values every time. Thus, one might have a function DrawString() that takes as arguments a string (bound to be unique for each call), a short integer font size (most often 12pt but other sizes getting used), an enumerated type specifying a style (almost always e_plain) and two short integers specifying horizontal and vertical displacements of the string's origin relative to the current pen position (almost always both zero). The function declaration might be:

```
void DrawString(char* txt, short fntsz, textstyle ts,
                          short dh, short dv);
```

with calls like:

```
DrawString("Hello World",18,e_plain,0,0);
DrawString("I said hello",12,e_plain,0,0);
DrawString("OK, sulk, I don't care",12,e_plain,0,0);
DrawString("Get lost",72,e_bold,1000,-2000);
```

Typing the extra arguments can become tiresome; particularly when it is obvious that the author of the library function would know in advance the sensible default values for the arguments.

C++ has a scheme that allows default values for arguments to be specified in a declaration. Trailing arguments, for which the defaults are acceptable, may then be omitted. Thus, it would be possible to have a declaration:

```
void DrawString(char* txt, short fntsz = 12,
            textstyle ts = e_plain,
            short dh = 0, short dv = 0);
```

This declaration would allow the above function calls to be rewritten as:

```
DrawString("Hello World",18);
DrawString("I said hello");
DrawString("OK, sulk, I don't care");
DrawString("Get lost",72,e_bold,1000,-2000);
```

Only trailing arguments can be omitted. So, the call

```
DrawString("Talk to me",12,e_italic,0,0);
```

would have to be written as:

```
DrawString("Talk to me",12,e_italic);
```

and not as something like:

```
DrawString("Talk to me",,e_italic);
```

It is possible to have several declarations of a function with different numbers of arguments being defaulted (possible, but a little baroque in style). So, one might have:

```
void DrawString(char* txt, short fntsz = 12,
            textstyle ts = e_plain,
            short dh = 0, short dv = 0);

void DrawString(char* txt, short fntsz,
            textstyle ts = e_bold,
            short dh = 0, short dv = 0);
```

Calls to `DrawString()` with a single argument would get the first version (12 pt, plain text, no offset); calls with two arguments would use the second set of defaults (bold text, no offset).

Although generally useful, the use of default arguments can sometimes be confusing. A maintenance programmer examining the text of a large program might note calls of the form `Compute(2);`, `Compute(i);`, `Compute(j + 4);` and infer the existence of a function `void Compute(int)`. Searches for that function would probably prove ineffective if it were actually defined as `void Compute( int, int x = 1, int y = 2, double z = 3.0)`.

4.5   AUTOMATIC TYPE COERCIONS

Most C programmers will have committed an error equivalent to that in the following program:

```
#include <stdio.h>
#include <math.h>

main()
{
    double x;
    x = sqrt(2);
    printf("x is %f\n",x);
}
```

The C system that I use runs this program and computes the value $\sqrt{2}$ as being 0.000000. That integer 2 was not the kind of value that function `sqrt()` was expecting. Function `sqrt()` wants a double as an argument.

The philosophy behind a C compiler is "the programmer is always right". Even in situations where compile time analysis will reveal that the programmer has almost certainly stuffed up, the actions of the compiler are premised on the basis that "the programmer is always right". Here, I asked for an integer argument to be pushed on the stack and call to be made to a function that will expect to find a double precision real. As far as a C compiler is concerned, I should get exactly the code that I asked for – although the simplest check on argument types would reveal an inconsistency.

In the ten years between C (early 1970s) to C++ (early 1980s) it was realized that even the best programmers make trivial errors that can be picked up and, sometimes, be corrected by a compiler. The C++ compiler will check the header <math.h> header file and determine that `sqrt()` expects a double but has been given an integer. The compiler could simply print an error message "You've forgotten to typecast from integer to double on line ...". Such error messages are extremely irritating when they terminate the program build process (one such error message from a current C++ compiler informs careless programmers "A semicolon was omitted at the end of line ..."). The compiler can determine from context that a conversion of integer to real is required, and can put in the code to perform the conversion.

C++ compilers essentially have tables defining known conversions. There are entries like (*integer available, real required, emit "float" instruction*), (*real available, integer required, emit "round" instruction*) and so forth. The data in these internal tables allow a compiler to generate the code for all these standard interconversions.

Explicit type conversions

A programmer is allowed to specify conversion explicitly. There are two styles of specification:

```
x = sqrt((double)2);// C style type cast

x = sqrt(double(2));// newer C++ type coercion
```

The C style casts must be used when the target type is something complex (e.g. "pointer to a function taking an integer and a pointer to integer and returning a pointer to a pointer to integer").

The next chapter, on programmer defined types, includes a brief explanation of how the standard type conversion tables of a compiler can be augmented with descriptions of functions that convert simple types, like integer, to user defined types such as "class Point". Once such conversion functions (actually, "constructors" for the classes) have been defined, the C++ compiler can use them as freely as the standard built-in functions that are used to convert integers to doubles etc.

Type conversions involving programmer defined types

4.6 CONST DECLARATIONS

In C, compile-time constants have to be declared with #define macros:

```
#define    SHRT_MAX    +32767
#define    M_LN2       0.69314718055994530492
```

There was nothing much wrong with such macro definitions for simple things like integer limits and special real numbers. But it was not easy to define something like a complete table of error strings (strings that could be used but weren't ever supposed to be changed). Further, the names of these constants disappeared after the macro preprocessing stage of compilation; this sometimes caused problems when checking compile time errors or when using a debugger to probe into a program's memory image.

In C++, one can have real constants. When a entity is declared a C++ const it retains all the other properties as required by its declaration (its type, its scope, its requirement for some memory locations, etc.) but the compiler makes certain that it is never changed. Examples of const declarations are:

```
const  short  NumGrades    = 6;
const  short  Marks[]      = { 0, 40, 50, 65, 75, 85 };
const  char*  GradeNames[] = { "Fail", "E", "D", "C", "B",
                               "Dean's List" }
```

A C++ compiler is wise to the devious ways of C programmers. Not only will it catch obvious attempts to change constants

```
NumGrades = 1;   // This will be stopped by the compiler
```

a C++ compiler will detect and stop some of the more underhand approaches:

```
int*      hack;
...
hack = &NumGrades;// You can't fool a C++ compiler this way.
*hack = 1;
```

A C++ compiler will reject this code fragment, reporting that an attempt has been made to take the address of a constant.

In C++, `const` is another modifier that defines an entity's type. Functions can return `const` entities. It is common to have functions returning `const` pointer types; this style will be used where a function is being employed to lookup some value in a private table.

Constant pointer

There are a few oddities associated with pointers. One can have:

- a pointer to some constant (the pointer is set, temporarily, to point to a data structure; the pointer can be used for read access to the data but cannot be used for write access);
- a constant pointer, i.e. a pointer that is effectively locked so that it always holds the same address (although the contents of memory at that address may change);
- a constant pointer to constant data.

Examples of these forms are:

```
// Pointer to constant data
const char* txt = "Hello World";
char* other = "Hi mum";
...
txt[0] = 'h'; // NO this pointer provides only 'read' access
txt = other; // OK, the pointer is a variable


// Constant pointer
char *const asmallbuffer = "01234567890123456789";
...
asmallbuffer[0] = ' ';   // OK, chars can be changed
...
asmallbuffer = &bigbuffer; // NO, can't reset pointer


// Constant pointer to constant data
const char *const dontchangeme = "chastity";
...
dontchangeme[0] = 'C';   // NO, data are constant
...
dontchangeme = &astring; // NO, pointer is constant
```

These declarations illustrate the `const` specifier qualifying the basic data type, `const char` ..., and qualifying the "pointer" aspect of the declaration, `char *const` ....

Cautions on const definitions in header files

Typically, most constants used in a program will be defined in a header file that is then "included" in many different modules. The situation is illustrated in Figure 4.1. As shown there, copies of `const` entities defined in a header file will be replicated in each compiled module that included that header file.

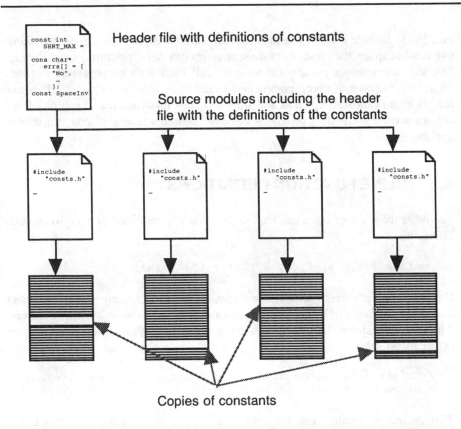

Figure 4.1   Const entities defined in a header file get replicated in compiled modules.

This replication doesn't matter if the header file defines only a few simple integers or a similar amount of data. However, it is obviously unwise to include the definitions of large data structures in an "included" header file. If an application has a large `const` array (e.g. a table generated by some parser generator, or a bit-map representation of an image), the data should be declared in the header file but defined somewhere else. So, a header file might read:

```
const int        MaxTokens = 512;
extern const int ParserTable[];
extern const unsigned char TheImage[];
```

(There are some differences between C++ and ISO C here. In C++ a `const` datum's scope is filescope unless its declared external.)

Start up cost of initializing consts

There can be other problems when there are `const` instances of programmer defined types. `Consts` must be initialized before entry to a main program; for things like integers and character strings, this initialization is trivial and is done by the compiler or linker. With programmer defined types, it may be necessary to execute an initialization routine (the "constructors" that are explained in Chapter 5). If a header file contains definitions of programmer defined types, then each module may get to include a set of routines for their initialization. Quite apart from any overhead in space, the presence of these routines can cause performance problems. The start up code for an application must call each such initialization routine. Therefore, the code of every module gets paged into memory at start up. This can lead to significant start up costs and slow launches for programs. Such problems are not serious with small programs but can become a matter of concern in real applications.

4.7 INLINE FUNCTION DEFINITIONS

The other main use of #define macros in C was the definition of code fragments. Things like:

```
#define max(A,B) (((A) > (B)) ? (A) : (B))
```

If a program really needs to compute a maximum value in many places, and can't afford the overhead of a function call, then these macros were quite convenient. Although convenient, they were a frequent source of bugs. They were easy to define incorrectly:

```
/* Distance(x, y) : distance of point(x, y) from origin */
#define Distance(x, y) sqrt(x*x+y*y)
```

This definition would work for calls like Distance(2,4), Distance(i,k). Sooner or later, a coworker on the project who didn't know that Distance() was a macro would try code like:

```
/* Need to know distance from origin of a point */
/* offset by (7,3) from current position (h,v) */
   d = Distance(h+7,v+3);
   ...
```

and then might have to spend hours trying to find "their bug".

Even if the macro author is careful and dutifully puts parentheses around everything, one can still have problems when there are calls with side effects. For example, the following code probably doesn't achieve quite the effect desired by its author:

```
/* Increment this signal count and record
   largest count value. */
newmax = max(sigcount++, currentmax);
...
```

Such macros are used when the author wants something with the semantics of a function call (evaluate the arguments, feed them to the "subroutine", get the result) but does not want to pay the time costs associated with a function call. Instead of paying for time (subroutine call instruction, build stack frame, ..., save result, free stack frame, return), the intention is to pay in a space overhead. Rather than a single copy of the code and subroutine call instructions at invocation points, the program contains multiple copies of the code – one copy at every point where a result must be computed. (There aren't necessarily any space overheads. Some of the "functions" commonly coded as macros, e.g. `max()`, require no more instructions when expanded than would be necessary for setting up a call to an out of line subroutine.)

With C++, if you want something with the semantics of a function call but with inline expansion then simply ask the compiler. It will sort out the code (and get it correct, unlike the average macro writer). Functions can be defined as being "inline". For example:

"inline" function definition

```
inline double Distance(double x, double y)
{
    return sqrt(x*x+y*y);
}
```

The C++ compiler will generate code that "evaluates the arguments" (x and y) and then performs the real computation; this code is then inserted at each point where the `Distance()` function is invoked. So, you get the correct answer for `Distance(h+7,v+3)` as well as for `Distance(2,4)`.

Compilers have limits on the complexity of functions that will be "inlined". In effect, the `inline` specifier is a hint from the programmer to the compiler (just as the `register` specifier is a hint). The compiler can ignore an `inline` specifier if it is inappropriate. (Compilers should generate a warning if a request to `inline` a function cannot be honoured.) Typically, compilers will not "inline" functions that have loops, or those that have multiple return statements, or those with very complex conditional statements.

Limits on inline functions

Only one definition is supposed to exist for a function declared `inline`. If modules containing `inline` function definitions are being compiled separately, there is no way for a compiler to check that the different modules in the program do in fact use the same function definition. Of course you may be able to "fool the compiler" and have different definitions for `max(,)`, or whatever, in different modules (though a smart compile/linker environment might notice and object). Unless you wish simply to cause pain and suffering to maintenance programmers, don't use different definitions for `inline` functions. Definitions of `inline` functions should go into a header file that is included by all modules that use those functions.

Fooling the compiler?

4.8 TYPES VOID AND VOID*

In old C, a program was built from "functions". Every routine was expected to compute and return a value; by default, the value would be of type integer. But often, procedures are required. Procedures don't return a value, they just produce side effects such as changes to data structures or outputs to a display screen. This feature of programs should be recognized by a programming language, which should therefore allow procedures to be declared.

Type "void" was introduced to allow declarations of procedures. Variables can't be declared to be of type `void`. The type can only be used in declaring routines. If a routine is declared to have type `void`, it does not return a value. A routine that requires no arguments can be declared as having a `void` argument list.

```
void Display(GraphicsContext*);
void SaveState(void);
```

Again, in C there was no general pointer type. Instead, the basic pointer type was `char*` (pointer to characters). The free storage allocation routines returned a `char*` pointer that was then type cast to `int*` or whatever other pointer type was required. At all other places in a program where a general pointer type was required, pointers had to be type cast to `char*`.

The style was a little misleading. In some place in a program a `char*` really was a pointer to a character string; elsewhere it would be used as a general pointer type. Type `void*` was introduced into the language to allow these two usages to be made distinct. Type `char*` should be used when referring to character strings; type `void*` should be used when the context simply requires a general pointer. (C and C++ don't have any true generic pointer type; a system can treat different pointers as being of incompatible types.)

Types `void` and `void*` are now a part of ISO C. Their use should be familiar to most readers. There are a couple of minor differences in usage between C++ and ISO C. As noted in section 4.2, ISO C expects a function taking no arguments to be specified as having a `void` argument list (e.g. `foo(void)`) whereas C++ will accept an empty argument list (e.g. `foo()`). In ISO C, a `void*` pointer value can be assigned directly to a pointer of defined type; C++ requires an explicit type cast in such assignments.

4.9 DECLARATIONS OF ENUMERATED TYPES AND STRUCTS

Support for types was limited in C. If a programmer wanted to introduce a new type, it was a two step process. First the form of the new type was given in a `struct` declaration, or `enum` declaration. Then, a `typedef` statement introduced a new type name; for example, in the Macintosh Quickdraw header files, one finds:

```
struct Point {
    short v;
    short h;
};

typedef struct Point Point;
```

Only after the `typedef` declaration can variables of type `Point` be defined.

In C++, a `struct` declaration, or `enum` declaration introduces a type name.

4.10 DECLARATIONS/DEFINITIONS AS STATEMENTS

In most block structured languages (C, Pascal, Modula2), all the variables used in a routine are declared at the head of that routine. C++ has relaxed this constraint and gone back to an older style where variables can be declared at the point where the programmer realized they were required. So, the following is legal C++:

```
void Receive()
{
/*
    Receive message, check for starting stx character,
    collect and buffer characters until etx character,
    verify checksum, copy validated message, …
    Set global Boolean 'ok' to true if valid message,
    Exit immediately, with 'ok' false, if find any error in
    input sequence.
*/
    ok = false;
    int ch = getchar(); // Need variable, ch, for input char
    const int stx = 02;
    if(ch != stx) return; // Bad start!
    int n = 0; // Better count the characters
    const int etx = 03;
    int space[5000]; // need some space for a buffer
    while(etx != ch=getchar()) space[n++] = ch;
    int checksum = 0; // their checksum is in last two chars
    for(int i=0; i<(n-2);i++) checksum += space[i];
    …
}
```

The advantage claimed for "declarations as statements" (and, therefore, being permitted in the body of a procedure) is that this style minimizes the risk of uninitialized variables. Sometimes, an appropriate initial value for a variable will only be determined after several statements have been executed; it is argued that in such cases, the declaration of that variable should be deferred until the required initial value is available.

The scope of a variable declared in a function extends from the point of declaration to the end of the block in which it is declared. In the example code fragment, the scope of the integer 'i' introduced in the for(;;) statement runs to the end of the procedure. (Some people expect, wrongly, that the scope of i would include only the body of the for-loop. There are languages where an iteration statement does introduce a local scope for the loop-control variable, but this is not the case in C++.)

It is not legal to jump into the scope of a variable without executing the declaration statement that brings that variable into scope. However, since the only way to achieve this effect is to use a "goto" statement and since no programmer would ever use a "goto", this is not a real issue.

4.11 REFERENCE TYPES

C++ introduces a new type specifier – "reference type". References have a couple of uses; the more important being the support of "call by reference" semantics for arguments to functions. "Call by reference" is Pascal's "var" parameter style (and is also the normal usage in Fortran).

In some ways, references are redundant. Pointers can be used to achieve almost anything that can be done with references. However, references do help make code clearer and simpler (and therefore, possibly, less error prone).

C's convention is that arguments are passed by value. The values of arguments are copied onto the stack before a routine is called. If the code of a routine modifies its arguments, the changes are made to the copies existing in the stack. On exit from a routine, the stack is cleared – and any changes made to those stacked arguments are lost. (C's conventions are always obscured by the case of arrays. The contents of an array won't be copied onto the stack because C interprets the use of an array name as meaning take and copy the address of the array's first element.) C's "call by value" style is appropriate if one wants to feed a little information to a function that will return a computed result. "Call by value" is inappropriate if one requires a procedure that is to modify the arguments that it is passed.

C's solution is to pass addresses around and make explicit use of pointers and pointer-dereferencing. The following code provides a simple example:

```
/*
C pointer style code.
*/

void swap(int*, int*);

main()
{
    int     x, y;
    x = 5;
    y = 6;
    swap(&x,&y);
    ...
}

void swap(int* a, int* b)
{
    int     temp;
    temp = *b;      // copy the value of the variable
                    //     that b points to;
    *b = *a;        // make the variable that b points to
                    //     equal to what a had;
    *a = temp;      // change the other variable
}
```

At the point of call, the addresses of the arguments are obtained and passed on the stack. The routine takes pointer arguments. At every point in the body of the code of the routine, these pointers must be explicitly dereferenced. The code looks ugly even in simple cases like procedure swap().

This pointer style puts the burden of a lot of hack, book keeping work on the programmer. One must remember to use the & operator to take addresses. One must insert the * dereferencing operator all over the place. Such hack work is better left to computers.

Using variables of reference type, the code becomes:

```
//  C++ Reference style code.

void swap(int&, int&);

main()
{   int    x, y;
    x = 5;
    y = 6;
    swap(x, y);
    ...
}

void swap(int& a, int& b)
{   int    temp;
    temp = b;
    b = a;
    a = temp;
}
```

Behind the scenes, the mechanism is probably the same. Addresses are pushed on the stack before the procedure call. In the code of the procedure, the addresses passed as arguments are dereferenced when accessing or changing the data. But all these operations are done by the compiler, and so do not have to be explicitly described by the programmer.

Reference parameters, like those in the argument list for swap(), are declared:

Declaring reference variables

```
// Declarations of reference variables as
// appropriate for the arguments to a function
<type>&          <variable name>;
int&             a;
unsigned char&   ch;
```

Declarations of reference variables, using &, are very similar to declarations of pointers, using *.

Often, functions need read access to large data structures. If structures are large, call by value is inefficient (it involves too much copying of information from an original structure to a stack-based copy) and may cause a program to abort with stack overflow. Call by reference is preferable. If only read access to the structure is required, the arguments for the routine should be declared as being of const reference types, for example:

const reference arguments

```
long CyclicRedundancyCode(const DataBuffer& buff);
```

This declaration states that `CyclicRedundancyCode()` is a function, returning a long integer, that expects a reference to a `DataBuffer` structure as argument and that it will require only read access to that structure. (Given such a declaration, when the definition is compiled the code is actually checked to verify that it does not modify the `const` argument.)

Specifying reference arguments as `const`, or leaving them implicitly non-`const`, helps those who need to use a library of functions. Client programmers can tell from a list of function signatures which functions will modify data structures that are passed as arguments.

Aliases?
Aliases!
References have a second use. They can be used to create aliases, i.e. alternative names for the same entity.

```
// Declarations of references and related variables as
// globals or local variables within a function
<type>&            <variable name> = <variable>;
int                jones;
int&               smith = jones;
```

Actual reference variables, that are globals or local variables, must be initialized at their point of declaration. Reference variables are initialized with the name of the variable for which they will serve as an alias. After the example variable definitions, there is only one integer variable occupying memory; but in the following code this variable can be referred to by using either of the names 'smith' or 'jones'. Aliases can be used to achieve effects similar to Fortran `EQUIVALENCE` statements that allow different arrays to be overlayed on the same storage locations.

The use of aliases is not illustrated in this text. No useful example could be found.

4.12 NEW DYNAMIC STORE MANIPULATIONS

In C, space allocation for dynamic, heap-based structures is handled through a set of run-time support routines and calls to the functions `malloc()` and `free()`. While C's free-store package can still be used, there is a preferred replacement. Programmer defined types, described in Chapters 5 and 6, will need to use the newer C++ free-store management system. In general it is unwise to mix two free-store management regimes, so it is best to make consistent use of the C++ package throughout a program. The C++ approach has the further advantage of having a cleaner syntax with no need for obscure type casts on pointers as are required when using `malloc()`.

'new' and 'delete'
operators
C++ has defined two extra operators – `new` and `delete`. The `new` operator applies to a typename; its effect is to create an instance of that type on the heap; possibly initialize the instance; and return a typed pointer. For example:

```
struct Point {
    short h;
    short v;
};
```

```
    Point*          nextPt;
    ...
    nextPt = new Point;
    ...
```

As described in Chapter 5, special initialization routines ("constructors") can be specified for programmer defined types. The processing effected by the new operator on such a type automatically entails a call to the initialization routine. If instead of being a simple `struct`, `Point` had been defined to have a constructor routine that took initial values for its coordinates, these would have appeared as arguments:

```
    nextPt = new Point(40,60);// Call assuming a "constructor"
                        //      for Points that initializes
                        //      the coordinates
```

The related new [ ] operator can be used to create arrays:

```
    char*       buf = new char[BUFFSIZE];
```

   It is always possible for a request for free-storage to fail. A free-storage manager may be set up to "throw an exception" (see Chapter 12 for exceptions) if allocation fails. Otherwise, code using free storage allocation should include checks to see that space was allocated. If an allocation request fails, the pointer returned will have value 0. A program that attempts to allocate an array of characters, assumes the allocation to be always successful and fills the array, will overwrite low memory and crash if the allocation step fails. It is possible to program defensively:

```
    char*       buf;
    ...
    buf = new char[BUFFSIZE];
    if(buf == 0) {
        error("Sorry, but you ran out of space.\n");
        exit(1); // hard fail, not the best solution.
        }
    ...
```

The implementation of the new operator supports more sophisticated, and partially automated styles of error handling and recovery. An application program can include a "new_handler" routine (new_handler is a global pointer, it has to be set to point to the routine that will handle storage allocation problems). A typical strategy is for a program to claim a temporary memory reserve at start up (e.g. by using new to allocate a block of ten thousand bytes). If a request for free storage fails, the routine identified by new_handler will be called; it should somehow warn the user of a low-memory situation and release some of its memory reserve, and then attempt to honour the new request.

'new_handler'
routine

   There are several other hooks for more complex free store management. It is possible to define type-specific variants of the new operator for programmer defined classes and structs. Other extensions to new allow a programmer to force

Class specific
versions of new and
delete

allocation of space at specific addresses. These features are only required in advanced applications of C++.

'delete'   Space allocated using `new` should be released using `delete`. Examples using `delete` are:

```
...
delete nextPt;   // release store for a single item
...
delete [] buf;   // release store for an array of items
...
```

If `new []` was used create an array, this should always be reflected in the `delete` step, as in the case illustrated for the character array `buf`. (This is a relatively recent change in C++, many texts show arrays being released with simply a `delete` operator rather than `delete[]`.)

After a `delete` operation, the value in the pointer variable is undefined. The pointer may still contain the old address; or, depending on implementation, it may have been set to zero. It is safer if the pointer is set to zero, for then it is obvious that the pointer is invalid. (If you mistakenly use a pointer that still contains a valid address, the program is likely to continue running. Eventually an error will become manifest but it may be long after the place where the incorrect pointer was used; this makes debugging difficult. If a program dereferences a zero pointer, there is a good chance of an immediate "bus error – core dumped" termination.) If the pointer is a local variable that is about to go out of scope it doesn't really matter; in other cases, it is sensible to program defensively and explicitly set pointers to zero after `delete` operations. (Setting the value to an odd number may work even better; almost all systems will get a hardware address error as soon as an attempt is made to use a pointer with an odd address.)

Programmer defined types may require to tidy up before they are destroyed. As described in Chapter 5, the author of a programmer defined type can include a "destructor" routine that will carry out such tasks. The processing of the `delete` operation will automatically entail a call to a destructor if one exists. (If an array of objects of a programmer defined type is deleted, a destructor routine would have to be invoked on each element of the array. The main reason for the `delete []` syntax is to help the compiler identify situations where it may be necessary to encode a loop through the array calling destructor routines for the individual elements.)

If a programmer chooses to redefine the `new` operator for a class, then the `delete` operator should also be redefined.

4.13 SCOPE QUALIFIER OPERATOR

C++ introduces another new operator – '::' "the scope qualifier operator". The most common use of the scope qualifier operator relates to classes. In a program that uses classes, like the Space Invaders program, one may have routines, e.g. a `Draw()` routine, for each of several different classes (`Saucer`, `Rock`, ...). When these various `Draw()` routines are defined, the compiler must be informed as to which class a particular definition belongs. The definition for a `Draw()` routine is qualified by the name of its class:

```
void Rock::Draw()
{
    PaintOval(fRect);
}

void Saucer::Draw()
{
    ...
}
```

As explained in Chapter 5, there can be variables that "belong" to classes but which have a kind of global existence ("static data members"). When such variables appear in the program text, their names will be qualified by the name of the class to which they belong.

The scope qualifier operator may appear without any prepended class name. This usage currently implies a reference to "global scope". One can have contrived examples of references to global scope variables:

```
int         Counter;      // a global variable
...
void Procedure1()
{
    int Counter, j, k;
    ...
    Counter++;    // use local, automatic, Counter variable
                  //          of Procedure1
    ...
    ::Counter++; // change the global Counter.
    ...
}
```

In this ugly code fragment, the declaration of `Counter` within `Procedure1` hides the global variable of the same name. The scope qualifier operator makes it possible to still refer to the global `Counter` from within the code of the procedure. There are no reasons why a sane programmer would choose to use the same name for a local variable as for a global that must be accessed.

There are sane uses of the scope qualifier operator to reference global scope. These are illustrated in Chapter 5. Essentially, the scope operator is used to flag global routines so that they stand out in program text. This usage makes it easier to read the code of class member functions.

4.14 STREAM INPUT AND OUTPUT

C++ is "C without the bugs". Where are many of the "bugs" in C programs? Often, they are in the input/output (i/o) sections.

The trouble with C's `stdio` library is that there really can be no compile-time checking. For example, the interface to the most common output routine is

```
int printf(const char *format, ...);
```

The format control string would be expected to have output conversion specifiers such as '%s' (output a string), '%x' (print as a hexadecimal value), '%e' (print as a floating point number in scientific notation). (These conversion specifiers may be further decorated with qualifiers that specify field widths, precisions, padding characters, and left/right justification.) But, there is no check that conversion specifiers are appropriate for the arguments that follow the format string.

Modula2 provides one form of "type-safe" output. In this style, separate type specific routines are used to output (or input) each datum:

```
WriteString('Image[');
WriteInt(i,10);
WriteString(',');
WriteInt(j,10);
WriteString('] = ');
WriteReal(img[i,j]);
```

At compile time, the arguments for the various Write routines can be checked. This approach is type secure. But it is very *verbose*.

An attractive approach for i/o would be one that is both type secure and concise. C++ provides for these requirements through its iostream objects.

<iostream.h>
<fstream.h>
<manip.h>

C++'s stream i/o facilities are made available through three header files. File <iostream.h> defines the main functions (replacements for printf(), scanf() etc.) and provides access to the predefined streams cin, cout, and cerr that correspond to C's stdin, stdout, and stderr. File <fstream.h> adds support for streams to and from files. File <manip.h> provides some additional formatting and other facilities.

An output stream is an object that can examine another run-time entity (an integer variable, a character pointer, a double precision real number, or whatever) and work out how to represent the information in that entity as a sequence of bytes that are buffered and eventually forwarded to a display, or to a printer, or to a file. Similarly, an input stream is an object that can be given a reference to a run-time entity of a particular type and which will then go searching in an input buffer for characters that can be converted into data that can be stored in the given entity.

Output streams "take data" from other run-time entities. Input streams "give data" to other run-time entities. The syntax of statements using streams tries to reflect these give/take semantics. Stream i/o statements are illustrated in this first complete C++ program:

```
#include <iostream.h>
#include <stdlib.h>      // needed for exit()

int main(int, char**)
{
    int    count;
    cout << "How many 'Hello Worlds' do you want?\n";
    cout.flush();
    cin >> count;
    if(cin.fail()) {
```

```
            cerr << "Could not read the input stream\n";
            exit(1);
            }
    if((count < 1) || (count > 1000)) {
            cerr << "An unreasonable number of Hello Worlds "
                        "was requested\n";
            exit(1);
            }
    for(int i=0; i < count; i++)
            cout << i << "\tHello World\n";
    return 0;
}
```

The first output statement:

```
cout << "How many 'Hello Worlds' do you want?\n";
```

is read "cout takes from the string ...". The '<<' operator is *not* a left shift. As explained in Chapter 5, C++ allows additional meanings to be given to standard operators. Operator '<<' retains its standard semantics of "left shift" when applied to integer arguments. But information in the <iostream.h> header file specifies alternative interpretations for '<<'. When reading this header file, a C++ compiler expands its internal table that defines the meaning of operators with these alternative interpretations for '<<'. The entries in the header file specify that when '<<' is found operating on a output stream (ostream) object (e.g. cout) and a string then a call to a "WriteString" routine should be generated. A call to a "WriteInt" routine should generated if '<<' is operating on an ostream and an integer; and a call to "WriteReal" should be generated by the compiler wherever it finds '<<' operating on an ostream object and a double precision variable. (In practice, the actual routine names are quite different, but the basic idea is as explained.)

*Overloaded <<
operator*

Because a compiler has this complete table of all the overloaded meanings for '<<', and details of the corresponding output routines, it guarantees type secure output. The output conversion function is guaranteed to be that appropriate for the data type whose value is to be emitted.

There are similar definitions of overloaded interpretations for the '>>' input operator. Operator '>>' means "right shift" if working on two integers. But, if '>>' is found working on an input stream (istream) object and an integer it means the compiler should encode a call to some ReadInteger(int&) routine; similarly, '>>' operating on an istream and a double will result in a call to a ReadDouble(double&) routine. Thus, the first input statement:

```
cin >> count;
```

is transformed into a call to an integer input routine that sets variable count.

The statement:

*Flushing output
buffers*

```
cout.flush();
```

causes the `cout ostream` object to execute its `flush()` procedure. This causes any pending output characters, those that are still in buffers in memory, to be sent to the output device so that a complete prompt can be printed before the program stops and waits for input. (The `iostream` package actually provides facilities for "tieing" input and output streams so that a call to flush output buffers is made automatically as the first step of every operation on a tied input.)

'fail()' and 'bad()'

Input/output operations can fail. If the input stream was empty (at end of file) or the next input characters were not digits, an attempt to read an integer would be unsuccessful. The read call would return, after first setting various error flags maintained by the `istream` object `cin`. The call to the `fail()` function of the `cin` object checks the flag that indicates the occurrence of an error of any kind during the last input operation. (There are other flags that can be checked to detect specific problems such as end of file. The "fail" flag indicates an error from which it may be possible to recover; a "bad" flag is set, check with `bad()` function, if an unrecoverable error has occurred.) If an error did occur, output is sent to the error stream and the program is terminated (using `exit()` as defined in the `<stdlib.h>` header file).

On using exit()

In the example just shown, and the examples in Chapters 5 and 6, error detection (like `cin.fail()` returning true) leads to a call to `exit()`. This is permissible in small demonstration programs and exercises, but such a policy would *not* be acceptable in a real program. Firstly, `exit()` does cause abrupt termination of the program (no attempt is made to unwind the stack or to close files etc.); side effects of such abrupt termination can cause problems in some environments. More generally, it is not appropriate for a program just to give up in disgust when it finds a situation it doesn't like. Although it may not be possible to recover from an error at the point where it is detected, an attempt should be made to tidy up and report the error to the calling environment. At the level of the calling environment, it may be possible to take remedial action; if not, further tidying can be done allowing the program to terminate gracefully. Section 3.16 introduced the idea of "exceptions" and "exception handling"; exceptions provide a means for implementing a disciplined approach to error handling. Chapters 12 and 13 contain a few small examples illustrating exception handling in C++ and Eiffel.

The error stream, `cerr`, is just another instance of class `ostream` and so had identical behaviour to the standard output stream `cout`. The two conditional statements check for different problems and, if errors have occurred, give messages to `cerr`. (The string "An unreasonable number of Hello Worlds was requested\n" has been broken because it was too long to fit onto one line. C++ compilers, and modern C compilers, are capable of joining successive strings into one long string. This feature may appear odd to those used to older C compilers.)

Type secure and CONCISE

The output statements in the for-loop illustrate the next feature of stream i/o. Operations can be concatenated:

```
for(int i=0; i < count; i++)
    cout << i << "\tHello World\n";
```

The statement `cout << i << "\tHello World\n"` is compiled into a call to a `"WriteInt(cout,i)"` and then a call to a `"WriteString(cout,"\tHello World\n")"` (those are not the true function names, just illustrations). Input and output statements that use this concatenation feature are as concise as any calls to printf() or scanf() but have the advantages of being type secure. (The way it is all made to work is illustrated in Chapter 5 when input and output of programmer defined types is explained.)

Most i/o operations using streams will use the '<<' and '>>' operators. But there are some other useful transfer routines. For example, for input there is "getline()" function that can read characters into a buffer, and an "ignore()" function that will skip input characters until a specified marker character is found (or a specified maximum number of characters has been skipped). File streams, defined in <fstream.h>, add capabilities such as being able to seek to a particular position in a file prior to performing an i/o transfer.

The file <manip.h> defines additional capabilities for setting formats etc. Using features from <manip.h> one can change output formats for numbers from the default decimal to either octal or hexadecimal. Field widths, numeric precision, padding characters etc. can all be specified.

Such features are illustrated in standard C++ texts. They are not explored here because most of the later example programs will be using graphic i/o functions as defined by various graphical user interface environments. Where streams are used (mainly when saving objects to files), the operations will all be very simple, using just the file-stream versions of those operations illustrated by the "HelloWorld" program listed above.

REFERENCES

C++'s extensions to C are covered in detail in all the numerous C++ text books. For the most part, these extensions are easy to learn. Most can be seen to be ways of adding compile-time checks, or as constituting alternatives to error prone constructs. Some reference material on stream i/o will probably be required; the appendix in Lippman's text is probably the most accessible overview.

Reference books

S.B. Lippman, *C++ Primer 2e*, Addison-Wesley, 1991.
   Detailed coverage of the C components of C++ and all the minor C++ extensions.
   Appendix illustrating use of stream i/o.
M.A. Ellis and B. Stroustrup, *The Annotated C++ Reference Manual*, Addison-Wesley, 1990.
   Explanations for why minor changes and extensions have been made to C (and why, in other cases, it was not reasonable to make changes). Details of arcane aspects, such as the way C++ mangles names. Justifications for the choices made when implementing the more significant extensions such as classes, inheritance mechanisms, access control mechanisms etc.

Programmer-Defined Data Types in C++

This chapter introduces C++ classes. The examples used as illustrations would commonly be referred to as "abstract data types". But, these 'points', 'rectangles', 'strings', 'linked-lists' and other data types are hardly "abstract", and, really, they are much more accurately described as "programmer-defined data types".

Pascal's `Type record … end` and C's `struct` definitions have in the past provided programmers with ways of constructing new data types from the built-in collection of integers, characters, reals etc. provided by the language. But variables of such programmer-defined types could never be employed with quite the same flexibility as could variables of the built-in types. Moreover, it was difficult to define secure data structures.

Limitations of C's structs

A programmer could specify a C `struct`, that was to represent a new data type, and provide a set of routines that manipulated such structures. However it was not normally possible both to allow such structures to be created and copied in arbitrary parts of a program while enforcing requirements that the data in the various fields only be accessed through the routines provided. A C `struct` declaration exposes its constituent data fields. Another programmer on the project could change a data field of any accessible instance of the structure type by a direct assignment statement. The direct manipulation of a field, rather than use of a procedure that kept different field values consistent, could leave a structure in an invalid state that would later cause errors.

The C++ language allows a programmer to add new types to those already built-in by the compiler-writer. New programmer-defined data types can be made "first class types", i.e. variables of these types can be used just as flexibly as can variables of the built-in types. Variables of standard types such as `short`, `long`, `unsigned char`, etc. can be created as statics, as automatics (on the stack), and as dynamics (on the heap) – so, it should be possible to have variables of programmer-defined types that can exist as statics, automatics or dynamics. Initial values can be defined for variables of standard types – so, it should be possible to define initial values for the variables of a programmer-defined type. Built-in variables can be assigned, defined as constants, passed as arguments, returned as

C++ classes as first class types

results from functions, and (where there is a defined semantics) combined using operators such as '+', '*', '==', '>>'.

If programmer-defined types are to be "first class" then they must also be capable of being assigned and combined using operators. Now, some of these requirements necessitate the use of fairly sophisticated programming techniques. The semantics of a '+' operation for integer operands is built into the compiler; if a programmer wants to define a '+' operation on strings to mean concatenation then these semantics must be specified. It takes a fairly sophisticated programming language (and a fairly sophisticated programmer) to define new semantics for basic operators.

C++: access controls

As well as allowing the definition of new data types, C++ allows the definition of controls on the usage of variables. The language supports the packaging together (encapsulation) of data fields and data manipulation procedures so that the individual data fields comprising a variable of programmer-defined type can only be accessed and manipulated through the nominated procedures. (There is no guarantee against fraud; it is still possible to perform address arithmetic and so obtain a pointer to some supposedly inaccessible data and then change those data by some operation that works indirectly through the pointer.)

"Object-based" rather than "Object-Oriented"

The language features introduced in this chapter are not "object-oriented" – there is no inheritance, no polymorphism, no real fun. But they do support an "object-based" approach to programming. The class declarations describe the forms of objects that will be created when a program runs. The programming style is of the form "Hey, object perform your XXX-function". Simple objects such as points and linked-lists are widely used in support roles in more advanced object-oriented programming; more complex objects such as Views (or Space Invaders) are likely to have, among their data fields, many instances of simpler object classes such as points or lists.

5.1 SIMPLE TYPES: POINTS AND RECTANGLES

Several of the simpler aspects of C++ classes can be illustrated by considering classes that allow the manipulation of 'points' and 'rectangles' in a two-dimensional plane. Points and rectangles are frequently required when programming graphics input/output on personal computers. Thus, the Apple Macintosh "Quickdraw" manager declares the structures:

```
struct Point {
    short v;
    short h;
};

struct Rect {
    short top;
    short left;
    short bottom;
    short right;
};
```

and provided a host of routines to manipulate actual instances of `Point` and `Rect` structures (the `pascal` keyword in the following function declarations instructs the compiler to pass arguments in the order required by Pascal rather than in the normal C order, this is necessary as the Macintosh Toolbox routines are implemented in Pascal):

```
pascal void AddPt(Point src, Point *dst);
pascal void SetPt(Point *pt, short h, short v);
pascal Boolean EqualPt(Point pt1, Point pt2);
pascal Boolean PtInRect(Point pt, const Rect *r);

pascal void SetRect(Rect *r, short left, short top,
        short right, short bottom);
pascal void OffsetRect(Rect *r, short dh, short dv);
pascal void InsetRect(Rect *r, short dh, short dv);
pascal Boolean SectRect(const Rect *src1,
        const Rect *src2, Rect *dstRect);
pascal void UnionRect(const Rect *src1,
        const Rect *src2, Rect *dstRect);
pascal Boolean EqualRect(const Rect *rect1,
        const Rect *rect2);
pascal Boolean EmptyRect(const Rect *r);
```

While such structures are perfectly usable, one does find inconsistencies in style:

```
struct Point p1, p2, p3;
...
p1.h = 5; p1.v =7;
...
SetPt(&p2,9,11);
...
p3 = p2; AddPt(p1, &p3);
...
if(EqualPt(p1,p2)) ...
```

Fields, like `h` and `v`, may be manipulated directly, or via the routines provided; operations combining points cannot be expressed as fluently as might be desired.

Such problems can be overcome by exploiting C++ classes. A useful specification of class `Point` will be developed in stages. First, the general form of a class declaration is illustrated; later, the class will be extended by defining semantics for operators, such as '+', that can be applied to points .

5.1.1 C++ class declaration

A simple class declaration differs from an old-style C `struct` declaration in two obvious ways. The routines that are to be used to manipulate class instances are nominated in the declaration. Access controls are defined on the data fields and routines. Here is the first declaration specifying a form for class `Point`:

Class declaration

```
class Point {
// First version of a point class
```

```
public:
    Point(short x = 0, short y = 0);
    Point            Add(const Point& deltaPt);
    Boolean          ZeroPt();
    Boolean          Equal(const Point& otherPt);
    ...
private:
    short            v;
    short            h;
};
```

Keywords class, public, private

The words `class`, `public`, and `private` are keywords in C++. `class` introduces a new data type declaration; `public` and `private` are terms that describe access controls. The form given above is the standard layout that will be used in this text. A class declaration provides a new type name. Variables of type `Point` can be defined, e.g. `Point p1, p2; Point* p; ...`. In term of the old C language, the class declaration combines both a `struct` declaration and a `typedef`; so that `class Point` is approximately equivalent to `struct Point { ... }; typedef struct Point Point;`.

Public interface

The class declaration starts by describing the "public interface" of the class; essentially, this is a summary of the ways in which points can be used in a program. Several routines are listed (here the ellipsis, ..., indicates the omission of additional functions for manipulating points that would actually be present in the class definition but are not required to illustrate its form). `Point(short x =0, short y = 0)` is a "constructor" i.e. a special function that will initialize a `Point` when it is created. Constructors are special in many ways, for example they do not have a return type. Constructors will receive more detailed consideration later.

The other functions used for illustration are `Add()` which calculates and returns as a result a new `Point` at some delta offset from a given `Point`; `ZeroPt()` which checks whether a `Point` is at the coordinate origin, and `Equal()` that tests whether two `Points` are at the same location. (It is assumed in these examples that "Boolean" is a type defined by some declaration e.g. `"enum Boolean {false, true};`.)

Private implementation

The keyword `private` separates the public interface from details of data fields and housekeeping functions that are required in the class being specified. A programmer who is using instances of class `Point` (the "client programmer") can stop reading the declaration at the `private` keyword; the subsequent details are only of concern to the programmers who first implement and subsequently maintain the code for the class. In the case of the example `Point` class, there are only two data fields in the private section. More typically, the private section would also contain declarations of additional functions. These would be "housekeeping" functions that are required in the implementation of a class.

Class members

The standard terminology for C++ uses the term *data members* for the various data fields declared in a class, and the term *member functions* for all the functions declared in a class (both the public interface functions and any private housekeeping functions). It is legal to include data members in the public interface, but this is rarely useful (if you want simple `structs` with unprotected data fields then declare a `struct`). The public interface will generally comprise constructors, functions that can be used to request information from an instance of a class, and procedures (functions returning `void`) that request actions by an instance of a class. The private members include all the data fields and any necessary housekeeping procedures.

There is a considerable flexibility to the form of class declarations. In earlier C++ texts, one would typically find something like class `Point` being declared as follows:

```
class Point {
    short           v;
    short           h;
public:
    Point(short x = 0, short y = 0);
    Point           Add(const Point& deltaPt);
    Boolean         ZeroPt();
    Boolean         Equal(const Point& otherPt);
    ...
};
```

*Older   style
layout for a
class   declaration*

This form of declaration relies on the fact that the default access specifier for members is `private`. Because `private` features of a class are not of interest to clients, most programmers now use the style where the `public` interface is defined first.

Even more complex layouts are legal. The following style might conceivably be produced by some CASE tool that automatically generated class declarations from some graphical input:

```
class Point {
public:     Point   Add(const Point& deltaPt);
private:    short   h;
public:     Point(short x = 0, short y = 0);
    ...
private:    short   v;
};
```

Although legal, this style of declaration would not be generally favoured.

C++ cannot completely separate the `public` interface of a class from its `private` implementation details. Such a separation is not possible because of the way in which variables of programmer-defined types can be used. In C++ one can have automatic instances of programmer-defined types, like the various `Point` variables in the following procedure:

*Complete
separation   of
public interface
and   private
implementation
is   not   possible*

```
void SketchLine()
{
    Point   startPt, endPt, currentPt;
    ...
}
```

When creating code for a procedure like `SketchLine()`, the compiler must determine a size for the "stack frame" that is to be allocated for the procedure's automatic variables. In this case, it must know the size of a `Point` variable and

consequently it must see the private implementation part in order to discover
what data members exist.

5.1.2 Definition of member functions.

Header file for
class declaration
and separate
implementation
file

Normally, the declaration of a class such as class Point will be contained in a
header file (e.g. "Point.h" or "Point.hpp"). This header file is included in all files
containing code with Point type variables. In addition to the header file, there will
be an implementation file that contains the code for the member functions of class
Point (e.g. "Point.C" or "Point.cp", naming conventions for both header files and
implementation files are system dependent). This arrangement of files is as shown
in Figure 5.1.

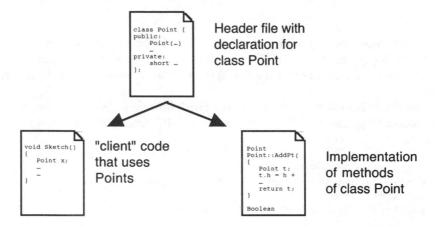

Figure 5.1 Class declarations are normally contained in header files that can
be included in files of "client" code and in the implementation files
that contain the code defining the class member functions.

It is not necessary to have unique files for individual classes. In practice, the
header file and implementation file would both typically contain details of several
classes – Point, Rectangle, Oval, etc.

The definitions of the member functions are all given with the class name
qualifying the function name. For the example Point class, some of the code with
member function definitions would be:

Member function
definitions

```
#include "Types.h"
#include "Point.h"

Point::Point(short x, short y)
{    // See later, section 5.1.14, for more efficient
     //  alternative way of initializing data members
     h = x; v = y;
}
```

```
Point  Point::AddPt(const Point& deltaPt)
{
    Point sumPt;
    sumPt.h = h + deltaPt.h;
    sumPt.v = v + deltaPt.v;
    return sumPt;
}

Boolean Point::ZeroPt()
{
    if((0==v) && (0==h)) return true;
    return false;
}
```

The construct `<class name> <scope qualifier operator><FunctionName>()`
(e.g. `Point::ZeroPt()`) identifies these function declarations as belonging to a
specified class (here, class `Point`).

Within a member function, all data members are accessible. As described in   *'this' pointer*
Chapter 3, each member function does have an additional implicit pointer argument
– a pointer to the actual record structure for which the function has been invoked.
The implicit pointer is named `this` and, for a member function of some class x is
of type "`x *const`" (i.e. a constant pointer to a variable of type x), here `this` is of
type "`Point *const`". Use of the pointer, `this`, when referencing data members is
legal in C++, so the following versions of the constructor are equally correct:

```
Point::Point(short x, short y)
{
    h = x; v = y;
}

Point::Point(short x, short y)
{
    this->h = x; this->v = y;
}
```

Generally, the preferred coding style omits the `this->` qualifier. However, it is a
question of style, and large C++ programs and class libraries can be found in which
the `this->` qualifier is used in all references to members.

5.1.3 Class as the unit of protection

In C++, access protection on members is "class based" and not "instance based".   *Access to data*
Consequently, in the code of the function `AddPt()`, the coordinates of the offset   *members*
`deltaPt` are accessed directly:

```
Point  Point::AddPt(const Point& deltaPt)
{
    Point sumPt;
    sumPt.h = h + deltaPt.h;      // or, sumPt.h =
                         //         this->h + deltaPt.h
    sumPt.v = v + deltaPt.v;
```

```
            return sumPt;
    }
```

The language could have been defined with access protection on a per-instance basis. Then, it would have been necessary to have defined accessor functions that returned the horizontal and vertical coordinates of a `Point` and the code for `AddPt()` would have to use these functions to obtain the coordinates of `deltaPt`. Modifier functions would be needed to set the coordinates in `sumPt`. The code would then have been something like:

```
short       Point::H()
{
    return h;
}
short       Point::V()
{
    return v;
}
void        Point::SetH(short newh)
{
    h = newh;
}
…
Point  Point::AddPt(const Point& deltaPt)
{
// Code style here is unnatural for C++, it is simulating
// the style of a language with instance-based rather than
// class based access controls.
    Point sumPt;
    sumPt.SetH(h + deltaPt.H());
    sumPt.SetV(v + deltaPt.V());
    return sumPt;
}
```

Since C++'s protection scheme is class based, there is no need to define accessor functions like `Point::H()`, or modifier functions like `Point::SetH()`, unless clients have a legitimate reason to read or change the individual coordinates of a `Point` object.

5.1.4 Constructors

Constructors "Constructors" are not normal member functions. Constructors can only be used to create new objects. Calls to constructors can appear in argument list for functions and in some expressions as well as in simple variable declarations. But one cannot ask an existing object to reexecute its constructor. So, although a constructor is typically used to put an object in some standard initial state, constructors cannot be used to re-initialize an object after it has been modified by some sequence of processing steps. If objects of some class will need to be reset to standard initial states, an "initialize" member function must be supplied. (A constructor can call such an initialization function; there is no need to duplicate initialization code in both constructor and the initialization function.)

The constructor for class `Point` was:

```
Point::Point(short x, short y)
{
    h = x; v = y;
}
```

with the class declaration specifying default values for x and y (`Point(short x = 0, short y = 0)`). Two arguments should be provided; if no arguments are given, or only one argument is given, omitted arguments are assumed to be zeros.

The compiler places a call to a constructor immediately after a variable comes into existence:

```
static Point gPoint0;      //  The compiler places calls to
Point gPoint1(1,1);        //  constructor functions to
                           //   initialize these variables in
                           //  the "preamble" code that is
                           //  executed prior to entry to main()

void Sketcher()
{
    Point startPt, endPt;// Code to build a stack frame
    short  i;            // of sufficient size for all
                         // local variables.

                         // Calls to constructors for
                         // startPt and endPt
    cin >> i;
    ...
    for(int j=0;j<i;j++) {
            Point current;
                         // Call to constructor for "current"

            ...
            Point* p;

            ...
            p = new Point;
                         // Call to constructor for (*p)

            ...
    }
    ...
}
```

In C++, there are a number of styles for variable initialization. For variables of simple arithmetic, pointer, and character types one has an "assignment" style of initialization:

Several styles for variable initialization

```
short i    = 99;
char* msg  = "Hello World\n";
float pi   = 3.142;
char starts = '\003';
```

Arrays and simple C structs can have initializer lists:

```
int divisions[] = { 1, 2, 4, 8, 16, 32 };

struct Monarch {
    char* name;
    char* address;
    short year_accession;
} UK_Queen = {
            "Elizabeth Windsor",
            "Buckingham Palace",
            1953
        };
```

Programmer-defined data types get initialized by their constructors; the notation looking a little bit like a function call:

```
Point gPoint1(1,1);
Point gPointm1(-1,-1);
Point gPoint4(4,4);
Point gPoint0;
Point x5(5);
Point other(79,-47);
```

The default arguments provided in the declaration of the Point constructor make gPoint0 become the Point (0,0) and x5 correspond to the Point (5,0).

Default constructor

Usually, a class should have a *default constructor*. This is a constructor that does not require any argument values. The definition of C++ has changed relatively recently with regard to default constructors. It used to be that a default constructor was one that *took* no arguments – e.g. "Point::Point()". The new definition is a constructor that *requires* no arguments – for example, a constructor with default values for all its arguments. The constructor defined above for class Point with its complete set of default values for arguments would, in the current definition, be accepted as a default constructor. If *no* constructors are defined for a class, then global instances of that class are "initialized to zero" (NULLs in pointer data members, 0s in ints, 0.0 in doubles etc.); other instances of the class are created with random bits left in their uninitialized data members.

Classes without default constructors

Classes do not in general *require* a default constructor. If a class is defined with one or more constructors that require explicit (non-defaulted) arguments, there will be no default constructor. All instances of such a class will have to be defined with appropriate argument lists for their constructors. There is only one restriction on such classes – it is not possible to declare arrays of variables of a class if it lacks a default constructor. An array declaration would, for class Point, be something like:

```
Point Corners[4];   // array of 4 Points, each
                    // initialized by default constructor
```

If such arrays of objects are required, a default constructor must exist. (Some current compilers cannot deal correctly with arrays when the default constructor takes defaulted arguments; these compilers can only handle arrays if the default constructor has no arguments – like a `Point::Point()` constructor.)

Most classes tend to have several constructors. Class `Point` is rather simple; after all, there are not too many different ways that one might wish to initialize a `Point`. Even so, a more completely defined, more practical version of class `Point` would probably define at least three constructors:

```
class Point {
// Slightly augmented version of a point class
public:
    Point();
    Point(short x, short y = 0);
    Point(const Point& likePt);
    Point           Add(const Point& deltaPt);
    Boolean         ZeroPt();
    Boolean         Equal(const Point& otherPt);
    ...
    void            SetPt(short newx, short newy);
    // SetPt with a change from global to local coordinates
    void            SetPt(const Point& globalPt);
    ...
private:
    short           v;
    short           h;
};
```

Class Point with multiple constructors

The first constructor, `Point()`, is a standard default constructor – as accepted by all C++ compilers. Its definition would be:

```
Point::Point()
{
    h = v = 0;
}
```

The following variable declarations would use this default constructor:

```
Point       start, end;
Point       target_coordinates[50];
Point*      p = new Point;
```

The second constructor, `Point(short x, short y =0)`, would take either one or two shorts as arguments. Its definition and use in variable definitions would be:

```
Point::Point(short x, short y)
{
    h = x; v = y;
}

Point       topleft(40,40), bottomright(400,300);
Point       mark1(10), mark2(20), mark3(40);
Point*      q = new Point(10,20);
```

Don't try to confuse the compiler by defining more than one default constructor

Note that it would be incorrect to have defined this constructor with two arguments both with defaults (e.g. Point(short x = 1, short y = 0)), because this would make it an alternative default constructor. Then, the compiler would be unable to determine which default constructor to use for a definition like "Point no_Point;" (is it (0,0) using Point::Point() or (1,0) using Point::Point(short x = 1, short y = 0)).

Copy constructor

The third constructor in this revised version of class Point is an example of a "copy constructor". Copy constructors initialize a new object with values copied from some given reference object. Once again, Points are rather simple objects so making a copy of the data is straightforward; but as will be illustrated in subsequent examples, there are a number of subtle issues related to copy constructors for more complex classes. But, for Points, it is all very simple; the h and v coordinates need merely be copied from the given reference Point. (Because Point is such a simple class, it isn't really necessary to define a copy constructor. The C++ compiler would have provided one automatically. The automatically provided function would have the same behaviour as the function actually defined here.) The code and usage of the copy constructor would be as in the following example:

```
Point::Point(const Point& likePt)
{
    h = likePt.h; v = likePt.v;
}
```

Code using copy constructor

```
void TrackMouse(const Point& mouse)
{
    Point          StartPt(mouse);
    Point*         tracker;
    ...
            {  // call global function "Point GetMouse()"
            tracker = new Point(GetMouse());
            ...
            }
    ...
}
```

5.1.5 Overloaded member functions

Overloaded member functions

The constructors Point(short, short) and Point(const Point&) "overload" the first Point() constructor. It is quite common to have overloaded member functions; the revised class Point has the further examples of the overloaded function(s) SetPt(). The definitions for these functions would be:

```
void Point::SetPt(short newx, short newy)
{
    h = newx; v = newy;
}
void Point::SetPt(const Point& globalPt)
{
    // convert global coordinates to local frame
    h = globalPt.h;
```

```
    v = globalPt.v;
    GlobalToLocal(&h, &v);
}
```

The second SetPt() function is somewhat contrived. It is included to provide a   *Assignments of*
simple example of an overloaded member function. There would be no need for   *class instances*
such a SetPt(const Point&) function if all it did was copy unchanged values of
data members. After all, the value of one Point can be simply assigned to another:

```
    Point a,b,c,d;
    ...
    a.SetPt(5,6);
    b.SetPt(7,8);
    c = a.Add(b);
    d = c;
    ...
```

In simple cases such as this, assignment is identical to assignment of structs in
C. The compiler emits code with a blockmove instruction that copies the correct
number of bits from the source to the destination. For Points, such a copy of the
h, v data fields is just what is required. As will be illustrated later with class
"Player", C++ provides a mechanism whereby the programmer who defines a class
can refine the meaning of "assignment" if a simple bit copying operation is
inappropriate.
    Some additional functions would be required before class Point had adequate
functionality; for example, one would probably want "SubPt(const Point&
subtrahend)". But the declarations and definitions of such functions are all
straightforward.

5.1.6 Simple version of class Point

Further refinement of the class declaration for class Point leads to:

```
class Point {
// Slightly refined version of a point class
public:
    // Constructors
    Point();
    Point(short x, short y = 0);
    Point(const Point& likePt);
    // Access functions
    Boolean     ZeroPt();
    short       H();
    short       V();
    ...

    // Modifier procedures
    void        SetPt(short newx, short newy);
            // SetPt with a change from global to
            //local coordinates
    void        SetPt(const Point& globalPt);
```

```
        void            SetH(short newval);
        ...

        // Combination functions
        Point           Add(const Point& deltaPt);
        Boolean         Equal(const Point& otherPt);
        ...
        ...
    private:
        short           fv;
        short           fh;
    };
```

In the public interface of a class, the member functions should be grouped (if there
are private implementation functions, these should also be grouped and listed before
data members). Grouping of member functions is not a requirement of the
language, it is just a matter of style. Constructors are given first. For a simple
class such as Point, the other member functions could be grouped according to
whether they are functions that return stored or computed data (e.g. ZeroPt() and
H()), or procedures that change the object (like SetPt()), or procedures that
combine data from more than one object (e.g. Equal() and Add()).

Naming
conventions for
class members
The data members have been renamed as "fh" and "fv". There are some naming
conventions for variables that are quite commonly used in the class libraries for
Macs and, to a lesser extent on PCs. Global variables have names that begin with
a small letter 'g'; data members in classes have names that start with 'f' (field?);
local variables in functions have names that start with any lower case letter other
than 'f' or 'g'; class names begin with a capital letter; constants may be fully
capitalized. Such conventions make the code of member functions more readable,
but are in no way standard.

Use of global
scope qualifier
"::"
Again, in order to make the code of member functions more readable, it is useful
to employ some convention to distinguish member functions from global
functions. For example, the code for SetPt() uses a function GlobalToLocal():

```
    void Point::SetPt(const Point& globalPt)
    {
        // convert global coordinates to local frame
        h = globalPt.h;
        v = globalPt.v;
        GlobalToLocal(&h,&v);
    }
```

A maintenance programmer extending class Point might not be familiar with all
the member functions of the class and have to check back to determine where this
function GlobalToLocal() comes from.

Make distinct
the use of
member
functions and
global functions
One of two conventions should be used to distinguish global functions from
member functions. Either all references to member functions should explicitly use
the this-> qualifier, or all references to global functions should use the explicit
global scope qualifier. In this text, the second convention will be used; global
functions will always use the :: global scope qualifier. The :: qualifier is
redundant; the compiler does not require it. It is used simply to help those who

must read the code. Using this style for function calls, and the preferred naming style for data members, the SetPt() function should be written as follows:

```
void Point::SetPt(const Point& globalPt)
{
    // convert global coordinates to local frame
    fh = globalPt.fh;
    fv = globalPt.fv;
    ::GlobalToLocal(&fh,&fv);
}
```

The latest proposals for C++ include support for "namespaces". At the *'namespaces'* moment, the names for global functions and variables, e.g. the GlobalToLocal(), EraseRect() and other example functions from the Maintosh's Quickdraw library, all go into a single global name space. The use of a single global name space raises the problem of collisions – it is all to easy to end up with one's own void ClosePicture() routine clashing with the void ClosePicture() that comes with Quickdraw, or one may get conflicts among separate imported libraries. The "namespace" feature will allow partitioning of the name space. For example, in future implementations all Quickdraw routines might be packaged in an explicit Quickdraw namespace:

```
namespace Quickdraw {
    void InitGraf(void *globalPtr);
    ...
    void ClosePicture();
    ...
};
```

The :: scope resolution operator would then be used in association with a specific namespace. The example call to GlobalToLocal(...) would be written as Quickdraw::GlobalToLocal(&fh,&fv). This more explicit naming scheme will further help to distinguish between calls to member functions and calls to service routines.

5.1.7 Enhancing class Point: inline functions improve efficiency

Class Point, with appropriate definitions for all its member functions, would be *Class Point is* usable. However, it does not yet realize many of the objectives. Programmer- *not yet a "first* defined data types should be able to be used as freely as built-in types – but the *class" data type* current version of class Point would not permit the definition and use of any constant Point objects. Where there are reasonable semantics, operators such as '+' should be used with programmer-defined data types; but the current class Point has only its clumsy Add() function leading to code like c = a.Add(b) rather than the preferred c = a + b. Standard built-in types can be output using C++ streams, but there is no mechanism allowing the current version of Points to be output to a stream. Finally, programmer-defined data types should not entail a performance overhead; but the current class Point involves function calls for even the simplest actions such as reading the value of a Point's vertical coordinate. All of these problems can be overcome through the use of additional features in the C++ language.

Efficiency and the cost of function calls

First, the "performance hit" fears of C programmers should be assuaged. C programmers have traditionally prided themselves on their abilities to achieve "efficiencies" at the micro-scale of individual statements in a program. Many of the C-hacks used to achieve "efficiency" were myths; a modern compiler can generate as good machine code from simple, straightforward source code as from some convoluted C-hack. (For example, C programmers usually prefer to access arrays using a pointer dereferencing scheme rather than by indexing. This was an appropriate choice in ≈1970 when C ran on a PDP-7 that had no index registers but did have auto-increment, memory-based pointers. However architectures have changed a bit since then and most machines can handle array indexing with reasonable efficiency.) Nevertheless, programmers could justifiably complain of a function call overhead on simple access to data members.

With a C struct declaration like:

```
struct Point {
    short fh;
    short fv;
}
```

the code generated for the statement

```
x = apoint.fv;
```

would be something like:

```
load an address register with base address of
                structure apoint
add in offset of fv data field
access memory at location specified in address register
```

The C++ class Point and member function V() would have the source code form:

```
x = apoint.V();
```

with an apparent translation into machine code being something like:

```
push address of apoint onto stack
jump to subroutine Point::V()

        // subroutine Point::V()
        // single argument 'this' Point const*
        load address register with argument
        add offset of fv data field
        access memory at location specified
        place result in "return" slot of stack frame
        return from subroutine
```

```
        get result from return slot
        (possible additional instructions clearing stack frame)
```

Obviously, there is considerable overhead in this code. Since a `Point` object is something that is likely to be heavily used inside tight loops, this processing overhead must be avoided.

This would be a place to use *inline* functions. If the function `Point::V()` were defined as:

Use of inline member functions

```
inline short Point::V()
{
    return fv;
}
```

then the code generated for

```
x = apoint.V();
```

would be identical to that produced for a simple C `struct`. There would be *no* performance hit.

C++ has two mechanisms that allow member functions to be made "inline functions". The first is to define the member function in the class declaration:

Defining inline functions in a class declaration

```
class Point {
// Illustrating one way to define "inline" member function
public:
    // Constructors
    Point();
    ...
    // Access functions
    ...
    short         V() { return fv; }
    ...

    // Modifier procedures
    void          SetPt(short newx, short newy)
                        { fh = newx; fv = newy; }
    ...
private:
    short fh, fv; // CAUTION, see next paragraph
}
```

This is *not* the preferred style. Defining member functions in a class declaration makes that declaration much less readable. The declaration should clarify the role of the class, and shouldn't be cluttered with detail of how it works. Further, this style really requires that the private data members be declared before any `inline` functions are defined because otherwise the compiler may not be certain what `fh` and `fv` refer to. If variables named `fh` and `fv` happened to already exist with global scope, the compiler might generate code for `SetPt()` that used these global variables, rather than the class data members appearing later in the class declaration. So, although perfectly legal, this style should be avoided.

Defining inline functions external to the class declaration

The second, and preferred mechanism is to simply declare the member function in the class declaration and define it elsewhere, *using the "inline" modifier at the point of definition*. There is one catch. The definitions of inline functions must be available in all translation units that use a class. A sensible arrangement of #include files will avoid any problems.

Three files:

A class will typically require three files. There will be a header ("class.h") file with the class declaration, a file with the current set of inline definitions that is named as an include file at the end of the header file ("class.incl.h"), and a file with the other function definitions ("class.cp"):

1) class declaration,

```
// File Point.h
// Start with compiler directives that deal with any
// potential problem of multiple inclusions:
#ifndef __POINT__
#define __POINT__
// includes for any header files needed to define points
#include <iostream.h>
class Point {
public:
    // constructors
    Point();
    ...
    // Access functions
    ...
    short  V();
    ...
}

#include "Point.incl.h"
#endif
```

2) inline definitions,

```
// File Point.incl.h
// Declaration of inline member functions of class Point
inline short Point::V()
{
    return fv;
}

...
```

3) implementation of other members.

```
// File Point.cp
#include "Point.h"

Point::Point()
{
    fh = fv = 0;
}

...
Point  Point::Add(const Point& deltaPt)
{
    ...
}
```

This organization of files makes it simple to change the status of a member function (at the cost of possibly lengthy recompilation). To make a function inline it can be moved from the ".cp" file to the ".incl.h" file and the `inline` qualifier added. Inline functions are meant to be simple. If, subsequently, additional more complex processing steps are added to an inline function it should be "demoted" and transferred back to being an ordinary member function defined in the ".cp" file. Simple copying operations between ".incl.h" and ".cp" files tend to be easier and less error prone than editing operations on a class declaration.

Even in simple projects, compiler directives should be used to eliminate the possibility of problems associated with multiple include files. The class `Point` might well get used later when defining classes "Window" and "Palette", and so the header files of both these classes would have inclusion statements `#include "Point.h"`. Later, another module might be written that required to reference both windows and palettes, it would have statements `#include "Window.h"` and `#include "Palette.h"`. During the preprocessing, file `Point.h` would end up being included twice – and this would lead to compilation errors. Such problems are avoided by using the `#ifndef Tag`, `#define Tag`, ..., `#endif` directives illustrated above for class `Point`. The tag, e.g. `__POINT__`, would initially be undefined, so the first include operation results in definition of the tag and declaration of the class. When the file `Point.h` gets included a second time, the preprocessor will find the tag defined and omit all text between the `#ifndef Tag` and the matching `#endif`.

Compiler directives to suppress multiple inclusion of header files

5.1.8 Constant instances of programer-defined types

"Inline" definitions for member functions overcome the performance problems noted above. The next issue to be resolved is that of "constant" objects. A program might want to define a constant `Point`, e.g. a `Point` that represents the origin of the coordinate system:

```
const Point        origin(0,0);
```

Defining a const object

The C++ compiler will accept this variable definition and `origin` will be correctly initialized by the constructor. The compiler will detect and reject any attempt to change `origin` by direct assignment or by assignment via a pointer:

```
{
Point b(7,0);
origin = b;         // Compiler will reject
...
Point* p = &origin;   // Compiler will reject
```

But what about member functions? It is certainly safe to have `origin.H()` or even `origin.Add(b)` as `H()` and `Add()` do not modify the object for which they are invoked. In contrast, the statement "`origin.SetH(101);`" would destroy the "constancy" of `Point origin` and so must be disallowed.

'const' member functions With a simple class like `Point`, the compiler could probably distinguish functions that change the state variables `fv` and `fh` from functions that leave them unchanged. In general, this is not possible. (Also, the compiler needs to know which functions preserve constancy when compiling client code and only has the class header file for reference.) The programmer must help the compiler by distinguishing functions that preserve constancy from those that can modify an object. The distinction is made by adding the keyword `const` to the declaration of those functions that do preserve constancy and so can be applied to `const` instances of a class. This `const` keyword comes after the function name and parameter list (a `const` keyword at the start of a member declaration describes a result returned by the function). Extending class `Point` to allow properly for `const` instances would result in declaration like:

Improved "const aware" declaration of class Point

```
class Point {
// Refined version of a point class, now "const aware"
public:
    // Constructors
    Point();
    Point(short x, short y = 0);
    Point(const Point& likePt);
    // Access functions
    Boolean        ZeroPt() const;
    short          H() const;
    short          V() const;
    ...

    // Modifier procedures
    void           SetPt(short newx, short newy);
            // SetPt with a change from global to
            //local coordinates
    void           SetPt(const Point& globalPt);
    void           SetH(short newval);
    ...

    // Combination functions
    Point          Add(const Point& deltaPt) const;
    Boolean        Equal(const Point& otherPt) const;
    ...
};
```

In the function definition (whether in the class declaration or in a separate file), the `const` keyword comes between the function template and the body, e.g. `short Point::H() const { return fh; }`. Const member functions that are defined externally to the class declaration must have the `const` keyword both with their declaration and with their definition. (Caution! Some compilers will accept code where a non-`const` member function is invoked by a `const` object. These compilers may mumble some kind of warning, but go straight ahead and emit code that changes `const` objects.)

5.1.9 Operator functions

The next step in the refinement of class `Point` is the replacement of a clumsy *Operator* function like `Add()` by a '+' operator so that one can have `Point a, b, c; … c =` *functions* `a + b;` rather than `c = a.Add(b)`. In order to do this, one must extend the compiler's understanding of the meaning of '+'.

In a compiler, the meaning of '+' will be defined by some table that specifies the code that is to be emitted when a '+' operator is found with particular arguments. For a simple computer, with no hardware floating point, a part of this table might read:

```
Operator + function table:
argument 1          argument 2          code to emit
integer             integer             move arg1,%r0
                                        move arg2,%r1
                                        add   %r0,%r1

float               float               push arg1
                                        push arg2
                                        jsr  floatoperatorPlus
```

If there is no hardware floating point, the code generated for the + operator will involve a function call to some floating point support routine (here named `floatoperatorPlus`).

The construct "`float + float`" is already handled by a function call; "`Point + Point`" should be dealt with in similar fashion. So the compiler's "+ operator translation table" must be extended to include:

```
Point               Point               push arg1
                                        push arg2
                                        jsr  PointoperatorPlus
```

The function "`floatoperatorPlus`" is already built into the run-time environment and is known to the compiler. The programmer-defining class `Point` must define the `PointoperatorPlus` function and describe it to the compiler.

Such operator functions are described in the class declaration. The somewhat clumsy `Add()`, and `Equal()` member functions would be replaced as follows:

```
    class Point {
// "Sophisticated" version of a point class,
// with operator functions
public:
    // Constructors
    Point();
    …
    // Access functions
    …
    // Modifier procedures
    …
    // Combination functions
    Point           operator+(const Point& deltaPt) const;
```

Class Point declared with some operator functions

```
Boolean          operator==(const Point& otherPt) const;
    ...
};
```

These member functions could have the definitions:

```
Point Point::operator+(const Point& deltaPt) const
{
    Point sumPt;
    sumPt.fh = fh + deltaPt.fh;
    sumPt.fv = fv + deltaPt.fv;
    return sumPt;
}

Boolean Point::operator==(const Point& otherPt) const
{
    if(fh != otherPt.fh) return false;
    if(fv != otherPt.fv) return false;
    return true;
}
```

When the compiler reads the class declaration, these additional meanings for
`operator +` and `operator ==` are added to its pre-existing translation tables.
Then, when it is dealing with the statement "`c = a + b`" in code like:

```
Point   a, b,  c;
...
a.SetPt(2,4);
cin >> i  >> j;  b.SetPt(i,j);
...
c = a + b;
```

the compiler will stack the left-value (`c`) and assignment operator, note and stack
the 'a' (type `Point`) and encounter the '+' operator. At this stage, it will check its
table of translations for operator '+' and realize that it can deal with "`Point +
Point`" constructs. Code is emitted to push the address of 'a' (the `this` pointer
argument of the member function) and 'b' (the argument) onto the stack and a
function is generated to `Point::operator+()`.

The assignment statement `c = a + b` could be written as:

```
c = a.operator+(b);
```

but this is even uglier than `a.Add(b)`. This style, though legal, is almost never
used and is really just a freak feature related to the implementation of these operator
functions.

Almost all of the standard C operators can be overloaded with class specific
meanings. So, class `Point` could acquire operator functions for '-', '+=', '-=', '/',
'*' and so on. There has to be some reasonable semantics; "`Point * Point`" yields
what?

An operator function can be defined with different argument types – "`Point
operator X <any type>`". With `Point`, operators '*' and '/' might be
meaningfully defined as scaling operations. So one could conceivably have:

```
class Point {
// "Sophisticated" version of a point class,
// with operator functions
public:
    // Constructors
    Point();
    ...
    // Access functions
    ...
    // Modifier procedures
    ...
    void        operator*(short scalefactor);
    void        operator/(short scalefactor);
    // Combination functions
    Point       operator+(const Point& deltaPt) const;
    Boolean     operator==(const Point& otherPt) const;
    ...
    ...
};

void Point::operator*(short scalefactor)
{
    fh *= scalefactor;
    fv *= scalefactor;
}
void Point::operator/(short scalefactor)
{
    fh /= scalefactor;
    fv /= scalefactor;
}
```

Operators precedence is unchanged; so "if(gPoint0 == a + b/2)" should halve
Point b, add the new value for Point b to Point a and then check whether the
resulting Point is located at the origin.

Unary operators can be defined. For example, it would be possible to define
operator ! ("not") to be a Boolean function that returned true if a Point was not
located at the coordinate origin:

*Example of
Unary operator
functions*

```
class Point {
// Version of a point class, with a surfeit of obscure
// operator functions
public:
    // Constructors
    Point();
    ...
    // Access functions
    Boolean     operator!() const;
    ...
    // Modifier procedures
    ...
    void        operator*(short scalefactor);
    void        operator/(short scalefactor);
    // Combination functions
    Point       operator+(const Point& deltaPt) const;
```

```
    Boolean          operator==(const Point& otherPt) const;
    ...
}

Boolean Point::operator!() const
{
    if(fh != 0) return true;
    if(fv != 0) return true;
    return false;
}
```

This definition for `operator!()` would allow code like:

```
cin >> i >> j; b.SetPt(i,j);
if(!b) {
        // deal with points not at origin
        ...
        }
```

Caveats re operator functions

Some C programmers take great delight in concise, even cryptic, code. Such programmers frequently approach operator functions with an excessive enthusiasm, and would consider the coding style `if(!b)` vastly superior to the alternative `if(!b.ZeroPt())`. Common sense is needed. Operator functions are acceptable when their meanings are obvious. They become a hazard to the software engineering process when used with abandon.

For example, class `Point` might need to be extended with new functionality that would reflect a `Point` in the origin (i.e. transform `Point` (2,3) to (-2,-3)). The author of class `Point` could equally readily define this functionality in terms of a standard member function `Point::Reflect()` or as an operator function, for example unary `operator-()` or unary `operator~()`.

```
void Point::Reflect()
{
    fh = -fh;
    fv = -fv;
}

void Point::operator-()
{
    fh = -fh;
    fv = -fv;
}
```

The class author is intimately familiar with the concept captured by a class and would find these alternatives as convenient, and possibly prefer the operator style because of its brevity. However, client programmers and maintenance programmers are likely to find code with statements such as `"b.Reflect();"` much more understandable than code with a statement like `"-b;"`.

Those readers still determined to define operator functions for their own classes should read the cautionary tales in Cargill's book on C++ programming style.

5.1.10Constructors as type conversion functions

The following code fragments represent perfectly correct usage of the `Point` class as defined above:

Some surprising behaviours with constructors

```
Point a,b(6),c(7,8),d(9,-5),e;
...
a = 17;    // Say what?
...
e = d + 11;  // What gives?
...
```

The assignment statement (with an integer apparently being assigned to a variable of type `Point`) and the arithmetic expression (that combines a `Point` and an integer) might have been expected to lead to compile time errors. However both are legal. Both involve the compiler arranging for a temporary `Point` object to be constructed and initialized with an integer value.

The compiler has in its symbol table the `Point` constructor:

Constructor ≡ type converter

```
Point::Point(short x, short y = 0)
```

and recognizes it as a *mechanism for converting a single integer argument into a* `Point`. If the compiler ever finds itself in a situation where it seems to need a `Point` but only has an integer, it remembers this constructor. The compiler realises that if it makes a temporary `Point`, using this constructor, it will be able to complete the translation of the code. So the compiled code that one gets is something along the following lines:

```
a = 17;     // Approximate translation
            make space in stack frame for a temporary point
            invoke constructor Point(short,short)
                    with arguments 17,0
            do a blockmove into 'a'
            free up space claimed for temporary
```

In the case of the assignment, the compiler's task is quite simple. It has already determined the type of the *lvalue* (the thing being assigned to) and knows its a `Point`. For class `Point`, assignment should entail a blockmove from another `Point`, so the right-hand side of the assignment statement must evaluate to type `Point`. Since the right-hand side is type integer, the compiler checks the declaration of class `Point` in its symbol table looking for a constructor that requires one integer argument.

The arithmetic expression "d + 11" (with d of type `Point`) involves a bit more work by the compiler. When parsing the expression, it first stacks 'd' and its type (`Point`), and then encounters the operator +. It checks its tables that define meanings for '+' and finds that operator + can be used with `Points`. It then reads the next token, the integer 11. The compiler would then check its translation table for a definition "operator+, arg1 = Point, arg2 = integer" – and would fail to find such a definition, finding only "operator+, arg1 = Point, arg2 =

Point". This function can be used provided the compiler can turn the integer 11 into a Point. So the compiler then searches its tables describing class Point and finds the appropriate constructor.

```
e = d + 11;         // Approximate translation
            make space in stack frame for a temporary point
            invoke constructor Point(short,short)
                    with arguments 11,0
            invoke Point::operator+ with arguments
                    d ('this') and temporary point
            do a blockmove of function result into 'e'
            free up space claimed for temporary
```

These automatic conversions allow mixed arithmetic expressions with Point + integer (e.g. d + 11), but what of integer + Point (e.g. 11 + d)?

With the definition of Point as given above, the expression 11 + d would cause a compile-time error, although d + 11 is legal. This asymmetry is possibly unattractive. A different definition of a function operator+() that can be applied to Points is possible in C++. This alternative definition would make both expressions 11 + d and d + 11 legal. In both cases, the code generated would entail the use of a temporary Point object, its initialization with the Point:: Point(short,short = 0) constructor, and then a call to the appropriately defined operator+() function. The techniques that allow this more general usage are described in most C++ text books but will not be explored here because this feature of C++ is not relevant to any of the later examples.

Beware the cost of constructors used as type converters! The compiler's ability to second guess the programmer and use constructors as type converters is often more of a pain than a benefit. Apparently innocuous expressions like those in the code fragment:

```
short i, j, currentoffset, deltaX;
Point       refPt, current;
...
for(i=0;i<10000;i++) {
    refPt = i;
    for(j=0;j<10000;j++) {
        Point mark(i,j);
        ...
        current = refPt + currentoffset + deltaX;
        ...
        }
    }
```

may involve a sufficient number of calls to constructors etc. as to make this seemingly simple code a major processing bottleneck. The source form of the code may give no real indication as to the amount of code actually generated. Points are fairly simple; their construction does not entail much work. This type of problem of concealed overheads is more serious with more complex classes.

5.1.11 Stream I/O with programmer-defined types

With `inline` functions for efficiency, `const` qualifiers for `const`-awareness, and operators functions for style, class `Point` is becoming almost usable. But, some additional mechanisms are still required if it is to be possible to use C++ stream style i/o for `Points`.

As noted earlier, C++ streams allow input and output to use forms such as:

"Stream" i/o for programmer-defined classes

```
int        i;
float      x;
...
cin >> i;
...
cout << "The result for i = " << i << " is " << x << "\n";
```

It would be attractive if `Points` could be read and written in the same way, allowing code such as:

```
Point      p1, p2, p3, p4;
Point      intersection;
cout << "Enter points on line 1\n"; cin >> p1 >> p2;
cout << "Enter points on line 2\n"; cin >> p3 >> p4;
...
cout << "The intersection of the given lines is at " <<
         intersection << "\n";
```

The standard streams `cout` and `cerr` are predefined objects of type `ostream`, `cin` is of type `istream`; other streams connected to files are of type `ifstream` (input files), `ofstream` (output files), or `fstream` (for files supporting both input and output operations). The classes `istream`, `ostream`, `ifstream`, `ofstream`, and `fstream`, are part of a large family of interrelated classes provided in the standard C++ libraries. These classes define overloaded operator functions; thus, class `ostream` declares the following member functions (along with many many others!):

```
class ostream : virtual public ios {
public:
    ...
    ostream&      operator<<(const char*);
    ostream&      operator<<(int);
    ostream&      operator<<(long);
    ...
}
```

(These functions return a reference to the `ostream` object for which they are invoked so as to permit the concatenation of stream operations.)

The way a compiler deals with operator functions was described above. After reading the declaration of class `ostream`, the compiler expands its translation table for the "<<" operator (which initially contains just the standard meaning for a left shift operator). The compiler will build something equivalent to the following table:

```
Operator  <<  function  table:
argument 1              argument 2              code to emit
integer                 integer                 move arg1,%r0
                                                move arg2,%r1
                                                left shift %r0,%r1


ostream                 char*                   push arg1
                                                push arg2
                                                jsr  ostream_charstar


ostream                 int                     push arguments
                                                jsr  ostream_long


ostream                 long                    push arguments
                                                jsr  ostream_long


ostream                 short                   push arguments
                                                jsr  ostream_short
```

The definition of these member functions has the code that converts these standard types to output data. The "ostream_charstar" function will copy characters from its argument string to some output-buffer, looping until a null character is found. The "ostream_long" function will check the current output format (hexadecimal, decimal, or octal), convert its integer argument into an appropriate character sequence in some temporary array, and then copy the characters to the output buffer.

The code that already exists in class ostream really has all required the functionality to print a programmer-defined data type. A programmer-defined data type is ultimately just built from integers, floats, character strings etc., and ostream and related classes already have the functions to deal with these. There is no need to change the ostream class (and one definitely wouldn't want to). All that is required is a function that will pick the various shorts, characters, floats etc. out of a programmer-defined type and pass them on via calls to existing functions of ostream. This function has to be defined as yet another overloaded meaning for the << operator. Like the functions that are members of class ostream, it should return a reference to its ostream argument as its result. The compiler must be told to emit a call to this routine that when it finds a << operator acting on an ostream and a variable of the programmer-defined class.

To print instances of class Point, a function something like the following would have to be defined:

```
ostream& operator<<(ostream& os, const Point& p)
{    // But see next section for why this isn't quite right
    os << "(" << p.fh << ", " << p.fv << ")";
    return os;
}
```

As explained in Chapter 4, the concatenated stream expression is interpreted as:

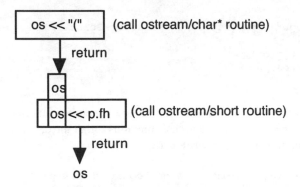

After reading this function definition, the compiler extends its table of meanings for << and can subsequently deal with code like:

```
Point p1(5,8);
cout << "Initial target position" << p1 << "\n";
...
```

The code generated would invoke the standard `ostream_charstar` function, the new function just defined, and then again the `ostream_charstar` function. The output produced at run-time would be the expected "Initial target position(5, 8)". (The argument that is a reference to a `Point` is declared as `const` because the function doesn't change this argument and, if the argument were not declared as `const`, it would not be possible to print any `const` instances of the `Point` class.)

5.1.12 Friend functions

Unfortunately, a C++ compiler would reject the code given at the end of the previous section for the function "`ostream& operator<<(ostream&, const Point&)`". The error message would point out that the `fh` and `fv` data members of `Point p` are "private" and not accessible to this function.

The function "`ostream& operator<<(ostream&, const Point&)`" is an independent, global function. It is *not* a member function of class `Point` (nor is it a member function of class `ostream`). Member functions of class `Point` are appropriate where the code expresses the idea "*Hey you point, do something*" e.g. `aPoint.SetH(49)`. The pattern "`output-stream variable, left shift operator, Point variable`" just does not match the style required for a member function.

As an independent global function, the function `ostream& operator<<( ostream&, const Point&)` should not have any privileged access to the data members of class `Point`. The data members `fh` and `fv` were deliberately declared as being `private` to prevent their being manipulated by arbitrary global functions.

If class `Point` already defined a public member function `void PrintOn( ostream&)` one could achieve the desired effect by encoding the global `operator<<` function as:

```
ostream& operator<<(ostream& os, const Point& p)
{
    p.PrintOn(os);
    return os;
}
```

In this case even without a defined PrintOn() member function, the problem could be avoided because there are access functions for all the data members of class Point. If the function had been written as:

```
ostream& operator<<(ostream& os, const Point& p)
{
    os << "(" << p.H() << ", " << p.V() << ")";
    return os;
}
```

then all would be well. But, this situation is special. Normally, there will not be access functions for reading *all* the data members of a class. Typically, most of the data members of a class are of no concern to clients and there is no reason to provide access functions. So, a more general mechanism that does not rely on access functions must be provided.

Friends   C++ provides the necessary mechanism through its concept of *friends*. Among other uses, the friend mechanism allows a global function to access directly the data members of an object of any class that has nominated that function as being a *friend*. (A global function can have many "friends" and be able to munge around inside objects of several different classes.) Here, function ostream& operator<<(ostream& os, const Point& p) must be nominated as a *friend of class* Point. The first coding for the function (the one accessing data members fh and fv rather than using the H() and V() functions) will be accepted by the compiler provided that the function has been correctly nominated as being a *friend* of class Point.

Class author   Quite obviously, the author of a class is responsible for nominating its friends.
selects its One cannot allow a strange function turning up later and proclaiming itself as a
friends friend – for this would immediately nullify the entire security mechanism provided for private data members.

Friend functions   Friends are nominated in the original class declaration. Here, a simple global function has to be nominated; later examples will illustrate friendship relations between different classes. For class Point and function ostream& operator<<(ostream& os, const Point& p), the declaration would be:

```
class Point {
// Refined version of a point class,
//          "const aware"
//          operator savvy
//          somewhat sociable
//    (and the implementation uses "inlines" so it is
//     reasonably efficient)
public:
    // Constructors
    Point();
    Point(short x, short y = 0);
```

```
        Point(const Point& likePt);
        // Access functions
        short           H() const;
        short           V() const;
        ...
        // Modifier procedures
        void            SetPt(short newx, short newy);
        ...

        // Combination functions
        Point           operator+(const Point& deltaPt) const;
        Boolean         operator==() const;
        ...
        // HERE  FRIENDS  GET  NOMINATED:
        friend          ostream& operator<<(ostream& os,
                                const Point& p);
private:
    short           fv;
    short           fh;
};
```

A friend can be nominated anywhere in the class declaration. Most often, friendships are an implementation detail not of real concern to class users but they are sort of public access functions. So, a suitable place for nominating friends might be after the public interface and before private implementation details (some programmers prefer a class to start with a list of its friends).

Code for a working version of class Point, and a test program that utilizes most of its functions, is provided. The files are in the sub-directory "pdt".

5.1.13 Class Rectangle

Class Point has been used to illustrate most of the features that C++ provides for simple programmer-defined data types. Class Rectangle does not introduce many new features. One new feature illustrated for Rectangle is data members that are of class types rather than built-in types like short etc. Rectangles are going to be defined in terms of two Point coordinates, a "top, left" Point and a "bottom, right" Point, they are also to have a "shade" that determines how they will be displayed.

Rectangles (at last)

```
class Rectangle {
public:
    // Whoever chose the following colour coding scheme
    // worked for Apple computer.
    enum RectangleShade {
            BLACK = 33, BLUE = 409, GREEN = 341,
            YELLOW = 69, RED = 205, WHITE = 30 };
    // Constructor(s)
    Rectangle();
    Rectangle(const Rectangle& refRect);
    ...
    // Access functions
    Point                   TopLeft() const;
```

A class with data members that are themselves instances of classes

```
        Point                   BottomRight() const;
        Boolean                 Valid() const;
        RectangleShade          Shade() const;
        ...
        // Modifier functions
        void                    Offset(const Point& delta);
        void                    SetRect(short top, short left,
                                    short bottom, short right,
                                    RectangleShade s = BLACK);
        void                    SetRect(const Point& tl,
                                        const Point& br,
                                        RectangleShade s = BLACK);
        void                    SetShade(RectangleShade s = BLACK);
        ...
        // Combination functions and others
        Rectangle               operator&(const
                                        Rectangle& otherRect) const;
        Rectangle               operator|(const
                                        Rectangle& otherRect) const;
        ...
        Boolean                 Contains(const Point& testPt) const;
        ...
        // Graphics
        void                    Draw() const;
        ...
        // friends and relations?
        friend ostream& operator<<(ostream& os,
                                    const Rectangle& r);
    private:
        Point                   ftopleft;
        Point                   fbottomright;
        RectangleShade          fshade;
    };
```

Class `Rectangle` will require the normal set of constructors, access functions (which should be declared as being applicable to `const` instances of the class), modifier procedures, a friend to handle printing, and various assorted member functions and operator-functions that combine and manipulate `Rectangles`. The code for these member functions is very similar to that for class `Point`.

Class `Rectangle` introduces two other new features of C++. Its constructors illustrate how information can be passed on and used in the initialization of constituent objects (such as the two corner `Points`). It has its own enumerated type "`RectangleShade`" and there are some conventions relating to the use of such an enumeration.

The type "`RectangleShade`" is not in the global name space. The type `RectangleShade` can only be used freely within the member functions of class `Rectangle`. Of course routines that create `Rectangles` will need to specify their shades, and so will need to use the enumeration `RectangleShade`. C++ has a naming convention that allows clients to use a class specific, non-global, enumerated type (provided, of course, that the enumeration is part of the `public` interface of the class).

5.1.14 Initializing component objects

When a class instance (e.g. a `Point` or a `Rectangle`) is created, its data members are initialized. Any data members that are instances of programmer-defined types are initialized using their default constructors.

The simplest constructors for class `Rectangle` are:

```
// default constructor
Rectangle::Rectangle() { fshade = BLACK; }

// copy constructor
Rectangle::Rectangle(const Rectangle& refRect)
{
    ftopleft = refRect.ftopleft;
    fbottomright = refRect.fbottomright;
    fshade = refRect.fshade;
}
```

If `Rectangles` are created and initialized using either of these constructors, the two `Point` data members will be initialized, using the `Point::Point()` constructor, prior to entry to the constructor bodies.

In the case of the copy constructor, the two `Points` are first set to (0,0) using the `Point::Point()` constructor, and then changed by assignment (i.e. blockmove copies) from the data in the `Point` data-members of the reference `Rectangle`. Obviously, this entails a little bit of redundant work. Once again, `Points` are atypically simple. With more complex classes, a considerable amount of work may be wasted by a process that involves first creating and then changing data members.

Sideline: const and reference data members

C++ provides a mechanism to avoid this redundant processing. In cases like this, use of the mechanism represents a minor optimization. However, the mechanism *must* be used when a class being constructed has data members that are a) instances of classes for which there are no default constructors, or b) "`const`", or c) references. (A class can have `const` data members. For example, it might be appropriate for `Rectangles`' shades to be fixed at creation time. If this were the case, the data member `fshade` should be declared as `const RectangleShade`. Classes can also have data members that are reference types. For example, one might have a class that represents some form of "graphic port" with a "bit map" data field. If the bit map structure had to be created and owned by some windowing environment, a "graphic port" object might be created with a reference to a pre-allocated bit map structure.)

Pre-initialization of data members

The information needed to initialize constants, references, or required by the constructors of data members, can be separated from other arguments of a class constructor and passed directly to the appropriate data members. This information is then used to initialize the various data members before entry to the body of the main constructor function. The C++ notation is very expressive of this concept of cut-out, and use before entry. If the definition of the copy constructor is rewritten to use this mechanism, it becomes:

```
Rectangle::Rectangle(const Rectangle& refRect)    :
          ftopleft(refRect.ftopleft),
          fbottomright(refRect.fbottomright),
          fshade(refRect.fshade)
{
    // Nothing left to do in body!
}
```

Here, the two `Points` are constructed using the `Point::Point(const Point&)` constructor. The enumerated type has also been pre-initialized. (A function call style initializer can be used for integers, enumerated types, `char*` pointers, and other built in types.)

The general scheme for these constructors with data member pre-initialization is illustrated diagrammatically in Figure 5.2.

Details of initialization of data members appear only in the definition. The class declaration just names the constructor and gives its argument list. In the definition, a colon (':') separates the constructor name and its argument list from a list with data members and their initial values. The compiler determines the order in which the various data members are initialized; this order is not related to their position in the definition's initialization list. The main body of the constructor follows the initialization list. Quite often, as in the `Rectangle` example, all the work is done prior to entry to the main body of the constructor which is then empty.

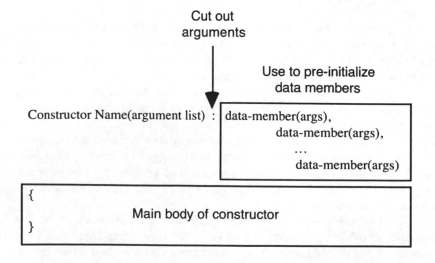

Figure 5.2 Layout of a constructor definition that uses the mechanisms for pre-initializing const, reference and other data members.

5.1.15 Using an enumerated type defined by a class

Rectangles will be created by client code, for example:

```
Rectangle  palette[4], testrect;
palette[0].SetRect(20,20,40,60,Rectangle::RED);
palette[1].SetRect(40,20,60,60,Rectangle::YELLOW);
palette[2].SetRect(60,20,80,60,Rectangle::GREEN);
palette[3].SetRect(80,20,100,60,Rectangle::BLUE);
…
if(Rectangle::YELLOW == testrect.Shade()) …
```

Using enumerated types defined (publicly) by classes

In client code, use of the enumerated type `RectangleShade` must always be qualified by its class name `Rectangle`.

5.1.16 const data members

The following fragments illustrate the changes in the class declaration, constructors, and member functions that would be required if the shade field were to be made `const`.

Rectangle with a const data member

```
class Rectangle {
// Rectangle with const shading
public:
    // Whoever chose the following colour coding scheme
    // worked for Apple computer.
    enum RectangleShade {
            BLACK = 33, BLUE = 409, GREEN = 341,
            YELLOW = 69, RED = 205, WHITE = 30 };
    // Constructor(s)
    Rectangle();
    Rectangle(const Rectangle& refRect);

    …
    // Access functions
    Point               TopLeft() const;
    RectangleShade      Shade() const;

    …
    // Modifier functions
    void                Offset(const Point& delta);
    void                SetRect(short top, short left,
                                short bottom, short right);
    void                SetRect(const Point& tl,
                                    const Point& br);

    …
    // friends and relations?
    friend ostream& operator<<(ostream& os,
                    const Rectangle& r);
private:
    Point               ftopleft;
    Point               fbottomright;
    const RectangleShade fshade;
```

A const data member

```
    };

    Rectangle::Rectangle() : fshade(BLACK) { }
```

Note that the default constructor *must* now pre-initialize the `fshade` field. The `fshade` data member cannot be set by assignment in the body of the constructor because by the time that code is being executed it already exists as a constant and shouldn't be modified. A compiler may take note of a `const` data member in a class declaration and refuse to allow assignment of instances of that class (older compilers have been seen to ignore the `const` declaration and generate code with assignments that then change the supposedly `const` member).

5.1.17 Examples of stream I/O functions for Point and Rectangle

Reading an object from a stream

The test code provided for class `Rectangle` (also in subdirectory "pdt") includes code for stream input of `Point`s and `Rectangle`s. Both `Point` and `Rectangle` classes have nominated friend input functions:

```
    class Point {
    public:
        ...
        friend ostream& operator<<(ostream& os, const Point& p);
        friend istream& operator>>(istream& ins, Point& p);
    private:
        ...
    };

    class Rectangle {
    public:
        ...
        friend istream& operator>>(istream& ins, Rectangle& r);
        ...
    };
```

Code for input functions is typically a little more long winded than that for output functions because it is always necessary to check for errors. In the following examples, the error checking is not comprehensive. Further, as noted in section 4.14, the response to an error (abandoning processing via a call to `exit()`) would not be appropriate in a production program. An error should be handled by restoring the state of the stream and signalling the error to the calling environment. Although essential in a practical program, such features go a bit beyond the scope of these simple introductory examples The somewhat primitive code for input of `Point`s is:

An operator>>() that works with istream& and Point&

```
    istream& operator>>(istream& ins, Point& p)
    {
        char* msg = "Trying to read a point but failed.";
        // Points are written in form (<integer>,<integer>)
        // It is possible that there will be leading whitespace
```

```
          // characters etc
          // so 1) scan input until left parenthesis and
          //           consume it
          //    2) get the h-coord as a short integer
          //    3) consume characters until have got comma
          //    4) read second integer
          //    5) consume characters upto and including final
          //           close parenthesis

          // ignore up to ~2000million characters while looking for
          //      a left parenthesis
          ins.ignore(LONG_MAX,'(');                                    istream::
          if(ins.fail()) {                                             ignore() - skip
                cerr << msg << " No leading left parenthesis\n";       characters
                exit(1);
                }

          short h;
          ins >> h;
          if(ins.fail()) {                                             istream::fail() -
                cerr << msg << " Error reading 1st coordinate\n";      check for errors
                exit(1);
                }
          ins.ignore(LONG_MAX,',');
          if(ins.fail()) {
                cerr << msg << " No comma.\n";
                exit(1);
                }

          short v;
          ins >> v;
          if(ins.fail()) {
                cerr << msg << " Error reading 2nd coordinate\n";
                exit(1);
                }
          ins.ignore(LONG_MAX,')');
          if(ins.fail()) {
                cerr << msg << " No closing right parenthesis\n";
                exit(1);
                }
          p.fh = h;
          p.fv = v;
          return ins;
     }
```

The code uses a number of member functions of the `istream` class in addition to the class `istream`'s member operator function `operator>>(short&)`. The function "`istream::fail()`" is used to check the success of the last input operation. Function `istream::ignore(int, char)` takes, as arguments, a count of a maximum number of characters to skip and a character that is to be sought. The function is most often used to remove trailing input to an end of line – e.g. `cin.ignore(SHRT_MAX,'\n')`. The operator function `istream::operator>> (short&)` is used to read the two short integer coordinate values for a `Point`. *istream functions ignore() and fail()*

The i/o code for class `Rectangle` illustrates use of streams and `Points`. `Rectangle`'s `tshade` field (the enumerated type) is read and written as an integer.

```
ostream& operator<<(ostream& os, const Rectangle& r)
{
    os << "(" << r.ftopleft << r.fbottomright << " "
                << r.fshade << ")";
    return os;
}

istream& operator>>(istream& ins, Rectangle& r)
{
    char* msg = "Error reading a rectangle.";
    ins.ignore(LONG_MAX,'(');
    if(ins.fail()) {
            cerr << msg << " No leading left parenthesis.\n";
            exit(1);
            }
    ins >> r.ftopleft >> r.fbottomright;

    short temp;
    ins >> temp;
    // I should really have included some code to
    // validate the shade codings, but never mind.
    r.fshade = (Rectangle::RectangleShade) temp;

    ins.ignore(LONG_MAX,')');
    if(ins.fail()) {
            cerr << msg << " No closing right parenthesis.\n";
            exit(1);
            }
    return ins;
}
```

5.2 A CLASS WITH RESOURCE MANAGEMENT RESPONSIBILITIES

Objects of classes `Point` and `Rectangle` are integral – each such object exists as a unique block of contiguous bytes on the stack or in the heap. The nature of these classes is such that class instances are going to have exactly the same form as ordinary C `struct`s. Each `Point` is just a small data record (of course, there would be some storage management overhead for `Point`s allocated on the heap using `new`). Although a `Rectangle` is built from two constituent `Point`s, each `Rectangle` is integral; a `Rectangle` would probably be represented by a block of a few bytes allocated as a single entity.

Quite often, classes are required where each object has some separately allocated storage structures. A simple example might be "class Player" – a new entity in some extended "Space Invaders" game. A `Player` is to have a position (this will be represented by an instance of class `Point`), an icon (actually, just an integer icon-identifier), and a name. Obviously, users will choose their own names, so a name is going to be a C string of arbitrary length. An initial declaration for class `Player` is as follows:

```
class Player {
    // This class does not allow for const instances.
    // This class does not support stream i/o
    // No default constructor.
public:
    // Constructors and related functions
    Player(const Point& startpt, short iconid, char* name);
    ...
    // Access functions
    Point        Position();
    ...
    // Modification procedures
    void         MoveBy(const Point& delta);
    void         MoveTo(const Point& place);
    void         ChangeName(char* newname);
    ...
    // Others
    void         Draw();
private:
    short        fIconId;
    char*        fname;
    Point        fp;
};
```

The application did not require const instances of class Player. This design decision is reflected in the class declaration. As the class declaration does not specify the access functions as being applicable to const instances, it will be impossible to make any use of "const Player" variables (const instances of the class can be defined, but they cannot be used in the code). Objects of this class would be ephemeral – existing only during a game; consequently, there would be no need to support stream i/o. There is no default constructor; there is only one defined constructor and it requires explicit, non-defaulted, arguments. Because a constructor has been defined, the compiler will not define its own default constructor. Consequently, it is not possible to have arrays of Player objects because for arrays one must have some form of default constructor. (One can have arrays of pointers to objects of class Player).

Some simplifications present in class Player

Various access functions might be needed by the application, e.g. a function returning the position of a Player. There would be some set of procedures for moving and otherwise modifying Players as the game progresses and some graphic functions. These functions and procedures are of no particular interest and will not be elaborated. It is only the constructor that involves new C++ features.

An instance of class Player would probably be created in some global routine such as the following:

```
Player*    GetNewPlayer(const Point& clikPoint)
{
    const int NAMEMAX = 65;
    char    buff[NAMEMAX];
    ::GDisplay("Enter your name (alias):");
    ::GReadString(buff,NAMEMAX);
    short id = ::GSelectIconDialog();
```

```
        Player* p = new Player(clikPoint,id,buff);
        return p;
    }
```

This routine uses a few global routines (`GDisplay()`, `GSelectIconDialog()` etc.) to obtain user input and then creates an appropriately parameterized instance of class `Player`. The name for the player is held in the character buffer that is an automatic, stack-based variable local to function `GetNewPlayer()`. When the function is exited, this buffer array goes out of scope and the space that it occupies is released.

Need to allocate space for associated data structures

Obviously, an instance of class `Player` cannot simply store the value of the `char*` pointer that it is given as an argument to its constructor. This `char*` pointer will either reference some global character buffer that may continue to exist but may get overwritten by other input data, or reference a temporary data structure that will shortly cease to exist (as in the `GetNewPlayer()` code). The `Player` object must allocate its own space to hold its name string. So, the constructor for a `Player` would be something like:

```
Player::Player(const Point& startpt, short iconid,
                    char* name) : fIconId(iconid), fp(startpt)
{
    fname = new char[::strlen(name)+1];
    ::strcpy(fname ,name);
}
```

The `Point` field does not have to be pre-initialized but it is more efficient to pre-initialize it, than to use a default constructor for `Point` and then change the value in the constructor body of `Player`. The constructor body of `Player` allocates a character array of sufficient size for the name and a trailing null character, and copies the given name into this array. (It would be inappropriate to call `strdup(,)` to create a copy of the name. The standard `strdup(,)` function will use `malloc()` to allocate space rather than `new`. Operator `new` may rely on `malloc()` but in some implementations it is completely distinct. In general, it is unwise to have two different heap management regimes employed in the same program.) The form of a `Player` object in the heap is illustrated in Figure 5.3.

5.2.1 Memory leaks and "destructors"

Pointer problems

A similar group of problems beset every class whose instances contain pointers to independently allocated memory. The problems are a) memory leaks, and b) dangling pointers.

The following code fragment creates a stack-based (automatic) instance and a heap-based (dynamic) instance of class `Player`:

```
void Game(const Point& p, short defltId)
{
    Player computerPlayer(p,defltId,'Computer rules OK');
    …
```

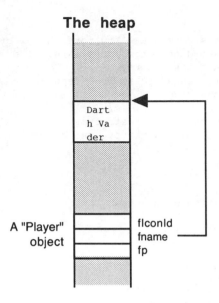

The heap

A "Player"
object

Dart
h Va
der

fIconId
fname
fp

Figure 5.3 Illustration of form of a Player object with its separately allocated
space for a name string.

```
Player*        user = GetNewPlayer(gPoint0);
...
delete user;
...
}
```

The dynamic instance, user, is explicitly deleted; the other instance,
computerPlayer is automatically freed on exit from the routine.

The spaces claimed in the stack frame for the computerPlayer instance, or on
the heap for the user instance, would each consist of a few bytes for the various
Point data members, icon-ids etc. It is this space that gets released when the
object is deleted or the stack frame gets cleared. The separately allocated heap space
used for name strings is *not* freed automatically.

If no explicit provision is made to clear up those name strings, they remain as   *Memory   leaks*
dead space on the heap. As the program runs, and instances of class Player are
created and destroyed, there is a steady decrease in the pool of available heap space.
Memory appears to "leak away". Eventually, an application may have to abort
when a request for a new heap space fails.

The author of class Player, and similar classes using separately allocated   *Destructor*
memory, must take part of the responsibility for releasing memory when an object   *routine*
is freed. The programmer must define a *destructor* member function. The C++
compiler then assists by arranging to place calls to the destructor (just as it placed
calls to the constructor).

A destructor has a fixed name: ~<classname>() e.g. ~Player(). There can
only be one destructor for a class; the destructor takes no arguments. The revised
class declaration and the destructor function definition are as follows:

```
class Player {
    // This class does not allow for const instances.
    // This class does not support stream i/o
    // No default constructor.
    // Destructor required to free separately allocated
    //      name string
public:
    // Constructor(s) and destructor
    Player(const Point& startpt, short iconid, char* name);
    ~Player();
    ...
    // Access functions
        ...
    // Modification procedures
        ...
private:
    short       fIconId;
    char*       fname;
    Point       fp;
};


Player::~Player()
{
    delete [] fname;
}
```

The code in the destructor function is concerned only with the separately allocated
memory. The array of characters pointed to by data member fname is deleted.

Destructors are never explicitly called from ordinary C++ code. (Almost
nothing is prohibited in C++, and there are mechanisms for explicitly invoking a
destructor. But these special features are not used in normal code.) The C++
compiler arranges the necessary calls.

```
void Game(const Point& p, short defltId)
{
                        // create stack frame of sufficient size
    Player computerPlayer(p,defltId,'Computer rules OK');
                        // call to the constructor
    ...
    Player*     user = GetNewPlayer(gPoint0);
    ...
    delete user;
            // Player::~Player() invoked for object user
    ...
    ...
            // object computerPlayer about to go
            // out of scope, call to Player::~Player()
}
```

With a destructor defined for the class, the separately allocated space for the name
fields does get freed. Memory leakage, the first of the two pointer-related problems,
can be avoided.

An object may acquire resources apart from separately allocated memory. For example, some objects open files when they are created (the file-open code forming part of their constructor routines), or they may claim a "device context" (a graphics resource in the world of Windows applications on the IBM-PC), or they may acquire a serial i/o line, or a Unix "port", or any other general "resource". Such resources need to be freed when no longer required. Such resource management aspects are often facilitated by having constructor ≡ "claim resource", destructor ≡ "free resource". The code of the destructor routine would include the instruction sequence to close the file, free the "device context", or perform whatever other steps are needed to release resources.

Generalize memory management to "resource" management

5.2.2 Dangling pointers and class specific assignment operators

The structure of class `Player` also raises the possibility of dangling pointers. Consider the following code:

Dangling pointers

```
void Display(Player d)
{
    ::Moveto(d.Position());
    ...
}

int main(int, char**)
{
    Player player1(gPoint0,SPACESHIPICON,"Captain Kirk");
    ...
    Display(player1);
    ...
}
```

Despite its innocent appearance, this code virtually guarantees that the program will terminate with "bus error – core dumped" (or an equivalent expletive from the operating system).

The same problem is present in the following code fragment:

```
Player*    players[10];
...
void Target()
{
    short   choice;
    Player x(gPoint0,NULLICON,"");
    ...
    for(;;) {
            x = *(players[choice]);
            ...
            }
    ...
}
```

```
int main(int, char**)
{
    ...
    players[0] = new Player(gPoint0,SPACESHIPICON,
                                "Captain Kirk");
    players[1] = new Player(gPoint100,TARDISICON,
                                    "Dr. Who?");
    ...
    Target();
    ...
}
```

The problem relates to the assignment of one object to another. This assignment can be an explicit one (as in the second fragment with the assignment to x) or an implicit one (as in the first code fragment where an assignment is involved in the copy constructor that would be used to create, on the stack, the argument d for routine Display()).

Figure 5.4 illustrates a possible sequence of events as might occur when the second code fragment is run. The three tableaux show the state of the heap before, during, and after a call to function target().

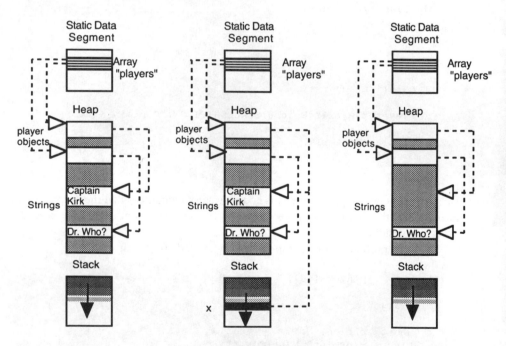

Figure 5.4 State of memory before, during and after a call to function target(). The assignment in target() leads to a shared data structure and eventually to a dangling pointer condition.

The various objects of class Player, for global area players[], are allocated on the heap, each having a pointer to some separately allocated string located elsewhere in the heap. On entry to function target(), object 'x' is allocated space on the stack (again, having a name pointer referencing some string space in the heap). Eventually, the assignment statement in `target()` is executed – for example, if the value of `'choice'` was 0, x would have its data fields changed by assignment from `players[0]` i.e. "Captain Kirk at gPoint0".

Before assignment, x and `player[0]` were:

```
x.fIconId   = 1000;        players[0]->fIconId  = 1001;
x.fname     = 0x9acc;      players[0]->fname    = 0x9c44;
x.fp        = (0,0);       players[0]->fp       = (0,0);
```

and after assignment they were:

```
x.fIconId   = 1001;        players[0]->fIconId  = 1001;
x.fname     = 0x9c44;      players[0]->fname    = 0x9c44;
x.fp        = (0,0);       players[0]->fp       = (0,0);
```

because the assignment is simply performing a blockmove from one location to another. Inevitably, after the assignment these two separate instances of class `Player` have pointers to the same "shared" string in the heap.

On exit from a function, all automatics are freed, and for class objects their destructors are called. The destructor invoked for x frees its name. The space originally allocated to the string "Captain Kirk" is now available for reallocation. Unfortunately, the object at `*(players[0])` still has a pointer to this now non-existent string. It is a *dangling* pointer – a pointer to nothing but trouble.

This is the trap that will eventually cause the program to crash.

The space on the heap will get reused for some other dynamic data structure (or, on some systems, a code segment). Subsequently, if the name of `players[0]` is changed, that other structure or code fragment is destroyed.

The problem is essentially the same in the other code fragment. There, the temporary argument on the stack gets to share a pointer to a name string. On return from the called function, the temporary is destroyed – and along with it so is the shared name. Again, a dangling pointer is left.

The sharing of the name string is incompatible with the usage intended for this class of objects. Each object must have a separate and distinct name. The compiler cannot know this restriction. The programmer must specify semantics for class `Player` that incorporates any such restriction.

Assignment and copy construction are the are two places where information in the data members will be copied from one object to another. So, the author of class `Player` must redefine the semantics of these operations.

Copy constructors and assignments

Copy constructors were illustrated earlier for the simple `Point` class. A constructor for class `Player` has to be a little more subtle; it must take into account its storage management responsibilities. When a new `Player` is created using data from an existing `Player` object it should get a separate, but identical, version of that name string. An appropriate constructor would be:

```
Player::Player(const Player& refPlayer)
{
    fIconId = refPlayer.fIconId;
    fp = refPlayer.fp;
```

```
            fname = new char[::strlen(refPlayer.fname) + 1];
            ::strcpy(fname,refPlayer.fname);
    }
```

This constructor allocates new space on the heap for the name for the object under construction and copies the original name into this space, so guaranteeing that each individual object has its own unique name string.

Defining an assignment operator

The examples for class `Point` showed how operators such as '+' could be redefined on a class specific basis. The assignment operator '=' is no different; a class can have a new meaning created for `operator=()` (the "operator equals function"). The assignment operator should be defined to copy values of simple data types (such as the `Point` coordinate and the short integer) and, again, create a duplicate string. These requirements lead to the following declaration and definition:

```
class Player {
        // This class does not allow for const instances.
        // This class does not support stream i/o
        // No default constructor.
        // Destructor required to free separately allocated
        //      name string
        // "Resource" (memory) aware copy constructor
        // Memory aware assignment operator
public:
        // Constructor(s) and destructor
        Player(const Point& startpt, short iconid, char* name);
        Player(const Player& refPlayer);
        ~Player();
        ...
        // Access functions
            ...
        // Modification procedures
        <return-type to be resolved>
                        operator=(const Player& rhs);
            ...
private:
        short           fIconId;
        char*           fname;
        Point           fp;
};
```

Partial implementation of Player:: operator=

```
<return-type to be resolved>
            Player::operator=(const Player& rhs)
{
        // some vital bits of additional code to go here
        fp = rhs.fp;
        fIconId = rhs.fIconId;
        fname = new char[::strlen(rhs.fname) + 1];
        ::strcpy(fname,rhs.fname);
        // and a bit more code here when the return value
        // has been decided
}
```

This version of the code is not satisfactory. One problem is illustrated in Figure 5.5.

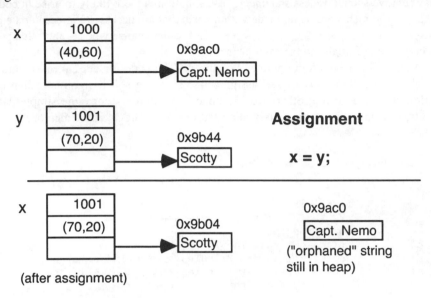

Figure 5.5 Assignment must create a copy of the string (avoiding problems with shared data) and, further, should release the storage associated with any pre-existing string.

Although this version of operator=() does correctly create new storage for the string, it does not release any previously allocated string. Once again, there is a memory leak. Any previous string must be freed.

```
<return-type to be resolved>
            Player::operator=(const Player& rhs)
{
    // some vital bits of additional code to go here
    fp = rhs.fp;
    fIconId = rhs.fIconId;
    delete [] fname; // release existing name space
    fname = new char[::strlen(rhs.fname) + 1];
    ::strcpy(fname,rhs.fname);
    // and a bit more code here when the return value
    // has been decided
}
```

operator=()
aware of memory
leakage   problem

Unfortunately, there is again a problem. Suppose an object gets assigned to itself. The code says "throw away the name string", "find the length of the other string", "copy the other string"; but, in a self assignment, it is the same string that gets thrown away and then used. Actually, on most systems such code will work; at least, it will work most of the time. There is only a very small chance that memory will get reshuffled as part of a particular delete operation. It only happens when programs are demonstrated to bosses or potential customers. When a memory reshuffle does occur, the program produces erroneous output or crashes.

(Self assignment of the form x = x; is admittedly unusual, and a good compiler may remove such a useless statement. The situation is more likely to arise in code where assignments are being made with objects that are accessed via pointers, e.g. Player* a; Player* b; …; *a = *b; …. Problems arise when a and b happen to be pointing initially to the same object.)

Handling the case of self assignment

A little defensive programming is required. The code for operator=() must check for the possibility of self assignment. If it is a self assignment, then no action should be taken; otherwise, the code is as above – copy the simple data fields, free any existing separately allocated memory, allocate new memory, copy data:

```
<return-type to be resolved>
            Player::operator=(const Player& rhs)
{
    if(this != &rhs) {
            // It is NOT a self assignment
            fp = rhs.fp;
            fIconId = rhs.fIconId;
            delete [] fname; // release existing name space
            fname = new char[::strlen(rhs.fname) + 1];
            ::strcpy(fname,rhs.fname);
            }
    // and a bit more code here when the return value
    // has been decided
}
```

The check for self assignment compares the address in the implicit parameter this with the address of the object on the right hand side of the assignment. If these addresses are equal, it is a self assignment. This check is satisfactory for all the simple examples that will be presented in this text. (There are situations, in more advanced uses of C++, where this form of check may be inadequate or inappropriate; see the discussions in Meyers' book *Effective C++*.)

Return value for the operator= function

It would be possible for the operator=() function to return void; but this would not allow the ordinary C style construct:

```
a = b = c = d = e;
```

Such a concatenated assignment should work for user defined types as readily as for integers. If operator=() returned void, such a concatenated assignment would be a syntax error. To make this construct legal, class X::operator=() must return X&. If Player::operator=() returns Player&, then the compiler will interpret:

```
Player a(…,…,…), b(…,…,…), c(…,…,…);
…
a = b = c;
```

as meaning:

```
a.operator=( b.operator=(c));
```

So, the final declaration and definition for the function are:

```
class Player {
    // This class does not allow for const instances.
    // This class does not support stream i/o
    // No default constructor.
    // Destructor required to free separately allocated
    //     name string
    // "Resource" (memory) aware copy constructor
    // Memory aware assignment operator
public:
    // Constructor(s) and destructor
    Player(const Point& startpt, short iconid, char* name);
    Player(const Player& refPlayer);
    ~Player();
    ...
    // Access functions
        ...
    // Modification procedures
    Player&        operator=(const Player& rhs);
        ...
private:
    short          fIconId;
    char*          fname;
    Point          fp;
};

Player& Player::operator=(const Player& rhs)
{
    if(this != &rhs) {
            // It is NOT a self assignment
            fp = rhs.fp;
            fIconId = rhs.fIconId;
            delete [] fname; // release existing name space
            fname = new char[::strlen(rhs.fname) + 1];
            ::strcpy(fname,rhs.fname);
            }
        return *this;
}
```

Once an `operator=()` function has been defined for a class, the C++ compiler will remember to use it in all situations. So, if a more complex class is defined:

C++ uses defined operator= functions as needed

```
class NetworkPlayer {
    ...
private:
    Player         fplayrec;
    short          fnode;
};
```

one can perform assignments:

```
NetworkPlayer p1, p2;
...
p1 = p2;
```

and rely on correct operation. If class `NetworkPlayer` does not introduce any additional data members with special semantics, there is no need to define a `NetworkPlayer::operator=()` function. The C++ compiler will handle the assignment by copying the simple data members, like `short fnode`, and use the defined `Player::operator=()` for the `fplayrec` data member.

5.2.3 Classes with separately allocated data members

It is all the fault of those pointers

If a class contains pointer-type data members, like the `char*` pointer in class `Player`, the author of that class should give careful consideration to the need to provide an appropriate "pointer aware" copy constructor, an assignment operator, and a destructor. Pointer data members often indicate the presence of separately allocated memory, or other resources, that must be properly managed.

It is not possible to say that a copy constructor, destructor, and assignment operator are *necessary* in a class with pointer members. Many subsequently developed examples will have classes with pointer members that serve as links to objects of other classes. These data members are used for object-to-object communication where one object requests action by another object. Several examples using the class libraries are constructed using instances of "class Document" (which basically stores data and organizes transfers to and from disk) and "class View" (which handles visual display of information). These classes will have links; documents need to ask views to re-display information after editing steps, views must ask documents for details of the current data.

```
class View;
class Document {
public:
    ...
private:
    View*   fmainview;
    ...
};
class View {
public:
    ...
private:
    Document*       fdoc;
};
```

Pointer data members like `fdoc` and `fmainview` are not part of the "state data" characterizing a particular view or a specific document. These pointers represent the physical realization of collaborative links that must exist among objects of varying classes. Other classes, like those that represent the various specialized data items stored in a document, will also need links to collaborators (e.g. each "paragraph", "shape" or other data item belonging to a document might need a link to the view in which it is displayed). The default bitwise copy constructor and assignment operator might suffice for such data items (the new copy probably requires a link to the same view as the original so the "sharing" is quite acceptable).

It is up to the author of the class to determine whether a pointer data member is "state" or "collaboration". If there are "state" pointer data members, and the class is to be used as a first class type (assignments allowed etc.), then the class author should provide a copy constructor, destructor and assignment operator. It may also be worth defining `private` copy constructors and assignment operators for those classes, e.g. `View` and `Document`, that are not intended to be used like normal data types (these `private` versions don't have to have any implementation). If the assignment operator for a class is `private`, the compiler will prevent any inappropriate attempt to assign instances of that class.

When to supply copy constructor, destructor, assignment operator

Sometimes it is desirable to initialize an object in two stages. Only part of the work is done in the constructor, the rest of the work is carried out in an explicitly invoked initialization routine.

Two phase initialization

The code given for the `Player` constructor did not actually check whether an array was allocated to receive the name string. The code simply says "give me a block of bytes; use ::strcpy to fill the block". Now, if the heap has been exhausted, the `new char[]` operation will return a NULL pointer (i.e. 0). As there is no check, the bytes from the name are duly copied into memory starting at location zero. The operating system will probably not appreciate an attempt to overwrite its interrupt vector and the program will be terminated with extreme prejudice. The code could be extended to check for this possibility, but what should it do if a NULL pointer is returned. Constructors have no return value. There is no simple way of conveying to the calling environment the fact that a failure has occurred. The only general solution to the problem is to cause an exception, or maybe abandon processing by a call to `exit()`.

Where a newly created object has to claim a resource (memory, a file, an I/O channel or whatever) there can be advantages in a scheme where initialization takes place in two stages. The code for the constructor forms the first stage. This code places the new object in a defined, safe state. The claim for the resource is then made in a separate, explicitly called startup/initialize/get-resource method. This get-resource method returns a status value. If the resource can not be acquired, the method returns a failure status that can be checked and acted on in the calling environment.

This type of two phase initialization will appear in some of the code of the framework libraries that are illustrated in later chapters.

5.2.4 Class "String"

Class `Player`'s handling of its name string makes it really a special, limited version of a "string handling class". Class `String` is a popular first example, or first exercise, in most C++ texts (see for instance the version in Shapiro's early text *A C++ Toolkit*). Class `String` has a private `char*` pointer to a dynamically allocated block of characters. Typically, class `String` will be defined with copy constructor and assignment operator analogous to those used in class `Player`. Functionality of class `String` will include `operator+()` as a concatenation operator

class String

(allocate a buffer for the longer string, copy existing bytes, append new bytes, release existing bytes), indexing (find character at position i, return NULL if i is out of range), determine length etc. Class `string` is left as an exercise for the reader.

5.3 WATCHING A PROGRAM WORK: SIDE-EFFECTS OF CONSTRUCTORS AND DESTRUCTORS

Constructors and destructors as debugging aids

One technique for debugging C++ programs exploits the automated calls to constructors and destructors of class instances. One can define the classes:

```
// Declaration of tracers
#ifndef __TRACERS__
#define __TRACERS__
```

Global function tracer

```
class FTracer {
// Tracer for simple (global) functions
public:
    FTracer(char*);
    ~FTracer();
private:
    char* ffunctionname;
};
```

Tracer for object creation and destruction

```
class OTracer {
// Tracer for object creation/destruction
public:
    OTracer(char*,void*);
    ~OTracer();
    void OTracer operator= () { };
private:
    char*  fclassname;
    void* fobj;
};
```

Member function tracer

```
class MFTracer {
// Tracer for member function invocation
public:
    MFTracer(char*,void*);
    ~MFTracer();
private:
    char* fmembername;
    void* fwho;
};

#endif
```

Implementation of tracers

```
// Implementation of Tracers
#include <iostream.h>
#include <iomanip.h>
#include <string.h>
```

```
#include "Tracers.h"

FTracer::FTracer(char* fn)
{
    cerr << "Entry to function " << fn << "\n";
    cerr.flush();
    ffunctionname  = new char[::strlen(fn) + 1];
    ::strcpy(ffunctionname,fn);
}


FTracer::~FTracer()
{
    cerr << "Exit from function " << ffunctionname << "\n";
    cerr.flush();
    delete [] ffunctionname;
}


OTracer::OTracer(char* cn,void* obj)
{
    fobj = obj;
    cerr << "Create a " << cn << ", address " << hex
                << fobj << "\n";
    cerr.flush();
    fclassname = new char[::strlen(cn) + 1];
    ::strcpy(fclassname,cn);
}


OTracer::~OTracer()
{
    cerr << "Destroy a " << fclassname << ", address "
            << hex << fobj << "\n";
    cerr.flush();

    delete [] fclassname;
}


MFTracer::MFTracer(char*fn, void* who)
{
    cerr << "Object at " << hex << who;
    cerr << " is performing " << fn << "\n"; cerr.flush();
    fwho = who;
    fmembername = new char[::strlen(fn) + 1];
    ::strcpy(fmembername,fn);
}


MFTracer::~MFTracer()
{
    cerr << "Object at " << hex << fwho;
    cerr << " finished " << fmembername << "\n";
    cerr.flush();
    delete [] fmembername;
}
```

If an instance of OTracer is defined as a data member of another class, and that class's constructors are extended to initialize the extra data member, then the creation and destruction of any instance of that class will be logged to cerr.

Similarly, defining an automatic variable of type MFTracer in a class member function, or of FTracer in a global function, will provide a log of entry to and exit from the traced function.

(The operator=() function for class OTracer is a little unusual. The values in OTracer objects essentially record identity, and should not be changed between construction and destruction. Since data members of type OTracer are to be added to classes that support assignment, such as class Rectangle, it might seem necessary to change the meaning of the operator=() function of the host class, i.e. class Rectangle, so as to prevent any change to OTracer data members. This would be too much trouble. It is much simpler to redefine the operator=() function of class OTracer so that it becomes a null operation. With the definition given, data members of class OTracer can be added to other classes without having to worry as to whether these classes support assignment.)

Tracers of these classes have been included in the code of class Rectangle, and program RectangleTest.C, as provided in the examples (in subdirectory "pdt"). Class Rectangle has an OTracer data member so that creation and destruction will be logged:

```
class Rectangle {
public:
    Rectangle();
    ...
private:
    ...
    RectangleShade        fshade;
    OTracer               ft;    // Tracer data member added
};

// Initialize extra data member
Rectangle::Rectangle(): fshade(BLACK) , ft("Rectangle",this)
{ }

Rectangle::Rectangle(const Rectangle& refRect) :
        ftopleft(refRect.ftopleft),
            fbottomright(refRect.fbottomright),
                fshade(refRect.fshade), ft("Rectangle",this)
{
}
```

(The tracer data member is initialized with the address of the object of which it forms a part (the this argument). Such usage allows the tracer to print the same address as will appear in the member function traces. An object tracer does not really require the address of the object of which it forms a part; it can print its own address in all outputs. However, the address printed by the object tracer would then be slightly different from that produced by member function tracers – the difference being the offset of the data member within the overall object.)

The member functions Rectangle::Valid() and Rectangle:: operator&() have automatic variables of type MFTracer:

```
Boolean Rectangle::Valid() const
{
```

```
    MFTracer        mftr("Rectangle::Valid()",this);
    if(fbottomright.H()<=ftopleft.H()) return false;
    if(fbottomright.V()<=ftopleft.V()) return false;
    return true;
}

Rectangle Rectangle::operator&(const Rectangle& otherRect)
const
{
    MFTracer mftr("Rectangle::operator&",this);
    ...
}
```

The test program is (letters, like **A**, mark segments and provide links to features in the output shown below):

```
#include "Rectangle.h"

A Rectangle gRect(0,0,340,512,Rectangle::WHITE);

int main(int, char**)
{
    B FTracer mft("main");
    cout << "Tests of Rectangle class\n";
    C Rectangle palette[4];
    palette[0].SetRect(20,20,40,60,Rectangle::RED);
    palette[1].SetRect(40,20,60,60,Rectangle::YELLOW);
    palette[2].SetRect(60,20,80,60,Rectangle::GREEN);
    palette[3].SetRect(80,20,100,60,Rectangle::BLUE);

    cout << "The first rectangles, array palette\n";
    for(int i=0;i<4;i++)
            cout << "palette[" << i  << "]" <<
                                palette[i] << "\n";

    cout << "How about some dynamic rectangles\n";
    D Rectangle* never_say_die = new
            Rectangle(300, 300,450, 650, Rectangle::WHITE);

    Rectangle* comes_and_goes = new
            Rectangle( 100, 100, 101, 1001, Rectangle::BLUE);

    cout << "and some more automatics, a,b,x,y and z\n";

    Rectangle a(0,0,100,100),
                b(40,40,60,200,Rectangle::WHITE);
    Rectangle x,y,z;

    E cout << "Do & and | operations on a, b "
                        "modifying x and y\n";
    x = a & b;
    y = a | b;
```

```
cout << "Rectangle a:\n\t"  << a << "\n";
cout << "Rectangle b:\n\t"  << b << "\n";

cout << " a & b gives " << x << "\n";
cout << " a | b gives " << y << "\n";

cout << "Now combine the two heap based Rects "
                    "to get a temporary\n";
cout << "\tand then invoke some functions "
                    "on that temporary\n";

// How's this for an ugly code fragment?
```
F `if(((*never_say_die) & (*comes_and_goes)).Valid())`
```
        cout <<
            "The two heap based rectangles must intersect\n";
else cout <<
            "The two heap based rectangles don't seem to"
                    " intersect\n";

cout << "Get rid of one of those heap based"
            " Rectangles\n";
```

G `delete comes_and_goes;`
```
cout << "Make some static rectangles.\n";
```

H
```
static Rectangle c(20,-20,100,120,Rectangle::BLUE);
static Rectangle d(40,40,20,30,Rectangle::YELLOW);
```

I
```
if(c.Valid()) cout << "The rectangle " << c
                                    << " is valid\n";
else cout << "The rectangle " << c << " is NOT valid\n";

if(d.Valid()) cout << "The rectangle " << d
                                    << " is valid\n";
else cout << "The rectangle " << d << " is NOT valid\n";
```

J
```
do {
        cout << "Enter a VALID rectangle\n";
        cout.flush();
        cin >> x;
        }
while (!x.Valid());

cout << "The rectangle read was " << x << "\n";

cout << "Enter a Point\n"; cout.flush();

Point p;
cin >> p;

if(x.Contains(p)) cout << "Rectangle " << x
            << " contains point "  << p << "\n";
else cout << "Rectangle " << x
```

```
                         << " does NOT contain point "  << p << "\n";
        return 0;
    K
}   L
```

The outputs are shown below with a commentary. (These outputs represent the interleaved output to streams cout and cerr; the cerr outputs have had * placed at the start of line to make them distinct.)

A
```
* Create an instance of Rectangle at 0x63a8
```
B
```
 * Entry to function main
```

(Note how the global variable is created prior to entry to the main program.)

```
Tests of Rectangle class
The global rectangle variable was ((0, 0)(512, 340) 30)
```

C
```
*Create an instance of Rectangle at 0xf7fffa98
*Create an instance of Rectangle at 0xf7fffaac
*Create an instance of Rectangle at 0xf7fffac0
*Create an instance of Rectangle at 0xf7fffad4
```

(Space may have been reserved in the stack frame on entry, but objects aren't actually constructed until the flow of control reaches their definitions.)

```
The first rectangles, array palette
palette[0]((20, 20)(40, 60) 205)
palette[1]((40, 20)(60, 60) 69)
palette[2]((60, 20)(80, 60) 341)
palette[3]((80, 20)(100, 60) 409)
How about some dynamic rectangles
```
D
```
*Create an instance of Rectangle at 0x91a8
*Create an instance of Rectangle at 0x91e0
and some more automatics, a,b,x,y and z
*Create an instance of Rectangle at 0xf7fffa78
*Create an instance of Rectangle at 0xf7fffa64
*Create an instance of Rectangle at 0xf7fffa50
*Create an instance of Rectangle at 0xf7fffa3c
*Create an instance of Rectangle at 0xf7fffa28
```

(Addresses of heap and stack variables are usually easily distinguished, as in these examples.)

E Do & and | operations on a, b modifying x and y
```
*Object at 0xf7fffa78 is performing Rectangle::operator&
*Create an instance of Rectangle at 0xf7fffa14
```

```
*Object at 0xf7fffa78 finished Rectangle::operator&
*Create an instance of Rectangle at 0xf7fffa00
```

(Here are tracers from the operator functions; note the creation of temporaries on the stack that hold the results being returned from the function. These temporary objects are destroyed a little later – at **E'**.)

```
Rectangle a:
        ((0, 0)(100, 100) 33)
Rectangle b:
        ((40, 40)(200, 60) 30)
 a & b gives ((40, 40)(100, 60) 33)
 a | b gives ((0, 0)(200, 100) 30)
Now combine the two heap based Rects to get a temporary
        and then invoke some functions on that temporary
```
E'
```
*Destroy the Rectangle at 0xf7fffa00
*Destroy the Rectangle at 0xf7fffa14
```

F
```
*Object at 0x91a8 is performing Rectangle::operator&
*Create an instance of Rectangle at 0xf7fff9ec
*Object at 0x91a8 finished Rectangle::operator&
*Object at 0xf7fff9ec is performing Rectangle::Valid()
*Object at 0xf7fff9ec finished Rectangle::Valid()
*Destroy the Rectangle at 0xf7fff9ec
```

(These entries from the trace log follow the execution of the statement

```
if(((*never_say_die) & (*comes_and_goes)).Valid())
```

The heap-based object at 0x91a8 starts executing its operator&() function; a temporary stack-based object is created at 0xf7fff9ec for the result; the temporary object executes its valid() function; finally, the temporary is destroyed.)

```
The two heap based rectangles don't seem to intersect
```
G
```
Get rid of one of those heap based Rectangles
*Destroy the Rectangle at 0x91e0
```

(The destructor for comes_and_goes is invoked for the explicit delete operation.)

```
Make some static rectangles.
```
H
```
*Create an instance of Rectangle at 0x67a4
*Create an instance of Rectangle at 0x67b8
```

(Statics are allocated space in the program's static data segment, and again have addresses that are distinct from both heap and stack variables. Although space is permanently allocated, these static variables are not constructed until the appropriate statements are reached. These statics outlive the main program. They are final destroyed at **H'**.)

I

```
*Object at 0x67a4 is performing Rectangle::Valid()
*Object at 0x67a4 finished Rectangle::Valid()
The rectangle ((-20, 20)(120, 100) 409) is valid
*Object at 0x67b8 is performing Rectangle::Valid()
*Object at 0x67b8 finished Rectangle::Valid()
The rectangle ((40, 40)(30, 20) 69) is NOT valid
```

(Several more invocations of assorted member functions for various objects.)

J

```
Enter a VALID rectangle
((20,40)(607,532) 33)
```

(Object input, then more tests of member functions.)

```
*Object at 0xf7fffa50 is performing Rectangle::Valid()
*Object at 0xf7fffa50 finished Rectangle::Valid()
The rectangle read was ((20, 40)(607, 532) 33)
Enter a Point
(67,89)
Rectangle ((20, 40)(607, 532) 33) contains point (67, 89)
```

K

```
*Destroy the Rectangle at 0xf7fffa28
*Destroy the Rectangle at 0xf7fffa3c
*Destroy the Rectangle at 0xf7fffa50
*Destroy the Rectangle at 0xf7fffa64
*Destroy the Rectangle at 0xf7fffa78
*Destroy the Rectangle at 0xf7fffad4
*Destroy the Rectangle at 0xf7fffac0
*Destroy the Rectangle at 0xf7fffaac
*Destroy the Rectangle at 0xf7fffa98
Exit from main
```

(On exit from main, all the local scope, non-static variables get destroyed.)

L (H')

```
*Destroy the Rectangle at 0x67b8
*Destroy the Rectangle at 0x67a4
*Destroy the Rectangle at 0x63a8
```

(This is the code executed after exit from main. Final tidying up operations are performed. Class instances allocated in the static data segment are destroyed. These include the global variable, and the statics – constructed at **H**, that are local to main() but which have outlived main().)

Traces such as those illustrated are a valuable aid to beginners at C++ who want to understand when temporary objects get created, how they are manipulated, and when they are freed. More generally, such traces can serve as a debugging aid.

Automatic analysis and display of traces

It is not really practical to read such a trace from anything other than a small demonstration program. The detail overwhelms and it is impossible to follow the events relating to a particular object. Instead of producing semi-readable text to the terminal, real tracers should direct more concise outputs to a file. After the traced program has run to completion, the data in the log file can be analyzed. A general purpose program can read trace files and provide a graphic display that illustrates the creation and destruction of objects and the processes by which they interact. Such display programs provide a helpful way of visualizing program behaviour and can help to find bugs (e.g. the failure to free data structures as in the program above).

5.4 A "CLUSTER" OF RELATED CLASSES: LINKEDLIST AND FRIENDS

The next group of classes form a very small "cluster", i.e. a group of classes designed to be used together. The various classes in a cluster may have many interdependencies, while presenting a much narrower interface to clients. The classes in this cluster are "LinkedList", "Link", and "LinkedListIterator". This example is revisited, with different versions being presented in the chapter on inheritance, and in the chapter on advanced C++ where "template" classes are introduced.

These classes illustrate the basics of "container classes". Inspired by Smalltalk, and guided by Gorlen's early work on the National Institute of Health's C++ class library (NIHCL), every C++ compiler product is now packaged with an extensive collection of "container classes". No C++ programmer ever *needs* to implement a "stack", or a "dynamic array", or "queue"; all such standard data structures and associated manipulation routines have been packaged and provided as properly encapsulated classes. Some of the class libraries contain much more elaborate containers – such as Btrees (a Btree class is provided in the Borland class library on the PCs and in some other libraries). Although these classes have all been written many times before, they still constitute *useful exercises for learning basics of C++*.

5.4.1 Class LinkedList

This class LinkedList is a doubly-linked list. Data items can be added at either end. Data items can be removed from arbitrary positions in the list. A data item is simply a void* pointer. Links are completely separate from the data items that are to be held on a list. Figure 5.6 illustrates the basic organization of these linked lists.

```
class Link;                    // "Forward reference"
```

class LinkedList
```
class LinkedList {
public:
```

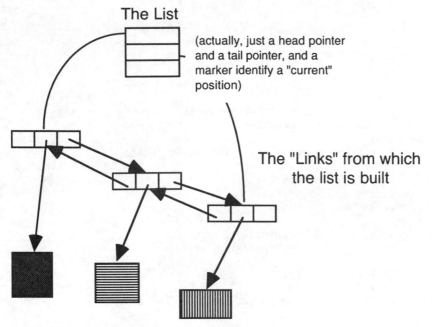

The List

(actually, just a head pointer
and a tail pointer, and a
marker identify a "current"
position)

The "Links" from which
the list is built

Assorted data items "stored"
in the list

Figure 5.6 An instance of class LinkedList has pointers to the head and tail of
a doubly-linked list of "Link" objects. In addition to
forward/backward links, the "Link" objects hold pointers to the data
items stored in the LinkedList container object.

```
    LinkedList();
    ~LinkedList();

    // Asking about list
    short           Length() const;
    ...
    // Put data in list
    void            Append(void* datum);
    void            Prepend(void* datum);

    void            PutBeforeMark(void* datum);
    void            PutAfterMark(void* datum);

// Removing stuff from list
    void*           RemoveFirstItem();
    ...

    friend class LinkedListIterator;
private:
    Link*           Head() const;
    Link*           Tail() const;
```

```
                    Link*          fhead;
                    Link*          ftail;
                    Link*          fmark;
              };
```

Class   Link
```
              class Link {
                    friend class LinkedListIterator;
                    friend class LinkedList;
              private:
                    Link(Link* prev, void* datum, Link* next);
                    Link(const Link& l);
                    Link*          Prev();
                    Link*          Next();
                    void*          Datum();

                    void           SetPrev(Link* newl);
                    void           SetNext(Link* newl);

                    Link*          fp;
                    Link*          fn;
                    void*          fd;
              };
```

The initial declaration "class Link" is, in Pascal terms, a "forward reference". It simply introduces the name Link so allowing the declaration of Link* data members in class LinkedList.

Make   a   habit   of declaring   access functions   as const

Class LinkedList has the typical public interface of constructors, destructors, access functions, and modifier procedures. There would be no purpose in defining const instances of LinkedList. Nevertheless, it is usually worthwhile declaring the access functions as being const – it helps distinguish these functions as simple accessors.

Pointers to data items are added to the list as void* pointers. Data retrieved from, or removed from, the list are also returned as void* pointers. Explicit type cast operations will be required in all code that uses these lists.

Friendly   classes

If a list iterator is to wander down a list, it must be able to follow the connectors from one link to the next and, consequently, is justified in having privileged access to Links. Similarly, if a LinkedList needs to append an item, it needs to create a new Link and then thread it onto the Link that was previously at the end of its chain, so it too needs to change fields within Link objects. The declarations of class LinkedList, and of class Link, both nominate other classes as *friends*. Class LinkedListIterator is a friend of both Link and LinkedList; class LinkedList is also nominated as a friend of Link. A data member such as LinkedList's head pointer fhead should not be accessed by ordinary clients but is required to be accessed in the code of class LinkedListIterator. Similarly, the various data member pointers in a Link have to be accessed by the LinkedList and LinkedListIterator classes. The friend relations define these restricted access privileges on a per class basis; any method of class LinkedList can contain code that accesses or changes Links. This usage is a small generalization of the previous friend declarations that nominated specific functions as having privileged access.

Because class `LinkedList` has been declared a friend of class `Link`, the code of a method of `LinkedList` could contain statements that directly modify a `Link`'s `fn`, `fp`, or `fd` data members. Such usage would be inappropriate. All code that changes data members should belong to the class. The privileged access being granted is really the use of private access and modification routines. Thus, in class `LinkedList`, private routines `Head()` and `Tail()` give access to the head and tail pointers. The code for class `LinkedListIterator` should use such functions rather than directly accessing the data members (since these functions will be "inlined" anyway, there will be no cost in terms of performance). Similarly, data members of class `Link` should only be changed via the private procedures such as `SetNext()`.

Use   private access   functions and   modifier routines

Class `Link` is unusual. It nominates its two friends, but declares that all other details of itself are private. Even the constructors are private. This is a useful defence mechanism. The author of this cluster of classes wanted to be certain that there would be no loopholes through which client programmers could accidentally get at, change, and probably disrupt the workings of the linked lists provided. Class `Link` is very private – instances can only be created and accessed by the code provided in class `LinkedList` and class `LinkedListIterator`. Client programmers may know of the existence of `Link`s but have no (legitimate) way of manipulating them.

A   very   private class

5.4.2 Planning interactions among objects

The earlier example classes like `Point`, `Player` etc. really do little more than encapsulate some data and provide services related to the data owned by class instances. Thus, an instance of class `Rectangle` owns some `Point` coordinates and a shading code, and will update these data, perform graphic output, and report values as requested. With the `LinkedList` cluster, one begins to get interactions among objects; for example, `LinkedList` objects create, destroy, and communicate with `Link` objects. As more complex classes are encountered, the networks of interactions among objects become more complicated. These networks of interactions must be diagrammed as part of a program's documentation.

Object-object interactions

The handling of an append operation of a linked list serves as a first example of the diagramming of such interactions. The `Append()` code, in file pdt/LinkedList.C, is:

```
void LinkedList::Append(void* datum)
{
    // create new link, its previous item is
    // whatever is current tail, its next is NULL
    // because its being appended to end of chain
    Link*  alink = new Link(ftail, datum,(Link*) 0);
    // update pointers
    // current last item (if there is one) now
    // has a next
    if(ftail != (Link*) 0)
            ftail->SetNext(alink);
    // tail should be changed
```

```
        ftail = alink;

        // and, maybe, head should be changed (if list was empty)
        if(fhead == (Link*) 0)
                fhead = alink;
        // set mark to reference this item
        fmark = alink;
}
```

The Append() member function creates a new Link object, initializing it with pointers to the data element that is to be stored and the Link that was previously the end of the chain. The Link object that was the old end of the chain is requested to update its fnext pointer. Then, the LinkedList object updates its own ftail, fhead, and fmark pointers .

Figure 5.7 shows a time sequence, progressing from the top to the bottom of the page, with control passing between objects as the result of method invocations.

5.4.3 Class LinkedList: Implementation details

Class LinkedList provides the following data storage and retrieval functions:

```
        void            Append(void* datum);   // add after last
        void            Prepend(void* datum); // add before first

        void            PutBeforeMark(void* datum);
        void            PutAfterMark(void* datum);

        void*           RemoveFirstItem();
        void*           RemoveLastItem();
        void*           RemoveMarkedItem();
```

In addition to functions for accessing the head and tail of the list (both to add and remove data) class LinkedList supports the notion of a "marker" with operations to add and remove data at the current position of this marker. The marker is set to the start of the list by operations that act at the head, or to the end of the list by operations that act on the tail, or to an arbitrary point in the list by actions that involve finding a particular entry.

As well as being able to Find() a data item, clients can query the length of the list or access the data elements at particular positions in a list. The functions available are:

```
        short           Length() const;
        Boolean         Find(const void* queryitem);
        void*           FirstItem();
        void*           LastItem();
        void*           MarkedItem();
```

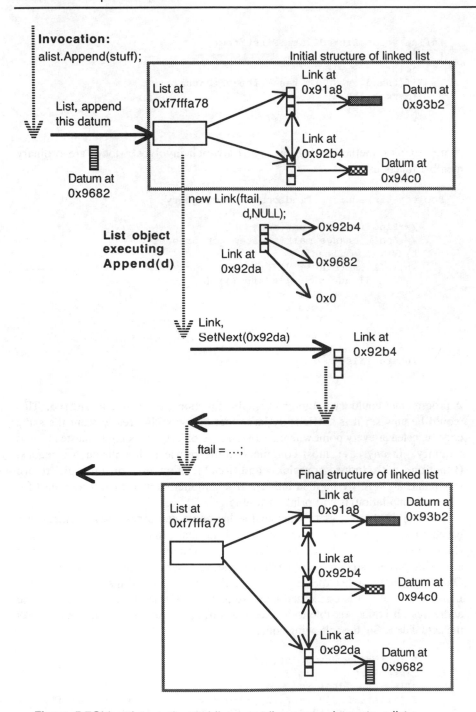

Figure 5.7 Object interactions while appending a new datum to a list.

Only `Length()` is const. The other methods all change the position of the marker for the "current element". Simple methods such as `FirstItem()` are defined as `inline`:

```
inline void* LinkedList::FirstItem()
{
    void*   res = NULL;
    if(fhead != NULL) res = fhead->Datum();
    fmark = fhead;
    return res;
}
```

More complex methods, such as `Find()` which involves a loop, are ordinary member functions:

```
Boolean LinkedList::Find(const void* query)
{
    // Find item and set mark
    // Don't change mark if item not present
    Link*   l = fhead;
    while(l != (Link*) 0) {
            if(query == l->Datum()) {
                    fmark = l;
                    return true;
                    }
            l = l->Next();
            }
    return false;
}
```

A programmer could try defining a complex function like `Find()` as `inline`. This would be unwise; it is sufficiently large that one wouldn't really want the same code repeated at every point where a `LinkedList` object was being requested to find an entry. In any case, most compilers would refuse to honour such a request (functions with loops, complex conditional statements, or multiple return statements are rarely compiled inline). If a compiler cannot honour a request for `inline` compilation, it will print a warning.

In `Find()`, the test as to whether the list contained a particular element was based on a comparison of addresses. That is all that can be done in this situation. Data elements are simply `void*` pointers. The structure of the data elements that are being placed is completely unknown and no other comparisons can be made. One might have a list of `Points`, or a list of `char*` strings – as far as the `LinkedList` object is concerned they are just data elements that exist at specific addresses. It is not possible to invoke a specific comparison function that checks the actual data. So, the following code:

```
Point* p1 = new Point(2,2);
LinkedList    alist;
alist.Append(p1);
if(alist.Find(Point(2,2))) cout << "(2,2) is in list\n";
else cout << "(2,2) is not in list\n";
```

will report that "(2,2) is not in the list". The report is correct – because it is a different instance of (2,2) that is in the list from the instance that was used in the `Find()` request. In the next chapter, another version of `LinkedList` will be

presented in which it is possible to have the Find() operation compare structural
equality rather than address equality.

The member functions of LinkedList are all simple and don't merit much *LinkedList's*
further discussion. The constructor simply sets all data fields to NULL. The *constructor and*
destructor has to tidy away any chain of Links: *destructor*

```
LinkedList::LinkedList()
{
    fhead = ftail = fmark = (Link*) 0;
}

LinkedList::~LinkedList()
{
    // only destroy links, do not do anything about
    // data structures pointed to by fields of Links
    Link*  l = fhead;
    Link*  temp = (Link*) 0;
    while(l != (Link*) 0) {
            temp = l;
            l = l->Next();
            delete temp;
            }
}
```

Again, in this situation, it is not possible to do anything about the data elements
referenced by the Links. This will change in the version of LinkedList presented
in the next chapter.

5.4.4 Class Link

Links themselves are very simple. All member functions are defined as inline:

```
inline Link::Link(Link* prev, void* datum, Link* next) :
    fp(prev), fd(datum), fn(next)
{
}

inline Link* Link::Prev()
{
    return fp;
}

inline Link* Link::Next()
{
    return fn;
}

inline void* Link::Datum()
{
    return fd;
}

inline void Link::SetPrev(Link* newl)
```

```
{
    fp = newl;
}

inline void Link::SetNext(Link* newl)
{
    fn = newl;
}
```

The copy constructor, Link(const Link& orig), could be written to copy the data members from the existing Link orig to the new Link; but there is no real need for this routine and it could be left undefined. It is declared as a private function so as to allow the compiler to prevent attempts to create copies of Links.

Caution on "inlining" all member functions of a class

Note, with more complex classes that have "virtual functions" and other more advanced features, there may be problems about having all the member functions declared as inline. A compiler needs some place to store information describing a class, information such as the "virtual table" used for accessing "virtual functions". Usually, a compiler will place this information in the load module that contains the first normal, i.e. non-inline, member function. If all member functions are inline, a compiler may become confused as to where it is to store this information. In such a situation, redundant copies of the same information may get replicated in many files. These problems are explored further by Shapiro in *A C++ Toolkit* and by Meyers in *Effective C++*. The problems won't arise for simple classes like Link, Point, Rectangle etc. and in these cases it is quite appropriate to declare all member functions as being inline provided that they don't involve loops.

The file LinkedListTest.C, in the "pdt" subdirectory, contains a test program that exercises most of the facilities provided by class LinkedList.

```
#include <iostream.h>
#include <stdlib.h>

const short SSIZE = 35;

#include "LinkedList.h"
static char*        shelf[SSIZE] = {
            "Effective C++",
            "Algorithms in C++",
            "OOPSLA 91 Proceedings",
            ...
            ...
            "The Annotated C++ Reference Manual"
            };

int main(int, char**)
{
    // Run a few tests on that Linked List class
    // using C-strings as the data types.
    LinkedList      thelist;

    thelist.Append(shelf[2]);
    thelist.Append(shelf[27]);
    thelist.Append(shelf[19]);
```

```
      thelist.Prepend(shelf[30]);


      cout << "The list has been made and now has " <<
            thelist.Length() << " items\n";
      PrintAList(thelist);

      if(thelist.Find(shelf[18])) cout << shelf[18]
            << " is in the list\n";
      else { cout << shelf[18] << " is NOT in the list\n";
            cout << "Adding it to the list\n";
            thelist.Append(shelf[18]);
            }

      …

      cout << "Removing " << (char*)
                        thelist.RemoveMarkedItem() << "\n";

      thelist.PutBeforeMark(shelf[7]);
      PrintAList(thelist);
      …
      …
      return 0;
}
```

The program creates an automatic (stack-based) instance of `LinkedList` and then loads it with some data. The data are C-strings, represented as `char*` pointers. Member functions, such as `Append()`, expect `void*` arguments – `char*` arguments are quite satisfactory. A member function such as `RemoveMarkedItem()` returns a `void*`; this must be explicitly type cast back to `char*` before it can be used.

5.4.5 Class LinkedListIterator

The routine `PrintAList()` uses an instance of class `LinkedListIterator`. Iterators are a "Lispish" notion; Lisp has long provided functions such as `mapcar` that iterate along a list, processing each element in turn. Since the early 1980s, iterators seem to have become an expected feature of a language supporting abstract data types.

LinkedList
Iterator

It would be quite simple to add a little extra functionality to the original `LinkedList` class and obtain a mechanism for iterating along a list examining each element in turn. The mechanism would extend the existing idea of a marker. Extra member functions would be needed. One would need a member, `Start()`, to set the mark at the start of the list. Another function, `Next()`, would move the mark to the next member and return a Boolean value – true if there was a next member, otherwise false. The access function `MarkedItem()` would be used to access the current item.

Although workable, an approach using a mark associated with the `LinkedList` class has some disadvantages. It is a little fussy – the client has to be aware of three functions: `Start()`, `Next()`, `MarkedItem()`. Only one list iterator can be active. So there are advantages in having a separate way of defining an iteration along a list.

The class LinkedListIterator is simple:

```
class LinkedListIterator {
public:
    LinkedListIterator(const LinkedList* alist,
                                    Boolean Backwards = false);
    void*           Next();
private:
    Link*           fl;
    Boolean         fBackwards;
};

inline LinkedListIterator::LinkedListIterator(const
            LinkedList* alist,Boolean Backwards)
{
    fBackwards = Backwards;
    fl = fBackwards ? alist->Tail() : alist->Head();
}

inline void* LinkedListIterator::Next()
{
    void*   res = NULL;
    if(fl != NULL) {
        res = fl->Datum();
        fl = fBackwards ? fl->Prev() : fl->Next();
        }
    return res;
}
```

An instance of class LinkedListIterator gets a Link pointer, fl, initialized with the head pointer (or tail pointer if it is a backwards iterator) of the list through which it is to iterate. Its Next() method returns the current datum and moves its fl pointer forwards (or backwards) through the list.

Use of the iterator class is illustrated by the code of procedure PrintAList():

```
void PrintAList(const LinkedList& alist)
{
    LinkedListIterator      Iter(&alist);
    void*                   Ptr;
    short                   i = 0;
    cout << "Contents of list:\n";
    while(Ptr = Iter.Next())
            cout << "\t" << i++ << "\t" << (char*)Ptr << "\n";
    cout << "(end of list)\n";
}
```

This code uses a stack-based instance of the iterator class which gets initialized with a pointer to the list (from which its constructor extracts the head pointer to the first Link). By default, a "forward" iterator is created. The iterator is used in the while loop where it returns successive elements that are then used in the print statements.

Class libraries with container classes normally provide appropriate iterators for all container types. So, if the library has binary-trees, one probably has a battery of iterators for various in-order, post-order etc. iterators. The coding of iterators becomes more complex if there is the possibility of the list being changed while iteration is in progress. Code in major class libraries, e.g. code in the ET++ library, should be used as an example of how to handle such intricacies.

Obscure uses of overloaded operators seem to be quite common in the context of iterator classes. Instead of a simply named member function like `Next()`, the iteration may be controlled by invocation of an overloaded `operator++()`. While it is all the same to the author of the iterator class, clients are generally better served when the functionality is clearly named.

5.4.6 Enhancement of the LinkedList class cluster

There will be overheads associated with the use of the `LinkedList` class. A significant part of this overhead will be contributed by the cost of creating and destroying the `Links`. These are just small data structures, but they have to be allocated on the heap. The heap-based allocation involves overheads in terms of both time and space.

Controlling space allocation: class specific new and delete operators

In C++, `new` and `delete` are both operators. Like any other operator, their meaning can be changed on a class specific basis. It would be possible for the author of the `LinkedList` cluster to take control of allocation and freeing of `Links`. For example, an array sufficient for 5000 `Links` could be pre-allocated. A `new` operator would be defined for class `Link` that would take the first unused entry in this array; the `delete` operator would mark a `Link` in the array as being reusable. Such complexities need not concern beginners. However, this option for additional control over storage allocation should be remembered because it does become relevant in more advanced work. Most C++ texts provide very similar examples that illustrate how to allocate space from a static array.

5.5 ASSOC: A CLASS "EMPLOYING A SUBCONTRACTOR"

Classes such as `LinkedList` provide standard capabilities that are required again and again. A `LinkedList` object manages a small storage depository and can respond to a request to keep items, or to find whether it is storing an item, or to give back an item that it had in store. Other classes that perform more complex tasks may need some storage. Rather than take on storage management responsibilities for themselves, it may be advantageous for such classes to "subcontract" their storage management requirements to a `LinkedList` object.

The next example class uses a `LinkedList` object as such a subcontractor. The class is "Assoc". It is loosely modelled on the Lisp concept of an association list. In Lisp, an association list is most commonly used to maintain a set of key/value pairs. Lisp's `assoc` function looks up a key in such a list and returns the corresponding value (or NIL if the key is not present). One can use these lists in a way which does not distinguish between key and value – either part of a pair can be used as search probe and will return the matching component. A typical early Lisp exercise uses such association lists to represent a dictionary that is used in extremely naive word-based approaches to English↔French translation.

The Assoc class implemented here is quite specific. It is intended to maintain a set of name↔number pairs. The example code given with the Assoc class illustrates its use in the construction of a table of pairs specifying a country and its international dialling code.

```
class Assoc {
// Class that maintains a form of Lisp "association list"
// It illustrates the use of composition; this class is
// really little more than a repackaged LinkedList.
// This version is written specifically to support
// associations between char* strings and long integers
public:
    // Constructors and destructors
    Assoc();
    ~Assoc();
    // Access functions, declare as const even though
    // don't intend to make the class truly const aware
    Boolean Contains(char*) const;
    Boolean Contains(long) const;
    Boolean Contains(char*, long) const;
    const char* Lookup(long) const;
    long Lookup(char*) const;
    short Size() const;
    // Modifier procedures
    void Add(char*, long);
    void Remove(char*, long);
    friend ostream& operator<<(ostream& os, const Assoc& a);
private:
    LinkedList      fl;
};
```

A LinkedList data member or a LinkedList* data member?

Class Assoc has an internal instance of class LinkedList (i.e. a sort of "in-house" subcontractor). (Of course, this instance of LinkedList comprises only its three pointer data members; all the actual Links forming any list will exist separately in the heap.) It would be possible to have a LinkedList* data member pointing to an instance of class LinkedList that has a separate existence in the heap (i.e. an externally located subcontractor). In this case, it is more appropriate to a have an "expanded" instance of LinkedList rather than a pointer.

Objects in these classes are small. Any object allocated in the heap involves some storage overhead. This storage overhead will depend on the implementation, but will be several bytes per heap-based structure. With these small classes, the storage overhead (and time overhead on initial creation) can become quite significant if the parts are separately allocated. In other cases, where the classes have many more data members, the instances will be larger, the overhead costs from separate allocation will be proportionately smaller and there may be advantages for the memory management system to work with several medium sized objects rather than a few large objects. Then, separate allocation of the "subcontractor" object might be appropriate.

Class Assoc does not advertise to clients the fact that it employs a LinkedList for its storage management. It provides its own unique interface with Add(), and Remove() procedures for changing data and a whole series of overloaded access functions for reading data. Thus, there are three overloaded Boolean Contains() functions; one checks whether a name is known, the next whether a numeric identifier is known, and the last determines whether a particular name number combination has been stored.

The two Lookup() functions return information (these are the functions that most closely correspond to the original assoc function of Lisp). Note that the function returning a name returns it as a const char* i.e. a pointer to a constant string. If a simple char* pointer were returned, there would be a loophole through which stored data could be changed from arbitrary points in a program. A client "intent on fraud" would obtain the address of the stored string and be able to write different information into that string. By declaring the return type of the Lookup() function as const, the compiler will try to prevent such actions (though it can't prevent a programmer "casting away" the "const" aspect of the pointer).

The implementation of class Assoc is straightforward. It uses a struct (AssocItem) to store (integer, name) pairs. The constructor is empty; it is not really necessary to define one but it is usually worth defining because this simplifies any subsequent extensions. The destructor tidies up, freeing any remaining AssocItems.

```
struct AssocItem {
    long    fint;
    char*   fstr;
};


Assoc::Assoc() {}

Assoc::~Assoc()
{
//  The list will free its links but not the data
//  So, here run along the list throwing away the data items
    LinkedListIterator    Iter(&fl);
    AssocItem*            ptr;
    while(ptr = (AssocItem*)Iter.Next()) {
            delete [] (ptr->fstr);
            delete ptr;
            }
//  Rest of structures should go when the destructor for fl
//  gets called.
    }
```

As this version employs a LinkedList of void* pointers, the LinkedList can only be responsible for storage. Searches based on content have to be the responsibility of the Assoc class. Consequently, a method such as Assoc:: Contains() uses an instance of class LinkedListIterator to retrieve each stored

AssocItem in turn, once retrieved the item is examined to determine whether it represents the desired data.

```
Boolean Assoc::Contains(char* probe) const
{
// If find a data item with a name equal to probe return true
    LinkedListIterator    Iter(&fl);
    AssocItem*            ptr;
    while(ptr = (AssocItem*) Iter.Next()) {
            char*            ref = ptr->fstr;
            if(0 == ::strcmp(probe,ref)) return true;
            }
    return false;
}
```

The file AssocTest.C, also in subdirectory "pdt", contains code exercising the capabilities of class Assoc. A compiler will come up with a series of warnings about the code of Assoc. Note that instances of class LinkedListIterator are initialized with *pointers* to LinkedLists. As Assoc has an internal data member fl of type LinkedList, the address of this member must be taken and passed as an argument. Here, the usage is quite safe and reasonable. However, in general, taking the address of a data member is a little unwise and the compiler is correct in reporting its concerns.

5.6 DATA CHARACTERIZING A CLASS AND SHARED BY ALL ITS INSTANCES

Static class members

There is another feature in C++, not illustrated in any of the earlier examples, that is important even to beginners employing simple programmer-defined data types. This feature – "static" members of a class – provides a mechanism for defining and manipulating properties that are characteristic of all instances of a class.

Imagine a program that displays molecular models (three-dimensional models of chemical structures) on a colour display. Different atom types need to be displayed in different colours. There are a few conventions; usually carbons are black, oxygen atoms are blue, sulphurs yellow, and often halogens are various shades of green. The user would naturally require control over the specific shades employed.

Molecular modelling and display programs are typically used by chemists involved in drug design. They need to be able to visualize and compare the shapes of different molecules (shape is often related to pharmacological activity). Such programs work with many molecules simultaneously – displaying each molecule in a separate window, or in a small pane within some larger window. The molecules will be instances of some class:

A Chemical Structure class with each structure having its own colour scheme

```
class ChemicalStructure {
public:
    ...
    void          Draw();
    void          SetColorsFromPalette(); // ???
    ...
```

```
private:
    FixedPoint      fcoords[MAXATOMS][3];
    Atomtype        ftypes[MAXATOMS];
    …
    RGBColor        fcolors[NATOMTYPES];   // ??? or a global??
};
```

The member functions of the class would provide the means to edit and display structures and choose colours for the different atom types. The data members would define the forms of individual molecules; there would be arrays with a few hundred entries defining the coordinates and types of the constituent atoms, and a small array defining the colours of the ten or so different types of atoms that can occur in these structures.

However, there is a major problem with the class declaration shown above. It makes the colours unique to each molecule. This would be unsatisfactory. Different molecules being compared by the user could be displayed using different shades and this would be a source of considerable confusion.

An alternative design would remove, from class ChemicalStructure, the data member fcolors[] and the member function SetColorsFromPalette(). These would be made globals:

A Chemical Structure class that makes use of global data for the colours

```
RGBColor    gcolors[NATOMTYPES];    // ??? only for use by
void        SetColorsFromPalette(); // ???ChemicalStructures

class ChemicalStructure {
public:
    …
    void        Draw();
    …
private:
    FixedPoint      fcoords[MAXATOMS][3];
    Atomtype        ftypes[MAXATOMS];
    …

};
```

This arrangement would have the required semantics. There is now one table of colours that can be used by all instances of ChemicalStructure.

Unfortunately, this solution is unsatisfactory from a software engineering perspective. Two new names have now appeared to "pollute" the global name space. There is nothing to associate these names with the class Chemical-Structure. It is quite likely that there will be other tables of colours, and other functions needed to manipulate these tables; these would be the colours used for screen backgrounds, borders, scroll-bars etc. A selection of many colour tables, and several variants of the function SetColorsFromPalette(), would be likely to lead to maintenance problems.

Avoid the "pollution" of global name space

In situations such as this, one requires a kind of global variable or global function that really belongs to a class. This requirement is satisfied by static members. Class ChemicalStructure would be declared:

Problems resolved by use of "static" data members

```
class ChemicalStructure {
public:
    …
    void          Draw();
    static void   SetColorsFromPalette();
    …
private:
    FixedPoint    fcoords[MAXATOMS][3];
    Atomtype      ftypes[MAXATOMS];
    …
    static RGBColor        fcolors[NATOMTYPES];
};
```

The `static` qualifier on data member `fcolors` identifies this array as being a "class data member" rather than an "instance data member". (There are suggestions that the distinction should be obvious in the name of the data member. Names with 'f', like `fcoords`, identify instance data members; names beginning with 'g' are globals. Possibly, static data members should have names beginning 'c', for class, or 's' for static. Later examples will generally use 's' names.) The declaration specifies that there is only one such colour table; further, this table is private to class `ChemicalStructure` and can only be accessed by member functions of this class.

Static member
functions

Static data members have a fairly simple interpretation – they are sort of global variables, but ones that belong to a class. A meaning for "static member functions" may be less obvious.

It is actually quite simple. "Static member functions", such as function `SetColorsFromPalette()` in the revised class `ChemicalStructure`, are functions that operate only on the static data members of a class. In the example, this function would be invoked when a user selected "Set Colours…" from some menu. The function would display a dialog that allows the user to change each of the entries in the unique `ChemicalStructure::fcolors` array.

The class declaration specifies that an array exists, this array has entries of type `RGBcolor` and is named `ChemicalStructure::fcolors`. The declaration does not define the array. No space has been allocated for this array of colours. The actual array must be defined elsewhere. Typically, it will be defined in the file that contains the definitions of most of the (non-inline) member functions of class `ChemicalStructure`. The definitions would appear something like:

Defining a static
data member

```
RGBcolor ChemicalStructure::fcolors[NATOMTYPES];
// In practice, the array elements would be initialized with
// default colours

void ChemicalStructure::Draw()
{
    …
}
```

(The array must not be *defined* as static – for that would give it the wrong scope!)

Invoking a
static member
function

The static member function, such as `SetColorsFromPalette()`, changes only the shared static data members. It does not work on an instance of the class. So code of the form:

```
ChemicalStructure aspirin;
...
aspirin.SetColorsFromPalette();
...
```

would be misleading – nothing is actually happening to structure `aspirin`. (The code is legal; it is just misleading.) For the code to be clear, there has to be a way of invoking the static member function that does not involve reference to any particular instance.

Static member functions are invoked using the class name. The preferred style for invoking `SetColorsFromPalette()` would be something like:

```
        if(menu.choice == cColors)
              ChemicalStructure::SetColorsFromPalette();
```

Unfortunately, the term `static` now has many subtly different shades of meaning. First, there is the general programming concept of "static data" as in the "static data segment". These are data structures that come into existence prior to entry to `main()` and remain in existence throughout the lifetime of a program. In C, all variables declared external to functions are placed in the static data segment as are local variables of functions that have been explicitly declared as being `static`. Those variables declared externally to any function default to having global scope; their names are available in the linking phase and can be referenced from other modules using `extern` declarations. In C, the qualifier `static` used on a non-local variable restricted it to file scope, hiding its name from the linking phase and preventing it from being referenced from other modules. Now, in addition to all previously existing meanings, `static` may convey information about the usage of a variable name in relation to a class.

Naming confusions

5.7 SUMMARY ON "PROGRAMMER-DEFINED TYPES"

Programmer-defined types are "components", or "subcontractors" or "concrete classes". These different names reflect slightly varied viewpoints.

Classes like those illustrated in this chapter are intended to be used directly. They represent almost tangible things – rectangles, assoc lists, chemical structures. .... They are not abstractions. Instances of these classes will be defined and used in programs. The term concrete seems to have been adopted to convey this reality (as opposed to the abstractness of classes such as "Message Items" or "Space Invaders" briefly illustrated in earlier chapters and explored in the next chapter). Further, these classes are not designed to be reworked and adapted to suit specialized situations.

Concrete classes

While some of these concrete classes may be essentially unique to specific applications, e.g. class `ChemicalStructure`, many others represent generally useful building blocks. A class such as `LinkedList` has potential for use in numerous different contexts within many varied programs. Thus, it constitutes a

Components

component from which one can construct more elaborate, more specialized entities. Large numbers of these components will normally be supplied with any object-oriented development environment. Component class libraries will include "collection classes" (`LinkedList` and all its more sophisticated cousins), "date/time classes", "display classes" (things like scroll-bars, icons, buttons etc. in graphic user interface systems), and possibly specialized mathematical classes (e.g. "rational numbers", "complex numbers", "indefinite length integers").

Subcontractors

While programmers may think in terms of components, designers may tend to think in terms of "subcontractors". When designing a class, a developer may decide to subcontract out a specific part of the overall work that instances of the new class are to accomplish. Thus, the designer of class `Assoc` decided not to bother with details of how data were to be stored, instead subcontracting this work to `LinkedList`. The information describing classes in a library can be viewed as "contracts" that instances of these classes can fulfil. Class `LinkedList` would advertise that it could accept contracts to save and return individual data items and to iterate through its store returning each element in turn. Several other classes in the library might accept similar contracts. The program designer reviews the contracts and selects a subcontractor. If it later proves that a particular subcontractor is too slow or otherwise unsatisfactory, a revised contract can be made some other class that has at least the same capabilities.

Declaring classes

A class will be declared in a header file (more than one class may be declared in the same file). The declaration should start with the public interface of the class – a version of the "contract" that class instances can honour. The first entries will declare constructor(s) and, if appropriate, a destructor. Then, access functions should be listed. It is *not* appropriate to have access functions for all data members. If access functions are listed for all members then the class is possibly being viewed as a glorified data structure and work that should be done by class instances is being left to the clients. It is appropriate to declare all access functions as `const`, even if no `const` instances of the class are to occur. If an access function returns a pointer to a data member, it should be returned as a `const` pointer .

Member functions and procedures that modify a class instance should then be declared. Any friends, either classes or global functions, should be listed.

The `private` keyword starts the specification of details of data members, any "housekeeping" functions, and any other specialized methods that are to be known to friends but not clients. Even friends should not change data members directly; instead, functions should be provided that can be invoked by friends to access/change data members that are otherwise private.

'inlines'

Any inline functions should be defined in a separate file that gets included, by a compiler directive, at the end of the header file. Inline functions have the normal limitations of no loops, no elaborate conditionals, no multiple return statements. With simple classes, it may be appropriate to have all member functions declared inline; but this style should be avoided in more complex classes with virtual functions.

Constructors

Most classes will have multiple, overloaded constructors. The important case of a copy constructor must be considered in all classes that manage separately allocated resources; otherwise, the default copy constructor created by the compiler can cause undesirable structure sharing. A constructor puts a newly created object into a known, safe state and may claim separate resources. Note that a two-phase initialization process is sometimes advantageous; in this style, the constructor only places an object in a standard state, resources are claimed in a separate method that can notify callers of any failure conditions.

Constructors involve an initialization step and then execution of the body of the constructor. In the initialization step, any internal instances of classes are constructed (e.g. the `Point` variables of class `Rectangle`) and any `const` or reference data members are set.

If instances of a class manage any separate resources, a destructor member *Destructor* function should be provided that frees up these resources.

Any simple class "X" should nominate, a globally defined function   *Stream i/o*

```
ostream& operator<<(ostream& os, const X&)
```

that can print instances of class X on a C++ stream. This function should be defined initially even if stream output is not going to be required in the final implementation. The function should then be included within conditional compilation directives. The ability to print class instances is a useful debugging tool during initial development.

A class whose instances are to be saved to file should provide the corresponding input function so that information can be read back from files.

```
istream& operator>>(istream&,X&)
```

It is quite common for the output function to save only the most important state data to file; other data members, e.g. links to instances of cooperating classes, are not saved. The input function must explicitly set all data members, not just those for which values are to be read from file.

Operator functions should only be defined when the semantics are obvious. *Operator* Remember, that what is obvious to you (the class author) is totally obscure to me *functions* (the client programmer). Generally, named member functions are preferable to operator functions.

In classes that manage separately allocated resources, it is almost certain that an `operator=()` assignment function will be required.

REFERENCES

Examples of programmer-defined types (or "abstract data types") are included in all *Further reading* C++ texts. There are also a number of books that explore the topics in more detail. In particular, the following are relevant:

J.S. Shapiro, *A C++ Toolkit*, Prentice Hall, 1991.
    This text presents a series of mostly fairly simple C++ classes. Because of the
    rapidity with which C++ has been evolving, some of the code already appears dated.

K.E. Gorlen, S.M. Orlow, and P.S. Plexico, *Data Abstraction and Object-Oriented Programming in C++*, John Wiley, 1990.

This text presents many ideas concerning simple abstract data types before exploring more advanced aspects of class hierarchies and class library design. The code is definitely dated.

S. Meyers, *Effective C++*, Addison-Wesley, 1992.

The C++ gotcha book. Explains why things are not quite the way you thought they were. Essential at every stage of learning (surviving?) C++.

T. Cargill, *C++ Programming Style*, Addison-Wesley, 1992.

Several cautionary tales relating to the use of C++. Read Chapter 5 before defining operator functions for your new class.

B. Meyer, *Object-Oriented Software Construction*, Prentice-Hall, 1988.

Explores the analogy of contracts and subcontractors.

C++ : *Inheritance and Virtual Functions*

6

The C++ features explored in Chapter 5 provide programmers with support for data abstraction. Coupled with the features described in Chapter 4 (that are for the most part error-suppressors that eliminate frequent bugs in C programs), these data abstraction facilities allow C programmers to continue using conventional procedural programming paradigms while guaranteeing the production of higher quality, more reliable software. For many programmers, that is all that is desired.

C++ was designed to be a practical language for a wide range of applications for programmers with a wide range of backgrounds. Many programmers making the transition from C to C++ have no desire to use object-oriented techniques and will limit themselves to the "abstract" (programmer-defined) data type subset of C++.

C++ does support an object-oriented programming style. While other languages may, by some measures, rate as "better OO languages", very few organizations can afford to make an abrupt switch to using a new programming paradigm and a different programming language. Continuity in development is also required. Currently, C++ is the most accessible route to using OO that is available to programmers.

The use of inheritance class hierarchies and polymorphism differentiates an object-oriented program from an object-based (abstract data type) program. Inheritance, the interplay between inheritance and protection, and provision for polymorphism in C++ were covered briefly in Chapter 3. In this chapter, the C++ realization of these features is further illustrated and two complete example programs are presented. Consideration of multiple inheritance, parameterized classes, and nested classes is deferred to Chapter 12.

Some aspects of inheritance are complicated in C++. The strong support for abstract data types can act to restrict flexibility. For example, in most OO languages the data members of a class are accessible in the code defined for derived subclasses. This is not normally the case in C++ where data members are private to a class. The class designer must make explicit provision for subclasses. If derived subclasses might have a legitimate reason for using private data, the class designer must provide protected access functions. Another complication in C++ relates to the redefinition of member functions in derived subclasses. C++ allows for two quite different kinds of redefinition, only one of which corresponds to the normal object-oriented idea of a subclass substituting new behaviour for an inherited behaviour. Once again, the onus of planning for an OO style falls on the designer of a base class who must choose the extent to which implementors of subclasses may refine inherited behaviours.

The next few sections illustrate some aspects of inheritance in C++. The first section deals with the simple but atypical case of a subclass simply adding behaviours to an existing class. In the example, such additions are almost orthogonal to the existing capabilities of the base class and so require no explicit provision by the author of the base class. However, in almost all cases, any extensions will require some degree of privileged access to the data members of the base class. This issue is considered in section 6.2 where the preferred solution, protected access functions planned for and provided by the base class programmer, is illustrated. Section 6.3 examines member function redefinition by subclasses and illustrates both variants that exist in C++. As illustrated, the programmer implementing the base class must plan for possible inheritance and define any member function that is to be open to change as being "virtual". Section 6.4 deals with "private inheritance", a quite distinct way of using inheritance to build new classes from existing available components.

The final three sections of this chapter include a reworking of the `LinkedList` class cluster of Chapter 5, the Space Invaders example, and an Operating Systems simulator. The Space Invaders example is a hybrid program combining normal procedural code for the "game" component and object-oriented code illustrating inheritance and polymorphism in the Invaders. The Operating Systems simulator is a more completely object-oriented program. It is a somewhat simplified discrete event simulation program addressing the type of problem for which Simula was originally devised.

6.1 STRICT INHERITANCE: EXTENDING AN EXISTING CLASS

The simplest use of inheritance is the creation of a subclass that merely *adds* features to an existing base class, but makes no other changes. For example, a program might have a class that defines a representation for organic molecules in terms of an array of atoms of defined types, and another array of bonds. The class could hide the actual representation, providing a public interface with methods like `AddAtom()`, `AddBond()`, `ContainsAsSubstructure()`, ... that allow structures to be created and manipulated in whatever ways are needed by a client application:

A base class
```
class ChemicalStructure {
public:
    enum AtomType { CARBON, HYDROGEN, OXYGEN, NITROGEN };
    ChemicalStructure();
    ...
    short       AddAtom(AtomType a = CARBON);
    short       AddBond(short a1, short a2);
    ...
    short       NumAtoms();
```

```
    AtomType      Atom(short i);
    Boolean       ContainsAsSubstructure(
                         ChemicalStructure* x);

    ...
private:
    AtomType      fatoms[kMAXSIZE];
    short         fbonds[2][kMAXSIZE];

    ...
    short         fnumatoms;
    short         fnumbonds;
    ...
};
```

Subsequently, a need might arise for a new specialization – a variant on Chemical-Structure in which each atom is characterized by its three-dimensional coordinates and which has the ability to draw itself as a "space filling model" (a collection of coloured spheres with no explicit indication of bonding). A new class ChemModel could be defined as "publicly derived" from ChemicalStructure:

```
class ChemModel : public ChemicalStructure {
public:
    ChemModel();
    int           SetCoords(short atomnum, double xa,
                            double ya, double za);
    double        X(short atomnum);
    double        Y(short atomnum);
    double        Z(short atomnum);
    void          Draw();
private:
    Point         ScreenCoords(short atomnum); // map 3d to 2d
    ...
    double        fx[kMAXSIZE],fy[kMAXSIZE],fz[kMAXSIZE];
    ...
};
```

A derived class that adds "orthogonal" features to a base class

"Public derivation" means that clients (the public) are aware that a ChemModel is a specialization of a ChemicalStructure. An instance of class ChemModel can be relied on to do anything that a ChemicalStructure can do, plus a little more. (The keyword public is required. The default inheritance mechanism in C++ is private – as explained in section 6.4.)

Public derivation

The extensions of class ChemModel are orthogonal to the existing functionality of ChemicalStructure. Their implementation requires use only of the public methods of class ChemicalStructure. For example, the method ChemModel::SetCoords() can use ChemicalStructure::NumAtoms() to check that it has been invoked to work on an atom that actually exists, and the method ChemModel::Draw() can use the public functions ChemicalStructure::NumAtoms() and ChemicalStructure::Atom() to obtain information required when drawing a structure:

```
short ChemModel::SetCoords(short atomnum, double xa,
    double ya, double za)
{
```

```
    // check valid atom …
    if(atomnum<0) return 0;
    if(atomnum>=NumAtoms()) return 0;
    …
    return 1;
}

void ChemModel::Draw()
{
    short  natoms = NumAtoms();
    for(int i =0; i < natoms ; i++) {
        ChemicalStructure::AtomType  a = Atom(i);
        switch(a) {
        case   OXYGEN:
            ::ForeColor(blue); break;
            …
            }
        …
        }
}
```

Cases like this are exceptional. Normally, there is coupling between extensions and the base class features.

6.2   PLANNING FOR EXTENSION: "PROTECTED ACCESS"

Extensions usually need access to more data than clients

Usually, code implementing extensions will require access to base class data that are not available through the public interface of the base class. For example, the data defining the bonds that exist among atoms is private in the example class ChemicalStructure presented in section 6.1. It might not be appropriate for this information to be public. Once bonds have been defined, the only other processes involving them might be coded best as methods of class ChemicalStructure (e.g. the method ContainsAsSubstructure() would involve comparing bonding arrangements in two structures). However, a different version of class ChemModel might need to draw structures showing atoms and bonds (e.g. as a "ball and stick" diagram). It would not be possible to implement this version of class ChemModel without some preplanned assistance by the author of class ChemicalStructure.

Protected access to base class data

In this sort of situation, the author of the base class needs to have made provision for subclasses to have some privileged access to private data (in the case of the example, the requirement is for access to the data defining the bonds). The required access is achieved using protected data (or, preferably, private data and protected access functions) as briefly outlined in section 3.4.

Figure 6.1 illustrates the general structure of a class designed as a "base" class. The class will have a *public interface* that defines the behaviours of class instances as seen by clients. These are the services that the class provides. So, a ChemicalStructure provides services for storing atoms and bonds and checking for containment of another ChemicalStructure; a SpaceInvader is something that can "Run" and check whether it might be damaged by gunfire; a LinkedList, as illustrated in Chapter 5, can append and remove `void*` pointers. The *protected interface* provides access to implementation details for programmers creating derived classes. The final part of a class declaration, the *private* part, specifies the types of the data resources that will be owned by each instance of the class; there may also be some private implementation functions.

C++ "base" class.

public interface:
functions that define the "services" that a class instance provides to clients.

protected interface:
functions that provide programmers of derived classes with access to the implementation of this class.

private data and functions:
data and specialized implementation functions

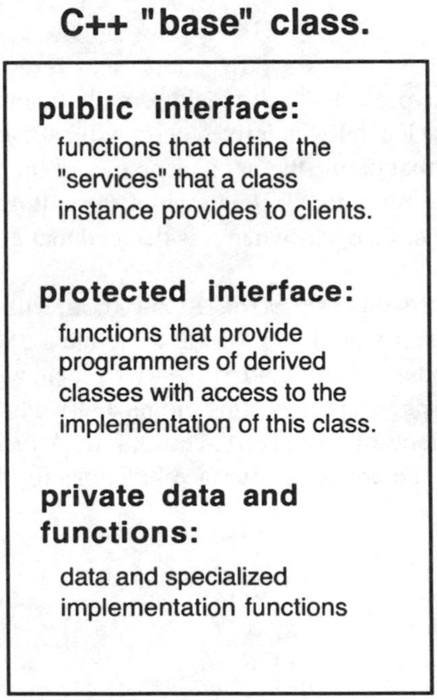

Figure 6.1   A C++ class designed as a base class for a set of specialized subclasses will provide a public interface for clients, a protected interface providing access to implementation for programmers of derived classes, and a private component describing data members.

```
class ChemicalStructure {
public:
    enum AtomType { CARBON, HYDROGEN, OXYGEN, NITROGEN };
    ChemicalStructure();
    ...
    short       AddAtom(AtomType a = CARBON);
    short       AddBond(short a1, short a2);
    ...
    short       NumAtoms();
    AtomType    Atom(short i);
    Boolean     ContainsAsSubstructure(ChemicalStructure* x);
    ...
protected:
    short       NumBonds();
    void        Bond(short bondnum, short& atom1,
                     short& atom2);
    ...
```

A base class supplying protected access functions for use by programmers of derived classes

```
private:
    AtomType        fatoms[kMAXSIZE];
    short           fbonds[2][kMAXSIZE];
    ...
    short           fnumatoms;
    short           fnumbonds;
    ...
};
```

In this example, the protected part would provide functions that allow data on bonds to be accessed. Generally, a derived class will have legitimate reason to use some of the functions that define the implementation of the base class. The author of the base class must move the declarations of these functions from the private part to the protected interface part when it is decided that a class may be used as a base class.

Given the revised version of ChemicalStructure, with the protected interface giving access to bonds, it would be possible to define a ChemModel subclass with different drawing modes. The standard Draw() behaviour would, as before, produce an image of spheres representing atoms and would utilize only functions from the public interface of ChemicalStructure. A Draw_Ball_and_Stick() method would rely on the protected interface functions for data defining the bonds that had to be drawn.

```
class ChemModel : public ChemicalStructure {
public:
    ...
    void        Draw();
    void        Draw_Ball_and_Stick();
private:
        ...
};

void ChemModel::Draw_Ball_and_Stick()
{
    // Draw the bonds
    short nbonds = NumBonds();
    for(int b = 0; b< nbonds; b++) {
            short a1, a2;
            Bond(b,a1,a2);
            ...
            }
    ...
}
```

Protected data members
It is legal to have data members declared in the protected interface. This gives programmers of subclasses direct access to these data members. This can be advantageous if there are many data members and they are "read" in numerous places in the code implementing methods of derived classes. However, it is often better to keep data members private to a class and provide access functions. If efficiency of access is a major concern. these access functions can be made "inline", as described in Chapter 5 (e.g. inline short ChemicalStructure:: NumBonds() { return fnumbonds; }). If a protected function interface is used,

then the data members can only be changed in those functions that are members of the base class. This limit on access can simplify debugging. It is not necessary to check through all of the code of a class and its subclasses to find where a data member might be changed.

6.3 PLANNING FOR EXTENSION: "VIRTUAL FUNCTIONS"

The other major requirement of subclasses is the ability to override inherited behaviours by extending, or completely replacing methods of their base class. The classes ChemicalStructure and ChemModel can again serve as the basis of an example. The first part of this example illustrates how a member function can be replaced and how, when appropriate, the replacement function can utilize the standard inherited behaviour. The second part of the example explains the two different forms of function redefinition in C++ and illustrates why, generally, the programmer of the base class must make explicit provision for redefinition of member functions through "virtual" function declarations.

6.3.1 Replacing an inherited behaviour

Class ChemicalStructure might provide a method PrintOn(ostream&) that allows an instance of the class to write itself to a disk file:

```
class ChemicalStructure {
public:
    ...
    void    PrintOn(ostream&);
    ...
protected:
    ...
private:
    ...
};

void ChemicalStructure::PrintOn(ostream& os)
{
    os << fnumatoms << "\n";
    for(int i  = 0; i < fnumatoms; i++) {
        ...
        }
    ...
}
```

The function ChemicalStructure::PrintOn() saves all the standard data defining a molecule, but it wouldn't be satisfactory for a program using instances of class ChemModel. All the laboriously entered data that define three-dimensional coordinate positions of atoms would be lost when the structure was written to a disk file. The programmer implementing class ChemModel needs to redefine the PrintOn method:

Redefining a (non-virtual) member function

```
class ChemModel : public ChemicalStructure {
public:
    …
    void   PrintOn(ostream&);
    …
};

void ChemModel::PrintOn(ostream& os)
{
    os << NumAtoms() << "\n";
    for(int i  = 0; i < NumAtoms(); i++) {
       …
       // and now write out the coordinates as well
       os << fx[i] << ", " << fy[i] << ", " << fz[i] << "\n";
       }
    …
}
```

This ChemModel::PrintOn() function saves the coordinate data along with the other information characterizing atoms.

Using inherited behaviour in replacement code

The programmer implementing class ChemModel could have reduced the amount of work done by exploiting the inherited PrintOn() behaviour. An instance of class ChemModel must save all the standard data defining a structure and, in addition, the extra data with coordinates. If a data layout of standard data followed by extra coordinate data were acceptable, the ChemModel::PrintOn() function could have been coded as follows:

```
void ChemModel::PrintOn(ostream& os)
{
    // Use inherited behaviour to save all the standard stuff
    ChemicalStructure::PrintOn(os);
    // Now save my coordinates
    for(int a=0; a < NumAtoms(); a++)
       os << fx[a] << ", " << fy[a] << ", " << fz[a] << "\n";
}
```

It is very common for the behaviour of an object of a derived subclass to involve doing everything that a base class object would do and then a little more, or to involve doing an extra processing step and then doing everything that a base class object would have done. All object-oriented languages provide some mechanism to facilitate such usage.

As illustrated in the second version of ChemModel::PrintOn(), C++ uses class names to disambiguate calls. The implementation of ChemModel::PrintOn() invokes the inherited behaviour via the explicit call to ChemicalStructure::PrintOn(). The C++ mechanism is completely general. In a three level hierarchy, a sub-subclass can call the method of the original base class or that of the intermediate level subclass. Similarly, in multiple inheritance hierarchies where there might be a choice of inherited behaviours, the programmer implementing a derived class can invoke a chosen inherited behaviour by using a qualified function name (e.g. in the DUKW example of multiple inheritance in section 3.6, the programmer could encode a call to `Truck::PrintOn()` or to `Barge::PrintOn()`). Of course, if it is to be called, the inherited function must be part of the base class's public or protected interface.

Languages using single inheritance, such as Smalltalk and Object Pascal, have extra syntactic structures that allow the code implementing a function in a derived subclass to invoke the inherited behaviour. Thus, Smalltalk has calls to SUPER methods; and Object Pascal uses calls like `inherited PrintOn(…)`. This has been introduced as an extension in some dialects of C++ (for example, in the version of C++ on the Macintosh). The extension is only applicable when used in a subclass created using single inheritance. The justification given for the extension is that it supposedly simplifies the recoding of classes if a class hierarchy is modified.

C++ dialects with "inherited::…" construct

6.3.2 Static or dynamic binding for member functions

The code just illustrated would work in some circumstances, but it does have problems. It might be satisfactory if a client application is working solely with instances of class `ChemicalStructure`, or solely with instances of class `ChemModel`. But, it is more likely for an application to require instances of both classes; for example, the application might work with a list of structures some of which have coordinate data (and so are created as instances of class `ChemModel`) and others that lack this information (and so are instances of class `ChemicalStructure`). It is also quite likely that code written to work with instances of class `ChemicalStructure` will be used with `ChemModel` objects. In these circumstances, the `PrintOn()` functions may exhibit anomalous behaviour.

For example, the application might have function along the following lines:

```
void CreateMolFile(ChemicalStructure& theMol)
{
    // Get name, and other info entered using a dialog.
    NameData      d;
    Name_the_Structure_Dialog(d);
    // Get a stream opened to a file
    ofstream      ofile;
    File_Open_Dialog(ofile);
    d.PrintOn(ofile);
    theMol.PrintOn(ofile);     // ?? something wrong here??
    ofile.close();
}
```

This function takes a "reference to `ChemicalStructure`" argument, uses dialogs to get some extra data and to open a file, and then writes the extra data and the structure to that file. Function `CreateMolFile()` might be called in code like:

```
ChemicalStructure mol1;
...
ChemModel         mol2;
...
```

```
CreateMolFile(mol1);
...
CreateMolFile(mol2);
...
```

Because an instance of class `ChemModel` is also an instance of class `ChemicalStructure`, it is quite correct to call `CreateMolFile()` with `mol2` as an argument. However, the programmer who implemented `ChemModel` might be disappointed to discover that this code would not save the coordinate data characterizing `mol2`.

Static binding in the method call

The problem relates to the original declaration of class `ChemicalStructure` and how this declaration will be used when the C++ compiler generates code for the function `CreateMolFile()`.

```
class ChemicalStructure {
public:
    ...
    void            PrintOn(ostream&);
    ...
};

void CreateMolFile(ChemicalStructure& theMol)
{
    ...
    theMol.PrintOn(ofile);
}
```

The class declaration basically says to the compiler "encode a call to the function `ChemicalStructure::PrintOn()` wherever the member function `PrintOn()` is invoked for a `ChemicalStructure` object". The code for `CreateMolFile()` says "here is a reference to a `ChemicalStructure`, emit a call to `PrintOn()` for it". The compiler dutifully encodes a call to `ChemicalStructure::PrintOn()`. The fact that the reference might be to a specialized variant of `ChemicalStructure`, a variant for which `PrintOn()` can have a different meaning, is never taken into account when the code for `CreatMolFile()` is generated by the compiler. (Information as to whether specialized variants of class `ChemicalStructure` might actually exist is never going to be available to the compiler when the code for `CreatMolFile()` is generated.)

Requesting the encoding of a dynamic call

The compiler must be informed as to when it is to encode dynamic calls that involve run-time lookup based on an object's class. It is the programmer of the base class who encodes this instruction to the compiler. If class `Chemical-Structure` had been declared as follows:

```
class ChemicalStructure {
public:
    ...
    virtual void PrintOn(ostream&);
};
```

then the encoding `theMol.PrintOn(ofile)` in function `CreatMolFile()` would have been quite different. The compiler would note that `PrintOn()` was virtual and so generate the longer code that involves extracting the address of the method table (virtual table) from the `theMol` data structure, indexing into this table to obtain the address of the appropriate `PrintOn()` routine and then calling this routine (as explained in Chapter 3).

In this form, the code will work. The example call `"CreatMolFile(mol1)"` has an instance of `ChemicalStructure` passed to the routine; `mol1` will contain a pointer to the virtual table of class `ChemicalStructure` where the appropriate entry points to `ChemicalStructure::PrintOn()`. In the other call, the argument `mol2` contains a pointer to the virtual table of class `ChemModel` where there is an entry for `ChemModel::PrintOn()`. In both cases, the correct routine will be selected when the code for the dynamic lookup is executed within `Create-MolFile()`. The coordinates will indeed be printed for `mol2`.

When a member function is not virtual, the compiler determines which of any possible variants to call based on the static type of the entity. If an entity has type `ChemicalStructure`, `ChemicalStructure&`, or `ChemicalStructure*`, the invocation of a non-virtual member function `PrintOn()` will be encoded as a call to `ChemicalStructure::PrintOn()`.

Static binding

When a function is initially declared as being virtual, the compiler encodes a dynamic lookup when the member is invoked via a reference or a pointer. So, for an entity of type `ChemicalStructure&`, or `ChemicalStructure*`, the invocation of a virtual member function `PrintOn()` will be encoded as "*hey referenced object, choose your very own PrintOn() function and execute it*". (If the entity is of type `ChemicalStructure`, the compiler can optimize and still encode a static call to `ChemicalStructure::PrintOn()`.)

Dynamic binding

When a program is built along OO lines, one generally requires dynamic binding throughout. So, why does C++ default to static binding and only provide dynamic binding when explicitly requested?

Static binding as default?

C++ was not developed to support solely the OO programming style. Developments using classes for abstract data types are equally valid in C++. In an abstract data type, or "object-based", approach there is no requirement for dynamic binding. The requirement for dynamic binding only arises when one has class hierarchies employing inheritance, in which derived subclasses are permitted to substitute new definitions for inherited behaviours. Given the user-pays philosophy in C++, it is inappropriate for those working with simple abstract data types to have to pay for compiler and run-time overheads associated with having dynamic function lookup implemented as the default.

C++ provides dynamic binding only when explicitly requested

Further, C++ classes are basically closed. Private data are not accessible in derived classes. If they were accessible, the keyword `private` would have no meaning because any programmer could change everything by simply deriving a subclass with a couple of extra methods that mucked around with the data. Private data only become accessible if the designer of the base class consciously chooses to declare the data members as `protected`, or provides protected access functions for the private data. Similarly, if a programmer can simply derive a subclass and replace any method, the concept represented by the original class becomes much less clearly defined. An apparent instance of the class, accessed via a reference or a pointer, may behave as in the original specification, or it may not. One would

C++ provides total security for data members unless more general access is specified explicitly

have to check the entire program to verify that there weren't any subclasses redefining methods.

You cannot meddle with a C++ class

Stroustrup expressed his view in a note published on the Internet:

"If I specify a concept/class, I don't want later programmers using my class to meddle except where I explicitly specified that overriding, access, etc. is allowed. You don't get "re-use" from individual language features, but from design."

C++ programmers must understand that the implementor of a derived class does *not* have the privilege of changing inherited behaviours. In C++, one must do as one's ancestors do unless they have, in advance, given an explicit written dispensation permitting changes to their behaviours.

Responsibilities of the programmer of a base class

The designer of a base class has to plan for all possible uses of that class. Each member function has to be assessed. If there is any possibility of a derived class having a legitimate reason for modifying the behaviour represented by a function, then that function must declared as being `virtual`. Each data member has to be assessed. If there is any reason why derived classes might need access to a data member, then that data member must either be made protected, or be given protected read/write access functions that allow changes to its value.

Disallow or warn on redefinition of non-virtual member functions?

Some warnings are generated by C++ compilers when derived classes redefine member functions in apparently dubious ways. For example, a warning is generated in the following situation:

```
class ChemicalStructure {
public:
    ...
    virtual void PrintOn(ostream&);
    ...
};

class ChemModel : public ChemicalStructure {
public:
    ...
    void    PrintOn(short, long);
    ...
};
```

The warning generated will point out that the `PrintOn(...)` member defined for class `ChemModel` hides the inherited virtual function (it is not a replacement because it has a different signature). An instance of class `ChemModel` (accessed as `ChemModel`, `ChemModel&` or `ChemModel*`) cannot be written to a stream using `PrintOn(ostream&)` – because this hidden function effectively doesn't exist for the `ChemModel` class (or so the ARM specifies, not all compilers comply).

Caution: no warnings when a non-virtual function is redefined

However, no warnings get generated when a derived class redefines a non-virtual member function of its base class. This means that it is possible for programmers to mistakenly think that they are overriding functions in an OO sense when in fact static binding is going to be used and their redefined functions will not necessarily get called. This could be a problem if conventional editors and compilers are used. Increasingly, C++ programs will be developed within language specific development environments. Such environments are cognizant of inheritance, rules for redefining virtual and non-virtual functions, and similar constraints. Different fonts, or other obvious visual indicators, are used to distinguish among the addition of functionality by a subclass, the redefinition of an inherited virtual behaviour, and the local replacement by redefinition of a non-virtual member function.

6.3.3 A standard form for a base class

The designer of a class that is to be used as the base class in some hierarchy has to plan for reuse. Certain features are typically required. Coplien, in his text on *Advanced C++*, goes into some detail on the canonical (i.e. standard) form for a class, and the canonical inheritance form. Briefly, a class designer should check that they have covered the following:

- Declared a class with
  a public interface;
  a list of any friend classes and/or friend functions;
  a protected interface for use by derived classes;
  a private implementation part.

  Form of declaration

- If instances of the class will create separately managed resources, a destructor must be provided that releases those resources. (Separately managed resources include auxiliary objects and ordinary data structures allocated on the heap, files, i/o ports etc.) A class using separately managed resources must also provide a "copy constructor" and an `operator=()` function for assignment. (If a class should *not* support assignment, an `operator=()` function can be declared as part of the private interface of that class. The compiler can then check for and prevent any attempts to use instances of such a class in assignment statements.)

  Resource manager classes require destructor, copy constructor, and assignment operator

- Any class that is intended to be used as a base class *must* provide a "virtual destructor":

  A virtual destructor is needed in any class that may be used as a base class

```
class TheBase {
public:
    virtual ~TheBase() { }
    ...
};
```

The base class may or may not have any separately managed resources. If the base class has such resources, it must provide a destructor; this destructor must be declared as virtual. If the base class does not own resources, it must still provide an empty virtual destructor function as shown.

Derived classes are likely to have additional separately managed resources and have destructor functions that need to be called to release those resources. In an OO style program, instances of those derived classes are likely to be accessed through variables of type `TheBase&` or `TheBase*`.

The C++ compiler will insert calls to destructors automatically at points where variables go out of scope or are freed using the `delete` operator. But, by default, the inserted call will be statically bound, so for variables of type `TheBase&` or `TheBase*` the calls will be to the destructor `TheBase::~TheBase()`. The code of the derived class's destructor would not be invoked and so its part of the resources would not be freed.

However, if the destructor of the base class is declared as `virtual` then a dynamic call is made. If a subclass has defined its own destructor, this is then invoked. A call to the base class destructor is made, automatically, at the end of a subclass's destructor.

Declare as virtual any member function that could conceivably need to be changed in a subclass
• Any member function for which redefinition by subclasses is permissible *must* be specified as `virtual`. Such functions can be part of the public, protected, or private interface.

It is meaningful to have `virtual` private functions; for example:

```
class Invader {
public:
    void   Run();
    …
protected:
    …
private:
    virtual void Move();
    …
};

class Saucer : public Invader {
public:
    …
private:
    virtual void Move();
    …
};

void Invader::Run()
{
    …
    Move();    //  virtual,
            // i.e. Move in a manner appropriate
            // to type of Invader executing Run().
    …
}
```

The `Move()` method is not part of the public interface of any class. `Invader::Move()` doesn't get explicitly invoked by code of a subclass so it does not need to be part of the `protected` interface.

If a redefined `virtual` function might need to invoke the standard implementation, then the original `virtual` function must be part of the `public` or `protected` interface of the class where it is defined.

- Data members of the base class may have to be declared as protected or protected access read/write functions may have to be provided. (Note, it is almost always an indication of a bad design if the public interface of a class has `"GetX(), SetX(value)"` read/write functions for each of its data members. If a class has such an interface, the author has probably been thinking in terms of a Pascal `record`, or C `struct`, rather than in terms of objects.)

Provide protected access to data members

6.4 "PRIVATE" INHERITANCE

"Private inheritance" is another mechanism that can be used in C++ to construct new classes from existing classes.

A possible use of "private inheritance" can be illustrated by considering the reworking of class `ChemicalStructure`. The original version, as used in the preceding sections, employs fixed sized arrays for storage of data on atoms. This would probably prove quite unsatisfactory. Most structures created in an application using class `ChemicalStructure` would probably be quite small (<100 atoms) but, occasionally, a user might want to create a protein (>1000 atoms). If large arrays are used, the space is wasted for most of the time. But, any fixed size array will eventually prove too small for some user. Further, the only information stored was the atomtype; this would probably prove inadequate. Instead of just an array of atomtype indicators, a useful `ChemicalStructure` would have to be composed of some collection of instances of class Atom.

The class designers would eventually have to substitute some form dynamic storage for the original array. Class `LinkedList`, from Chapter 5, might appear as a suitable candidate. Why not reuse `LinkedList` as the basis of a better `ChemicalStructure`?

```
class Atom;

class ChemicalStructure : public LinkedList { // ?? public??
public:
    ...
    void    AddAtom(Atom*);
    ...
};

inline void ChemicalStructure::AddAtom(Atom* a) { Append(a); }
```

With other appropriate modification to the class this would be practical and would result in a much more useful version of the `ChemicalStructure` concept where there was no arbitrary size limit.

However, public inheritance is inappropriate. It says that a `Chemical-Structure` *is* a `LinkedList`. Consequently, the following code is valid:

Public inheritance permits inappropriate use of instances of class

```
ChemicalStructure       mol1;
```

```
char                    buff[100];
int                     i;
ChemicalStructure*      pm = new ChemicalStructure;
…
mol1.Append(buff); mol1.Append(&i); mol1.Append(pm);
…
```

Why not? Class `ChemicalStructure` clearly states that its instances are just useful containers for storing any kind of junk.

Even if class `ChemicalStructure` is to be built on top of a `LinkedList` base, it is clearly inappropriate for clients to know this. "Private inheritance" could be used:

Declaring a private inheritance relationship

```
class ChemicalStructure : private LinkedList {
public:
    …
    void    AddAtom(Atom*);
    …
};

inline void ChemicalStructure::AddAtom(Atom* a)
{ Append(a); }
```

The keyword `private` is optional. If the inheritance mechanism is not specified it defaults to `private`. However, it is good practice to use an explicit declaration specifying `private` inheritance.

When class `ChemicalStructure` is declared as privately derived from class `LinkedList`, clients cannot see any relation between these classes and therefore cannot ask a `ChemicalStructure` to perform as a `LinkedList`. Consequently, code like "`mol1.Append(&i)`" is simply illegal and gets rejected by the compiler. But the relationship between the classes is apparent within the code of class `ChemicalStructure`. Within the implementation of class `ChemicalStructure`, the normal base-class/derived-class relations hold. The code of member functions of `ChemicalStructure` can involve calls to any public or protected functions defined in its `LinkedList` base. Class `ChemicalStructure` is reusing the implementation of dynamic storage provided by the designer of class `LinkedList`.

Inheritance for reuse of implementation

The relation between class `ChemicalStructure` and class `LinkedList` is an example of "inheritance for the reuse of implementation". Usually, inheritance has some associated semantics; inheritance relations among classes define "*is_a*" relationships – like "a Saucer is a kind of Invader". Here, there is no such meaning. Inheritance is being employed as a crude "hack" – it saved the implementor of class `ChemicalStructure` from having to design a dynamic array structure.

A preferable alternative: employ a subcontractor

Inheritance for implementation reuse is not necessary. The implementor of class `ChemicalStructure` could have obtained the data storage services of a `LinkedList` in a more natural way. Instead of this phony adoption of `LinkedList` as a putative parent, an instance of class `LinkedList` could be employed as a subcontractor:

```
class ChemicalStructure {
public:
```

```
    …
    void    AddAtom(Atom*);
    …
private:
    …
    LinkedList      ftheAtoms;
};

inline void ChemicalStructure::AddAtom(Atom* a)
{
    ftheAtoms.Append(a);
}
```

Class ChemicalStructure requires a data member that is an instance of class LinkedList. All actions involving storing and subsequently finding atoms can be delegated to this LinkedList object. (The code generated for this version probably costs one extra indirection cycle each time the list of atoms gets used. However, the effect on program performance is obviously negligible. If performance were an issue, one wouldn't be using such a list in the first place.)

It is possible to have the list object allocated separately; in that case the data member would be a pointer of type LinkedList* and the constructor for class ChemicalStructure would have to create an instance of class LinkedList. It is a design choice as to whether to have the auxiliary object as a data member or have a pointer data member and a separately allocated object. Either might be appropriate; it depends on how the auxiliary object gets used. Such issues are covered briefly in the chapter on design. In this case, it is probably more appropriate for the list to be an actual data member.

Class ChemicalStructure is conceptually much cleaner when seen as an employer of a LinkedList than when seen as the "inheritor" of an artificially adopted LinkedList parent. This is the normal situation. Private inheritance for purposes of implementation results in unnecessarily complicated class structures.

Arguments are sometimes presented justifying the use of private inheritance. Some arguments appeal to "efficiency". More realistically, it is argued that an advantage of private inheritance is that the derived class can, when necessary, access protected functions or data members of its private parent class and can redefine any inherited virtual functions. This obviously doesn't apply for simple classes like LinkedList which has no provision for extension – there are no protected members and no virtual functions. But, in a more general case, one could have a class DynamicArray, with virtual functions etc., that provides more or less the functionality required. The implementation of class DynamicArray could be reused, through private inheritance, with virtual functions redefined as needed.

Private inheritance when need to override virtual functions?

Essentially the same effect can be achieved by creating a new server class (class MyDynamicArray), publicly inherited from DynamicArray, which appropriately redefines functions and provides a public interface that meets requirements. Then, an instance of this new server class can be used as a "subcontractor" and the use of private inheritance can still be avoided. If one didn't want class MyDynamicArray to appear in the global name space, it would be possible to define it as a nested class within class ChemicalStructure; nested classes are touched on briefly in Chapter 12.

6.5   CLASS LINKEDLIST AND FRIENDS REVISITED

The `LinkedList` class cluster illustrated in section 5.4 was somewhat limited by the fact that it worked with `void*` pointers. The implementation of the code of `LinkedList` could not do much with the things that were to be stored in the list. The class did provide a `Find()` method to check whether a specific entity was stored in the list, but this worked by checking addresses and not the values of the stored items. There was no provision for making an ordered list, because the `void*` items could not be compared in any meaningful way.

Simple use of single inheritance allows the creation of a more useful `LinkedList` cluster. The revised version illustrated here is in the general style of classes found in class libraries. A `LinkedList` will store pointers to dynamically created instances of class "Item" (or instances of any class derived from class "Item"). "Items" are going to be things that can compare themselves (< , = , >), and can check for equality of content (as well as for identity). The new `LinkedList` class can rely on Items having such behaviours and can, consequently, itself be more elaborate providing additional functionality such as the ability to keep its contents ordered.

6.5.1   A design for class Item and its subclasses

Of course, an "Item" is just an abstraction; any real program will work with instances of specific subclasses derived from class `Item`. For example, a program using names would have a class `Name` based on `Item`; a program working with point coordinates might have a class `IPoint` derived from `Item`. (The definition of class `Item` is similar to, though much simpler than that of the root class `Object` as used in libraries such as Gorlen's NIH class library.)

Capturing general ideas in Analysis diagrams

Figures 6.2 and 6.3 illustrate ideas about the nature of an "Item" and possible specialized subclasses. These figures are in the style that will be used for object-oriented analysis, as presented in Chapter 11 and introduced here to give a first simple illustration of the ways in which ideas about objects and classes can be documented.

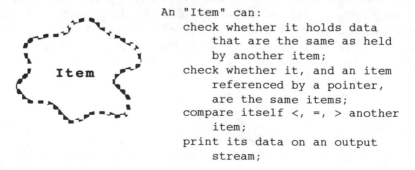

```
An "Item" can:
   check whether it holds data
      that are the same as held
      by another item;
   check whether it, and an item
      referenced by a pointer,
      are the same items;
   compare itself <, =, > another
      item;
   print its data on an output
      stream;
```

Figure 6.2   An initial idea for an "Item".

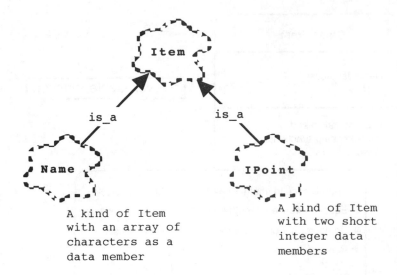

A kind of Item
with an array of
characters as a
data member

A kind of Item
with two short
integer data
members

Figure 6.3   A simple class hierarchy as identified during the analysis phase of some project.

The fuzzy blob labelled Item indicates that the program will use instances of some "class Item". At an early stage, as in preliminary analysis, the nature of this class is still somewhat obscure; it doesn't have any firm boundaries. The blob representation is stolen from Booch's *Object-Oriented Design with Applications*.

Figure 6.3 indicates that, even at the analysis stage, it is known that the program will require specializations of Item. These will occur as objects that are instances of "class Name" and objects that are instances of "class IPoint".

"Name" objects are to hold character strings and presumably provide some member functions for manipulating strings. "IPoint" objects are to be some variant on the "Points" illustrated in Chapter 5.

During a design phase, these initial ideas about "Items" would become more concrete. A design level class diagram for class Item is illustrated in Figure 6.4. The basic form of the diagram is similar to the class diagrams proposed by Coad and Yourdon, or by Henderson-Sellers (but is more C++ specific than their notations).

Representing detailed models in design diagrams

At this stage it has been decided that class Item is not a pure abstract class; it can instantiated even though Items are not much use because they have no data fields. (If class Item were intended to be pure abstract, i.e. not instantiatable, it would have been labelled as such on the diagram.) Default implementations are to be provided for all its member functions.

Abstract or concrete class

The diagram names the class, lists the data resources that will be owned by each instance of the class (in this case, there are none), and lists member functions that perform internal housekeeping tasks (again, there are none of these for class Item). In the case of a more complex class, it would be quite likely for additional data members and "housekeeping" functions to be identified in later more detailed design steps.

Private resources and implementation details

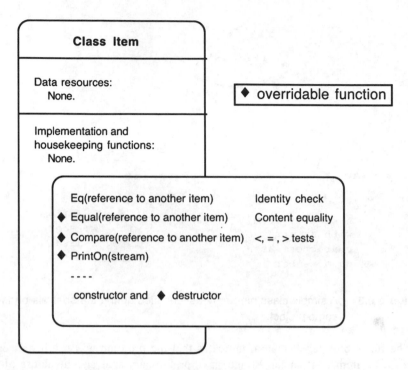

Figure 6.4   Initial design diagram for class Item.

Public interface defining services

    The public interface for the class is made distinct. This should already be pretty much complete by the design stage. The various services provided by an instance of the class are listed as a series of functions with some brief comments. If a class is intended to be used as a base class in some hierarchy, then functions that can be modified in subclasses should be identified.

Member function signatures

    The arguments for member functions should be indicated. However, the function signatures will be refined later during detailed design or during implementation. In the example in Figure 6.4, functions such as Compare() are shown as having "a reference to another item" as an argument; the implementors would have the choice of using C++ references or C++ pointers for such arguments.

Constructors and destructors

    The constructor and destructor are included in the public interface because "Items" can be created and destroyed in client code. They are separated from the other services defined in the public interface. Creation and destruction are things that get done to objects, rather than things that objects can be asked to do. The destructor is marked as overridable, as would be expected for any class that is to act as a base class in a hierarchy.

Friends

    When designing for C++, it is appropriate for the diagrams to identify any friend classes or functions that violate normal access restrictions. Figure 6.6, the design for class Link, has example friend relationionships with class LinkedList and class LinkedListIterator.

Implementation

    The declarations for class Item, and simplified versions of Name and IPoint, are:

```
class Item {                                                            Class Item
    // Item:
    //     intended to be used as a base class for
    //     things that get used with Lists etc.
    //     A few virtual methods are provided - compare,
    //     equal, ... (default implementations are provided
    //     to avoid the need for classes to implement these
    //     methods even if not relevant)
    //     There is a virtual destructor.
    //     There is a non-virtual "eq" identity test method.
    //     It is expected that usually Items will be dynamic
    //     entities allocated on the heap, so arguments to
    //     compare etc. are of type Item*
public:
    Item();
    virtual ~Item();
    // Equality tests, as in Lisp
    //     eq is an identity test (same address)
    //     equal is a content equality test
    short               Eq(const Item*) const;
    virtual short       Equal(const Item*) const;
    // Compare: ordering of objects
    //          this less than other -1,
    //          this equal other 0,
    //          this greater than other 1
    virtual short       Compare(const Item*) const;
    // Output
    virtual void        PrintOn(ostream& os) const;
};

// Somewhere will need definition of global output function

extern ostream& operator<<(ostream&,const Item&);

// This function doesn't need to be a friend of class Item
// because it uses PrintOn() from Item's public interface

class Name : public Item {                                              Class Name
    // Incomplete class declaration
    // Only for use in the example illustrating lists
    // Still needs copy constructor, assignment operator
    // and some functions that do something with names
public:
    Name(char*);
    ~Name();
    virtual short       Equal(const Item*) const;
    virtual short       Compare(const Item*) const;
    virtual void        PrintOn(ostream& os) const;
private:
    char*               fstring;
};

class IPoint : public Item {                                           CLass IPoint
    // Incomplete class declaration
    // Only for use in the example illustrating lists
    // Needs some functions defining "semantics" of Point
public:
```

```
            IPoint(short h = 0, short v = 0);
            virtual short        Equal(const Item*) const;
            virtual short        Compare(const Item*) const;
            virtual void         PrintOn(ostream& os) const;
        private:
            short                fh,fv;
        };
```

In an OO program there will usually be two versions of "equality", corresponding to Lisp's "eq" and "equal" functions. Since most objects will be dynamic entities created in the heap, they will be accessed via pointers. Sometimes it is necessary to check whether two pointers are referencing the same object; this is equivalent to Lisp's "eq" predicate. The base class can define Eq() as a non-virtual function:

Eq: a test for identity

```
short Item::Eq(const Item* other) const
{
    // Eq : identity test, i.e. same address
    return (this == other);
}
```

Note the use of const qualifiers; they indicate that this function modifies neither the object that is executing the code, nor the object accessed via the pointer argument. It is possible that some const IPoint or const Name objects might be used; if these are to be permitted, several of the member functions must be defined as const. Even if it is not intended to have const instances of any class derived from Item, the const qualifiers are still helpful as they allow the class declaration to make a clear distinction between functions that can/cannot modify an object.

Equal: a test for structural equality

The second version of equality, Lisp's "equal" predicate, tests whether two objects contain the same data. Obviously, the implementation will be (sub)class specific. The base class Item can define a default, virtual implementation that makes "equal" synonymous with "eq":

```
short Item::Equal(const Item* other) const
{
    // Default Equality test,
    //     they are only "equal" when they are the same object
    return Eq(other);
}
```

(Note, the keyword "virtual" appears only in the class declaration; it is not repeated when the function is defined.) Member Equal() will be redefined in subclasses, e.g.:

```
short IPoint::Equal(const Item* x) const
{   // Oh, bother. a type cast is needed.
    // Trust me compiler, I know what I am doing.
    IPoint* p = (IPoint*) x;
    return ((fh == p->fh) && (fv == p->fv));
}
```

Here, there is a potential problem with the type system. The compiler cannot guarantee that `IPoints` are always going to be used consistently. If it is to override the inherited `virtual Item::Equal()` function, function `IPoint::Equal()` must have the same signature. So, `IPoint::Equal()` takes a `const Item*` as its argument. This argument must then be type cast to a `IPoint*` before the data members of the two points can be checked for equality. This opens a hole in the type checking system; because the following code is legal:

```
IPoint*     p1;
Name*       n1;
...
p1 = new IPoint(1,1);
n1 = new Name("Point at 1,1");
...
IPoint* nl = new IPoint(hl,vl);
...
if(p1->Equal(n1)) ...; // IPoint == Name ? !; did you mean nl
```

Obviously, there will be no situation where a programmer intentionally checks a point and name for equality. But it is possible that the use of the wrong identifier will result in such a check being made. The function `IPoint::Equal()` would be invoked with anomalous results or, possibly, even a program abort.

An attempt to avoid this problem by the definition of a *different* function, `IPoint::Equal(const IPoint* x)`, results in much more subtle errors. This definition would hide the inherited virtual function. When `IPoints` were compared (using variables of type `IPoint`, `IPoint&`, or `IPoint*`) the function `IPoint::Equal()` would be used. If `IPoints` were compared when accessed as `Item&` or `Item*` variables, function `Item::Equal()`, with quite different semantics, would get used. The behaviour of a program with such a function might seem quite arbitrary and mysterious. Fortunately, a C++ compiler would generate a warning message ("hides virtual function") if it encountered a definition of `IPoint::Equal(const IPoint* x)`; such warning should be heeded.

The problem of unsafe type casts is well known to designers of class libraries. In some libraries, e.g. the ET++ library, mechanisms are provided that can catch type-casting errors at run-time. So, if an `IPoint` is given a `Name` and told to check it for equality, a run-time error will be generated. This checking requires some class identification data that are recorded in each object (usually as some kind of integer tag field). Instead of a simple type cast, a "guarded" type conversion operation is performed. This conversion operation includes code that checks the compatibility of the destination type (e.g. `IPoint*`) and the tag field of the object that is being type-cast (`IPoint`, `Name`, `Item`).

The output functions (virtual member `PrintOn` and a global `ostream&` `operator<<`) that are defined in class `Item` are modelled on those of common class libraries. Together they allow stream output to be used conveniently with classes descended from class `Item`.

It is advantageous if code of the form:

```
Item*       iptr;
...
```

```
Name*       n1 = new Name("Test text");
IPoint*     p1 = new IPoint(10,10);
...
short       userchoice;
...
iptr =  (userchoice == 1) ? n1 : = p1;
cout << "You chose " << *iptr << "\n";
...
```

can work intelligently and print either the string or the point. Global functions, like
ostream& operator<<(ostream&,Item&), cannot be virtual. After all, such
functions are global, they are not members of a class. There is no point defining
similar global functions for the individual classes like IPoint and Name because
those functions wouldn't be called for code like cout << *iptr.

Stream i/o can be made to work using a global function that invokes the
virtual PrintOn() member function of class Item

```
ostream& operator<<(ostream& os,const Item& thing)
{
    thing.PrintOn(os);
    return os;
}
```

PrintOn() has an empty definition in class Item, but is open for redefinition in
subclasses:

```
void Item::PrintOn(ostream& /* os */) const
{
    // Nothing to print in the default
}

void IPoint::PrintOn(ostream& os) const
{
    os << "(" << fh << ", " << fv  << ")";
}

void Name::PrintOn(ostream& os) const
{
    os << fstring;
}
```

The other member functions of Item, IPoint, and Name are all straightforward.
Class Name allocates memory on the heap for a character array and so has a defined
destructor function to release the space. The code for the functions is included in
files Items.[Ch] and DemoItems.[Ch] in the oocc/LinkedList directory.

6.5.2 Extended LinkedList class

The structure of the class cluster of LinkedList, Link, and LinkedListIterator
from section 5.4 is unchanged. Classes Link and LinkedListIterator are
essentially the same except that they work with Item* elements rather than void*

elements. Class `LinkedList` has additional functionality. The relations among instances of these classes were described in section 5.4 and are summarized schematically in Figure 6.5. (Most notational schemes for documenting relations among classes use different types of arrow to represent different kinds of relation. One type of arrow will indicate an *is_a* relation such as those in Figure 6.3; an arrow with a different form of arrowhead or shaft will indicate a "uses" relation; an arrow with various different fletches will signify create and destroy operations. In this introductory treatment, it seemed more appropriate simply to label the arrows with details of the relationships.)

Class `Link` remains an oddity in that its interface is private. The interface is diagrammed in Figure 6.6. It is the same as that described in section 5.4 with changes to use `Item*` pointers and the addition of a `SetDatum()` member.

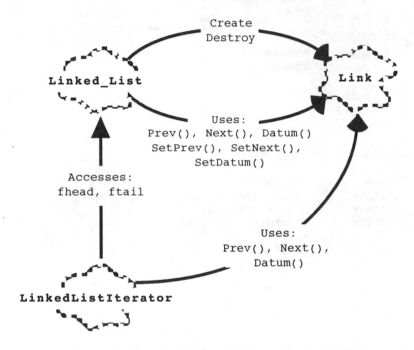

Figure 6.5 Interactions among instances of classes in the LinkedList cluster.

The revised and extended declaration for class `LinkedList` is:

```
class LinkedList {
    public:
LinkedList();
    virtual ~LinkedList();
    // Asking about list
    int          Length() const;
    short        FindPosition(const Item* queryitem) const;
    // Methods finding/retrieving items modify "mark" so
    // are not const
    Boolean      Find(const Item* queryitem);
    Boolean      FindSimilar(const Item* queryitem);
```

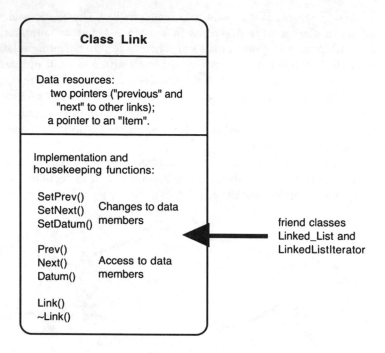

Figure 6.6   Class Link: a very private class.

```
        Item*           FirstItem();
        Item*           LastItem();
        Item*           MarkedItem();
        Item*           GetAt(short position);

        // Put data in list
        void            Append(Item* datum);
        void            Prepend(Item* datum);
        void            PutBeforeMark(Item* datum);
        void            PutAfterMark(Item* datum);
        void            PutItemAt(Item* newitem,short position);
        Item*           ReplaceItemAt(Item* replacement,
                            short position);
        void            Insert(Item* datum);

        // Removing stuff from list
        Item*           RemoveFirstItem();
        Item*           RemoveLastItem();
        Item*           RemoveMarkedItem();

        friend class LinkedListIterator;
protected:
        Link*           Head() const;
        Link*           Tail() const;
        Link*           Mark() const;
        void            SetMark(Link*);
        void            SetTail(Link*);
        void            SetHead(Link*);

private:
```

```
    Link*       fhead;
    Link*       ftail;
    Link*       fmark;
};
```

The class has been defined with some limited provision for subclassing. Thus, there is a virtual destructor; the base implementation, LinkedList::~ LinkedList, releases the links and any data items referenced by the links. Further, there is a group of protected access functions for modification of the private data members fhead, ftail, and fmark.

Actually, the overall structure of the cluster limits the scope for extension through subclassing. Any modified version of class LinkedList would need privileged access to links. However, friendships are not inherited nor are they transitive (your friend's friend is no friend to you). A new child subclass of class LinkedList does not share its parent's friends and would not have the required access. (Are you friends with your parent's close friends?) The implementors of a new form of list would have to have the right to change the existing cluster by adding their new class as an additional friend of class Link.

Friends are neither inherited nor shared

The extended class LinkedList attempts to provide all the facilities that might be required by any likely extension. Thus, it has the capabilities of a queue:

```
    Item*       FirstItem();
    Item*       RemoveFirstItem();
    void        Append(Item*);
```

and of a dequeue (double ended queue) through the addition of:

```
    Item*       LastItem();
    Item*       RemoveLastItem();
    void        Prepend(Item*);
```

and of a generalized list through the addition of the "mark" operations; and, finally, of a priority queue through the Insert(), FindPosition(), GetAt(), Put-ItemAt() and similar functions.

Such an overly broad and diffuse interface makes this LinkedList unsuitable as a library class for direct use by clients. Client programmers require classes that express individual and distinct concepts – like class Queue, class DEQueue, class PriorityList, and so forth. If the implementation of class LinkedList did prove satisfactory, it would need to be repackaged. The specialized variants would provide a restricted interface to a LinkedList subcontractor (an alternative would be to use private inheritance). For example, class Queue might be defined along the following lines:

```
class Queue {
public:
    Queue();
    ~Queue();
    Item*       Get();
    void        Put(Item*);
    Boolean     Empty();
```

```
private:
    LinkedList    fstuff;
}

inline Item* Queue::Get()
{
    return fstuff.RemoveFirstItem();
}
```

Dependence on
behaviours of Items

Several of the member functions for the revised class LinkedList depend on
the behaviours defined for class Item. Thus, there are now two "Find" functions,
one use 'eq' equality and the other uses 'equal' equality:

```
Boolean LinkedList::Find(const Item* query)
{
    // Find item "eq" to query and set mark
    // Don't change mark if item not present
    Link* lptr = fhead;
    while(lptr != (Link*) 0) {
            if(query->Eq(lptr->Datum())) {
                    fmark = lptr;
                    return true;
                    }
            lptr = lptr->Next();
            }
    return false;
}

Boolean LinkedList::FindSimilar(const Item* query)
{
    // Find an item "equal" to query and set mark
    // Don't change mark if item not present
    Link* lptr = fhead;
    while(lptr != (Link*) 0) {
            if(query->Equal(lptr->Datum())) {
                    fmark = lptr;
                    return true;
                    }
            lptr = lptr->Next();
            }
    return false;
}
```

and the member function FindPosition() depends on Items being able to
compare themselves:

```
short LinkedList::FindPosition(const Item* query) const
{
    // move along list until find an existing item that is
    // "larger" than the query item.
    // Return an index for the position,
    // (start count at 1, not zero).
    Link* lptr = fhead;
    short pos = 1;
    while(lptr != (Link*) 0) {
```

```
            Item* other;
            other = lptr->Datum();
            if(other->Compare(query) > 0) break;
            pos++;
            lptr = lptr->Next();
            }
      return pos;
}
```

The remaining functions for class `LinkedList` and friends are included in the "oocc" subdirectory along with a simple test program that creates and manipulates a list of `IPoints`, a list of `Names`, and a list with elements of mixed types.

6.6   EXAMPLE: SPACE INVADERS

As described in Chapter 3, the Space Invaders program is a hybrid program consisting of some procedural code that handles windows, menus, mouse tracking etc., and classes for the `Gun` and the `Invaders`. The responsibilities of class Invader are summarized in Figure 6.7. Instances of class `Invader` will be able to "Run" and check whether they've been destroyed by gunfire (in which case they are deleted). The class is also to encapsulate all details of its subclasses. Individual invaders, as `Saucers`, `Bombs`, or `Rocks` will be created by requesting creation of a random invader by the class itself (i.e. it is in Smalltalk terms a "class method" and will be implemented as a static member function of the class).

An initial design diagram for class `Invader` is shown in Figure 6.8. Here, there are some class responsibilities in addition to instance responsibilities. The class provides a `RandomInvader()` method that will create a randomly chosen invader. The class could also have held the data fields that define the limits on invader movements. These have been made instance variables in the current design; in the implementation, all invaders are constrained to the same limits. These limits define an area slightly smaller than the window in which the program runs. Having these limits as instance variables allows flexibility, maybe different individual invaders or different types of invader should be constrained to move in different regions. However, if all invaders do use the same limits, the data members defining the limits should be made static (i.e. shared class variables).

```
The class "Invader" can:                         An "Invader" can:
    create a randomly                                Run (advance
    chosen instance of                                 on gun, display
    one of its specialized                             itself etc.)
    subclasses.                                      CheckDestroyed
                          Invader                      (was it lasered
                                                       by the gun?)
                                                 and
                                                   can be destroyed.
```

Figure 6.7   Class Invader: class and instance responsibilities.

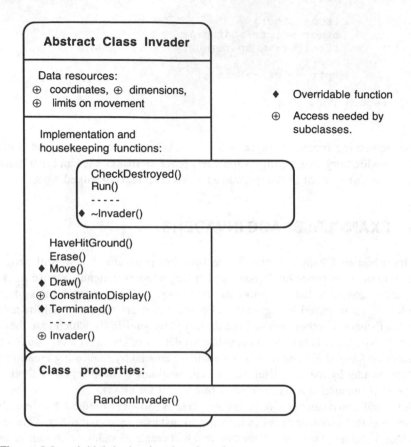

Figure 6.8   Initial design for class Invader.

The other instance variables define an invader's position and its dimensions. These data members will have to be accessible in the code of subclasses. As usual, it would be best to make the variables private to class `Invader` and provide protected access functions; however, for simplicity, they were made protected. When planning for a C++ implementation, the class designer must consider such details of access control.

Class `Invader` has a very narrow public interface with just the three methods. Invaders can be told to run, check for destruction, and can be deleted; since invaders can be deleted, the destructor function must be public (it must also be virtual because class `Invader` is a base class for a hierarchy).

The code for `Invader::Run()` is to be:

```
Boolean Invader::Run(Point gunPosition)
{
    Erase();
    Move(gunPosition);
    Draw();
    return Terminated();
}
```

Subclasses have no reason to change the basic Run() behaviour. Similarly, the methods HaveHitGround(), Erase(), ConstrainToDisplay(), and Check-Destroyed() can all be defined at the level of class Invader. There is no reason for these functions to be made virtual. Method ConstrainToDisplay() is to be called at the end of an invader's Move() function; its task is to constrain movement so that invaders don't move off the screen. It will have protected access status (as will the Invader constructor which will be invoked as part of the process of creating any specialized kind of invader). The other non-virtual functions could all be private as they are only invoked from within code of class Invader.

There are three other virtual functions – Move(), Terminated(), and Draw(). Default implementations can be defined for Move() and Terminated(). The default move behaviour is to move downwards slowly wobbling from side to side. The default Terminated() behaviour is to check whether the invader has hit the ground (this would need to be changed for any invaders that advance along the ground or that can fire at the gun). Function Draw() will be a pure virtual function; there is no default graphic representation for an invader.

The class hierarchy for the invaders is shown in Figure 6.9. Class Invader is a pure abstract class (it cannot be instantiated). All actual invaders will be instances of one of the subclasses Saucer, or Bomb, or Rock (or other specializations as defined in the exercises). Subclasses of class Invader must override Invader::Draw() and may redefine the other virtual functions.

The complete code for class Invader and its subclasses is in files in the MacInvaders subdirectory. (The conventions for Macintosh programs have all class names starting with a 'T'; consequently, in the Macintosh version of the code class Invader appears as TInvader etc.)

The declaration for class Invader is:

```
class       Invader       {
public:
    static Invader*     RandomInvader(short hmax,short vmax);
    Boolean             CheckDestroyed(Point gunPosition);
    Boolean             Run(Point gunPosition);
    virtual             ~Invader();

protected:
    void                ConstrainToDisplay();
    Invader(short hmax,  short vmax);
private:
    enum INVADERS { ROCKTYPE = 1, SAUCERTYPE = 2,
                    BOMBTYPE = 3 };
    enum INVCOUNT { INVADERTYPES = 3 };
    virtual Boolean     Terminated();
    Boolean             HaveHitGround();
    void                Erase();
    virtual void        Move(Point gunPosition);
    virtual void        Draw() = 0;
protected:
    short               fhsize, fvsize;
    short               fh, fv;
    short               fmaxh, fmaxv;
};
```

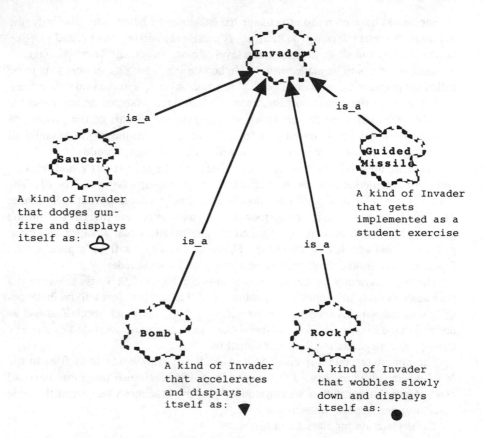

Figure 6.9   The Invaders class hierarchy.

The enumerated types "INVCOUNT" and "INVADERS" define the number and types of specialized invader. Enumerations in a class declaration are quite often used in this way to define integer constants that specify properties of that class. These enumerations would have to be changed if additional subclasses of Invader were to be defined.

Getting class Invader
to create specific
invaders

The static member function used to create invaders is:

```
Invader* Invader::RandomInvader(short hmax,short vmax)
{
    Invader*      vader;
    int           choice;
    choice = ::RandomFun(1,INVADERTYPES);
    switch (choice) {
case ROCKTYPE:
            vader = new Rock(hmax,vmax);
            break;
case SAUCERTYPE:
            vader = new Saucer(hmax,vmax);
            break;
case BOMBTYPE:
            vader = new Bomb(hmax,vmax);
            break;
```

```
        }
     return vader;
}
```

The arguments `hmax` and `vmax` define limits on movement of invaders. Additional branches would have to be added to the `switch() {...}` statement to deal with other kinds of `Invader`.

The declarations of the various subclasses are straightforward. For example, the Saucer class is declared as:

```
class Saucer       : public Invader {
public:
     Saucer(short hmax, short vmax);
private:
     virtual void        Draw();
     virtual void        Move(Point gunPosition);

     short               fhspeed, fvspeed;
     short               fcolor;
};
```

Defining Saucer as a
subclass of Invader

A `Saucer` keeps track of its current speed, and it also has a colour. The virtual keyword is repeated in the declarations of the overridden `Draw()` and `Move()` member functions; this is not strictly required, but often helps the reader. Some authors advise that the class declaration should contain comments distinguishing between overridden functions and declarations of new members.

Examples of constructor routines for the base class and one subclass are:

```
Invader::Invader(short hmax, short vmax)
{
     // Arguments hmax ... ...
     fmaxh = hmax;
     fmaxv = vmax;
     // the initialization of fhsize and fvsize should
     // be done by the subclass.
     // These values define size of a bounding rectangle for
     // the invader image (2*fhsize by 2*fvsize).
     fhsize = 0;
     fvsize = 0;
     // Initialize the coordinates of invader as random in
     // horizontal range and a bit above the top of the screen.
     fh = ::RandomFun(30,fmaxh-30);
     fv = -::RandomFun(0,60);
}

Saucer::Saucer(short hmax, short vmax) : Invader(hmax,vmax)
{
     fhsize = 5;
     fvsize = 3;
     int n = RandomFun(1,4);
     switch (n) {
case 1:
          fcolor = greenColor;
          break;
```

```
case 2:
        fcolor = magentaColor;
        break;
case 3:
        fcolor = cyanColor;
        break;
case 4:
        fcolor = redColor;
        break;
        }
    fvspeed = RandomFun(1,4);
    fhspeed = RandomFun(-8,8);
}
```

The Invader part of a Saucer is initialized first, then its size and colour are set and horizontal and vertical components of its speed are selected.

Code for graphics operations is, naturally, filled with calls to globally defined procedures. In this example, they are calls to Macintosh Quickdraw Toolbox functions. Although not strictly necessary, the global scope qualifier, ::, is used to highlight all calls to global functions. In future, when "namespaces" are used, the scope resolution operator would qualify a namespace, e.g. Quickdraw:: EraseRect().

```
void Invader::Erase()
{
    // If invader is within visual range
    if(fv>0) {
    // Take position of invader, initialize a zero sized
    // rectangle at that point,
    // grow rectangle to correct size, clear it.
        Rect   arect;
        Point  where;
        where.h = fh;
        where.v = fv;
        ::Pt2Rect(where,where,&arect);
        ::InsetRect(&arect,-fhsize,-fvsize);
        ::EraseRect(&arect);
        }
}
```

In addition to the invaders, the game requires an instance of class Gun. The role of the gun is summarized in Figure 6.10. The gun will be created and destroyed by the main program so its constructor, and any destructor, will be part of the public interface. (Class Gun would probably not require a destructor; it does not own any separately allocated resources and is not a base class so does not need to make provision for subclasses.)

The Gun will have a Run() method that will move it and, if appropriate, perform any "fire" action. This Run() method can return a boolean value indicating whether it fired; this result can be used by the main program to determine when invaders need to check if they have been destroyed.

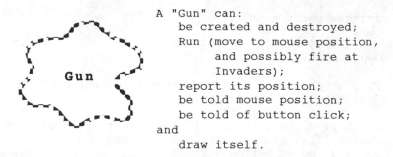

```
A "Gun" can:
    be created and destroyed;
    Run (move to mouse position,
         and possibly fire at
         Invaders);
    report its position;
    be told mouse position;
    be told of button click;
and
    draw itself.
```

Figure 6.10   Analysis diagram for class Gun.

The controlling game code will be tracking the mouse and recording button presses. Each time the mouse moves, or the button is pressed, the gun object should be informed. So the public interface of class Gun will have additional members allowing these data to be passed in. Another function in the public interface will be one that returns the current position of the gun, for this information is required by some invaders.

The code for class Gun is in files MacInvaders/Gun.[cp/h]. Its declaration is:

```
class Gun            {
public:
    Gun(short hmax,short vmax);

    Boolean      Run();
    void         Trigger();
    void         RecordCursorPosition(short a,short b);
    void         Draw();
    Point        Position();
private:
    void         Move(short dh);
    void         Erase();
    void         LaserOn();
    void         LaserOff();

    short        fh,fv;
    short        fch,fcv;
    long         flastshot;
    long         ftriggerpulls;
};
```

A class not open to extension!

Note the complete absence of any support of extension; there are no virtual functions, no protected data, and no protected access functions. It is not intended that the game be extended by the provision of different types of gun.

Some examples of the methods of class Gun are:

```
Gun::Gun(short hmax,short vmax)
{
    fh = hmax / 2;
    fv = vmax - 20;
    flastshot = 0;
```

```
        ftriggerpulls = 0;
}

Boolean Gun::Run()
{
    // gun can always move. It moves to cursor position
    if(fch != fh) Move(fch-fh);
    // But can't always fire.
    // It is allowed to fire if sufficient "reload" time
    // from last shot, and trigger has been pulled
    // (i.e. mouse button clicked)
    Boolean        fired = false;
    long           t1;
    t1 = ::TickCount();
    if ((t1>(flastshot+ShootInterval)) && ftriggerpulls) {
            // Shoot, and record time of shot so as to limit
            // firing rate
            LaserOn();
            LaserOff();
            flastshot = t1;
            ftriggerpulls = 0;
            fired = true;
            // Clear any other mouse clicks, each shot is
            // to be deliberately aimed
            ::FlushEvents(mDownMask,0);
            }
    return fired;
}

void Gun::LaserOn()
{
    // Line representing Laser blast is drawn in short
    // segments. Experimentation showed that this gave a
    // slightly better visual effect.
    ::MoveTo(fh,fv-15);
    for(int x = fv - 20; x > 0; x -= 5) ::Line(0,-5);
}
```

Much of the rest of the code is simply "Mac hacking" – creating windows from
a resource, handling a simple event loop etc. The remaining portions of interest are
the sections where the code creates and uses invaders. The program uses a single
instance of class Gun, and a global array of Invader* pointers:

```
Gun*        gGunBase;
Invader*    gInvaders[MAXVADERS];
```

The game starts by creating a gun and a set of randomly chosen invaders:

```
void InitialiseGame()
{
    // Get limits from screen information.
    gHeight = qd.thePort->portRect.bottom -
                    qd.thePort->portRect.top;
    gWidth = qd.thePort->portRect.right -
                    qd.thePort->portRect.left;
```

```
        gLimit = gHeight - 32;

        // Create the gun
        gGunBase      = new Gun(gWidth,gHeight);
        // and the invaders
        for(short i = 0; i < MAXVADERS; i++)
            gInvaders[i] = Invader::
                RandomInvader(gWidth,gLimit);
    }
```

The code is independent of the number of different types of invader because all such information has been encapsulated in the `Invader` class.

The main part of the game has a loop in which the gun gets a chance to run, as do the invaders:

```
...
Boolean      gGunBase_destroyed;
...
do {
        ...
        ...
        for(i = 0; i < MAXVADERS; i++) {
            gGunBase_destroyed = gInvaders[i]->Run(
                                      gGunBase->Position());
            if(gGunBase_destroyed) break;
            }
        ...
    }
```

Each invader executes its run procedure; because the `Move()` and `Draw()` functions called from `Run()` are virtual, each invader acts according to its own type.

The form of the display on a Macintosh screen is shown in Figure 6.11.

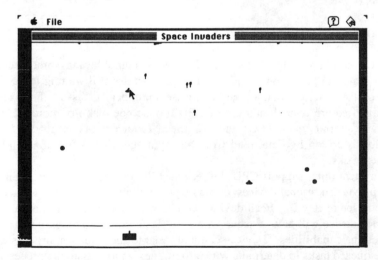

Figure 6.11   Space Invaders on the Macintosh

6.7 EXAMPLE: OPERATING SYSTEM SIMULATION

This example was originally an assignment for students. It involves building a small discrete event simulation system. The system being simulated is a large old-fashioned mainframe computer running multiple programs taken from a "batch stream". The intent is that a basic framework be created that can simulate the execution of jobs in such an environment and provide timing details that allow comparison of different scheduling algorithms. Parts of this framework are to be open for further extension. For example, it should be possible to substitute alternative memory management schemes into the simulated operating system, or make a change to one of the scheduling algorithms (e.g. to change the disk scheduling algorithm from the default first-come/first-served algorithm to a sweep algorithm).

The example is considerably more object-oriented than Space Invaders. In this case, the final main program is a typical OO one – "create the main object, tell it to run". It provides further illustration of interacting systems of classes and demonstrates simulation – a very natural domain for object-oriented techniques.

6.7.1 The assignment

Write a C++ program that provides a framework for student exercises on scheduling policies such as those used in Operating Systems. The program will involve components representing:

- the operating system (OS),
- a central processing unit (CPU),
- a disk (DISK)

and
- a Peripheral Processor Unit (PPU).

OS, jobs, and tasks The OS component has a list of the devices available and some record of memory usage, and maintains a queue of "jobs" that are still waiting to be loaded and run on the system. Each job will comprise a number of "tasks". For example, a job might require a small amount of CPU time, some disk i/o, more CPU time and finally output via a PPU; each of these requirements is a "task". The simulation is driven by data, read from an input file, that define jobs and their constituent tasks.

Devices The system must support CPU, DISK and PPU devices. Later versions may need to provide additional devices, or may have to allow for multiple instances of a particular type of device. Each device can either be idle or working on a current task. Each device will have a queue whose entries correspond to tasks that require its processing capabilities. Devices can add tasks to their queues, process a task, examine queued tasks to determine which to run next, and initiate the processing of a selected task.

Jobs and their constituent tasks Each job is characterized by a name, details of (maximum) memory requirements, an approximate estimate of total processing time, and a priority. The estimated total time and priority can be used by a scheduling algorithm in the OS that selects the next job to add to the group of currently running jobs. The simulation relies on "job traces". When a job is run on a computer one can (at least in principle) trace its activity; e.g.

- a particular job is queued for the CPU;
- when run on the CPU, it uses 5 milliseconds (ms) of CPU time before requesting a disk transfer;
- it is then placed on the disk queue;
- when run, it requests a particular track with the transfer requiring some number of milliseconds of seek and transfer time;
- the job is then requeued for the CPU;
- when run, it uses 17 ms of CPU time before requesting output to a peripheral i/o device like a terminal or printer;
- it is queued for the PPU, which it uses for 4 ms;
- it is then requeued for the CPU; ...

(Note, the system uses non-pre-emptive multiprogramming; a job only loses a CPU when it makes a request for an i/o transfer.) Details of jobs and their task traces are stored in job records in the input file provided for this assignment.

Example job records are:

```
Student72366:CS      40    600   1
   C 5 P 18 C 2 D 77 C 5 … P 10 C 2 Q
Bennett:CS          150 10000   5
   C 20 P 11 C 5 P 18 …  C 100 C 2 Q
Operator             30    500   7
   C 3 P 16 … C 222 P 17 C 2 Q
```

The first part of each record defines the job name, memory and time requirements and priority; the rest of the record is the trace of tasks. For example, the job for "Student72366" requires 40 pages of memory, a maximum of 600 ms CPU time, and is to run at the lowest possible priority; the job for "Bennett" requires 150 pages of memory, a total of ≈ 10000 ms of CPU time, and is to run at priority 5. When job Student72366 runs, it will use 5 ms of CPU time before requesting i/o; its first i/o task requires 18 ms of time on the PPU; the job's next task involves the CPU for 2 ms; a disk transfer (referencing disk track 77) is then required; once the disk transfer is complete, the job will use 5 ms of CPU time before requesting another i/o transfer. Eventually, the job terminates with a 10 ms data transfer on the PPU and a final 2 ms burst of processing (which would correspond to overheads incurred as a job exits from the system).

The OS component in the simulation program controls all activity. It will start by creating and initiating the various devices, and then read a set of jobs from a file, appending them to its main job queue. The OS will then execute a "Run" procedure that will simulate activity for a finite period of simulated time. The Run procedure cycles until the time limit is exceeded; on each cycle the OS will:

Organizing the simulation

- Determine whether another job can be added to the set of jobs running in memory;

- Check whether it is time to generate an "operator's report" summarizing current job loads and queues (such reports should be generated after approximately 1000 ms of simulated time);
- Advance the system time to the time at which the next device task is completed.

When a device completes its current task it should return a reference to the job of which that task was a part. The OS will then identify the next device that that job requires, and will transfer the job to the appropriate queue. The device that completed the subtask will reschedule its queue and initiate its next task.

OS: Job Scheduling and Memory Management:

The OS will attempt to maximize the number of currently running jobs. Memory limitations restrict the degree of multiprogramming that can be achieved. Memory usage by the OS itself is ignored. The basic simulator should use a simple form of paged memory. There will be a fixed number of memory pages available (this should be a parameter defined when starting the simulation, it should default to 500 pages of memory); each job specifies a memory requirement in terms of pages. Paging is *not* being used to provide "virtual memory"; a job must be loaded in its entirety. The paging scheme exists to simplify memory management. If there is a total of 100 free pages, the OS can start any job requiring ≤ 100 pages of memory; there is no need to look for a contiguous block of free pages. The OS may restrict consideration to the first job on its input queue, adding it to the running set if there is sufficient memory; alternatively, the OS may scan the queue until it finds a job that fits within the available free memory, or it may rank all jobs that fit within free memory and select the best. (The job ranking is determined by a formula combining size, estimated total time, and priority; a simple formula is size*time/priority. The OS should favour jobs for which this formula yields a low score.)

Disk: Task scheduling:

The disk should maintain a record of the track where its heads are located. The time to complete a task is based on the distance that the heads must move from their current location to the required location. The disk is a large old slow disk with 200 tracks (0-199) and a track-to-track time of 5 ms; rotational latency can be ignored. When the simulation starts, the disk heads are located over track 0. The disk can determine current head position and the track number required for each task, and so could attempt to optimize the order in which requests are handled. The default implementation is to be a first-come/first-served queue; but it should be easy to modify the scheduling algorithm.

CPU and PPU Task scheduling:

Both CPU and PPU devices should use first-come/first-served queues for scheduling queued tasks (again, alternatives could be explored in exercises that are part of an operating system course.) The CPU is to maintain a record of the total number of milliseconds for which it has been idle.

Outputs required from the simulator

Reports identifying jobs are to be generated when a job is first loaded into memory by the OS and when it finally terminates. At regular intervals (≈1000 ms of simulated time), an "operator's report" should be printed. This report should identify the status of all queues and show the amount of time for which the CPU has been idle. When the limit on simulated time is reached (or when all jobs in the input file have been processed) a final report should be printed giving statistics such as CPU idle time and number of jobs completed. Part of the output from one implementation of the simulator was:

```
================
Report at time 42174:
Jobs started: 12, jobs aborted: 0, jobs finished: 7
Device : C:
    Currently working on:Job 11, Operator
    Size 30, TimeLimit 500, Priority 7
    Queue length : 3
---
Total idle time for this cpu is now 4325
Device : P:
    Currently working on:Job 8, Fairly:ElecEng
    Size 100, TimeLimit 11000, Priority 2
    Queue length : 0
---
Device : D:
    Idling
    Queue length : 0
---

Currently using 490pages of memory
========
--- finished job Job 11, Operator
    Size 30, TimeLimit 500, Priority 7

--- finished job Job 6, Boys:TheoChem
    Size 170, TimeLimit 15000, Priority 2
+++Starting job Job 13, Wesley:Physics
    Size 150, TimeLimit 8000, Priority 4
    at time 42286.
--- finished job Job 8, Fairly:ElecEng
    Size 100, TimeLimit 11000, Priority 2
+++Starting job Job 14, Lawler:Civil
    Size 100, TimeLimit 10000, Priority 5
    at time 42293.
```

6.7.2 Finding the objects and their classes

Many of the objects that will be needed in the simulation are quite obvious from
the assignment specification. There will be an OS object, one or more CPU
objects, one or more PPU and DISK objects, numerous JOB objects, and a variety
of lists and queue objects. Classes defining these different types will have to be
defined and, where appropriate, formed into class hierarchies. One hierarchical
arrangement of classes is obvious from the specification; there will be some general
concept of "Device" with "Device/CPU", "Device/Disk", and "Device/PPU"
specializations.

"Jobs" look relatively simple, and so might as well be the first things analyzed   *"Jobs": how do they*
in any detail. Job objects are going to be created by the OS when it reads the file   *get created and*
with the simulation data; they are going to be stored in a list by the OS; moved   *destroyed*
around and finally destroyed by the OS. It would be reasonable to make "Job" a

specialization of `Item` as defined in section 6.5, because then the `LinkedList` class can get used to store jobs.

Jobs: What do they do?

The first thing that a `Job` will do is read its parameters. It will need a `char*` data member for a name, a few integer fields for priority and memory requirements, and another list. A `Job` probably needs a `LinkedList` to hold its tasks. Once created and put in the main input list by OS, the next thing that will happened to a `Job` is that it will be asked its size and, possibly, its score (size*time/priority) when the OS tries to start some jobs. So, a `Job` better provide functions returning these data. If a `Job` does get picked to run, it is going to be asked the device that it requires next; then it will be asked for additional information – a CPU or PPU will ask how much time, a DISK will ask for block number; actually, as far as Job is concerned it is just being asked for the integer device control parameter associated with its next task. Obviously, if a `Job` owns a list of Tasks, it had better keep track of them; it can remove and delete the first Task in its list when it is told that the work has been completed. A `Job` needs to be able to print out some data.

Task: another class

"Task" had better become a class. Again, it may as well be derived from class `Item` so that Tasks can be used with `LinkedList`. Probably only Jobs need to work with Tasks; there is potential here for friend relations etc. if efficiency becomes important. Tasks don't actually have much to do. They are really just storage structures for a device identifier, and integer parameter – those "C 10" and "D 77" entries in the data file. Initial ideas for "Jobs" and "Tasks" are summarized in Figure 6.12. (Jobs also interact with LinkedLists. But this interaction is an implementation detail and not an intrinsic part of the problem. So, it is not necessary to show the interaction in an analysis diagram.)

Devices

The CPU, PPU, and DISK seem generally very similar so it would be worth starting by describing a general class Device and then considering subclasses where behaviours are specialized. A Device must own a list of things waiting for its attention; in the specification, these "things" were identified as tasks, but really they will be `Jobs` with a `Device` asking a `Job` for task-details when necessary. A Device needs to report its current status, add a job to its queue, schedule (choose the next job to get a task run), identify the job for which it has just completed a task, and work out when it will finish its next task. Very few of these behaviours will have to be overridden in subclasses; the only obvious one is the calculation of the time a task will take.

Devices are going to have to be stored in lists. It would be possible to implement the assignment using one instance of a CPU, one DISK, and one PPU and have the OS interact with them explicitly; however, this would not allow for the required extensions with other devices or multiple devices of a particular type. So it would be more sensible to start with a dynamic array or list of devices.

Actually, devices may have to be in more than one list. The OS would have a list of all the devices it is controlling; in addition, an ordered list might be used to organize the timing of the simulation. When a device starts work on a task, it gets put into this timing list at a position determined by the time at which it is due to complete the task.

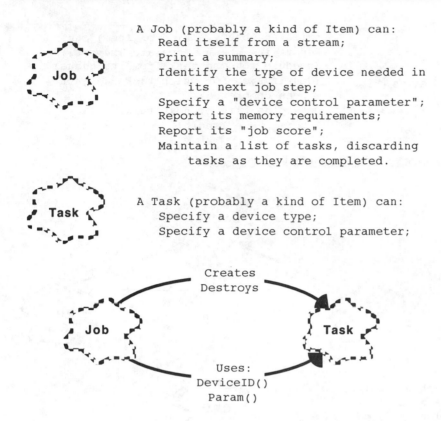

```
                        A Job (probably a kind of Item) can:
                           Read itself from a stream;
                           Print a summary;
       Job                 Identify the type of device needed in
                              its next job step;
                           Specify a "device control parameter";
                           Report its memory requirements;
                           Report its "job score";
                           Maintain a list of tasks, discarding
                              tasks as they are completed.


                        A Task (probably a kind of Item) can:
       Task                Specify a device type;
                           Specify a device control parameter;
```

Figure 6.12 Jobs and Tasks.

If the OS is going to have multiple devices of each of several types, it is going to have to have some way of identifying them and allocating work to them. In the example input for the assignment, identifiers like C (for CPU), D (for DISK) and P (for PPU) were used. Individual CPUs in a set of CPUs are probably going to be interchangeable; a job won't care which CPU it runs on. Similarly PPUs will be interchangeable. So there may well be several "C" devices and several "P" devices. When a job wants disk i/o it will definitely require a specific disk unit; so an extended system would probably have several instances of class DISK each with a distinct identifier. When allocating a job to a device the OS may be able to identify the device uniquely, or it may face a choice. If it does have a choice, e.g. it controls three CPUs and has a job that needs CPU-cycles, it will have to select one; presumably the one with the smallest existing workload (which might approximate to the one with the shortest queue). So, devices are going to have to be able to report on their overall workload.

The roles identified for Devices are summarized in Figure 6.13. Figure 6.14 shows the device class hierarchy. Subclasses will have to be defined for CPU and DISK. The PPU does not have any extra behaviours. If suitable defaults are defined in class Device, it may not be necessary to have a specialized PPU subclass.

A Device (probably a kind of Item) can:
Report its status (operator report);
Schedule (choose task from its queue);
Identify job for which task has just
    been completed;
Determine time of next task;
Report queue length;

Figure 6.13   The responsibilities of Devices.

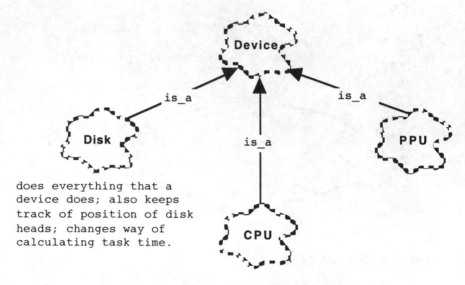

does everything that a
device does; also keeps
track of position of disk
heads; changes way of
calculating task time.

does everything that a device
does; also keeps track of time
that it is idle.

Figure 6.14   Class hierarchy for devices.

class OS   The os will also be defined as a class. The main program will simply create an instance of class os, get it to load the job records from a file, and tell it to run. The implementation of the main program can be completed by the analyst (who likes to demonstrate that she still has the ability to hack out code):

Typical OO main
program!
```
int main(int, char**)
{   OS      OS0;
    OS0.LoadJobs();
    OS0.Run();
    return 0;
}
```

Apart from LoadJobs() and Run(), the only other service that class os should provide to clients will be SystemTime(); the os will keep track of simulated time, devices will need access to the time (e.g. when a CPU is calculating how long it has been idle). Really, SystemTime() is a kind of global entity; it can reasonably be associated with class OS but it doesn't logically belong to an instance; this function will be a class responsibility.

The other responsibilities of class OS will all be private. There will be a method for generating operator reports, a method that tries to fit a new job into memory (this has to be overridable so that different memory management schemes can be explored), and various routines for maintaining lists of devices, lists of jobs, lists of completion times for tasks and so forth.

There will be many other classes in the system. But these will all be variants on lists and queues. They are not the concern of the analyst. The designer will determine what list structures to use – probably selecting them from a library of reusable components.

Other classes

6.7.3 Design and implementation detail

The first design decision was to reuse the LinkedList and Item classes from section 6.5 as the basis for all storage structures. "Tasks", "Jobs", and "Devices" are all going to be stored in linked list structures (instances of class LinkedList) and therefore must be derived from class Item.

Following this decision, the design level class hierarchy is as shown in Figure 6.15.

The class hierarchy is "a forest of little trees". There are three quite separate level-0 classes: Item, LinkedList, and OS (along with the support classes Link and LinkedListIterator which are not shown). Class Item is parent to the more specialized level-1 classes Task, Job, and Device.

Class Device will be further specialized in the derived classes CPU, DISK, and PPU. The "forest of little trees" hierarchy is typical of most OO programs other than those constructed using an elaborate "framework" style class library (in those programs, it is typical for all classes to br derived from some standard base class provided as part of the library and for the resulting class hierarchy to form a single, bushy tree).

Class Task is the simplest of the new classes. As noted in the analysis, a Task is little more than a data structure that identifies a Device and provides an integer parameter (milliseconds of time for a CPU or PPU, track-number for a DISK).

CLass Task

"Devices" have to be identified in some way consistent with the data given in the job traces used as input. The assignment specification does not provide any detail here; the example traces all show single character device identifiers: "C" for CPU usage, "D" for DISK etc. In a generalized system with multiple CPUs, PPUs, and DISKs, any CPUs should suffice for a processing task and any PPU could handle terminal i/o so one character device type identifiers should suffice for CPUs and PPUs; however, the DISKs would need to be distinguished (the contents of track 77 on disks A and B would, after all, be different). DISKs would presumably appear in generalized traces as "D1", "D2" etc., a "char*" string would be more appropriate than a single character for a device identifier. Both Tasks and Devices will have to use "char*" device identifiers.

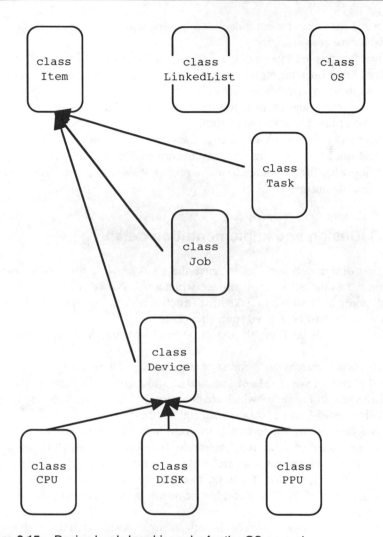

Figure 6.15   Design level class hierarchy for the OS example.

The form of class `Task` is indicated in Figure 6.16. A `Task` will have two data members – its parameter and its device identifier. There will need to be a constructor which will take as arguments a string (device identifier) and integer value and will need to create a copy of its string argument. Making a copy of a device identifier string will require space to be allocated on the heap, so a destructor routine will have to be provided to free that space when a `Task` gets destroyed. In addition to the constructor and destructor, class `Task` should provide two access functions; one that returns the value of its integer parameter and a second that returns a `const` pointer to the device identifier string.

There would be no necessity to redefine the virtual functions `Equal`, `Compare`, and `PrintOn` that are inherited from class Item because these are not required in this application. However, it might be worth redefining `PrintOn` for debugging purposes.

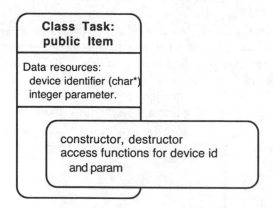

Figure 6.16 Design for class Task.

The implementation of class `Task` is trivial. The code is provided in the file Job.C in the OS subdirectory.

As noted in the analysis section, an instance of class `Job` will need to be able to: *Class Job*

- read its set of tasks and other data from a stream;
- print a summary;
- identify the device required for its next task;
- provide the integer parameter associated with its next task;
- provide details of its memory requirements;
- provide its job score

and
- maintain a list of tasks, discarding tasks as they are completed.

Most of these requirements map very simply into the design; however the maintenance of the list of tasks requires a bit more detailed consideration.

When a `Device` completes a processing step, it should inform the `Job` for which that step was completed. The `Job` can then remove and discard the first entry in its list of `Tasks`. A `Device` completing a task should also identify to the operating system the job that has been advanced. If the `Job` has completed, the `OS` should get details printed and then delete the `Job`; otherwise the `OS` should get the `Job` to identify the next device required and then queue it appropriately. (If a `Job` asks for a non-existent device, the `OS` should presumably abort the job.) The `OS` needs to be able to ask a `Job` whether it is completed. So, a `Job`'s maintenance of a list of tasks will involve two interface functions: a "Complete a step" function (for use by `Devices`) and a "Completed?" predicate (for use by the `OS`).

A `Job` will have as data the user name, details of size, priority, time limit (not actually used in the assignment as specified) and a list of tasks. It would probably be useful to associate a unique identifier with each job. This unique identifier could be an integer that is initialized from some "global variable" that gets incremented each time an instance of class Job is created. Obviously, a static class data member would be used for this "global" in a C++ implementation.

The design structure proposed for class `Job` is illustrated in Figure 6.17. The linked list used to store `Tasks` could be created as a separate structure in the heap (in which case, a `Job` would have a `LinkedList*` data member) or be a constituent data member of a `Job`. Since it is not intended that any other entity have access to the task-list, there is no reason for it to have an independent existence and it would therefore be more appropriate to use a `LinkedList` data member. The "class data" (the counter used to provide unique job identifiers) should be made distinct from instance data (as in Figure 6.17 where a separate box is used).

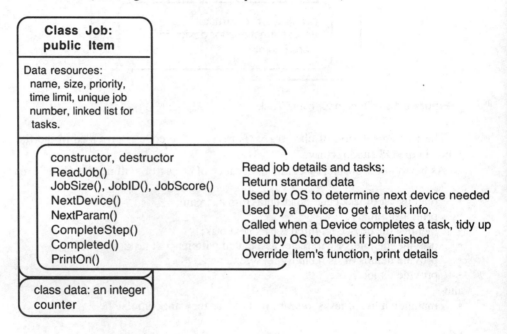

Figure 6.17   Design for class Job.

Neither class `Task` nor class `Job` need make any provision for extension, so there are no virtual functions and no protected functions giving access to private data. The assignment requires some provision for extensions – but these relate only to the "Device" classes and the `OS` class.

The implementation of class `Job` is straightforward. The class declaration comes directly from the design structure:

```
class Job : public Item {
public:
    Job(char* name);
    ~Job();
    void        ReadJob(istream&);
    Boolean     Completed();
    const char* NextDevice();
    short       NextParam();
    void        CompleteStep();
    short       JobSize();
    short       JobID();
    long        JobScore();
    virtual void PrintOn(ostream& os) const;
                // Override PrintOn inherited from Item
```

```
private:
    char*        fName;
    short        fSize;
    short        fPriority;
    long         fTimeLimit;
    short        fJobNumber;
    LinkedList   fTaskList;
    static short sJobCounter;
};
```

A name (the user name data from the trace) is associated with each job. Naturally, this will be represented as a separately allocated string in the heap, with the space claimed in the constructor function. Necessarily, there will be a destructor function that will release this space:

```
Job::Job(char* name) : fSize(0), fPriority(0), fTimeLimit(0)    Constructor
{
    fJobNumber = ++sJobCounter;
    fName = new char[::strlen(name) + 1];
    ::strcpy(fName,name);
}

Job::~Job()                                                     Destructor
{
    delete [] fName;
}
```

The destructor function as written does not include any code to get rid of any remaining "links" or "tasks" in any incompletely processed task list. Such tidying up is handled automatically; the C++ compiler will invoke the destructor for the `LinkedList` data member `fTaskList`.

Most of the member functions of class `Job` return data concerning the state of   *Member functions of* the task-list, details of the first task in the task-list, or other simple data. Such   *class Job* functions could be made "inline"; but there is no real need for such optimizations. (If it were really necessary to optimize performance of the OS-simulator, it would be more appropriate to focus on a redesign that avoids the relatively inefficient linked-list structures rather than to attempt minor optimizations by "inlining" functions.) Note that member functions of the task-list return `Item*` pointers; consequently, type casts to `Task*` are required whenever details of a `Task` are required:

```
Boolean Job::Completed()
{
    return (0 == fTaskList.Length());
}

const char* Job::NextDevice()
{
    Task*  t;
    t = (Task*) fTaskList.FirstItem();
    return t->DeviceID();
}
```

```
short Job::NextParam()
{
    Task*  t;
    t = (Task*) fTaskList.FirstItem();
    return t->Param();
}

short Job::JobSize()
{
    return fSize;
}

long Job::JobScore()
{
    return (fSize*fTimeLimit/fPriority);
}

short Job::JobID()
{
    return fJobNumber;
}
```

The CompleteStep() member function removes the first Task from the task-list, and then deletes it. (This design, where each individual Task is a separately allocated object in the heap, makes very heavy demands on the memory management routines. A more efficient program could be written that used an array of simple data structures, with an index identifying the data that define the current task. The entire array could be allocated and freed as a single entity.)

```
void Job::CompleteStep()
{
    Task*  t;
    t = (Task*) fTaskList.RemoveFirstItem();
    delete t;
}
```

The inherited PrintOn() member needs to be replaced with code that outputs summary data defining a Job:

```
void Job::PrintOn(ostream& os) const
{
    os << "Job " << fJobNumber << ", " << fName << "\n";
    os << "\tSize " << fSize << ", TimeLimit " << fTimeLimit;
    os << ", Priority " << fPriority << "\n";
}
```

The definition of operator<<() for class Item as a call to PrintOn() allows the following stream style output operations:

```
Job*   j;
j = ...;
...
cout << "The job at : "<< hex << j << " is " << *j;
```

This output will print the hex address of a Job and the data that characterize it.

The principal function in class Job is ReadJob() which reads the data characterizing a Job. This design assumes that OS::LoadJobs() routine will read the user-name data from the trace input file, create an instance of class Job, and then request that this new Job read the remaining input data. This input function is fairly typical, with checks for input conversion errors and statements that skip over any trailing characters at the end of an input line. Termination on input error is reasonable in this context; the user will have to edit any incorrectly formatted data file.

```
void Job::ReadJob(istream& is)
{
    // Read simple arguments
    // Size, MaxTime, Priority
    is >> fSize >> fTimeLimit >> fPriority;
    if(is.fail()) {
        cerr << "Bad header data for job:";
        cerr << "Job " << fJobNumber << ", name "
                    << fName << "\n";
        exit(1);
        }

    // next should get a series of (one or two character)
    // device names and integer parameters terminated
    // by a "Q"
    for(;;) {
        char devicename[10];
        is >> devicename;
        if(0 == ::strcmp(devicename,"Q")) break;
        short data;
        is >> data;
        if(is.fail()) {
            cerr << "Bad trace info for: ";
            cerr << "Job " << fJobNumber << ", name "
                << fName << "\n";
            cerr << "last inputs : " << devicename << ", "
                << data << "\n";
            exit(1);
            }

        Task* t;
        t = new Task(devicename,data);

        fTaskList.Append(t);
        }

    is.ignore(SHRT_MAX,'\n');
}
```

Finally, somewhere in the implementation the static data member sJobcounter should be defined:

```
short Job::sJobCounter = 0;
```

Although an instance of class Job will be handled as a single entity, it is (as illustrated in Figure 6.18) a composite made up of many separate objects.

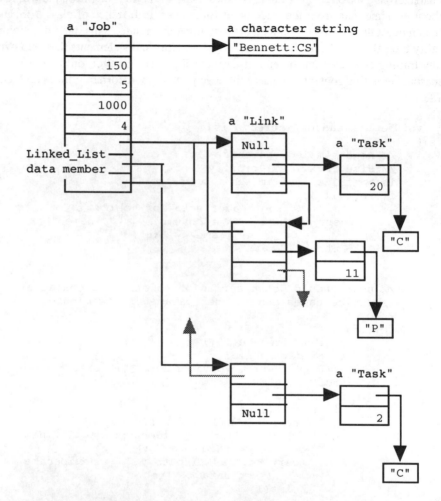

Figure 6.18   Structure of a run-time instance of class Job.

Class Device and
class OS The behaviours of the OS and various devices require further analysis before the designs for these classes can be completed. The heart of the simulation will be the OS::Run() function. This function will entail a loop of the following general form:

```
repeat
    try to load an additional job into memory;
    possibly generate another operator report;
    identify the device that completes next;
    move the simulated system time on to the time of
            completion of the device that is completing a task;
    identify the job for which a task has been completed;
    get the device that finished to start on its next task
            (if it has another task queued);
    determine type of device required in the next task of
```

```
                the job that made progress;
        identify the device with least workload that can handle
                the next task of the job being considered;
        add the job to the queue associated with that device;
    until time-to-stop
```

The simplest way to handle the active devices is to keep them in a list ordered by their completion times. The device that completes next will be the first entry on this list; it should be removed from the list of active devices. It should have a record of the time that it will complete its current work; this time can be interrogated and used to update the simulated system timer. When a device starts on a new task, it should get re-inserted in the list of active devices in a position determined by its calculated completion time. Instances of the LinkedList class can be used as ordered lists provided that the "Items", that are to be inserted, provide a suitable Compare() function. "Devices" will need to implement a Compare() function that uses their completion times in the comparison operation.

The assignment specification would permit the allocation of jobs to devices to be hard coded:

```
next_device_needed = thejob->NextDevice();
if(0==::strcmp(next_device_needed,"CPU")
        theCPU.AddJob(thejob);
else
if(0==::strcmp(next_device_needed,"PPU")
        thePPU.AddJob(thejob);
...
```

But this is just the kind of code that one wants to avoid. It freezes the original specification, so making it much more difficult to add extra devices. A more sensible approach would be for the OS to have a list of all available devices. When a job must be allocated to a device, the OS should iterate along its list of available devices asking each whether it can handle the job and, if so, asking the current workload. The job can then be queued on the device of appropriate type that has the minimum workload. If that device were idle, it should also be told to schedule some work and then get added to the list of active devices. The code for allocating a job would be along the following lines (code from OS::PlaceJob()):

```
Device*         d;
Device*         chosen = 0;
long            workload = LONG_MAX;
LinkedListIterator     Choices(&fDevices);
// Iterate through devices
while(d = (Device*)Choices.Next())
    // Identify those that are of correct type
    if(0==::strcmp(devicewanted,d->DeviceClass())){
        // and have smallest workload
        if(d->WorkLoad() < workload) {
            chosen = d;
            workload = d->WorkLoad();
            }
        }
```

```
        if(0 == chosen) {
            // Job has requested a non existent device
            cout << "Request for non-existent device\n";
            AbortJob(j);
            return;
            }
        chosen->AddJob(j);
        if(chosen->Idling()) {
            if(chosen->Schedule()) fBusyDevices.Insert(chosen);
            }
```

This design entails the OS maintaining two lists of devices, as illustrated in Figure 6.19. Of course, this raises a potential difficulty. A device may appear in two lists; therefore, there is a risk that an attempt will be made to delete a device twice if both those lists are destroyed. The lists would be destroyed when the program terminates. Consequently, the OS code should include a "close down" routine that avoids this problem – for example, the CloseDown() routine could remove any devices still in the list of active devices at the time when a simulation run gets terminated.

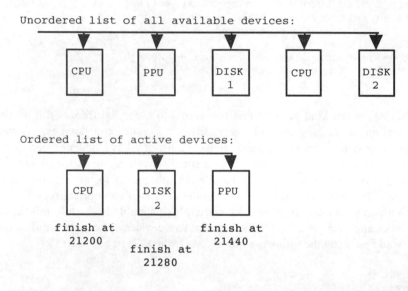

Figure 6.19   OS needs to maintain a list of all available devices and an ordered list of active devices.

Behaviours of   The preliminary considerations in the analysis phase identified the behaviours
"Devices"   of devices as including: reporting of status, scheduling, device identification, calculating and identifying completion time of a task, and reporting on workload. Additionally, "devices" must be able to add jobs, identify a job for which a task has been completed (and at the same time notify that job that it has advanced), report whether it is idle, and provide a Compare() routine that allows devices to be ordered according to their completion times. These roles are summarized in Figure 6.20.

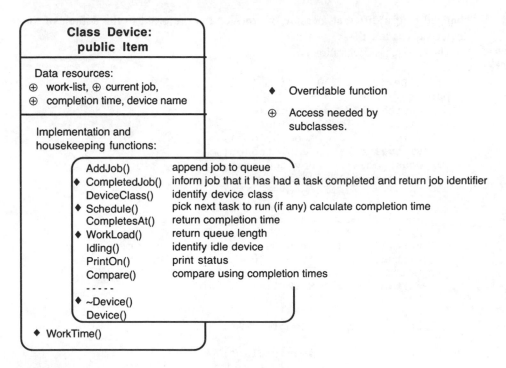

Figure 6.20   Design diagram for class Device.

Devices will have to own data that specify the job for which a task is currently being processed, the calculated completion time of that task, provide a device identifier, and list the other queued jobs.

Unlike class `Task` and class `Job`, class `Device` must include provision for extension in derived subclasses. Apart from the device name, all the data members will need to be accessible in any derived subclasses and should therefore be made "protected" or have protected access functions. (In this simple case, protected data members should suffice.) Additionally, several of the member functions must admit redefinition in derived subclasses. For example, a "Disk" will calculate the completion time of a task in a quite different manner from a "CPU" or a "PPU". For a "CPU" or a "PPU", the parameter associated with a task is the number of milliseconds of processing time required and so the completion time can be determined by simply adding this value to the current system time. For a "Disk", the parameter specifies a track-number and the processing time is determined by the seek-time needed to move the disk-heads from their current position to the required track. Consequently, a "virtual" `WorkTime()` function is required.

Provision for extension in class Device

The processing required when a task is completed will also vary according to the type of device. A record of idle time is required for CPUs; Disks need to keep a record of the position of the disk heads. Consequently, the `CompletedJob()` method should be virtual so that these different types of data can be updated when a `Task` is completed for a `Job`. Two other member functions have been marked as virtual: `Schedule()` and `WorkLoad()`. These have been made virtual to facilitate modification if the standard first-come-first-served scheduling policy is to be changed or if a different measure of workload is preferred to the default of the device's queue length.

The C++ class declaration is:

```
class Device : public Item {
public:
    Device(char* deviceclass);
    virtual        ~Device();
    void           AddJob(Job*);
    virtual Job*   CompletedJob();
    const char*    DeviceClass();
    virtual Boolean Schedule();
    long           CompletesAt();
    virtual long   WorkLoad();
    Boolean        Idling();
    // Override Item ...
    virtual void   PrintOn(ostream& os) const;
    virtual short  Compare(const Item*) const;
protected:
    virtual long   WorkTime();
    LinkedList     fPending;
    Job*           fCurrent;
    long           fCompletesAt;
private:
    char*          fDeviceClass;

};
```

The code for class Device is all straightforward. The constructor and destructor look after the device name:

```
Device::Device(char* deviceclass)
{
    fDeviceClass = new char[::strlen(deviceclass) + 1];
    ::strcpy(fDeviceClass,deviceclass);
}

Device::~Device()
{
    if(fCurrent)
            delete fCurrent;
    delete [] fDeviceClass;
}
```

Several functions provide access to data members or computed properties:

```
const char* Device::DeviceClass()
{
    return fDeviceClass;
}

long Device::CompletesAt()
{
    return fCompletesAt;
}
```

```
long Device::WorkLoad()
{
    return fPending.Length();
}

Boolean Device::Idling()
{
    return (0 == fCurrent);
}
```

As an override of the inherited `Item::Compare()` member, `Device:: Compare()` takes an `Item*` argument that has to be type cast to `Device*`. It has been written to return a short 0/1 flag which suffices here, but if more general usage were intended it would be better to have it return an enumerated type e.g. `enum COMPAR { LESS = -1, EQ = 0, MORE };` The result of `Compare()` will get devices ordered according to their completion times:

```
short Device::Compare(const Item* i) const
{
    // Devices get inserted into a work queue
    // based on completion times
    Device*     d; = (Device*) i;
    return fCompletesAt - d->fCompletesAt;
}
```

Default implementations can be defined for the virtual functions `CompletedJob()`, `Schedule()`, and `WorkTime()`:

```
Job* Device::CompletedJob()
{
    // current task completed
    // may have extra work to do, by default just identify
    // the job for which work has been done
    Job*   j = fCurrent;
    j->CompleteStep();
    fCurrent = 0;
    return j;
}

Boolean Device::Schedule()
{
    // If have no work in Pending list return false
    // i.e. (fail to schedule)
    if(0 == fPending.Length()) return false;
    // Otherwise, pick an item, get working
    fCurrent = (Job*) fPending.RemoveFirstItem();
    fCompletesAt = OS::SystemTime() + WorkTime();
    return true;
}

long Device::WorkTime()
{   long   jobsteptime;
    jobsteptime = fCurrent->NextParam();
    return jobsteptime;
}
```

Definitions for the other member functions of class `Device` are in file Device.C in the oocc subdirectory of ftp-able files associated with this text.

Subclasses of Device   Three subclasses have to be defined:

```
class PPU : public Device {
public:
    PPU(char* dc);
};

class CPU : public Device {
public:
    CPU(char* dc);
    virtual Job*        CompletedJob();
    virtual Boolean     Schedule();
    virtual void        PrintOn(ostream& os) const;
private:
    long                fIdleTime;
    long                fFinishedLastJob;
};

class DISK : public Device {
public:
    DISK(char* dc);
    virtual Job*        CompletedJob();
protected:
    virtual long        WorkTime();
    short               fHeadPosition;
};
```

Class `PPU` is not really necessary as there are no additional or changed behaviours. It is defined simply to facilitate subsequent extension. Class `CPU` requires extra data member to record idle time and the time at which it was last working; and class `DISK` needs a data member to record a track-number.

The redefined `CompeteJob()` and `Schedule()` member functions of class `CPU` both invoke the inherited member function from class `Device` as well as performing additional steps:

```
Job* CPU::CompletedJob()
{
    // record time at which job finished so that
    // can compute idle times
    fFinishedLastJob = OS::SystemTime();
    // Then do same as class Device does
    return Device::CompletedJob();
}

Boolean    CPU::Schedule()
{
    // Use standard class Device code to determine whether
    // can schedule a job
    Boolean        canSchedule = Device::Schedule();

    // If scheduling a job, adjust idle time
    if(canSchedule) {
```

```
                    fIdleTime += OS::SystemTime() - fFinishedLastJob;
                    }
          return canSchedule;
     }
```

Class `Disk`'s redefinition of `WorkTime()` is a complete replacement of the inherited behaviour:

```
long DISK::WorkTime()
{
     // Disk head move times 5
     short delta = fCurrent->NextParam() - fHeadPosition;
     delta = (delta < 0) ? -delta : delta;
     return delta*5;
}
```

Figure 6.21 summarizes the design for class OS. The easiest way of extending *Class OS*
the simulation system would be to define a subclass of class OS in which some of
the member functions are redefined. Given that arbitrary extensions may be
necessary, it is appropriate for all data members of class OS to have protected
access (so that they may be used in subclasses) and for the majority of the member
functions to be virtual so that they may be redefined. The main program requires
only a very simple interface to class OS so few of the members are public. Most
members are "protected"; this allows any derived classes ("class OS1", "OS2" etc.)
to call the default implementations in any redefined versions of these member
functions.

The most likely refinement of class OS would be one with a more complete
memory management model. The version defined simply has integer data
members defining the total number of pages of memory and the number currently
in use. The total number of pages would be initialized in the constructor routine.
Redefinable (virtual) functions are provided for claiming and releasing memory.
These member functions would need to take a job-identifier as argument so that a
more complex memory management system could record the allocation of pages to
specific jobs. The default implementation would simply involve updating the count
of pages in use.

The `LoadJobs()` member function would read the trace input file, creating
instances of class `Job` and getting these jobs to read their own data. Another
`LinkedList` could be used to queue the `Jobs`. The `Run()` method was outlined
earlier. It is the main loop in which time is advanced as devices complete work on
individual tasks. The system's record of time should be defined as a static variable
accessible by a static member function. This allows devices etc. to access this
information (via `OS::SystemTime()`)without the need for a `OS*` pointer data
member in each device as would be required if the time were held as an instance
variable.

The members `StartJob()`, `AbortJob()`, and `FinishJob()` would print details
to the record of the simulation program. `StartJob()` would also involve a call to
`ClaimMemory()` and would add the job to the queue associated with the device
required in the job's first task. `AbortJob()` and `FinishJob()` would call
`ReleaseMemory()` and update counts of jobs processed.

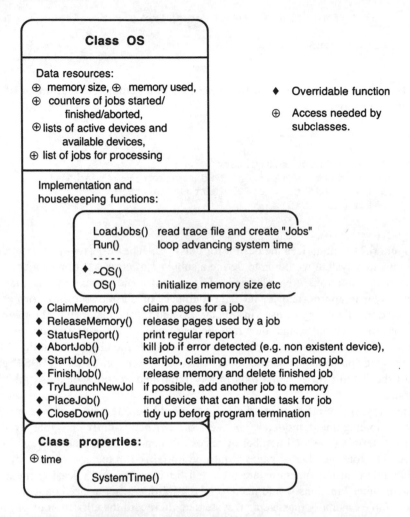

Figure 6.21   Design diagram for class OS.

The default implementation of `TryLaunchNewJob()` would simply consider the first job in the input queue and determine whether there was sufficient free memory. This would be another function likely to be replaced in any student exercises exploring more realistic scheduling policies.

Implementation of
class OS

The C++ class declaration is:

```
class OS {
public:
    OS(short sz = 500);
    virtual ~OS();
    static long         SystemTime();

    void                LoadJobs();
    void                Run();
```

```
protected:
    virtual void       ClaimMemory(Job* j);
    virtual void       ReleaseMemory(Job* j);
    virtual void       StatusReport();
    virtual void       AbortJob(Job* j);
    virtual void       StartJob(Job* j);
    virtual void       FinishJob(Job* j);

    virtual Boolean    TryLaunchNewJob();
    virtual void       PlaceJob(Job*);
    void               CloseDown();

    short              fmemsize;
    short              fmemused;
    short              fJobsFinished;
    short              fJobsStarted;
    short              fJobsAborted;
    static long        sOStime;
    LinkedList         fDevices;
    LinkedList         fBusyDevices;
    LinkedList         fJobQueue;
};
```

Most of the member functions are quite straightforward. Definitions for all are in the example files provided.

The constructor creates the devices as required in the assignment:

```
OS::OS(short sz)
{
    fmemsize = sz;
    fmemused = 0;
    // OK, told that we have at least one CPU, one PPU,
    // and one DISK
    CPU*   c = new CPU("C");
    fDevices.Append(c); // CPU 1
    PPU*   p = new PPU("P");
    fDevices.Append(p);
    DISK*  d = new DISK("D");
    fDevices.Append(d);
//   c = new CPU("C");
//   fDevices.Append(c); // CPU 2
    fJobsFinished = fJobsStarted = fJobsAborted = 0;
}
```

Member LoadJobs() handles the opening and closing of the input file of trace data (jobfilename is a global name string), and the creation of Jobs:

```
void OS::LoadJobs()
{
    ifstream jobfile;
    jobfile.open(jobfilename,ios::in);

    char namebuf[100];
    while(jobfile>>namebuf) {
            Job*    j;
```

```
                    j = new Job(namebuf);
                    j->ReadJob(jobfile);
                    fJobQueue.Append(j);
                    }
            jobfile.close();
    }
```

The general form of member `Run()` was given earlier; the definition of this function is:

```
    void OS::Run()
    {
        long nextreport = 0;
        for(;sOStime<RUNLIMIT;) {
                // Get as many jobs in memory as possible
                while(TryLaunchNewJob());

                if(sOStime >= nextreport) {
                            StatusReport();
                            nextreport = sOStime + REPORTING;
                            }

                if(0 == fBusyDevices.Length()) {
                        cout << "Machine idle. Go home early\n";
                        break;
                        }

                Device*      d;
                d = (Device*) fBusyDevices.RemoveFirstItem();

                // adjust time forward
                sOStime = d->CompletesAt();

                // Identify job that has progressed
                Job* j;
                j = d->CompletedJob();

                // Get device working on something else
                if(d->Schedule())
                            fBusyDevices.Insert(d);

                // And look at job
                if(j->Completed()) FinishJob(j);
                else PlaceJob(j);

                }
        cout << "Simulation ended\n";
        StatusReport();
        CloseDown();
    }
```

This code is much more "object-oriented" than the earlier Space Invaders example. Almost every statement is of the form "*hey object, do something*".

EXERCISES

1. Extend the Space Invaders example by implementing class GuidedMissile. A GuidedMissile is an Invader, with a suitable graphic image, that moves directly toward the current position of the gun accelerating slowly as it approaches.

2. Extend the "Gun" class, and the interface between the "Gun" and "Invader" classes (i.e. the CheckDestroyed method of Invader), to allow a Gun to fire sideways at Invaders advancing along the ground. Then implement Class Dalek, an Invader that enters at extreme left or right of the display area and advances along the ground until close enough to exterminate the gun.

3. In Chapter 3, the discussion of the desirability of automatic garbage collection used an example of a Transport class with a crew of Daleks. Implement the version of the Space Invaders game described in that section. Try to find a generalized way of dealing with memory management for the Daleks.

4. This exercise allows you to explore "efficiency".

   a) First, time the execution of the "OS simulation" program as provided.

   b) Then, make all suitable member functions "inline" so as to enhance their efficiency and again time the execution of the program.

   c) Finally, take the original code and replace some of the LinkedLists with more efficient data structures and again time the execution of the program.

   This exercise should show that playing with "inline" produces only small improvements in performance and that if there are performance problems with a program it will normally be necessary to explore alternative algorithms or data structures.

5. Write a simulation program for an "electronic office":

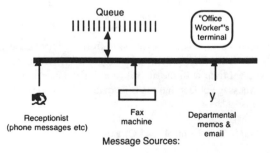

The electronic office will provide executives with a display of incoming documents that require their attention.

It will handle three classes of document. These are:
    Office Memoranda
    Facsimiles received on the fax machine
    Phone messages that have been recorded by secretarial staff.

The system must make provision for additional classes of document.

Incoming documents are to be stored in a queue.

The system will provide mechanisms for reviewing the contents of the document queue, and for selecting a document from this queue. The selection mechanism will take either a chosen document (identified by its position in the queue) or, by default, will take the document at the front of the current queue.

Once a document has been selected, the following processing options are to be enabled:

Discard – the document is to be expunged from the system
Requeue – the document is to be requeued for subsequent
consideration (added at end of existing queue).
Summarize – a summary of the document is to be presented.
Display – the contents of the document are to be displayed.

All documents have the same summary data. These consist of a time stamp that specifies when the document arrived at the executives workstation, an identification of the source of the message, and a 'Subject' header. The subject header will be a single word, or short phrase, that identifies the subject matter covered in the contents of the document.

The summary display should be the same for all classes of document.

Document contents vary. The current document classes have the following data:
Office Memoranda : a block of text.
Facsimiles: a picture.
Phone messages: a sound recording
Appropriate display mechanisms are to be chosen for these different document classes.

The document sources are to be "scripted". An input file should be provided for each separate input source. This input file should contain a sequence of messages ordered by arrival time. Each message has the standard data (time of arrival, summary etc.) and a record with the text, pictorial or encoded sound data.

The document sources are obviously "objects" from various classes. Each should have a Run() member function that gets invoked regularly. A source's Run() function will compare the current time with the arrival time for the next message, and if appropriate generate a new message object that reads its data and then gets appended to the queue.

6. Implement an "Air Traffic Control" (ATC) game.

The ATC game is to give users some feel for the problems of scheduling and routing aircraft that are inbound to an airport. Aircraft are picked up on radar at a range of ≈150 miles. They are identified by call sign, position, velocity, acceleration, and details of the number of minutes of fuel remaining. (A plane's velocity and acceleration are reported in terms of x', y', z' and x'', y'', z'' components.) The user (game-player, trainee air-controller, or whatever) must direct planes so that they land successfully on the single east-west runway.

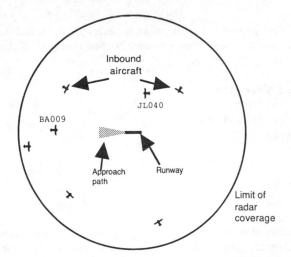

The game is to work with a main loop, each cycle of which represents one minute of simulated time. In each cycle, a function is called that may add another plane to the controlled air space (planes can be generated from an array of prototype examples), all planes in the controlled space update their positions etc, and then the program waits for input from the player. The player may request a status display and can give commands to individual planes.

A simple status display and command line interface for user input will suffice. The player should be able to get a list showing the status of all planes, e.g.

```
BA009   (-128,4.5,17000) (4,-0.5,-1000) (0,0,0) fuel = 61
JL040   (20,90,22000) (-8,0,0) (0,0,0) fuel = 110
```

Each line would identify a plane's call sign; the first triple gives its x, y, z position (horizontal distances in miles from the airport, height in feet); the second triple gives the x', y', z' velocities (miles per minute, feet per minute); the third triple is the accelerations x", y", z" (miles per minute per minute, feet per minute per minute); finally the number minutes of fuel remaining is given. For example, the BA plane is flying at 17000 feet, currently 128 miles west of the airport and 4.5 miles north of the runway, its total ground speed is ≈4.03 miles per minute (241mph), it is descending at 1000 feet per minute, it is not accelerating in any way, and it has 61 minutes of fuel remaining. The JL plane is north-north-east of the runway heading west and maintaining its height and speed.

The player can send new commands to planes; a command will specify the accelerations (as x", y", z" components) that should apply for a specified number of minutes (subsequent directives will override earlier settings) e.g.

```
BA009   0 0 100 3
```

This command instructs plane BA009 to continue with no change to horizontal velocities, but with a small positive vertical acceleration that is to apply for the next three minuts (unless changed before then in a subsequent command). This will cause the plane to reduce its rate of descent.

Controlling planes by specifying accelerations is unnatural. But it makes it easy to write
the behaviours of the planes. You need simply remember your high school physics
equations:

```
u    initial velocity
v    final velocity
s    distance travelled
α    acceleration
t    time
```

$$v = u + \alpha * t$$
$$s = u * t + 0.5 * \alpha\, t^2$$

There are a number of potential problems that the player should try to avoid. Planes may
collide. If a plane flies to high, its engines stall and it crashes. If a plane flies too low it
runs into one of the hills in the general vicinity of the airport. If it runs out of fuel, it
crashes. If its (total) horizontal speed is too low, it stalls and crashes; if the horizontal
speed is too high, or if it is descending too fast, the wings fall off (and, again, it crashes).
(The parameters defining speed and height limits etc should be "consts" declared in one
of the header files for your program.)

At each cycle of the simulation, every plane recomputes its position and velocity.
Normally, one unit of fuel is burnt each minute. If a plane has a positive vertical
acceleration ("*Climb!*", "*Pull out of that dive!*", ...) or its horizontal accelerations lead to
an increase in its overall horizontal speed, it must have been burning extra fuel; in such
cases, its remaining fuel has to be decremented by one extra unit.

Planes must be guided so that they line up with the approach path to the runway. If they
are flying due east at an appropriate velocity, within certain height limits and with some
remaining fuel, they may be handed over to an automatatic landing system. These
planes are considered to have been handled successfully, and are deleted from the
simulation. For a plane to achieve a safe landing i t must satisfy the following
constraints:

Approach height	(600 ... 3000) feet
Distance west of the runway	4 miles ... 10 miles
Distance north/south of runway	< 0.5 miles
Fuel remaining	> 4 minutes
x' (x velocity, miles per minute)	+2 ... +3 mpm (i.e. flying east 120-180mph)
y' (y velocity, i.e. north/south)	-0.1...+0.1 mpm (minimal transverse speed)
z' (vertical ascent/descent)	-500 ... 0 fpm (descending, but not too fast)
x", y", z"	≈0 (no accelerations)

The program should provide an entertaining game environment where the players are
challenged to land as many planes as possible, with the game code detecting and
reporting misadventures like collisions, planes flying into hills or descending so fast that
their wings fall off.

The "objects" you will need obviously include planes, but there will also be
assorted lists, user-interface objects etc. Start by identifying likely candidates.

Tools and Class Libraries

$\mathcal{E}$xploiting Reusable Code

One of the main claims of the object-oriented programming approach is that it facilitates the development of complex programs by allowing "reuse" of components. Such "reuse" must be demonstrated.

Most compilers for OO languages on personal computers are now supplied with "class libraries". In addition to those provided with compilers, there are many other class libraries that are either available from commercial suppliers or which exist as public domain code (usually accessible from archive sites on the Internet). Most of the public domain class libraries are for C++ on Unix, with a few for C++ on IBM-PC operating systems.

Some of these class libraries are simply collections of supposedly useful components. The classes in these libraries will typically define "abstract" (programmer defined) data types. Instances of these classes that can be used to manage particular resources. For example, a library might provide a class that represents a kind of "sparse array manager". An instance of this class could be created, in a client program, to provide an initially empty array to which elements could be added, accessed, modified, or removed without the client programmer having any need to be concerned with storage management issues or the coding of efficient access methods. As another example, a library class might simply provide a "wrapper" around some system data, for example the data defining the time and date. The library class would provide several alternative ways of accessing the time data (e.g. as a string, as millisecond count, or whatever).

Class libraries

The majority of classes in such a library will be "concrete" (i.e. directly instantiatable) with minimal or no provision for further extension or specialization. As well as libraries providing standard utilities (e.g. a "date" class), and data structures (e.g. a "sparse array" class), there are similar but more specialized class libraries. For example, a library might contain a set of classes that facilitate the use of "sockets" in a distributed Unix environment.

Concrete classes

Although such class libraries are useful, they do not make a dramatic impact on program development. (The last time that creating a linked-list was a significant part of your work was probably back in first year university when you were writing your first programs in Pascal.) While it may take you only an hour or so to write your own "class date" and "class LinkedList", it is often better to use the classes from a library. The library classes will have been thoroughly tested and shown to perform correctly whereas it is quite likely that a hurriedly implemented version will contain errors in the handling of less common cases.

Partially implemented abstract classes

Libraries with classes that are designed to be extended or specialized by client programmers do have a more significant impact on the costs of developing new applications. The classes in these libraries will for the most part be "partially implemented abstract classes". Each class will define some concept (abstract type) and specify its behaviours. Default implementations of many of the behaviours may be provided, while the implementation of others will be "deferred" (i.e. left as "pure virtual functions" in C++). In some cases, the designer would have been able to define sensible default implementations for all the behaviours of a class. Any such class will be instantiable, but will usually have provision for further specialization by subclassing. The designer of each library class will have given careful consideration to the needs of programmers who must implement derived classes. So, protected access functions would typically be provided for most (maybe for all) private data members (or the data members might be given protected access status). The majority of the member functions would be declared as being virtual; and, of course, the library classes would all have virtual destructors.

GUI class libraries

Such libraries are most commonly encountered as "GUI libraries" (Graphics User Interface libraries). These libraries are inherently platform specific (though some commercial suppliers can provide equivalent libraries for popular platforms such as Unix/X-Windows and IBM-PC). The classes define the various types of visual element from which an interface can be constructed. Thus, there might be a set of classes that help handle scrollable views that are too large to be displayed in their entirety. Such a set would include classes that can be instantiated to provide various forms of scroll-bar that control movement, and a "scroller" class that maintains a coordinate frame and interacts with the scroll-bars. A programmer employing such a class library would probably be able to use the scroll-bar class unchanged. The scroller class might be abstract, with some member functions left as pure virtual functions. The programmer creating a new application would then create a specialized subclass of class `Scroller`, providing definitions for all pure abstract functions as well as adding new functionality and, possibly, replacing some inherited virtual functions.

Class libraries like these GUI libraries can significantly enhance programmer productivity. The ways in which instances of different classes interact will all have been sorted out, thus saving the application programmer from having to do a lot of design work. Typically, there are default implementations for the vast majority of the member functions of the various classes. Consequently, instead of having to code up an entire interface, the application programmer need only implement those functions that are inherently application-specific (the pure virtual ones) along with any others whose default implementations are considered to be unsatisfactory. Furthermore, the use of standard classes enhances consistency among different products because applications built using the same GUI classes will tend to have the same "look and feel".

Framework class libraries

Code reuse can be maximized through the exploitation of "framework class libraries". A framework can be viewed as providing a skeleton application that can be extended and specialized through class inheritance. This skeleton will be constructed using instances from a large set of classes. These classes will involve many collaborations; they are much more strongly interrelated than the relatively simple classes in a basic GUI library. Typically, a program built using a framework will involve an "application" object that works with "document" objects that interact with "views". The collaborations among instances of these classes are often complex. For example, the application might detect that a window was moved and determine that it has to request a view to update itself; the view would find that it needs to communicate with the document to get data drawn; the document must then obtain, from the view, details of the display area so that it can identify the particular data that are visible and must be redrawn.

The "application", "document" and "view" classes in a framework library will be partially implemented abstract classes. While many of the behaviours of these classes, and patterns of interactions among instances of these classes, can be identified, actual details of implementation of some behaviours have to be deferred and left for definition in derived subclasses. It is not uncommon for these classes to have one hundred or more different member functions; most of which will have default implementations.

MacApp-1.0 (circa 1986) was one of the earliest of these frameworks. It was created to facilitate the implementation of programs (applications) for Macintosh computers. It had been noted that, excluding games and other special programs, all the early Macintosh programs could be conceived as involving: i, an "application" that handled interaction with users and managed "documents", ii, one, or more, documents that managed the run-time data structures representing the information manipulated by the user, and that organized data transfers to and from disk files, iii, views of the data, and iv, application specific data structures (e.g. "paragraphs" in a word-processor, "records" in a database, or graphics elements in a MacDraw-like graphics editor). The first three of these four parts were essentially the same irrespective of the application. Inevitably, very similar code was being reimplemented in each new program.

MacApp as an example framework

It was realized that if suitable abstractions could be found, much of the standard code could be provided in a library (with the proposed standard Macintosh look and feel actually implemented correctly for once). For example, every Macintosh application had had to include an "event-loop" that received mouse-down, key-stroke, update, and other "events" from the underlying operating system and which distributed these to other components in the code for handling. This standard code could be provided as a "Run()" method for an "application" class. This Run() method would sort out the events, passing updates to appropriate "view" objects, resolving mouse-downs into menu requests (that might be handled by "views", or "documents", or by the application object itself) and drawing actions (that would be handled by views). The general behaviour of an "application" object, and its modes of interaction with "views" and "documents" could be almost completely characterized with default implementations provided for all required functions.

The general behaviours of "documents" could also be characterized. Documents handle some menu requests, determine how much disk storage is needed for saved data, and transfer data to files. Of course, it is not always possible to provide a useful default implementation. A "document" must be able to accept a "DoWrite" message – but what it writes is obviously program- specific. A document class could be defined, but many of its methods were, necessarily, deferred. A programmer implementing a new program for the Macintosh would have to create a specialized subclass of "class document" that provided definitions for all these deferred methods.

MacApp has evolved substantially between version 1 and the current version 3 (and its implementation language has switched from Object Pascal to C++). As well as proving a fairly useful development aid for the Macintosh platform, MacApp served as a model for many other framework libraries.

ET++, OWL, MFC

ET++ is a public domain class library for C++/Unix environments. Its initial structure was based on MacApp-2 (with C++ substituted for Object Pascal) but it has evolved into a functionally-richer and generally more sophisticated system. (Like much public domain software, there are bugs in ET++ and it will occasionally crash or freeze; but the same can be said of several costly commercial software packages.) Both Borland's OWL class library and Microsoft's Foundation Classes (MFC) are based in part on MacApp concepts as reimplemented for IBM-PC with the Windows operating system. These libraries, particularly Borland's, are now generally more advanced than MacApp in the use that they make of C++.

"Programming by difference" with the framework class libraries

When applications are created using these frameworks, the programming style is very much one of "programming by difference". A "chemistry editor application" is just like a standard application except that it creates "chemistry-documents" rather than plain "documents". So, the programmer changes the DoMakeDocument() method in a ChemApplication class derived from the standard Application class. A "chemistry-document" is just like a standard document, save that it has some data structure representing a molecule, its DoRead() and DoWrite() methods actually involve data transfers, and it creates views of its data rather than blank views. So, the programmer creates a new ChemDocument class derived from the library class Document, adds some data fields and data manipulation methods, and provides implementations for deferred functions like DoRead(). The amount of code that must be written for the new application is significantly reduced. (Most of these class libraries seem to come with an example of a quite sophisticated text-editor. The code for this editor program will comprise less than one hundred lines; everything will be done by creating instances of standard window classes, text-handler classes etc.) A further advantage of using the libraries is that the programmer is able to focus more on, and spend more time on application-specific issues because so much the user interface, i/o, storage-management, and other components are provided by classes from the library.

These frameworks differ in detail. But, they are sufficiently similar in overall structure that basic principles learnt from one framework will be relevant to the others. Of course, mastery of any one framework only comes with months of usage.

Becoming familiar with a class library

Although framework class libraries, and the slightly simpler GUI libraries, can be extremely useful, their size and complexity makes usage difficult for new programmers. Before a programming team can create a new application, the individual programmers must become familiar with the classes in the library and the ways in which they interact.

For example, a programmer might choose to have a pointer to a "view" in the objects that represent data (molecules, paragraphs, pictures or whatever), and therefore would write code that places such links in data objects as they are read in from a file. This would work – provided that the required views existed before the data objects were read from the file (otherwise an invalid pointer value would be recorded in each data object). How could the programmer know whether the code is correct? Somehow, that programmer must first realize that there is a potential problem, and then be able to determine whether "Document::DoMakeViews()" is executed before or after "Document::DoReadData()".

Conventional subroutine libraries can be documented by just listing the arguments and result types of each routine and providing a paragraph of explanation. Similar documentation can be provided for class libraries with a note describing the class as a whole and short paragraphs describing each member function. But such documentation doesn't really help that much because it does not reveal the pattern of dynamic interactions among instances of different classes.

Documentation of class libraries

More specialized documentation tools are required. The most common tool is a "class browser". A browser allows the source code of a class library to be examined. While a conventional editor can be used to view code, it presents information in terms of files containing program statements; the presentation from a browser is in terms of the classes. A browser can display hierarchical relations among classes, present class declarations, and show method definitions. In complex class hierarchies, a browser can show where a virtual method is introduced in a hierarchy and where it has been redefined in derived classes. The display of a class can show just the features that it defines, or all its features as obtained through inheritance. The browser will typically be able to find all places where a particular method is invoked. Using a browser, a programmer who has some basic familiarity with the class library can rapidly tease out details of how instances of particular classes might interact and so be able to determine how to extend the interactions when additional processing is desired.

Class browsers and other tools

The next four chapters explore some of the class libraries and their support environments. Chapter 7 provides a brief overview of some of the available libraries and includes pictures illustrating features of typical OO development environments. Chapters 8 illustrates simple use of class libraries to create a program that allows some data to be viewed in different ways. Implementations using Borland's OWL library, ET++, Symantec's Think Class Library, and MacApp are available; most of the code fragments shown are from either the MacApp ot ET++ implementations. Chapter 9 reviews a number of the more advanced facilities commonly available in framework class libraries; while Chapter 10 presents the analysis and design for a specialized cut-and-paste graphics editor that can be built using these libraries.

Tools and environments
Examples using the class libraries

Development Tools and Libraries

7

This chapter provides a brief overview of available class libraries, tools, and environments used for the development of OO programs. The tools include "class browsers", "object-aware" source level debuggers, and environments that integrate an editor, possibly some form of source code control system, and a "make" facility.

7.1 CLASS LIBRARIES

7.1.1 Component libraries

There are now a number of component class libraries that are either available commercially or that can be obtained freely having been contributed to the public domain. These libraries provide "useful building blocks" for the construction of other programs. These "building blocks" are concrete (i.e. directly instantiatable) classes defined to represent commonly required abstract data types. The classes in such a library may be independent; but often most will be formed into a class hierarchy. Any such hierarchy will tend to be a quite simple tree-structured hierarchy with a broad but shallow branching structure. There are sometimes bushy clusters of related classes within the overall tree; classes in these clusters have to be instantiated together as they involve interdependencies and interactions. (Although more sophisticated, these clusters are like the `LinkedList` cluster introduced in Chapter 5.)

In the mid-1980s, it was argued that a "software components subindustry" would start. Such a subindustry has not really materialized yet, and perhaps it never will have a significant role. "Building blocks" like a class `LinkedList` or a neatly encapsulated class `Date` etc don't have much commercial value. Such classes simply don't give sufficient leverage to the programmer. They may be convenient, but one wouldn't pay much for a commercial package because there are public domain versions of such classes available for free, and one can always write one's own versions.

Software components subindustry?

Class libraries that capture a company's expertise are proprietary

At first glance, it might appear that companies could make profits from specialized class libraries tailored for narrow vertical markets. For example, a company with significant experience in a special field (e.g. computer-aided circuit design, or computer representation of chemical structures) could compose a special purpose class library that would be extremely useful to any programmer implementing a new program in that domain. But it would not be profitable for that company to market its class library. The value of that library would be that it captured the appropriate abstractions for the specific domain as they had been identified through years of work in that domain. A specialized consulting company that worked in given vertical domain would obviously continue to seek consulting work and could not expect to survive on sales of its published library. Even if released without source code, the classes defined in the library, and their patterns of interactions, would help competitors create products rivalling those of the original software company. Usually, purchasers are only interested in class libraries that include source code, and the availability of the source code further compromises the originating company's intellectual property rights.

Limited public domain and commercial libraries

Most class libraries developed in the near future are likely to be in-house efforts that capture the particular expertise of individual companies. Such libraries are not intended for distribution; they are created to act as aids to subsequent developments within the originating companies. There will be some public domain and commercial libraries. The public domain libraries will typically be the products of universities, or government research institutes, that have research funding which requires any products of the research be put into the public domain. The commercial libraries will for the most part be licensed versions of the same basic code with some extensions, improvements, and support justifying the fee.

Learning to create class libraries

Of course, programmers working on in-house developments will have to learn how to create the specialized class libraries that their companies require. Contrary to popular myths, one cannot build a useful class library by scavenging bits of code from past projects. Those past projects may suggest useful abstractions that might be developed into reusable classes, but that development will require time and effort.

Using existing class libraries as examples

One of the major values of the better public domain class libraries, and low cost commercial libraries, is that they provide programmers with a model of how to organize a library and how to write classes that really are reusable.

NIH class library

The first significant component library for C++ was the NIH class library developed by Gorlen and his colleagues. The library is documented in the book by Gorlen, Orlow, and Plexico. This is a public domain library that is available from several archive sites on the Internet (try ftp from alw.nih.gov, 128.231.128.251, file pub/nicl-3.0.tar.Z). A version for the IBM-PC can be purchased with Gorlen's text. Sightly modified and extended versions of the library are also available commercially.

The NIH class library includes a number of simple concrete classes such as class `Date`, class `Time` (time of day), class `String` (packaged character strings), a specialized class to help with regular expression matching, class `Point` and class `Rectangle` (coordinates etc), and class `Fraction` (for rational arithmetic).

Figure 7.1 illustrates another part of the NIH class library – its "Collection" classes. These include various `LinkedLists`, `Sets`, and `Arrays`; these classes form

a rather more elaborate hierarchy than any previously illustrated. Class
Collection is itself derived from class Object. (Class Object, not shown in
Figure 7.1, defines some standard member functions for input and output,
comparison operations, etc.)

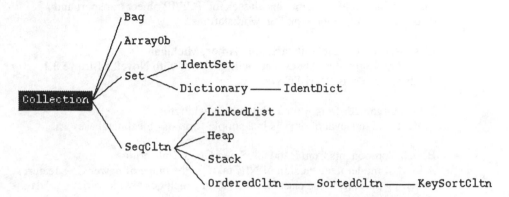

Figure 7.1   Part of the NIH class library showing the "Collection" classes.

Class Collection defines (pure) virtual functions that identify various
behaviours that are common to all kinds of collection, such as the ability to add an
object (add()), remove an object (remove()), report the current number of items in
the collection (size()) etc. Class SeqCtln imposes an ordering on its members
and provides additional member functions such as first(), last(), and at().
Effective implementations of these functions are provided in the concrete classes at
the leaves of the tree.

The NIH class library also includes a variety of classes that implement
coroutine process structures (somewhat similar to the basic control structures of the
original Simula language).

Including specialized classes that support object i/o (discussed in Chapter 9) the
library contains almost 60 classes with approximately 275K bytes of source code.
It is basically, a single-inheritance, tree-structured hierarchy with most classes
derived from a base class Object. Classes from the NIH library may be of
immediate value in some cases. A more general use is probably as an example of
how class libraries can be organized and how code should be written to handle
difficult tasks like "cloning" and object i/o. (The actual code of the library classes
is dated; the exploration of issues in class design, as provided in the accompanying
text book, is of more lasting interest.)

Companies do market component libraries. Some recent offerings include:

• Tools.h++ and Linpack.h++ from Rogue Ware Software, Cornwallis, Oregon.
  These libraries are for Unix or IBM-PC platforms. The Tools.h++ library
  contains the usual date, linked list, stack, queue, and similar classes along with
  a few more exotic items such as tokenizer and regular expression classes (these
  help implement simple parsers) and a B-tree class that can help organize
  records on disk. The classes are claimed to be "bullet proof" (which suggests
  an unusual approach to quality control or, possibly, staff management). The

Linpack.h++ library includes classes for vectors and matrices, statistics classes, and classes for handling signal processing tasks.

• NetClasses from PostModern Computing Technologies, Palo Alto, California.
The NetClasses library contains classes for TCP/IP object transport and distributed programming on Sun workstations.

• BTrv++ from Classic Software, Ann Arbor, Michigan.
This library provides classes that facilitate work with Novell's Btrieve 3.1 database system on IBM-PC systems.

• Object Organizer from Raima, Bellvue, Washington.
Storage and retrieval of objects in a simple form of object database.

• Booch Components from Rational, Santa Clara, California.
A kind of modernized version of NIH taking advantage of newer C++ features including parameterized classes. This library includes a wide variety of "data structure" classes (queues, rings, sets, graphs, trees, lists, etc.), "tools" (classes that package algorithms for searching, pattern matching etc.), along with subsystems supporting some forms of concurrency (coroutines etc.), and exception handling.

• Mejin++: Network Integrated Services, Santa Anna, California.
This is apparently a specialized library for MS-DOS that includes classes for statistics, mathematics, and simulations.

• GreenLeafComm++: Greenleaf Software, Dallas, Texas
Specialized library classes for asynchronous communications.

7.1.2 Graphics User Interface (GUI) class libraries

"User friendly" programs

GUI libraries represent one specialized form of class library that has attracted a lot of interest. GUIs are popular as they can help the construction of "user friendly" programs.

Early programs were for engineers or accountants; such programs could be quite "user-hostile", demanding their data in precise formats and requiring arcane command languages to control their operation. Such "user-hostility" did not matter (because the users of such programs were neither perceptive enough to notice nor assertive enough to complain). Nowadays, every program is expected to be "user-friendly" and to indicate this "friendliness" through graphics displays replete with button controls, pop-up menus, and scrollable views.

Such user interface requirements add extra complexity to programs. As well as implementing the data structures and algorithms required for some computational problem, developers must implement a *friendly* user interface.

User friendliness => consistency

Fortunately, one facet of "user friendliness" is *consistency*. All programs running on a particular platform are expected to work in similar style. Their windows should have the same functionality and controls (drag region, grow box, close box, etc.); a menu bar should be placed at the top of each window or at the top of the screen; dialog boxes are to be used to select processing options and should use a variety of standardized controls such as check boxes and groups of radio buttons. Such requirements for consistency really makes the reuse of interface components a part of the specification of a program. Reuse becomes mandated, it is not simply a matter of enhanced productivity.

The GUI libraries provide individual building blocks and complete subassemblies for the construction of user interfaces. They include relatively simple things like action-buttons, and groups of radio buttons. The "sub-assemblies" are provided in the form of clusters of related classes. Typically, a GUI will have a cluster of interrelated "scroll-bars" and "scrolling region" that can be used to used to move a window over a large diagrammatic representation of data. Another common cluster or sub-assembly would consist of a set of classes for the display and editing of text using multiple fonts and styles.

The classes in a GUI library will be a mix of concrete classes and partially implemented abstract classes. Action buttons, scroll-bars, and text-displays will all be fully implemented classes that can be instantiated and used in a program without further refinement. However, unlike the simple concrete classes in the component libraries, these GUI classes will have been designed to admit further refinement. If a programmer needs a specialized action button (e.g. one that provides some audible response in addition to the normal visual response to selection) then the library class `ActionButton` can be used as the base for some program specific class, e.g. `class MyActionButton : public ActionButton`. The majority of methods in the library class will have been defined as `virtual`, and data members will either be `protected` or have associated `protected` access functions.

Library classes open to further specialization

The libraries will usually have some abstract classes that can only be used through the creation of specialized subclasses. The majority of methods in such classes will have default implementations with only a few `pure virtual` functions that *have* to be overridden.

The GUI libraries are more complex than the component libraries. Part of the complexity results from more extensive use of inheritance. Although the inheritance structures are usually just trees, they may be quite deep. A specialized element, such as a radio button, may be six or more branches away from the root of the tree.

More elaborate class hierarchies

Most of the additional complexity results from the fact that there are many interactions among instances of the different classes in these libraries. A program built using such a library may have to have an instance of some "graphics world" class, and an instance of some "event dispatcher" class. All the other objects created from the GUI classes will depend on services from these components. There are also numerous interactions among instances of classes in the clusters; handling a text edit operation may involve a "text command" object interacting with a "text storage" and a "formats" object, with these in turn interacting with a "text view" object that has to communicate with a "scroller" object.

Multiple interactions among classes defined in the library

The simpler classes in a GUI library, the action buttons etc., can be understood in isolation just like the classes in a simpler "components" library. A programmer can read the description of an action button, instantiate one or create a new class by replacing just one method.

It is more difficult for a would-be user of such a library to understand the classes in the more tightly knit clusters. It is usually necessary to study such classes, particularly their patterns of interactions, before instances can be used really effectively in programs. Detailed study is essential before the classes can be extended through the creation of new subclasses.

Thus, the GUI libraries do present something of a challenge in the form of a steep "learning curve". There is a lot to master before a programmer can fully exploit such libraries. The return on this investment of effort is worthwhile. The user interface components of new programs can be created with a fraction of the effort that would have been required using just the low-level graphics primitives and system calls of a host platform. The interfaces that are developed are much more likely to fully support the expected look-and-feel of a specified environment.

Interviews   Once again, one of the first of these libraries was the product of a research group that contributed it to the public domain. The "Interviews" library developed at Stanford by Linton *et al*. is now in its third version. It is available by Internet ftp from interviews.stanford.edu and other archive sites. Interviews has been evolved into a complex and powerful tool for the construction of programs running in a Unix/X-Windows environment. Its sophistication and complexity make it unsuitable for novices.

The class hierarchy is complex. Interviews V. 3.0 is somewhat exceptional among current C++ class libraries in that it makes reasonably extensive use of multiple inheritance e.g. the class `Button` inherits from both class `MonoGlyph` ($\approx$ something that draws itself) and from class `PointerHandler` (a class that handles pointer, i.e. mouse, initiated input events). "Virtual base classes" are used (see the discussion of multiple inheritance in section 3.6); classes defined deep in the hierarchy may inherit from these base class via more than one inheritance path. Overall, the class hierarchy is a kind of forest with a few small tress of semi-independent classes along with a more complex group of classes joined by multiple inheritance into an elaborate directed acyclic graph. Other sophisticated features include support for "reference counting" as an aid for managing shared objects (see the discussion of memory management in section 3.15).

Interviews separates the input and output aspects of an interactive environment with "handler" and "dispatcher" classes for input, and "glyphs" and display-management classes for output. As just noted, a component such as a button that combines input and output functions will use multiple inheritance to combine features as needed. The display management classes include class `Display`, class `Window` (with specializations such as class `IconWindow` and class `TopLevelWindow`). The "glyphs" introduce the most complexities. Glyphs are basically things that can draw themselves, examples include class `Character`, class `Space`, and class `Rule`. However, they have quite complex behaviours because Interviews provides controls on graphics layouts that are almost as complex as those found in the TEX composition language.

Other branches of the Interviews library include "style" classes and "kits". Styles can be used to specify attributes such as the main foreground and background colours. Kit classes define the look-and-feel for components such as buttons, or scroll-bars (Interviews comes with kits for Motif and Open-Look).

There are also classes packaging operating systems components (e.g. class `Directory`), and useful utilities such as a class `String` and a class `List`.

In addition to its main class libraries, the Interviews package contains other components such as Unidraw, and ibuild. The Unidraw library is somewhat similar to the frameworks described in the next section; it provides basic abstractions for building graphic editors. "ibuild" is a program for building user interfaces by arranging standard kinds of interactive component. Once a proposed interface has been constructed, ibuild will generate the corresponding C++ code thus providing a starting point for the implementation of the final application.

GUI libraries provide much leverage for the programmer, eliminating a lot of tedious coding. The benefits of consistency from using a standard GUI library outweigh any advantages that might accrue from using an in-house interface library. Consequently, GUI libraries are potentially of significant economic value. Several companies are competing, each trying to establish its GUI as the standard for developments on particular platforms.

Commercial products

Current GUI products on the market include:

- C++_Views: CNS Inc, Chanhaussen Minnesota.
  GUI tools for Windows 3 together with a Smalltalk-like set of "collection classes".

- CommonView 3: Glockenspiel, Dublin, Ireland.
  C++ GUI classes providing cross-platform compatibility for Windows, OSF/Motif and Presentation Manager.

- Zinc: ZINC Software, Pleasant Grove, Utah.
  A GUI library in Turbo C++ for the IBM-PC.

7.1.3 Framework libraries

The framework libraries are really GUI libraries that have been specialized to meet the requirements of a particular type of program. As discussed in Chapter 8, the frameworks provide a kind of skeletal version of the "classical Macintosh program" – a program that allows a user to interactively edit data in some "document" that can be saved to file, restored, extended, printed and displayed in various views. A typical example would be a "spreadsheet" application that creates spreadsheet documents; these documents containing tables of numeric data and formulae. The data in a document can be displayed as a spreadsheet form, or views of selected data can be shown as pie-charts etc.

There are two "framework" style libraries for the Macintosh – MacApp from Apple Corporation, and TCL (Think Class Libraries) from Symantec. Somewhat unusually, both these frameworks support development in multiple languages. The current version of MacApp is written in C++ but the program specific code can be implemented in Object Pascal, or in a special dialect of Modula2, as well as C++. The Think Class Libraries work with Think's object-extended dialects of Pascal and C and with Symantec C++. (Think's Object Pascal is almost the same dialect as the Apple version; the C dialect incorporates a limited level of support for OO

Macintosh frameworks

programming similar in style to Object Pascal.). Members of the MacApp and TCL teams were working together on a new multi-platform class library ("Bedrock") but this seems to have run into problems.

Unix   There aren't many framework libraries for Unix. Interviews' Unidraw component was noted above. Brown University has also got some packages available. The illustrative code used in Chapters 8 and 10 comes from ET++ implementations of the example programs. This framework was developed by Weinand, Gamma, and Marty with support from the Swiss National Science Foundation and the Union Bank of Switzerland. Inspired in part by MacApp's design, ET++ takes advantage of a more powerful platform (typically a Sun workstation) and provides many extensions as well as having a more consistent design and implementation. The source is available by ftp from iamsun.unibe.ch [130.92.64.10].

Frameworks for MS-DOS, Windows, etc.   Currently, the most active area for framework development is for Windows on IBM-PC architectures. The leading product is Borland's OWL (Object Windows Library) which provides most standard framework components; in addition to the main framework classes, the Borland product includes an extensive set of collection classes some of which use the latest C++ features such as parameterized types. (Borland also supplies the "Turbovision" class library for MS-DOS based developments, but this product is not actively supported.) Microsoft entered this market relatively late but has introduced the MFC (Microsoft Foundation Classes) library. Another player is Vleermuis Software Research whose GUI_Master product is a MacApp-like framework for IBM PC, OS/2 or Windows 3.

7.2   CLASS BROWSERS

While the framework libraries, and the slightly more limited GUI libraries, can be of great value to developers, they also present formidable problems. Something of these problems can be realized from a few statistics concerning the ET++ framework library. Table 7.1 lists the files in the principal directory of the ET++ system; there are many other files in various subdirectories (e.g. another subdirectory holds all the files containing code of ET++'s "container classes"). Table 7.2 shows some statistics that were published concerning an early version of ET++ library when it incorporated >40000 lines of C++ code:

Considerable time and effort would be required of a programmer who had to read, study, and understand 43494 lines of C++ code. But if the contents of the library are not known, they cannot be reused.

New support tools – *class browsers* – are essential. Browsers do not solve all the problems associated with large libraries, but they do provide a means for conducting explorations through the code. Browsers were pioneered in the Smalltalk research environment. Most of the browsers that are now available are recognizably derived from the original Smalltalk browser, and all work in much the same manner.

AccessMem	EditTextItem	NumItem	ShadowItem
Alert	EnumItem	ObjArray	ShellTView
Application	Error	ObjFloat	Slider
BackgroundItem	EvtHandler	ObjInt	SortedOList
Bag	Expander	ObjList	Splitter
BitSet	FileDialog	Object	StaticTView
Bitmap	FileType	ObjectTable	Storage
BlankWin	Filler	OneOfCluster	StreamConnection
BorderItems	FindDialog	OrdColl	String
Buttons	FixLineTView	Panner	StyledText
ByteArray	FixedStorage	PathLookup	System
CType	FloatItem	PictPort	Text
ChangeDialog	Font	Picture	TextCmd
CheapText	Form	Point	TextFormatter
Class	GapText	PopupItem	TextItem
ClassManager	GotoDialog	Port	TextView
ClipBoard	GraphView	PrintDialog	Token
Clipper	Icon	PrintPort	TreeView
Cluster	IdDictionary	Printer	Types
CmdHistDoc	ImageItem	ProgEnv	VObject
CodeTextView	Init	PttyConnection	VObjectPair
CollView	InitCol	Rectangle	VObjectTView
Collection	Ink	RegularExp	VObjectText
Command	Iterator	RestrTView	View
CycleItem	LineItem	Root	Window
DevBitmap	ManyOfCluster	RunArray	WindowPort
Dialog	Mark	ScrollBar	WindowSystem
Dictionary	Menu	Scroller	
Directory	MenuBar	SeqColl	
Document	Metric	Set	

Table 7.1   The files in the main directory of the ET++ framework library.

Figure 7.2 shows a view of the MacBrowse class browser being used to explore   *MacBrowse*
the MacApp library. Like the Smalltalk browser illustrated in Figure 2.4, the
MacBrowse class browser uses a window divided into several separate panes. Pane
1 presents a list of all the known classes, either listed alphabetically or grouped
hierarchically with class/subclass relations indicated by indentation.

Once a class has been selected in pane 1, panes 2 and 3 list the class's member
functions and data members. These listings show only the additional members (or,
in the case of overridden functions the replacement members). A small region of
the window is used to display type information, as in Figure 7.2 where it shows the
inheritance hierarchy for class TDialogTEView (this region can also used to display
the types of data members or to identify the source files containing member
functions).

Component	Classes	Methods	Lines
Basic building blocks	60	604	11157
Application framework	16	228	3792
Graphics	103	908	14632
Programming environment	9	75	1472
System dependent parts			
Abstract interface	16	273	4270
SunOS	5	38	1435
PostScript	1	17	586
NeWS	6	51	575
SunWindows	6	35	3284
X11.3	6	57	1300
Server	6	56	991
Totals	234	2342	43494

Table 7.2   Sizes of components in an early version of the ET++ framework library.

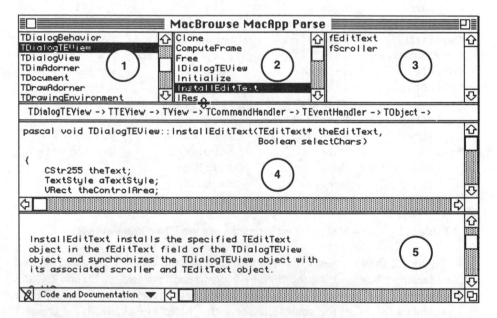

Figure 7.2   The main window of the MacBrowse class browser.

The rest of the window can be used to display code or documentation on individual member functions. The MacApp library comes with reasonably good documentation at the member function level. Each function has some associated explanatory text together with a commentary suggesting when (if ever) it might be appropriate to override the implementation in some specialized derived class. In Figure 7.2, this part of the window is split between the documentation and code browsers.

Of course, a class is part of a hierarchy. It may add just one data field or one method to an existing base class. The view of the class in isolation can be misleading – one does not see all its methods and data fields. Again, a browser will (usually) help – it can *flatten* the class hierarchy and list all the methods of a class (indicating the ancestor from which they are inherited). Figure 7.3 illustrates some of the other displays of class details etc. that can be obtained from MacApp, e.g. a list of all methods and all data fields for a class.

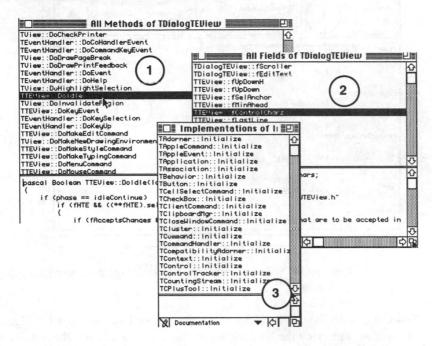

Figure 7.3 Collage of views from MacBrowse: 1) All methods of class TTEView, 2) All data members of class TTEView, and 3) all implementations of function Initialize().

MacBrowse can retrieve other information – such as references to particular functions, or lists of all classes that have an implementation for a method with a given name.

A class browser forms part of the "Programmers Environment" that is automatically linked with an ET++ program. Special command key combinations can be used to force the browser window to be displayed; its form is shown in Figure 7.4.

ET++'s PE browser

The same basic style is used with a scrolling list showing known classes displayed in pane 1 of the window; pane 2 lists the methods defined for a selected class; pane 3 is used to display the results of queries e.g. a request for details of all classes that implement a particular method. The bottom part of the window displays the implementation of methods or the class definition. Unfortunately, there is no documentation browser – documentation is one of the poorer aspects of the ET++ library.

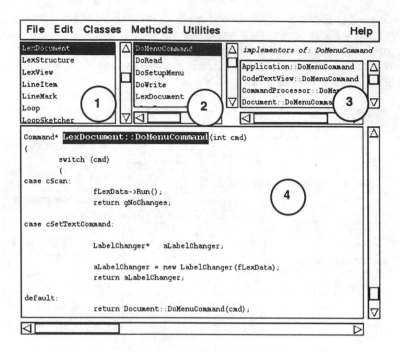

Figure 7.4 The class browser from ET++'s integrated programming environment.

ET++ can also display its hierarchy in a more graphic form, as shown in Figure 7.5. As well as displaying the hierarchical relations among classes, this view can be used to explore other relationships – e.g. client-server relations (where a method in one class invokes some method of another class), or composition relations (where one class has an instance of another class as a data member).

Some systems use a graphic form of display as the main view for their browsers. Figure 7.6 shows the browser for the TCL library. If a class is selected, a pop-up menu will appear listing its methods. Separate windows can be opened to look at implementation of a method or the definition of a class. One of the browsers for the Borland environment on IBM-PCs uses a similar style of interaction. Sometimes browsers can only be applied to executable programs and can show only those classes used in the viewed program. This is a severe limitation – it makes it difficult to explore the library as a whole.

7.3 DEBUGGING AIDS

Good debugging environments for OO programs are typically extended versions of conventional source-level debugging systems that have some additional capabilities for inspecting objects at run-time. The primary debugging environment for MacApp programs is "Sourcebug"; most of the class libraries used on Unix come with some modified form of the standard Unix "gdb" debugger.

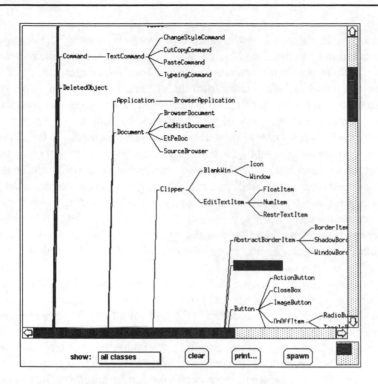

Figure 7.5   A graphic view of ET++'s class hierarchy.

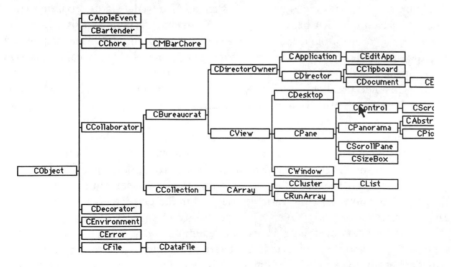

Figure 7.6   The class hierarchy of the "Think Class Library" for the Macintosh
                  as displayed by its class browser.

If a generated executable is to be analyzed using one of these debuggers, the *Special compile time*
source code must have been compiled with special compile time switches set. *options cause*
When such debug options are set, a compiler will insert extra instructions into the *generation of*
generated assembly language and will save many of its internal symbol tables for *information for the*
debugger

later use in the debugger. The symbol table data can be used by the debugger to analyze individual `structs` and instances of classes. Given a pointer specifying the address where a `struct` or class instance starts, a debugger can interpret the subsequent bytes using information from the original symbol table. The symbol table would have recorded the form of a `struct` by specifying its various fields, e.g. "two integer fields, an array of ten characters, an address pointer, ...". The field names and contents can be printed out in suitable formats. A typed pointer, e.g. `Link*`, can be interpreted by the debugger. The value in a pointer data field of one structure, together with type information concerning that field, can be used to switch focus and display a related structure. Thus, it is quite easy to use a debugger to inspect a run-time structure such as an actual linked list existing in memory.

The additional instruction sequences put into the executable code obviously vary somewhat with the compiler and debugger system. Quite often, the added instructions would include code at the start of each function that sets some global pointer to the address of a character string containing the name of that function (if the function is a member of some class, the saved name will include its class e.g. `LinkedList::Append`). Sometimes, extra instructions are inserted after the code generated for each individual statement in the body of a function; these instructions move an integer line count to another global variable that will be utilized by the debugger. With access to such run-time data, and other symbol table information defining the files containing the various functions comprising a program, a debugger can arrange to display the source text corresponding to instructions as these are executed.

Figure 7.7 illustrates the Sourcebug system applied to the Cards program that is used as an example in Chapter 8. When Sourcebug is launched, it allows the developer to scroll through the source text of the program and place preplanned breakpoints; in the example shown in Figure 7.7, one of these breakpoints was placed in the function `TCardView::DoMouseCommand()`.

Once planned breakpoints have been set, the program being debugged can be started. It will run normally until either an error occurs or a preplanned breakpoint is reached, at which time control will return to Sourcebug. Sourcebug will update a display showing the call sequence leading to the code with the breakpoint or error. If no disastrous errors have occurred, program execution can be continued either normally or in some step-by-step fashion controlled from Sourcebug.

When execution is stopped at a breakpoint, the developer can open additional "inspector" windows to view variables defined in the current stack frame (or any earlier frame in the call sequence). The ability to inspect objects can greatly simplify the process of finding errors. For example, the absence of links between supposedly collaborating objects is the cause of many program crashes; but such errors are easily tracked down if one can stop a program at preplanned points and inspect class instances. If at some point one finds that a link is `NULL`, or contains the address of something that is obviously not the intended collaborator, then one can scroll back through the code of the last few functions executed and try to identify where a step setting a link has been omitted or a point where the wrong address may have been used.

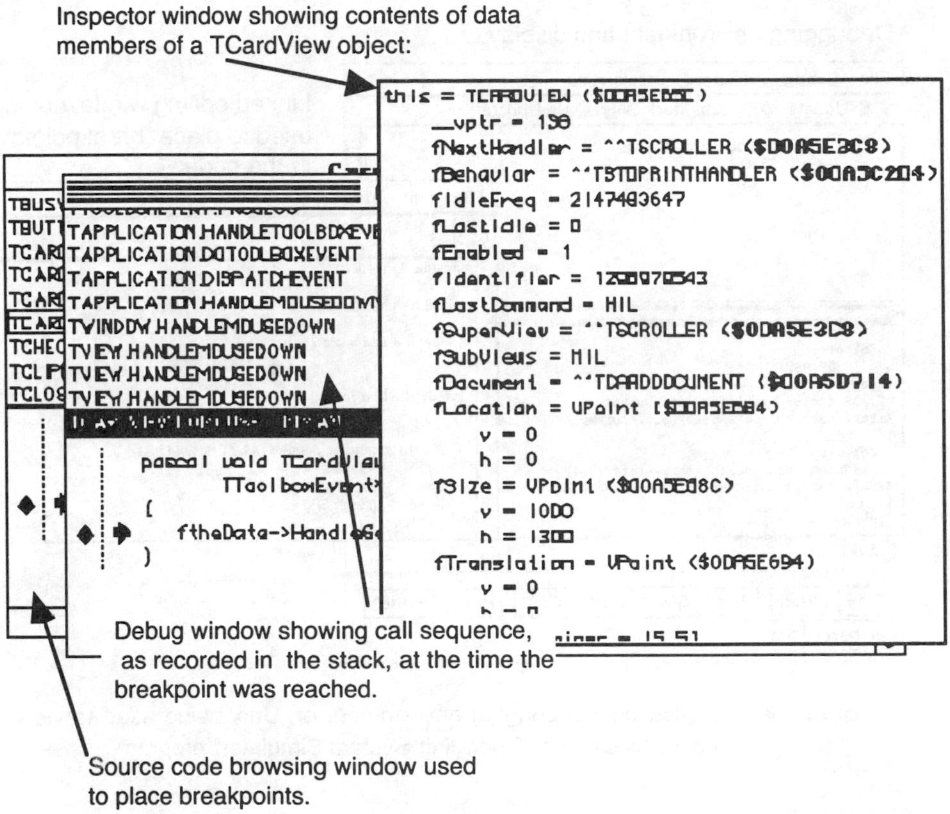

Inspector window showing contents of data
members of a TCardView object:

```
this = TCARDVIEW ($00A5EA5C)
    __vptr = 198
    fNextHandler = ^^TSCROLLER ($00A5E3C8)
    fBehavior = ^^TSTDPRINTHANDLER ($00A5C2D4)
    fIdleFreq = 2147483647
    fLastIdle = 0
    fEnabled = 1
    fIdentifier = 1200070543
    fLastCommand = HIL
    fSuperView = ^^TSCROLLER ($00A5E3C8)
    fSubViews = HIL
    fDocument = ^^TCARDDOCUMENT ($00A5D714)
    fLocation = UPoint [$00A5E884)
        v = 0
        h = 0
    fSize = VPoint ($00A5E08C)
        v = 1000
        h = 1300
    fTranslation = UPoint ($0DA5E694)
        v = 0
        h = 0
                        timer = 15 51
```

```
TBUSY
TBUT  TOOLBOXEVENT .NODE88
TCAR  TAPPLICATION.HANDLETOOLBOXEVE
TCAR  TAPPLICATION.DOTOOLBOXEVENT
TCAR  TAPPLICATION.DISPATCHEVENT
TCAR  TAPPLICATION.HANDLEMOUSEDOWN
TCAR  TWINDOW.HANDLEMOUSEDOWN
TCHEC TVIEW.HANDLEMOUSEDOWN
TCLP  TVIEW.HANDLEMOUSEDOWN
TCLOS TVIEW.HANDLEMOUSEDOWN

         pascal void TCardVIe
                TToolboxEvent
         {
             ftheData->HandleG
         }
```

Debug window showing call sequence,
 as recorded in the stack, at the time the
breakpoint was reached.

Source code browsing window used
to place breakpoints.

Figure 7.7   Sourcebug source level debugger being applied to an example
program.

Objectworks

Figures 7.8 and 7.9 illustrate aspects of a rather similar tool in the Unix
environment. These figures show the application of the Objectworks version of
gdb to the analysis of the "Operating System" example program from Chapter 6.
Again one window is used to display the run-time stack and debugging information
while another window can be used to view the source code and place breakpoints.

As shown in Figure 7.9, Objectworks uses a rather more graphic style to
represent the structure of a class instance as it exists in the program at run time; the
diagram is taken from an Objectworks inspector window associated with the "OS"
object in the example program.

The ET++ system has its own interface to gdb allowing the use of breakpoints.
A basic object inspector tool is automatically linked into each ET++ program
anyway. Command key combinations can be used to call up the inspector at run-
time. Figure 7.10 illustrates ET++'s inspector display.

Debugging environment and display of

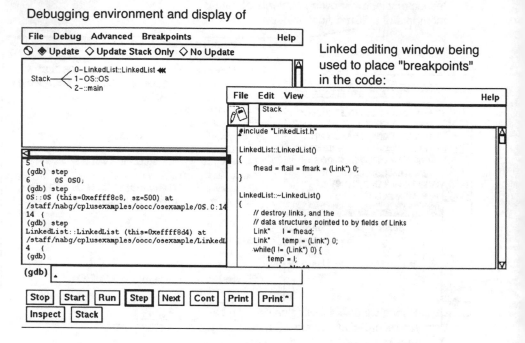

Linked editing window being
used to place "breakpoints"
in the code:

Figure 7.8   Objectworks debugging environment on Unix being used to view
the working of the "Operating System Simulator" program.

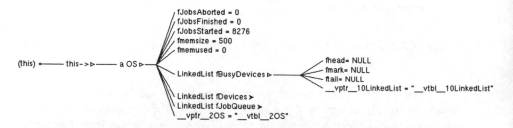

Figure 7.9   Objectworks' display of an instance of a class as it exists at run-
time.

Pane 1 of the inspector window (Figure 7.10) has a scrolling list that contains
the names of those classes currently instantiated in the program; any class of
interest can be selected by mouse-click. Pane 2 of the window will then display a
list of all instances of that class; in this case there was only one instance. If an
instance is selected from this list, its data members will be displayed in pane 3.
Pointer types, such as `fNodeList` in figure 7.10, act like hypertext buttons –
clicking on their values will switch focus to the referenced object. The arrow
buttons in the centre of the window permit one to move back and forth among
viewed objects so one could start with a particular object, follow a link to some
other referenced object, e.g. the `ObjList` referenced by the `fNodeList` pointer,
view some of the data items in that list and still get back to the original object.

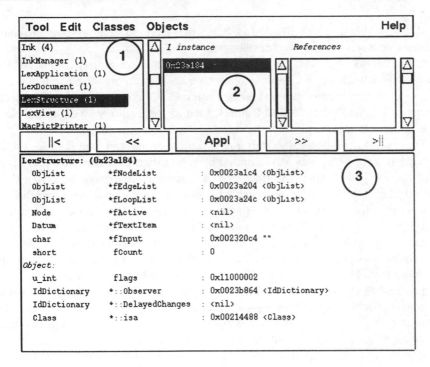

Figure 7.10 ET++'s object inspector used to view the data members of an object at run-time.

7.4 DEVELOPMENT ENVIRONMENTS

A conventional development environment usually comprises a number of separate tools. Thus, on a Unix system, a developer might use "vi" or some other text editor, the Source Code Control System (SCCS) for version management in a project, "make" for organizing compilations, and one of the C++ compilers. Similarly, the Macintosh MPW environment provides an editor, the "Projector" tool for version management, the MABuild script for compilation and, again, a choice of C++ compilers. A class browser is yet another separate tool that a programmer working on an OO program must utilize.

The increasing availability of workstations with greater memory and faster processors makes the use of integrated development environments a little more practical. These systems may never be quite a flexible and supportive as the top-range Lisp or Smalltalk environments. But they can provide significant support to the development programmer and so facilitate the development process.

Some of the currently available environments are experimental systems that have been put in the public domain. Others are commercial products. The brief examples in this section illustrate some of the features in the Objectworks environment (a commercial product from ParcPlace) and the "Sniff" environment (a public domain system created by Bischofberger at the Union Bank of Switzerland, Informatics Laboratory in Zurich, soon to be released as a commercial product). Although not illustrated here, the Borland IDE system provides comparable levels of support to developers.

Objectworks

Objectworks has some hybrid features. It is primarily intended for programmers working on C++ developments but has been organized so that it can also support conventional projects written using C. Of course, ordinary C projects cannot take advantage of the OO features, such as the facilities for exploring class inheritance structures.

Objectworks can be used to start a new project or one can load the files for an existing project. The files are compiled by a modified C++ compiler that saves much of its symbol table data (just like the source code debugging systems) in special files. (The compiler generates a file with a ".cyi" extensions for each C++ or C source file; these files tend to be rather large.) The data characterizing individual compiled modules are combined to provide a form of data base that describes the structure of the program that is being developed. When files are subsequently edited, or other files added to the project, the database is updated appropriately. The database is used by all the other tools in the system.

Objectworks control window

Figure 7.11 shows the main control window of the Objectworks program (pane 1) together with displays of data produced using some of the components. Objectworks provides access to a selection of standard editors through the three control buttons in the top-left of the window (the chosen editors are "vi", an editor similar to a Macintosh style cut-and-paste editing environment, and emacs). Most of the other control buttons invoke various specialized browsers. Two browsers are for exploring *Contains* relationships for files; one reveals the contents of files and directories, the other can expand and clarify #include dependencies that exist between files. Another group of browsers are used to view *Usage* relations – such as calls to functions, references to globals (there shouldn't be many in an OO program but Objectworks is also intended for use with C sources), and declarations of variables of particular types or classes. There are other browsers for viewing class hierarchies and client server relations among classes.

Contents browsers to find elements

Pane 2 in Figure 7.11 shows a *Contents* browser being applied to a directory to view the files in a project (the Operating Systems Simulation project used as an example in Chapter 6). This is the normal starting point for exploring code. Files and directories are shown as a kind of branching tree; the individual files can be displayed by name or can be "opened" to show their contents in terms of class, struct, and enum declarations and function definitions. By expanding individual files to reveal their contents, it is possible to find any particular class declaration of or member function implementation. This emphasis on files as a starting point is quite suited to C programs; but is less satisfactory for class based programs.

Class hierarchy and other browsers

Once classes and methods have been revealed using the contents browser, other browsers can be applied by dragging a tool icon from the palette to the data element that is to be displayed. This process opens a new browser window focussed on the selected data element. The third pane in Figure 7.11 was obtained using the class structure browser to view class Device in the OS example code. The fourth pane shows relationships between class Device and other classes (class LinkedList, class PPU, class OS etc) as diagrammed by one of the *Client-Server* browsers.

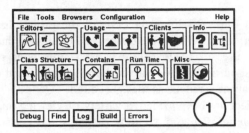

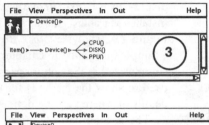

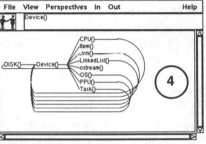

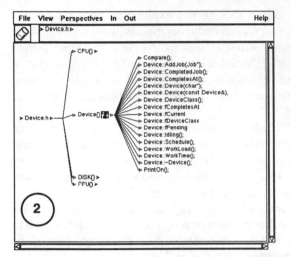

Figure 7.11   Collage of views showing Objectworks browsers applied to the OS Simulation program (example in Chapter 6).

Integrating documentation

Objectworks can help integrate documentation with code. The information browsers will retrieve text documentation associated with different program elements as can be displayed by the contents browser. There are facilities for generating documentation files automatically from comments embedded in the source text (provided that these follow specified conventions).

Other tools in Objectworks

In addition to the many different browsers, Objectworks has a "build" tool for running makefiles, and the associated debugging environment illustrated in the section 7.3.

Sniff

Sniff is an environment for editing, browsing, and cross referencing classes in object-oriented projects implemented using C++. It is a public domain system, generally available from the same sites as hold the ET++ class library (e.g. Internet ftp from iamsun.unibe.ch [130.92.64.10]). Sniff consists of a group of interacting programs that can run on different workstations in a work group. Each user would run a separate copy of the visualization and editing environment, but would normally share other tools such as Sniff's parser.

Like Objectworks and other systems, Sniff builds a "database" defining a project. Conceptually, a project can consist of several subprojects. A shared library, such as the ET++ framework, can be defined as a non-editable subproject of many different projects.

Sniff's "fuzzy C++ parser"   The various files containing the code of the project have to be analyzed to extract the information defining the classes and other data that a user might wish to review. Sniff uses a "fuzzy C++ parser" – this does not analyze the code completely, just sufficiently to extract the more important details The parser is fast, and the analysis is redone as files get edited.

Symbol Browser, and Class Browser   Once a project has been loaded, a variety of viewing and editing tools can be applied. A *Symbol Browser* window displays the classes in the project and subprojects (or any chosen group of subprojects). A class can be selected for display in a *Class Browser* window; this allows exploration of the methods, instance variables, enumerated types, friends and other aspects of the selected class. Figure 7.12 shows a *Class Browser* window focussed on class PPU from the OS example program of Chapter 6. The inheritance hierarchy for class PPU is summarized in the bottom part of the window; the top part of the window is being used to show the methods for the class, these are listed in alphabetical order with details of where they are defined.

Hierarchy Browser   The second part of Figure 7.12 shows the contents of a *Hierarchy Browser* window that was opened to display class PPU within the overall hierarchy defined in the program. As well as being used to view a chosen class like this, *Hierarchy Browser* windows are used to display information in response to queries seeking information on all classes that implement a particular method.

Retriever tool and editor   Sniff's *Retriever* tool performs "grep"-like regular expression searches for specific strings in the files of a project or specific subproject. The editor, shown in Figure 7.13 permits code to be modified; the symbol list on the right part of the editor window speeds navigation within and between files.

Sniff remembers   One little feature of Sniff that should be adopted by other environments is its use of a history list of browsing contexts. When exploring a class library, a programmer typically starts reading the code of one specific class, follows a call to some other class that is partially reviewed, then a link is followed to some third class – at which point it becomes necessary to get back to the original context. In most browsers this is quite difficult – it is up to the user to remember the class, method, and line where such a search started. Sniff maintains a history list of previously viewed contexts in one of its menus; restoring a previously viewed context is a trivial matter of menu selection. (Sniff was obviously designed by programmers with practical experience of the problems of exploring a large and complex library such as ET++.)

Other facilities   Sniff creates makefiles automatically, manages builds of programs, can interpret compiler error messages to select erroneous statements for editing, and can link to a "gdb"-based debugging environment.

Sniff is affordable   Sniff is considerably more economical in its use of computing resources than alternative environments. The file space required to maintain a record of a project is substantially smaller than that used by Objectworks. Run-time requirements for CPU time and main memory are also significantly less.

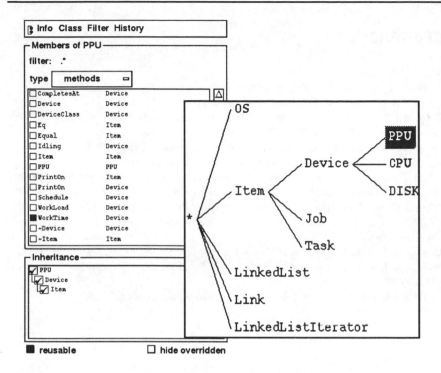

Figure 7.12   Exploring the OS example program with the aid of Sniff's browsing tools.

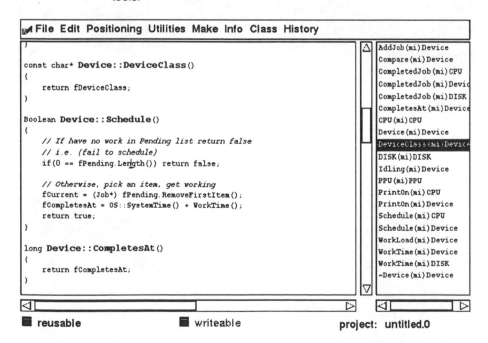

Figure 7.13   Editing the source of a class within the Sniff environment.

REFERENCES

References W.R. Bischofberger, *Sniff – A Pragmatic Approach to a C++ Programming Environment*, Usenix Association C++ Technical Conference, pp.67-80, 1992.

K.E. Gorlen, S. M. Orlow, & P.S. Plexico, *Data Abstraction and Object-Oriented Programming in C++*, Wiley, 1990.

S. Holzner, *Microsoft Foundation Class Library Programming*, Brady, 1993.

S. Holzner, *Borland C++ Windows Programming*, 3e, Brady, 1994.

N. Rhodes and J. McKechan, *Symantec C++ Programming for the Macintosh*, Brady, 1993.

A. Weinand, E. Gamma, and R. Marty, *Design and Implementation of ET++, a Seamless Object-Oriented Application Framework*, Structure Programming, **10**, 1989.

D. A. Wilson, L.S. Rosenstein, and D. Shafer, *C++ Programming with MacApp*, Addison-Wesley, 1990.

N.C. Shammas, *What Every Borland C++ 4 Programmer Should Know*, SAMS, 1994.

D.A. Wilson, L.S. Rosenstein, and D. Shafer, *C++ Programming with MacApp*, Addison Wesley 1991.

M.J. Young, *Mastering Microsoft Visual C++ Programming,*, Sybex, 1993.

Simple Use of a Framework Class Library

8

This chapter illustrates how a typical "Framework Class Library" can be used to construct a simple application program. The frameworks – MacApp and TCL on the Macintosh; Borland's Object Windows library, Microsoft's MFC library, GUI_Master, etc. on the PC, and ET++ and others on Unix – all embody essentially the same model for a program. Although names for classes and member functions vary, and responsibilities for classes may be allocated in somewhat different ways, programming with these different libraries can be approached in a generally similar fashion.

8.1 FRAMEWORK CLASS LIBRARIES

8.1.1 The problem domain addressed by framework libraries

The main classes in these frameworks provide a model of a particular kind of program – it is a classic "Macintosh" application controlled largely through user-commands entered as mouse-actions (data entry, selection, movement etc.) or by menu selection. The prototypical programs are graphics programs and, to a lesser extent, word processing programs. They are "editors". These programs allow a user to build, incrementally, some complex data structure. In a graphics editor, this structure is essentially a list of shapes each characterized by information defining position, line-styles, and fill-patterns. A word processor's data-structure might be a "list of paragraphs", each containing some text and having associated formatting information. Usually, this data structure, or rather a printed representation of the structure, is the "product" created by the user of the program.

Frameworks for building "editors"

Sometimes, a data structure may have to "do something" once it has been created. For example, a program might provide an editing facility for building a representation of an AC electrical circuit that can be "activated" resulting in a display of voltages, currents, phase-relations etc. Any such data processing capabilities would involve simple algorithmic coding that, generally, would not represent a major portion of the development costs of the software. Such coding is necessarily application specific and completely independent of the framework. The framework classes exist to help build the "editing" component of these programs.

User-controlled, interactive programs

These "editor" programs are highly interactive. The user selects a data entry "tool" and then uses this tool to add another data element to the overall structure; or, the user may select an existing data element and enter a command such as "*change the font used here to Courier*", or "*change the fill pattern to diagonal stripes*". Other user commands may affect overall organization or behaviour rather than individual data elements. For example, a "File/PageSetup" command that reduces the page size could, in a case where data elements have associated page positions, result in a change to every member element of the entire data structure. A "File/Save" or "File/Print" command obviously induces behaviour that involves the complete data structure.

Modeless programs

Individual user commands should be independent; the processing of a command is completed and then the program waits for the next command. Commands rarely set "modes" that define the subsequent allowed processing steps. Different processing options can be selected in arbitrary orders.

Program-controlled input is simpler

Older style scientific and business programs tended to have a modal structure with very rigid control flows: 1) get the input data, 2) process the data, 3) print the results. Any data input mode would be under program control, with the program prompting the user for successive data elements that had to be provided in a fixed order. When interactions with users are under program control, the programmer's task is simplified. Because the sequence of any interactions is known in advance, the programmer can plan to have all required data structures created and correctly initialized prior to use.

Complex control flows

With user control, things get more difficult. A characteristic of these user-driven programs is complexity in the flow of control. User commands such as a change to the "page setup", the resizing of a window, or an editing action that adds or removes data, can occur in any order. Typically, each such command will affect many different aspects of the run-time environment – aspects such as menu options, data displays, and window/view structures. Since each of the listed user commands affects the "size" of the data, they all necessitate updating the amount of "travel" in any scroll-bars that might be associated with a window displaying the data. Consequently, code determining this travel may have to be called from any of several quite different contexts. It is in cases like this that problems arise. A programmer may fail to perceive an implicit consequence of some user command; this failure could result in programs with aberrant behaviours (e.g. a program whose scroll-bars appear oblivious to data editing actions and only respond when a window is explicitly resized). Alternatively, the programmer produces some ad hoc solution with a complex flow graph.

Complexity of the control flow often results in program errors. After all, routines are typically written assuming a particular call pattern, with concomitant prior creation and initialization of all necessary data structures. When other call sequences are imposed later, initialization steps can be forgotten and so errors may arise. (Often these errors seem intermittent and so are difficult to detect. The required data structures will usually have been set up by some normal call sequence made prior to the occurrence of one of the extra "sideways" calls. Problems only arise with non-standard patterns of interaction.) If the programmer implementing one of these interactive editors lacks a strong model, the code tends to become convoluted, unreliable, and difficult to maintain or extend.

A model for interactive programs

The framework libraries were developed starting with a particular model for these interactive programs. The model envisages many different "objects" being present at run-time. Each of these objects owns some part of the program's data and handles any commands that relate to their data. User actions, such as menu selections, keyed input, or mouse-based interactions, are translated into commands that get sent to particular command-handling objects for processing. (The term "event-handler" is often used rather than "command-handler". However "events" tend to be low-level things, e.g. signals from the operating system that there is a keystroke to process or that some window has been activated. "Events" are translated into semantically richer "commands" that get sent to the various handler objects. Consequently, command-handler seems to be a better general term than event-handler.)

"Command-handler" objects

The authors of the framework library started by identifying a set of "command-handler" objects that could be conceived of as being present in any executing program. The various tasks that a program has to perform were then partitioned amongst these objects, and so the forms of specialized kinds of "command-handler" classes evolved. For example, there was a "document" command-handler. A document "owned" the data structure that was being manipulated by the program's user, it looked after changes to that structure, and it arranged transfers between disk files and main memory. An "application" command-handler was responsible for overall organization of the objects that made up a running program, and it performed much of the work involved in translating low-level operating system events into "commands" that could be sent to other objects. "View" command handlers provided a x, y-coordinate framework for the display of data, methods for drawing data, and methods for responding to mouse-based data selection and editing actions.

Application, Document and View command-handlers

The authors of the frameworks sorted out all the typical patterns of interaction that occur among these principal command-handler objects. These interactions were then encoded in the member functions that were defined for these classes. For example, if an `Application` object found that it was dealing with a "Quit" command, it would remember to warn any `Document` objects that were present. Each `Document` object could then check to determine whether its data structure had been altered. If the data had been altered, the `Document` would display a standard alert asking the user whether the data should be saved before the program terminated. Once it had either saved or abandoned its data, each `Document` would proceed by disposing of any `Window` objects that it had created to display its data. Here, there is a complex pattern of interactions among instances of the `Application`, `Document`, and `Window` classes. But the normal responses can all be defined and provided as methods `Document::DoClose()`, `Application::`

Standard patterns of interactions among command-handlers

DoQuit() etc. Of course, these methods would be specified as being "virtual" – a programmer should have the opportunity of changing standard patterns of behaviour if really necessary.

Command-handler classes are quite strongly coupled

In a typical framework, there are hundreds of such patterns of interactions associated with standard user-commands and run-time events. Each interaction may involve several objects; the work performed by each object may entail a series of nested method calls; and there may be some cycles in the call patterns. For example, the Application object might ask a View to handle a mouse action and, in doing so, the View might ask the Application to perform some function such as adding a "mouse-command object" to a list. The various classes of command-handler are consequently quite strongly coupled.

Partially implemented "Abstract" classes

The principal command-handlers are "abstract". Thus, the methods of class Application are simply an expression of the standard behaviours present in every program. Similarly, the library's Document class will epitomize a standard "document". Many of the methods defined for these classes will have been provided with default implementations that do "the standard things". But, a few methods will be incomplete, or undefined. Thus, the Document class in the library will contain a DoWrite() method that transfers data from memory to a disk file; however, Document::DoWrite() might be a pure virtual function, more likely it will have an *empty* implementation, and at best all it can do is save any "housekeeping" information that might be useful to the framework. The authors of the framework cannot know what data should be saved, for that is a program-specific issue. Each of the main command-handler classes has a few such methods that *must* be given effective definitions in subclasses. (These methods could have been defined as C++ pure virtual functions, so making the principal command-handler classes genuinely abstract. However, "do-nothing" definitions are provided in these libraries even for those methods that must be overridden. Consequently, these library classes can be instantiated, although they are not much use as actually defined.)

Building a program based on a framework library

Program-specific subclasses must be defined for command-handlers

Specific subclasses have to be created for each of the major command-handler classes; so one gets class MyDocument : public Document – a specialized document that really does hold data and can transfer data between memory and a disk file. A real program is built up by having a specialized Application that creates specialized Documents that are shown using instances of specialized kinds of View classes.

Prebuilt components based on command-handlers

Although each new program must define its own specialized versions of class Application, class Document and a few other command-handlers, the framework libraries supply many pre-built specialized forms of command-handler. These will include things like "check boxes", "scrollers", "windows", "text", and so forth. Class CheckBox might define a specialized kind of command-handler that owns a boolean variable, can draw a little box (containing some on/off indicator), and can respond to a mouse-entered command that sets or clears the boolean and then gets the box redrawn. An instance of class Text might be quite an elaborate entity – an object that owns a block of characters (possibly with auxiliary formatting data), that can display these characters, and which can respond to keyboard commands that add more characters or replace existing characters. Class `Window` would (as one might expect) specify the form of standard windows that a program could create for the display of its data. Class `Scroller` would be some kind of command-handler that deals with all user-actions that necessitate moving a large display view so that different portions fit inside a window. Although such classes will normally be used as concrete classes and get directly instantiated, they will have been written to permit further extension. Consequently it is possible to define new specializations – such as buttons that produce some sound when selected.

A framework library may offer other standard components – simple concrete classes that can be instantiated and used as needed. Typically, there are a few "collection classes" – the usual lists, dynamic arrays and similar useful data structures for storing items. In some cases, these collection classes are based on examples provided originally in the NIH class library. Then there will often be classes like `Point`, `Rectangle` etc. that may get used when manipulating coordinates while creating displays of data.

Lists and other useful data stores

Problems relating to the use of frameworks

Many experienced programmers feel quite unhappy when first starting to use these elaborate frameworks. They are upset by their loss of control. They end up writing methods such as `MyDocument::DoWrite()` and `MyDocument::DoMenuCommand()` but have no idea of how these functions get invoked. Of course, this is exactly what one wants. It really shouldn't be necessary to have to redo all the code that responds to a menu selection and eventually calls a menu-handling routine; it is only the menu-handling routine that is new, all the other bits are standard. While it may take some time to get used to the frameworks, programmers should appreciate the fact that the frameworks can save them from having to write lots of code.

Problems with the command-handler classes

There are some real problems with the highly connected command-handler classes. Some steps, e.g. opening an old document, involve lots of interactions among objects of different classes. If any non-standard processing is required in a particular program, the developer must be able to intervene in these interactions and arrange for additional processing. Unfortunately, it is often difficult for a developer to know exactly where and how to intervene. The developer may have to work through the framework code and laboriously map the interactions among objects in order to determine where to override and extend the standard behaviours of the framework classes.

8.1.2 Handling a user's commands

Principal command-handlers

Figure 8.1 shows some of the "command-handler" objects that will be present during the execution of a typical program. The program illustrated, ChemEdit, is a graphics editor with a palette of tools for drawing and manipulating structural diagrams of organic molecules. At run-time, there will be a single instance of a ChemApplication object, one or more ChemDocument objects, and for each ChemDocument there will be a PaletteView object and a ChemView object.

Of course, there will be many other objects present. These will include instances of some of the component classes in the framework library. Thus, each ChemDocument will have a Window object that frames its two views. A Window object is itself a highly specialized command-handler that deals with a few user actions; for example Window objects organize response to the use of controls for window-resizing.   There might be a Scroller object, with its associated Scrollbar objects, assisting each ChemView object; these components would be used if a ChemView object was likely to be too large for the screen. Finally, there would be program-specific objects, such as a ChemStructure object. A ChemStructure object might itself make use of List objects or instances of other collection classes provided with the framework library.

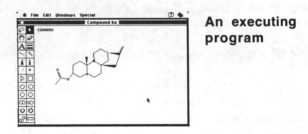

An executing program

some of the command-handler objects

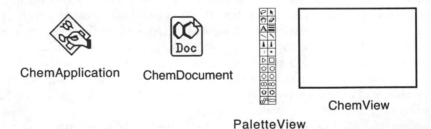

ChemApplication   ChemDocument

PaletteView

ChemView

Figure 8.1   An executing program and some of the "command-handler" objects that will be present.

The development programmer working with a framework library has to define these principal command handlers as specialized subclasses of the framework's Application, Document, and View classes:

```
class ChemApplication : public Application {
public:
        // Replace a few virtual member functions
        …
    };
```

```
class ChemDocument : public Document {
public:
    // Replace a few virtual member functions
    ...
    // Add some program specific functions manipulating
    // ChemStructure data structures
private:
    ChemStructure*       fStructure;   // Add some data fields
    ...
};

class ChemView : public View {
public:
    // Replace a few virtual member functions.
    ...
};
```

At the simplest level of use of the frameworks, very few member functions need be redefined in these subclasses.

The responsibilities of the principal control handler classes are illustrated in the following examples that show how instances of framework classes might cooperate to process user-entered commands.

Example: handling an "File/Open..." command

Figure 8.2 illustrates some of the steps that would typically be involved when a ChemApplication object handles an "Open ..." command (as initiated by menu selection or maybe a command-key). (The exact sequence of steps is framework dependent; the example illustrates the general approach.) An "Open ..." command would (by some devious call sequence) eventually result in the invocation of the Application::OpenOld() member function with data passed as arguments identifying the file. (Member names like Application::OpenOld() are "generic"; each framework will offer similar functionality, but the names of member functions vary, and there may be differences in the allocation of responsibilities to classes.) The Application::OpenOld() member function calls a DoMakeDocument() member function. This is one of those member functions that *must be overridden*. The default implementation does nothing, and here one would want to create a ChemDocument object:

A standard "OpenOld()" calls a specialized "DoMake-Document()"

```
class Application : public CommandHandler {
public:
    ...
    virtual void OpenOld(...); // standard OpenOld behaviour
    ...
    virtual Document* DoMakeDocument() {
            /* Make what kind of document? Override this! */
            }
    ...
};
```

Application is_a specialized CommandHandler

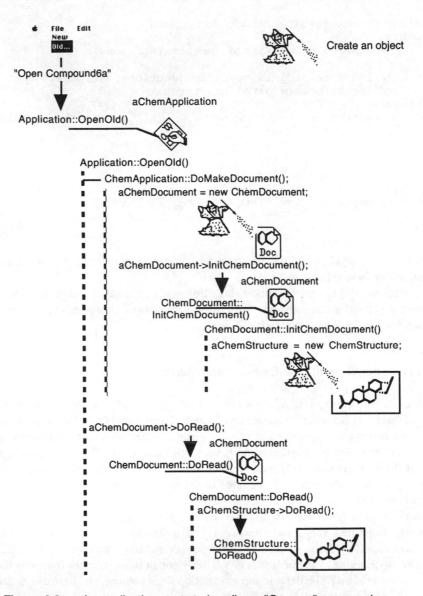

Figure 8.2   An application starts to handle an "Open ..." command.

ChemApplication
is_a specialized
Application

```
class ChemApplication : public Application {
public:
    ...
    // implement a real DoMakeDocument()
    virtual Document* DoMakeDocument();
};
```

With a unique way of
making Documents

```
Document* ChemApplication::DoMakeDocument()
{
    ChemDocument* aChemDocument;
    ...
    aChemDocument = new ChemDocument;
```

```
        aChemDocument->InitChemDocument();
        ...
        return aChemDocument;
    }
```

The "magic" of virtual function dispatch guarantees that the replacement `ChemApplication::DoMakeDocument()` method, rather than the default `Application::DoMakeDocument()` method, will be called from within `Application::OpenOld()`.

A newly created `ChemDocument` object would then create an instance of the `ChemStructure` class to hold the program specific data. This might be done as part of the `ChemDocument`'s constructor routine; or, as shown in Figure 8.2, it might be done as a part of a separate initialization routine called from `ChemApplication::DoMakeDocument()`.

Once `ChemApplication::DoMakeDocument()` had completed and returned, the `Application::OpenOld()` function would continue with other initialization steps and eventually ask the new document to read its data from the existing file. `Document::DoRead()` is another of the virtual functions that *must be overridden* in a derived class. An effective implementation of `ChemDocument::DoRead()` could consist simply of a call asking the new `ChemStructure` to read its own data.

DoRead() reads data

At some stage, during the execution of `Application::OpenOld()`, the new document would be asked to create the objects needed to display its data. These objects would be instances of the various specialized `View` classes created for the program along with things like a `Window` to frame these views and, possibly, additional support objects such as a `Scroller` and its `Scrollbars`. `Document::DoMakeViews()` is another of those virtual member functions that *must be overridden*. The display structures are necessarily program-specific. The programmer implementing the `ChemDocument` class must provide an effective function, `ChemDocument::DoMakeViews()`, that creates the objects that will display its data.

DoMakeViews() creates display structure

Once the window has been opened, its contents need to be drawn. This will be achieved by the `Application` object asking the `Window` object to redraw itself. The `Window` object will work through the list of views that it owns (in this example, that is the `PaletteView` and the `ChemView`) telling each to redraw itself.

Windows and Views get redrawn

`View::Draw()` is another of those virtual member functions that *must be redefined* in subclasses. The default implementation draws nothing. The specialized implementation for `PaletteView::Draw()` might involve getting a bit-map image of the tool palette and copying this to the screen. The code for `ChemView::Draw()` has to get the data structure drawn; its implementation would probably consist of a request to the associated `ChemDocument` asking it to organize the drawing of the data. `ChemDocument` will simply forward the "draw yourself" message on to the `ChemStructure` object.

Example: handling a "mouse command"

Once the windows and views had all been displayed, the program would wait for further actions by the user. The user's next action might be to select a tool from the

*Application
determines event
relates to a Window*

tool palette by positioning the cursor over the image of the tool and causing an event by clicking the mouse-button.

The processing of such an event is relatively easy, see Figure 8.3. The low-level mouse-down event is picked up by the operating system. This event is translated into a message to a `View` object requesting that it deal with the user's mouse-based action. The process of event translation involves identifying the appropriate `View` object that is to handle the action, and the creation of a data structure containing information such as the position of the mouse and associated data such as time, details of any keyboard modifier keys being set etc. The responsibility for event translation is shared between the operating system, the run-time support routines of the framework, and the `Application` object. While different frameworks do vary markedly in their mechanisms for event translation, it happens "behind the scenes" and is rarely of concern to the application programmer.

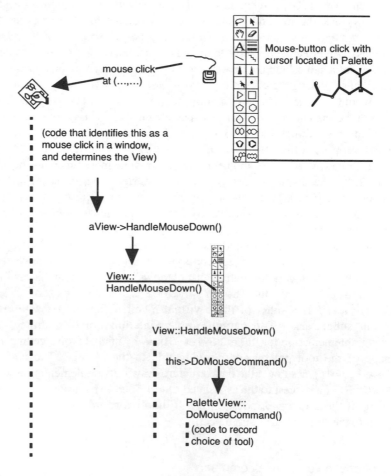

Figure 8.3   Dealing with a mouse click in a View.

*View deals with
mouse command*

Obviously, what a `View` object does in response to a mouse command is program specific. The virtual function `View::DoMouseCommand()` is again one that *must be replaced* in any effective class. The implementation for `PaletteView::DoMouseCommand()` would consist of some code that identifies and records the selected tool and then probably arranges to get the palette redrawn with the new tool highlighted. `ChemView::DoMouseCommand()` would be a lot more complex. A mouse action in the `ChemView` would indicate that the user wanted to edit the displayed structure; the code in `ChemView::DoMouseCommand()` would have to determine the type of editing action (e.g. add bond, erase bond, change atom, etc.) and arrange to create an "editor-object" or "command-object" that could handle the editing process.

Example: Dealing with menu and keyed commands, the "handler-chain"

Dealing with mouse commands has one simplifying aspect. In this case, it is fairly easy to identify the object that should be responding to the user – it will be the smallest `View` enclosing the cursor position (or, sometimes, an enclosing view). While the position of the cursor in a mouse-based editing action essentially identifies the `View` object that is to deal with the user's action; the low-level event data that the OS provides to characterize a keystroke or menu selection action cannot identify the particular object that has to deal with these user requests. Consequently, some more elaborate, general purpose mechanism has to exist to relate command-handlers and commands.

The basis of this mechanism is illustrated in Figure 8.4. The various command-handlers like `Views`, `Windows`, `Documents`, and `Application` objects are linked together. At any particular time, one particular command-handler object will be identified as the primary target for incoming menu commands (and also for any keyboard input). This target object (known in some frameworks as the "gopher") will be identified by some global pointer or by some pointer data member in the `Application` object. Usually, the target will be one of the views contained in the frontmost window belonging to the program.

The target chain

The `Application` object converts low-level event data into some form of higher level user-request. (For example, a simple data table might be used to map menu selections onto integer identifiers for the various user commands handled by the program's objects.) The target object is given first go at dealing with the user-request (using its `DoMenuCommand()` method for menu-based requests, or a `DoKeyCommand()` for handling keyboard input). If the target cannot deal with the request, it passes the request on to the next command-handler in the chain.

If, for example, a user selects an "About Application" menu option, the `Application` object would determine the equivalent command number (e.g. `cAbout = 1`). It would then send a request to the target object *"please execute your DoMenuCommand method to handle command number 1"*. The target might be the `ChemView`. Its `DoMenuCommand()` function cannot handle command 1, so it would arrange to pass the request on to its enclosing `Window`. The `Window` would similarly pass the request on to the `ChemDocument` object that had created it. The `ChemDocument` object would also defer on handling a command 1, and pass it back to the `ChemApplication` object. The `Application::DoMenuCommand()` code, that would be executed by the ChemApplication object, handles command 1 by arranging to display a dialog containing information about the program.

Handling standard
menu commands

Standard code in the framework libraries will handle a few menu-based requests. Thus, on the Macintosh with the MacApp framework, class Application contains code to handle the menu requests: /About, File/New, File/Open, File/Close, and File/Quit. A Document object has code to handle File/Save and File/Revert; this standard code invokes the virtual DoWrite() and DoRead() functions.

Program-specific
menus

All other commands will be program-specific. The development programmer provides data defining additional menu options and their mappings into command numbers. The developers must also implement the necessary DoMenuCommand() routines for specialized Views, and possibly provides a DoMenuCommand() routine for the specialized Document type that extends the normal Document:: DoMenuCommand() function. The definition of such functions is illustrated in Chapter 9. The default menus, and associated default processing suffice for simple examples like the program in this chapter.

Identifying the
"target"

Most of the work involved in the maintenance of the command-handler chain is done by standard code in the framework. If a program has a window containing multiple views, more than one of which is capable of handling menu commands, then the development programmer may have to include code that identifies the current target view within that window.

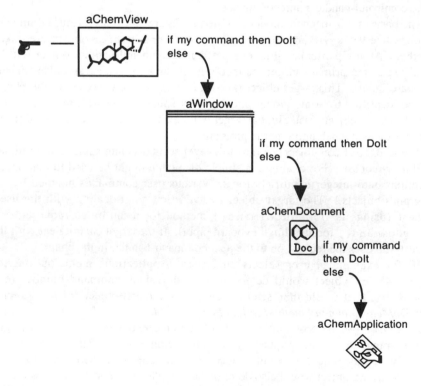

Figure 8.4   The target chain of command-handlers.

8.1.3 Example Frameworks

Figures 8.5 (for MacApp) and 8.6 (for ET++) illustrate two of the frameworks used to implement the example program from this chapter. The figures show only small portions of these frameworks. MacApp has more than 200 classes (though several are just for internal use within the framework) while ET++ has over 500 classes.

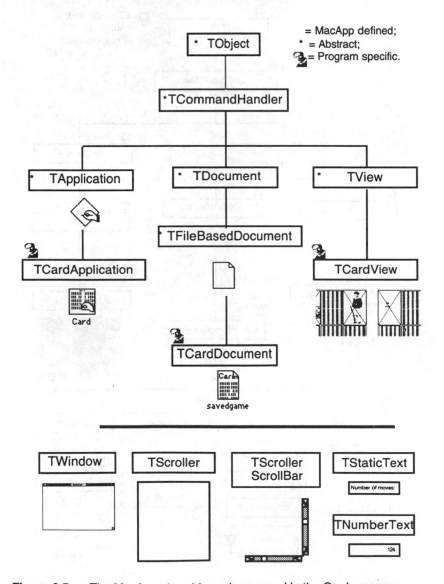

Figure 8.5   The MacApp class hierarchy as used in the Cards program.

Both these frameworks, and some other current frameworks, define tree-structured hierarchies for their main classes with some base class `Object` at the root of the tree. (Some frameworks, e.g. Borland's OWL and MacApp, have

Tree structure hierarchy with class Object as base class

naming conventions in which all class names starting with 'T' –TWindow , TButton, etc.) The framework's base class will have few or no data members, but may have several member functions. The functionality defined for the base class varies among frameworks; it may include support for debugging, along with functions for "object i/o" and "dependency handling" (both illustrated in Chapter 9).

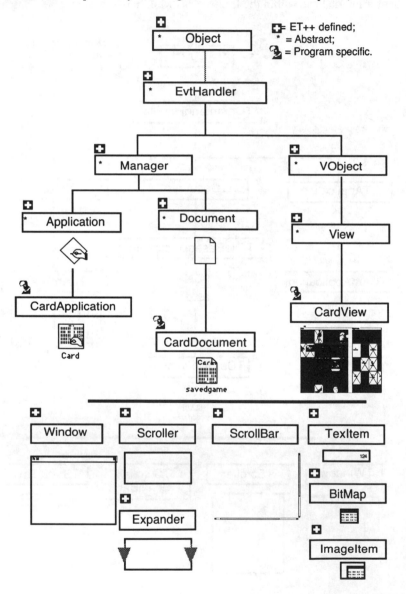

Figure 8.6   The ET++ class hierarchy as used in the Cards program.

Some of the more recent frameworks have started to exploit multiple inheritance. For example, "object i/o" facilities (possibly referred to as "streamability") may be defined in a class Streamable that is distinct from the base class `Object`. Such an arrangement allows greater flexibility when defining classes, only those classes requiring i/o support need inherit from class `Streamable`.

Apart from various minor classes, e.g. "iterators" for lists like class `LinkedListIterator` in Chapter 6, a framework class will be derived from its base `Object` class. For example, a `TScrollerScrollBar` (in Figure 8.5) is a kind of `TScrollBar`, which is a kind of control manager (`TCtlMgr`), which is a kind of `TControl`, which is a `TView`, and consequently it is a `TObject`.

Program-specific classes, e.g. a `ChemStructure` class in the `ChemEdit` program or a `CardData` class in the Cards program developed in this chapter, can be derived from a framework's base class. Such derivation is necessary if the new class is to take advantage of features like object i/o provided in the framework. In some frameworks, the collection classes have been written assuming that all objects stored in a collection will be derived from that framework's base class; this forces use of the base class if collections are needed.

Use of the framework's base class is optional for application-specific classes

Figures 8.5 and 8.6 provide some indication of the degree of similarity among these frameworks. Both have `Application`, `Document`, and `View` classes that are "event/command"-handlers derived from a base `Object` class. As illustrated in the figures, these classes serve as the bases for program-specific derived classes. The frameworks provide similar concrete classes for standard forms of user interaction and data display.

There are differences in library structures. Thus, MacApp distinguishes a document's run-time responsibility for data and display structures from its role as an interface to the file system. An extra class, `TFileBasedDocument`, is used to add file-handling to a basic `TDocument` class. Similarly, ET++ introduces an extra level of abstraction – "Visual Object" (`VObject`) – into its class hierarchy; a `VObject` in ET++ is something that can handle user-commands and can draw itself. Most of ET++'s standard user interface components are derived from this `VObject` class rather than the more specialized `View`.

Many concrete classes from these frameworks can be used without any detailed knowledge of either their ancestry or their behaviours: *"This program requires a window to frame its views – we'll use a standard one"*, or *"The view needed to display the xxx-data will be too large to fit in the window – we had better nest it inside a scroller"*. In other cases, e.g. using MacApp's `TNumberText` class, the development programmer has to be able to find out about the particular services that a class provides to its clients. These classes can be adequately described by an entry in a library manual that includes a brief descriptions of the class's intended role and a list of member functions that will be of interest to a client programmer – e.g. `TNumberText::GetValue()` and `TNumberText:: SetValue()`.

Because they must create their own derived classes, development programmers require a somewhat greater understanding of their framework's `Application`, `Document`, and `View` classes. Such classes are complex. For example, there are more than 120 member functions defined in MacApp's `Application` class (≈ 60 methods inherited from ancestors with the rest defined in class `Application` itself). But unless some unusual behaviour is required of a new program, or it is to handle more advanced features such as inter-application communication on a network, most of the member functions can be used unchanged. Basic use of a framework can be achieved once a development programmer knows which member functions of these classes must be given effective definitions and has developed some understanding of the principal ways in which instances of these classes can interact.

Specializing
Application

An effective specialization of a framework's class `Application` can be created by providing:

- an initialization routine or constructor. (This should invoke the initialization routine or constructor for the framework's class `Application` and then perform any special processing required e.g several frameworks permit the use of some form of "signature" that can uniquely identify files created by a particular program; the setting of this "signature" might form part of this initialization work.)

- an effective implementation for the virtual member function `Application::DoMakeDocument(...)` (or its equivalent)

Specializing View

In simple programs, like the examples in this chapter and Chapter 10, `View` classes will have definitions like:

```
class ChemView : public View {
public:
    // Provide definitions for essential virtual methods
    // inherited from class View

    // Need an effective "Draw" method:
    virtual void Draw(...);             // Override
    // Need to handle the mouse:
    virtual void DoMouseDown(...);      // Override
    // We have a menu-command to deal with:
    virtual void DoSetupMenus(...);     // Override
    virtual void DoMenuCommand();       // Override

    // Now define any new functions unique to this class
    void EstablishLinks(PaletteView* itsPalette, ...);
    …
private:
    // Extra, program specific data members
    …
};
```

Minor extensions to
the framework's View
class

It is rare for specialized `View`s to need lots of new member functions. The role of a `View` is not being changed; it is still just something that can respond to mouse actions and menu commands, and can display data. There may be a need for one or two extra data members, e.g. in the ChemEdit program the main `ChemView` class would probably define a pointer member so that each `ChemView` object can have a link to its corresponding `PaletteView` object (such a link might be needed to allow the `ChemView` to find out which of the editing tools was to be used). A function to set such links, like the `EstablishLinks()` function in the class declaration given above, would be one of the few additional member functions that would then be defined .

A few member functions of View *must* be given effective implementations. Obviously, the inherited View::Draw() function must be replaced because the default implementation draws nothing, while every real View exists to display some data. Views generally react to mouse events; consequently they usually provide an effective definition for the function DoMouseDown() (or DoClick() or whatever the function may be called).

Specialized Views must define functions for drawing and mouse handling

If menu commands are to be handled, the relevant member function(s) must be defined. A command-handler typically has two main functions for working with menus. A DoMenuCommand() function contains the code that deals with menu-based commands (in some frameworks, this may be subsumed by a more general DoCommand() function). The DoSetupMenus() function (or UpdateMenus() or ...) handles the enabling/disabling of menu options. Just before menu options get shown to a user, code in the framework arranges to call the DoSetupMenus() function of each command handler in the target chain. Consider for example a ShapeView that shows some graphic shapes and handles a Shape/Reshade command that should only be enabled when one or more shapes are in a "selected" state. The function ShapeView::DoSetupMenus() would enable the Shape/Reshade option if it had a record of selected shapes, and disable the option if nothing were selected.

Views may need to handle menu commands

Similarly, if a View must respond to keyboard events, it will need to provide a replacement implementation for the DoKeyCommand() function or its equivalent. (The default implementation for this function will simply pass any keyed-data request on to the next command handler in the target chain.) However, it is somewhat unusal to have to define one's own routines to handle keyboard events. When text needs to be handled, one can generally use a more specialized text-view class from the framework; such a class will define how to handle keystrokes (and, probably, how to handle other related things like the menu options Edit/Cut Text, Edit/Copy Text, and Edit/Paste Text).

Keyboard events?

A framework's View class is likely to have more than 100 member functions. More sophisticated programs may need to modify or extend more of these functions. For example, the Views used in the example programs in this chapter and Chapter 10 are fixed in size; but Views often grow as more data are added (e.g. a text-view might start with a size sufficient to display just one line of data, but would then grow as more lines were entered). Programs with such resizeable Views must redefine more of the member functions from their framework's View class.

100+ routines that you usually don't have to implement!

It is the framework's Document class, or its equivalent, that requires most refinement by the developer of even a simple program. A Document's responsibilites include organization of data in memory, transfer of data to and from files, and the construction of displays.

Specializing "Document"

"Documents" and "Windows" for data display are intimately related. A framework can use this relation to provide a slightly different organization in its class hierarchy. The role of the "window" can be enhanced to include many of the responsibilities that are here accorded to a Document class. Rather than creating a Document, an Application written for one of these frameworks would create an instance of some ProgramWindow or FileWindow class. This ProgramWindow

class Document ≈ class "ProgramWindow"

would then own the program's data and organize its display, possibly using additional "child windows".

While the framework's Document class can define prototype functions identifying its various responsibilities, their implementation is necessarily program specific. A typical declaration for a specialized Document class might be:

```
class MyDocument : public Document {
public:
    // Provide definitions for essential virtual methods
    // inherited from class Document

    // Create display structure that shows data
    virtual void DoMakeViews(…);              // Override
    …

    // Functions for saving and restoring data
    virtual void DoRead(…/* file info */);  // Override
    virtual void DoWrite(…/* file info */);// Override
    …

    // Functions for getting document into a standard state,
    // freeing space etc
    virtual void DoInitialState(…);           // Override
    virtual void Revert(…);                   // Override
    …

    // Functions for handling any special menus
    virtual void DoSetupMenus(…);             // Override
    virtual void DoMenuCommand(…);            // Override

    // Program specific functions
    …
private:
    // and some special data members ...
};
```

Functions that must be provided

The development programmer must provide a function that creates the display structure – some approaches are described in the next section. The transfer of data to/from files will require effective implementations for at least two functions – here called DoRead() and DoWrite(). A particular framework may require additional functions, e.g. MacApp relies on having an effective implementation for a function that calculates the size of a file prior to its being written to disk. Documents typically handle some menu options and so will normally have to provide their own versions of the menu handling functions.

Framework-specific features of Documents

Functions like the DoInitialState(), Revert() etc. are more framework specific. They will embody particular choices as to a system's "look and feel". For example, most frameworks only support the idea of "reverting" a file in situations where a program's data have been saved to a file and then been further changed; but on the Macintosh, programs are expected to be capable of "reverting" even if the data they are operating on have never been saved to file – the data are then "reverted" by being put in some "initial state". Most such functions have default, do-nothing, implementations. If these defaults are used, a simple application will usually run though it may not have quite the expected "look-and-feel". Keeping track of whether a document's data have been changed since last saved to file is another responsibility that has to be handled by the document – the way it is handled tends to be rather framework specific.

In addition to its classes, framework libraries may provide additional support to *Failure handling* the development programmer through the inclusion of some uniform mechanism for failure handling. Programs written using these frameworks are typically creating numerous structures on the heap, reading and writing files, and maybe loading predefined data resources such as bit-mapped pictures. Problems can arise anywhere; heap space may be exhausted, a file may be improperly terminated and cause a read error, a required "resource" may not be present at the expected location. In the face of such difficulties, programs are supposed to retreat gracefully rather than collapse or quit. Graceful retreat requires a fall-back position, and a means for tidying up any partially accomplished work. For example, suppose the ChemEdit program is asked to open a second file but finds that it cannot read the list of bonds recorded in that file, then it needs to get back to the state where only one document is open, any new window made for the second document has been closed, and any partially built ChemStructure has been freed.

Most frameworks provide at least some support for this kind of failure handling. In the newer frameworks, failure-handling mechanisms are based on " exceptions" as defined for the forthcoming standard for C++ (C++ exceptions are illustrated briefly in Chapter 12). However, current failure handling mechanisms are all to at least some degree framework specific.

8.2 DISPLAY STRUCTURES

Programs built using these frameworks typically have complex visual displays, with windows that contain instances of program-specific View classes along with instances of standard framework classes like Scrollers, RadioButtons, TextItems etc., see Figure 8.7

When a display structure is built, the programmer must specify the physical layout of the various Views, define behaviours that describe how this layout may change, and specify the command-handling chains. The physical layout of a display can usually be varied. For example, if the window containing the structure shown in Figure 8.7 was enlarged, it would probably be appropriate for components 6, 7 and 8 to remain unchanged in their original positions, for component 5 to grow vertically but not horizontally, for the scrollbars to move to the edge of the enlarged window (both growing but only in one dimension each), and for views 1 and 2 to enlarge both horizontally and vertically. There have to be mechanisms for specifying these behaviours. The "command-handler" chains are almost as complex to set up. Each View has to identify a "next handler" to which it can pass any commands it may receive but can't handle.

Display structures like that in Figure 8.7 can be built using procedural code. Such code would be along the following lines:

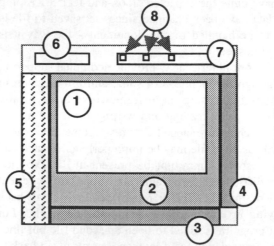

1:      A variable size main view
             for the display of data;
2:      A Scroller;
3, 4: Scrollbars for the Scroller;
5:      A palette view;
6:      A Statictext item;
7:      A Cluster;
8:      Radio buttons in the
             Cluster.

Figure 8.7   A complicated set of nested views.

```
Window* BuildDisplay()
{
    RadioButton* aRadioButton;
    …
    Window*       aWindow;

    aButton = new RadioButton(...);
    bButton = new RadioButton(...);
    cButton = new RadioButton(...);

    aCluster = new Cluster();
    aCluster->AddSubView(aButton,
            40, 40, /* coords for first button  */ ...);
    aCluster->AddSubView(bButton, ...);


    …
    theVertBar = new Scrollbar(HFIXED, VADJUSTABLE, ...);
    theHBar = new Scrollbar(HADJUSTABLE, VFIXED, ...);

    theScroller = new Scroller(...,theVertBar, theHBar, ...);

    …

    aWindow = new Window(WINDOWHPOS, WINDOWVPOS,
            WIDTH, HEIGHT, HRESIZEABLE, ...);

    aWindow->AddSubView(...);

    …
    return aWindow;
}
```

Such code is difficult to read, and even worse to write. Any change in desired layout necessitates recoding.

Many frameworks have support programs that are intended to ease the creation of display structures. These support programs are themselves usually simple graphics editors, constructed using the framework. They allow a development programmer to sketch out the form of a desired display structure by selecting standard elements from a list or palette and placing them in a window. Once the desired display structure has been composed, the support program can generate a suitable representation. This may mean generating the actual C++ code for a `BuildWindow()` function or it might entail the creation of a data structure that defines the display layout. If code is generated, it can simply be linked to the rest of the program. Frameworks that use data structures to define display layouts, e.g. MacApp, include some `InterpretDisplayStructure()` function in their run-time support routines. This run-time function can create the required instances of the different `View` classes and thread them together to establish both the physical nesting and command-chain relations. (With frameworks like MacApp and Symantec's Think Class Library, TCL, that use data structures to define display layouts, the programmer may have to do some special setup work prior to linking their program. Otherwise, the linker may fail to include code to handle a special kind of `View`, e.g. a `RadioButton`, that is needed for the display.)

"View editor"
support programs

8.3   CARDS PROGRAM

The example program, "Cards", is required by an imaginary software company that sells simple computer aided learning (CAL) aids and games targeted at children. The company wanted a CAL program that could be used for teaching languages.

The program required is a kind of "flash cards" teaching aid. There are two decks of cards; one deck having pictures on the card faces, the second deck having corresponding words or symbols (these too might be "pictures" if the language is iconographic). The idea is that the child using this CAL aid would see a picture card and find the matching word card (or vice versa). Each deck has 36 cards. The trial version of the program is to work with just a single pair of card decks. These are illustrated in Figure 8.8; they happen to be a set for teaching French.

The matching process is organized as a simple card game, similar to pelmanism. As illustrated in Figure 8.9, the two decks of cards are laid out face down. The player turns over one card from each deck to reveal their faces. Each turn of a card counts one move. If the cards match, they are to be marked, by the program, as having been paired. An unmatched card will be replaced face down when another card is subsequently selected from the same deck. This process is repeated until all cards have been paired. The program is to keep and display a count of the moves. The objective of the game part of the CAL lesson is to get the cards paired in the minimum number of moves.

Of course, the cards are positioned differently in each game. Once a game is over, with all cards paired, the program has to shuffle the cards and start over with all cards again face down. The program has to allow the user to save a partially completed game.

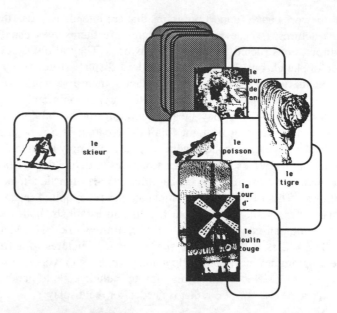

Figure 8.8   The matching picture and word cards.

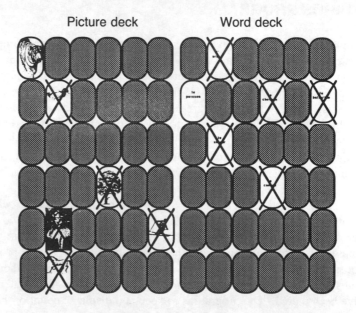

Figure 8.9   The playing tableau for the card game.

The program is to use a single window that displays the playing tableau and the number of moves used so far. A card is to be selected by the user clicking the mouse button when the cursor is above its image. The selection of a card that is already face up, or which has been taken, counts as a (wasted) move, resulting in an update of the number moves but causing no other changes to the display. (Clicks on the window between cards are also to count as moves.)

8.4 CARDS: ANALYSIS AND DESIGN

There are three main aspects to the design of such a program. These are the data-model, the display structure, and the overall organization. The data-model is necessarily program specific; in this example, the data-model will specify how the playing cards are organized. The display structure will define a set of views; different aspects of the underlying data will be displayed in the various views. As several different views must be nested together in a display window; part of the view definition will involve a specification of this nested arrangement. Finally, there are components that organize the program – things like the "application" object and any "document" objects.

The next three sections examine each of these main parts in turn. This initial examination provides some general ideas about what classes might get used. The level of detail varies; while some parts can be specified fairly completely, others can only be roughly characterized. Section 8.4.4 uses simple scenarios that illustrate particular interactions among objects that might exist during the execution of the Cards program. This kind of scenario analysis helps one clarify the roles of the various objects, and hence complete the design of their classes.

8.4.1 A data model for the Cards program

Cards seem to be good candidates for being objects. In fact, because they are simple, and obviously correspond to something having some physical reality, it is possible to proceed immediately to a fairly detailed design for a "Card" class. Each card obviously owns some data. These data include i) the card's state – face up, face down, or "taken", ii) some reference to a bit-image that represents the picture that appears on the card's face, and iii) some key value that allows cards in the "picture" deck and "word" deck to be matched easily. Cards also have some obvious responsibilities. They need to be able to draw themselves in a manner appropriate to their current state (of course, they will have to be told exactly where to draw themselves); they will have to "flip-over" (changing state from face up to face down, or vice versa); at the start of a new game they will be told to reset themselves to face down; the matching process that checks for card pairs will have to be able to ask a card for its "value"; and, there may be some other responsibilities. These basic ideas about a "card" are summarized in Figure 8.10.

Class Card?

In Figure 8.10, a default constructor is defined for class Card along with a separate Initialize() routine. As explained in the next paragraph, it seemed convenient to have arrays of cards. Since an array requires a default constructor, the cards will be identical to start with. They will therefore have to be individually initialized with values for their pictures etc.

It would be possible to make the data model work directly with cards, but the notion of a "card deck" seems to be a useful abstraction. Cards do belong to decks. A card deck can determine its layout and hence the position occupied by each card. If the cards are held as an array that forms a part of an instance of a CardDeck class, then memory management should be simplified a little (because the program will be working with one large composite object instead of many small objects).

Class CardDeck?

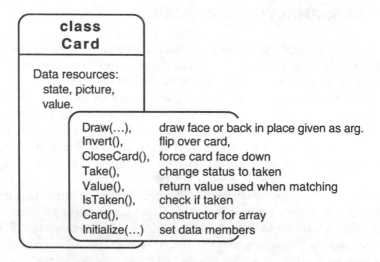

Figure 8.10   Initial design ideas for class Card.

If a CardDeck manages the layout of cards, it should probably take responsibility for determining which card has been selected when the user tries to select a card with the mouse. Since a CardDeck owns the cards, it had better be responsible for writing them to a file when a game is saved, and for reading back information from a file when a game is restored. At the start of a game, a CardDeck had better shuffle its cards to make each game unique.

It is not possible to characterize fully the responsibilities of a CardDeck at this stage. Interactions among the various objects will have to be explored in a little more detail. Scenarios will have to be constructed that show how something like a mouse click is handled. These scenarios will identify the various objects involved in a particular approach to the handling of an event and show their manner of interaction. It may be necessary to consider several different approaches involving different objects or different forms of interaction. It is only after such diagrams have been sketched that it becomes possible to partition responsibilities among classes and to determine the data that must be passed between instance of classes during the handling of a given event.

Consequently, class CardDeck is left defined only by a fuzzy "analysis" level class diagram showing responsibilities, and a rough "class composition" diagram that records the fact that a CardDeck object is a composite containing an array with many instances of class Card.; see Figure 8.11.

Class CardData   The program needs to keep two separate card decks, a count of moves made, and a count of matched cards. Actually it needs such a set of data for each "document". Obviously, there will be advantages in having some "record structure" that groups these data elements. This "record structure" is easily promoted to an "object" by identifying some responsibilities.

If a "CardData" object is created, it can act as a kind of manager for the CardDecks and related information. Such an organization should simplify interactions between the program-specific components and the instances of the framework classes.

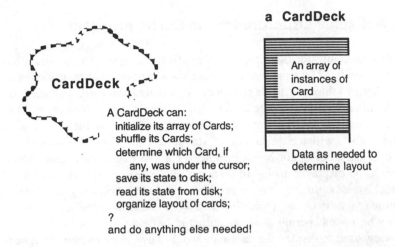

Figure 8.11 Analysis diagram for class CardDeck and a composition diagram revealing that an instance of CardDeck is composed from many Cards.

For example, the "document" object won't need to know about multiple CardDecks etc.; it need merely know that there is a CardData object that manages application-specific data. If the document requires data to be written, it can simply send a "do write" message to this data manager which will then have the responsibility of passing on the request to each data object that needs to be saved. Similarly, a "view" doesn't need to communicate directly with different CardDecks; when a view must handle a mouse click it can simply pass the cursor position to the data manager and rely on it to forward a "handle click" message to an appropriate data element.

So, a CardData class emerges as an appropriate abstraction. An instance of this class will own two CardDecks and a few other data elements. The CardDecks might as well be made constituent components rather than separately allocated entities.

The responsibility of class CardData is the "management" of the two CardDecks. This will mainly entail the forwarding of messages from other run-time objects. Details have to be deferred until some scenarios have been used to map out expected interactions. At this stage, the concept of class CardData is left very fuzzy, as illustrated in Figure 8.12.

Figure 8.12 Analysis diagram for class CardData.

8.4.2 A display structure for the Cards program

Concrete classes from the library display standard data items

The display structure for this program is relatively simple. The program has to display two kinds of information: a counter showing the number of moves used, and the playing tableau with its two decks containing the cards in their various states. Additional views could be used to display labels, e.g. a "Number of Moves" text-label beside the move count. The views used to display the count, and any associated label, will use instances of standard view classes provided in the framework; the label might be an instance of a `TextItem` view-class while the count could be displayed using an instance of a `NumberText` view-class. (The framework's class `NumberText` might be intended primarily for input and verification of user data; but, typically, there would be a "disabled"-data member in the class which would permit use as an output-only field.)

Custom designed views for complex data

The view used to display the card data would have to be custom designed. A class `CardView` would be derived from a standard class `View` provided in the library. Class `CardView` would have to provide an implementation for a virtual `Draw()` function inherited from class `View`. The implementation of `CardView::Draw()` would probably consist of a request to the `CardData` object to arrange that each card draw itself.

The view used to display the card decks will have to be nested inside a scroller. The two card decks could only be fitted into a window that would fit onto the screen if the cards were drawn on a very small scale; so small that the user would not be able to distinguish them.

The scroller, and associated scroll bars, will be placed inside a window, along with the other views needed for the display of the number of moves etc. as illustrated in Figure 8.13.

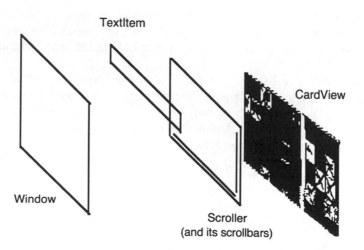

Figure 8.13   A nested arrangement of Views for the Cards program.

8.4.3 Application and Document classes

The Cards program naturally has its own Application and Document classes. These are minor specializations of the framework's class Application, and class Document.

Class CardApplication would need to define a constructor, or maybe a special *CardApplication* initialization routine, and would have to replace the default DoMakeDocument() method. The replacement DoMakeDocument() function would create an instance of class CardDocument. The total code for the CardApplication class would amount to something like:

```
CardApplication::CardApplication()
{
    IApplication(...);  // deal with inherited extra inits.
    this->SetSignature(...); // do any work of our own;
    ...                      // anything else? e.g initialize
                             // a random number generator
}

Document* CardApplication::DoMakeDocument()
{
    CardDocument* aDocument = new CardDocument;
    // Possibly ask the CardDocument to execute an
    // initialization routine that extends the work of
    // its constructor
    aDocument->OtherInitialization(...);
    return aDocument;
}
```

A CardDocument would have a pointer data-member that would reference an *CardDocument* instance of class CardData. As usual, the CardDocument class will provide a minimal set of redefined virtual methods: DoMakeViews(), DoRead(), and DoWrite().

A CardDocument will create an instance of class CardData in its constructor or other initialization method, and eventually get rid of it in its destructor or some separate Free() method. Its DoMakeViews() method would either read in and interpret a data structure that defines its display structure, or construct the display by explicitly creating and initializing the various components. Either way, instances of classes CardView, TextItem, Scroller, etc. would be created and interlinked to build the run-time display structure.

Several links will be required between views and data elements. These links would have to be constructed during the process of creating and initializing a new document. The setting of these links would probably be done in class CardDocument's DoMakeViews() method. The following section identifies the required links through an analysis of various scenarios where interactions among different objects are explored.

8.4.4 SOME SCENARIOS FOR OBJECT INTERACTIONS

The most important aspects of the Cards program are its abilities i) to allow manipulation of cards with mouse clicks, ii) to draw a picture of the playing tableau, and iii) to save/ restore a game to/from a file. It is in these aspects that most of the more complex interactions will occur between the data model and instances of standard framework classes.

These interactions should be explored so as to identify where collaborations occur, and therefore where instances of different classes need to know about each other. This information is needed before the design of the classes can be completed. Classes whose instances are involved in collaborations will have to have data members that are links (pointers) to their collaborators and member functions that can be used to set such links. The links will have to be set in the code where class instances are created − such as the code that handles the initialization of newly created documents.

Handling a Mouse Click

The initial steps involved in handling a mouse click are performed by the operating system, the framework's run-time support routines, and by the CardApplication object which will be executing standard Application member functions. Once the event has been properly identified, it will result in a request to the CardView object to perform its DoMouseCommand() function. Frameworks vary here; some may be able to send such a request directly to the target CardView; in other frameworks, the request to the CardView might be transmitted via intermediaries such as the Window and Scroller objects.

New responsibility for CardData: HandleSelection

Things become program specific only at the point where a specialized View starts its DoMouseCommand() function (see Figure 8.14). In the Cards program, things are simpler than usual. There is no need to track mouse movements while providing sophisticated visual feedback, nor is there any need to set up a mechanism that allows the user to reverse changes to the data that will result from the handling of the mouse click. A mouse click should cause a card to turn over and then a check should be made for a match between exposed cards in the two decks. This work can be organized by the CardData object. So, the interaction between the view and data model is trivial. The CardView object tells the CardData object "*you handle a selection attempt at point ...*". The code for CardView::DoMouseCommand() would be something like:

```
void CardView::DoMouseCommand(const Point& theMouse)
{   // argument is a structure defining the cursor position
    ftheData->HandleSelection(theMouse);
}
```

Need a link CardView → CardData

Of course, the CardView object must have a pointer (ftheData) to the corresponding CardData object. This pointer must be set when a new CardDocument object and its associated CardView and CardData objects get created.

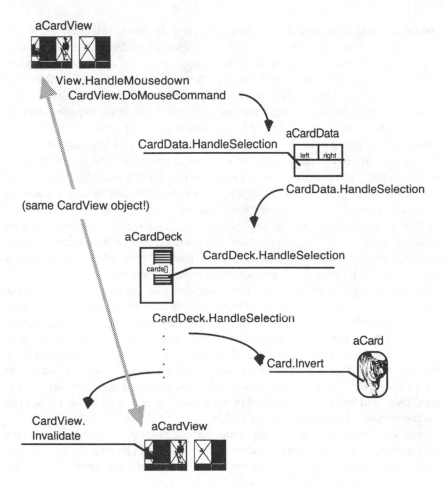

aCardView

View.HandleMousedown
CardView.DoMouseCommand

aCardData

CardData.HandleSelection

(same CardView object!)

CardData.HandleSelection

aCardDeck

CardDeck.HandleSelection

CardDeck.HandleSelection

aCard

Card.Invert

CardView.
Invalidate

aCardView

Figure 8.14   Handling a mouse click.

Each mouse click counts as a move; so, the first thing the `CardData` object should do is increment a record of moves and arrange for a `NumberText` view to change to show the new move-count. Class `CardData` will need a data member for the count, and a link to the view used to display this count. The link should also be set during the start up sequence where the `CardView` was linked to the `CardData` object.

CardData needs a link to display field for move count

The `CardData` object then has to identify the `CardDeck` affected. Since the layout for the playing tableau is fixed, it should be sufficient for the `CardData` object to use some static constant `Rectangles` to define the positions of the `CardDecks`. A simple "point in rectangle" test should suffice to identify the active `CardDeck`. This deck will then be asked to complete the card-selection process. Since `CardDecks` are component parts of `CardData` objects, there are no problems in communication – no need for extra links.

New responsibilities for CardDeck: HandleSelection

A `CardDeck` object will handle a selection attempt by identifying the card located under the cursor when the mouse click occurred. If that card was already taken, no further processing would be needed. Otherwise, any card already face up in the deck would have to be flipped face down, and the newly selected card would have to be set face up. Class `CardDeck` can probably determine the position occupied by each of the cards in its array (it does not seem worth making this a responsibility of each individual `Card`); there had better be a function that calculates the area occupied by a `Card` at a specific (row,column) position in the layout. Class `CardDeck` will need another (private) function for iterating through its `Card` array to determine which `Card` includes a specified point.

Extra data members in class CardDeck

Each deck is going to have to keep track of which `Card`, if any, is "open" – i.e. is face-up and still unmatched. Some data members will have to be added to record this information. The `CardDeck` object must obviously collaborate with the `Cards`, e.g. to tell a `Card` to invert itself. Again, there is no difficulty in communication because the array of `Cards` is a component part of a `CardDeck`.

Don't draw unless asked to

When a `Card` inverts itself, it just updates its internal record of its state. It does not redraw itself. In GUI and application frameworks, data objects normally draw themselves only on specific request. The framework will try to schedule its screen updates so as to minimize the frequency of drawing operations.

Need a link CardDeck→ CardView

The `CardDeck` object that is flipping over cards has to be able to inform the `CardView` that some portions of the existing display are now invalid, and will have to be redrawn when the underlying system next gets round to scheduling screen updates. (A `View` handles an "invalidate" notification by passing this information on to the run-time system that keeps track of the parts of the screen may need to be redrawn.) Just as there was a need for a pointer in class `CardView` to an instance of class `CardData`, there will need to be a link from class `CardDeck` to `CardView`. Again, there will have to be a method in class `CardDeck` that allows this link to be set when instances of this class are created.

CardData object checks for matches in its HandleSelection routine

Once the affected `CardDeck` has completed the process of identifying the new open `Card`, the `CardData` object can resume the checks in its `HandleSelection` routine. The `CardData` object has to be able to determine whether there are matching open `Cards` in the two decks.

More responsibilities for CardDecks: report "open" cards, handle requests to mark cards Another link to a collaborator

Class `CardDeck` will have to provide routines in its interface that can be used to ask whether a `CardDeck` object has an open `Card` (`HasSelection()`), to report the identity of this `Card` (`SelectedCard()`), and to mark a `Card` (`TakeCard()`) as matched.

If the open cards do match, the `CardData` object will have to increment the count of matched-cards, tell the decks to mark their cards as matched, and then tell the `CardDocument` that it has been changed. Marking the document as "changed" will result in it enabling the File/Save and File/Revert options. Since the `CardData` object must collaborate with the `CardDocument`, it will need a pointer data member for this collaboration link. (If the document is marked as changed at each move, it is too easy for a player to cheat. They can save their current game, inspect all the remaining hidden cards, revert their game to its saved state, and complete the matching process with no further mistakes. The temptation to cheat is reduced if the player only gets a chance to "save" or "revert" after a successful move.)

If all cards are matched, the `CardData` object should provide suitable visual and audible feedback to the player, and then reinitialize the game, the move and match counts, the count display etc.

Handling Drawing

When the OS/framework gets round to updating the screen, a message will be chanelled to each `View` requesting that it redraw any part of itself that intersects with a specified "update rectangle". Once again, the `CardView` object will simply pass the draw request on to the `CardData` object. This will check whether the left deck, the right deck or both card decks intersect with the update-rectangle. If a `CardDeck` does intersect with the update-rectangle, it will be asked to draw itself, see Figure 8.15

Extra responsibilities: CardData & CardDeck "Draw" functions

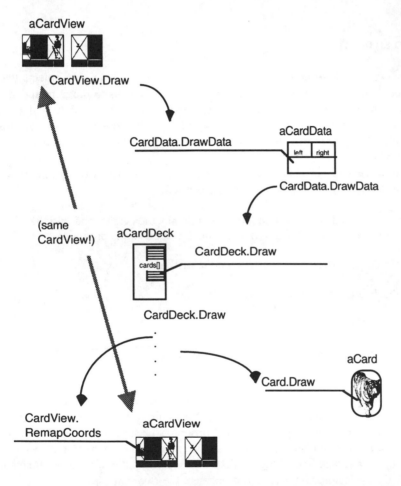

Figure 8.15 The main steps involved in the handling of a draw request to a CardView object.

The `CardDeck` will iterate through the cards that it manages, getting each one that intersects with the update-rectangle to redraw itself. The scheme is straightforward apart from one possible glitch. Normally, one works in a coordinate system defined by a `View` (probably using 32-bit coordinates); thus, here the positions of `Cards` are in terms of the `CardView`. But, when actually calling

Changing coordinates

low-level graphics routines, one is more likely to have to work in the coordinate system defined by an enclosing `Window`, or maybe even in screen coordinates (possibly, in 16-bit coords). `Views` maintain data that can remap their coordinate systems onto the coordinate system that must be used for actual drawing operations. It *may* be necessary to ask the appropriate `View` to remap coordinates before a drawing operation. The `CardDeck` object might again need to use its link to its associated `CardView` to get coordinates changed. The documentation on a framework's `View` classes will specify if/when such coordinate changes may be needed.

Opening an old file

The framework will define the overall pattern of interactions among objects that occur when a file is opened. This sequence will involve calls to program-specific routines to build display structures and to read data. Objects that will later collaborate need to be linked; these links should also be constructed during this file-opening phase (or during the equivalent processes involved in the creation of a new document).

The basic sequence of operations involved in opening a file are as summarized in Figures 8.16 and 8.17. The application object will:

- create and initialize a new document, which may then create and initialize other objects, such as the `CardData` object needed in this program;

- get the document to read its data from the disk file;

- get the document to create its display structure.

and

- set collaboration-links once all the data objects and `Views` have just been created.

The exact order of these operations depends on the framework – for example, the display structure may be created before data objects are restored from file.

Extra responsibilities: ICardDocument, and CardData:: ReInitialize

The Cards example is simpler than most programs because it has a fixed amount of data. An instance of class `CardData` can be created either in the `CardDocument` constructor, or as illustrated in Figure 8.16 in a separate `ICardDocument` function. No other data allocations are required because the two card decks, and all their constituent cards, are component parts of the `CardData` object. The `CardData` object will have to initialize itself properly – getting both card decks to place all cards face down, and resetting the move count to zero. It might be useful if these operations were packaged as a `ReInitialize()` method, because this would facilitate the setting up of a repeat game.

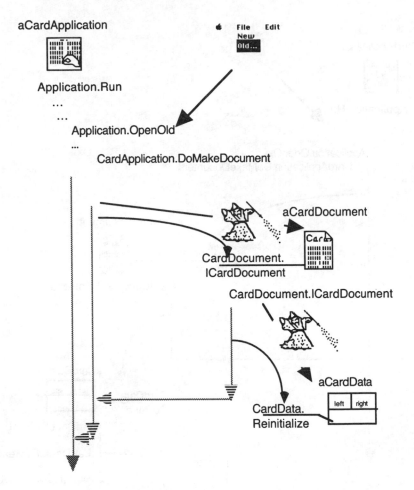

Figure 8.16 First steps in opening an old file – creation of document and data model objects.

The `CardDocument` object will keep a pointer referencing its own `CardData` object; this is obviously required so that it can pass on messages such as "DoRead". As noted earlier, a `CardData` object will need a reciprocal pointer so that it can keep its document informed about changes. So, at some stage during this initialization process, a pointer to the document should be placed in the `CardData` object.

Links needed
CardDocument
↔CardData

Once the document has been created and initialized, the `OpenOld` method of class `Application` gets the document to read its data (Figure 8.17). The `DoRead` method of class `CardDocument` might have to start by invoking the `DoRead()` function of its parent class. (Some frameworks use the first few bytes of the files that they create to hold their own "housekeeping" data; if the framework used does have such a policy, all specialized `DoRead()` and `DoWrite()` functions should start by calling `Document::DoRead()` and `Document::DoWrite()` respectively.) Once any such housekeeping chores have been completed, the `CardData` object can then be told to read its own data.

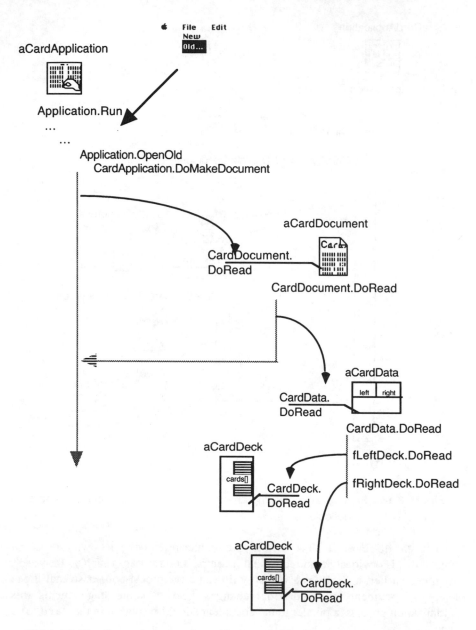

Figure 8.17   Second phase of opening an old file – restoring data.

CardData and CardDeck get responsibility for reading/writing their own data   The `CardData` object would read information such as the number of moves that had been made when the game was saved, and then get each card deck object to read its own data. In this simple situation, there wouldn't be any benefit from having cards etc. explicitly represented as "objects" in the data file. The `CardDeck::DoRead()` and `DoWrite()` functions could use binary block transfers to copy bytes to/from their arrays of cards (so the code would be implemented using low-level i/o calls equivalent to the `read()` and `write()` system calls on Unix).

In the final stage of opening a file, the application's OpenOld method would tell the new document to create its views. This may be done procedurally, or by reading and interpreting a data structure. Then the collaboration links would be set.

The CardData object needs links to both the main CardView and the Number-Text used to display a move count. The CardDocument object will have pointers to these two Views (it will either have just created them procedurally, or will have some mechanism for extracting such pointers from the data structure that it loaded). So the CardDocument can pass these pointers to the CardData object that it has created; class CardData will have to provide some SetLinksToViews() member function. This will take as arguments the pointers to the collaborating Views and would use these data to set private pointer data members. The same function can also be used to set the pointer to the CardDocument.

Getting pointers to view objects

The individual CardDecks also require links to the associated CardView (as shown in Figures 8.14 and 8.15, a CardDeck may have to tell its CardView to "invalidate" some region or may need to ask it to remap some coordinates). The pointer to the CardView will have to be passed to each CardDeck from somewhere in the code for CardData::SetLinksToViews(); class CardDeck will also have to define some LinkToView member function.

A CardView needs a pointer to the associated CardData object so that it can forward the requests for drawing and handling of mouse-down events. Class CardView has to provide some function that can be called from class CardDocument and will be used to pass the pointer to the CardData object. The framework may already define a suitable function in its View class, and have a call to this function in its standard processing sequence. After all, one almost always has to set up these collaboration links so the process might as well be standardized. For example, MacApp's View class has a PostCreate() method that takes as argument a pointer to the Document that created the View – this function is called for each View object as it is created from a display data structure. The default implementation of PostCreate does nothing; but it can be extended, in a class specific manner, to allow a View to get from its Document object any required links to other collaborators . If the framework doesn't provide such a function, the developer implementing a specialized View class will have to define a SetLinksToData(...) member function and arrange for this to be called at some suitable point in their DoMakeDocument() or DoMakeView() functions.

Linking views to the data model CardData

8.5 DETAILED DESIGN

The little scenarios in section 8.4.4 more or less suffice to complete the characterization of the various classes that are required for the Cards program. A great many such scenarios have to be worked out when analysing and designing a more realistic application. Each scenario explores the ways in which objects might interact when the program is performing a particular task. These scenarios help the designer identify additional responsibilites for each class and establish where collaborations will be needed. This analysis allows initial "fuzzy" class diagrams, like that in Figure 8.12, to be made a little more explicit.

Designs for the classes are summarized in Figures 8.18 – 8.21.

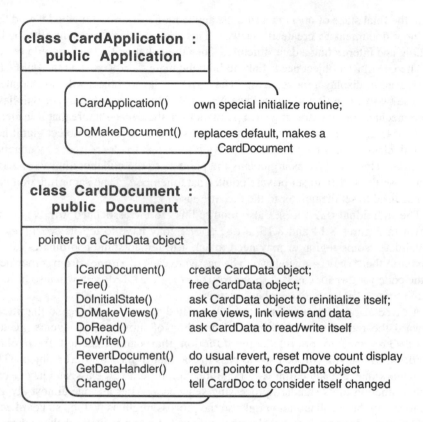

Figure 8.18   Design diagrams for classes CardApplication and CardDocument.

CardApplication and CardDocument

Design diagrams   The final design for the CardApplication and CardDocument classes is shown in Figure 8.18. As expected, class CardApplication requires only minor changes from the framework's Application.

Class CardDocument requires one extra data member – its pointer to its CardData object. The framework will probably provide a pointer data member that references the main Window; this can be used to obtain references to any Views nested within the Window, and there is no need for the CardDocument to keep separate pointers to these Views. The function that creates the display structure and sets the links between the CardView and CardData objects can use local pointer variables to reference the Views.

Design diagrams, like Figure 8.18, distinguish the public interface and private implementation aspects of a class. All the functions identified for class CardDocument are public; the only private information is the pointer data member. Obviously, class CardDocument is not going to be used as the base class for any more specialized class, so there is no need to consider things like a protected interface.

Design outlines for   In addition to providing an overview diagram like Figure 8.18, the designer
classes   would provide outline explanations for each of the member functions. Usually,

these outlines would specify the signatures for the routines. Such details have been suppressed here because they are too framework dependent. The names and responsibilities of framework defined member functions will also vary; but the following outline should be more or less compatible with several frameworks.

The responsibilites identified for class `CardDocument` are:

- `ICardDocument()` and `Free()`
  Depending on requirements of the framework used, these may be separate functions or may be subsumed into the class's constructor and destructor.

  `ICardDocument()` completes any document initialization steps required by the framework, then creates an instance of class `CardData`. This `CardData` object should be reinitialized (a process that randomizes the card decks etc.). The `Free()` method should release the `CardData` object and then invoke any inherited `Free()` method provided in the framework's `Document` class.

- `DoInitialState()`
  Need only be implemented if required by the framework. This function would act to reinitialize the game: get the `CardData` object to `Reinitialize` itself, and get the `View` that displays the move count reset to show 0.

- `DoMakeViews()`
  Construct the display, then establish links between the `CardData` object and its collaborating `Views` (the `CardView` and the move count).

- `DoRead()` and `DoWrite()`
  Perform any housekeeping work required by the framework (e.g. call inherited read/write function), then pass read or write request on to the `CardData` object. These functions will have arguments defining the "stream" to be used for i/o; pass this data on in the call to the read/write method of `CardData`.

- `RevertDocument()`
  Invoke the standard `Document::Revert()` method which should arrange for data to be again read from file. Then reset the display of the move count and arrange for the main `CardView` to be redrawn.

- `GetDataHandler()`
  Return a pointer to the `CardData` object. (Here, it is assumed that the member function of class `CardView` that is used to set a link to its collaborating `CardData` will actually be passed a pointer to the `CardDocument`. It will then need to ask the `CardDocument` for a pointer to the data object.)

- `Change()`
  Mark the `CardDocument` as changed or unchanged since last saved. (This allows the standard `Document::DoSetupMenus()` function to enable or disable the Save, Revert and similar options in a File menu.) The implementation of this responsibility varies considerably between frameworks. Some frameworks have a public data member in their class `Document` that is either an integer change count or a boolean "dirty" flag. In such cases, there is no need for a `Change()` function, the `CardData` object can update its document state

directly. However, some form of Change() function will be needed in frameworks that use more elaborate mechanisms to record their document's state.

Scenarios like those in Figures 8.15 should be kept as part of the documentation of a design, with each class being characterized by a list of the scenarios where its instances are shown.

Class CardView

As illustrated by the design shown in Figure 8.19, class CardView is only a minor specialization of the standard View class. It adds a private data field and overrides two or three of the virtual methods. The member function descriptions on the diagram pretty much summarize the class's responsibilities.

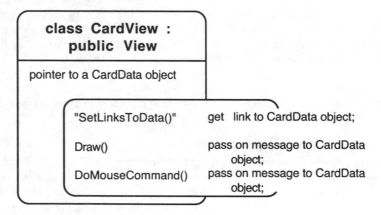

Figure 8.19   Design diagram for class CardView.

Class CardData

A design for class CardData is shown in Figure 8.20. It has private data members for counts of cards matched and moves made, and has two instances of class CardDeck as component data members. In addition, it has three pointer data members; these are for the links to the CardView, NumberText, and CardDocument objects with which a CardData object must collaborate. All the member functions are public – getting called in functions in class CardDocument or class CardView. In a more complex program, the design documentation would have to include information identifying the client class(es) that use each of the member functions.

Again, the designer would provide an outline for each member function to supplement the class design diagram. The functions of class CardData are:

*   CardData()
    Constructor. Set counts of moves and matched cards to zero, and set collaborator-links to NULL.

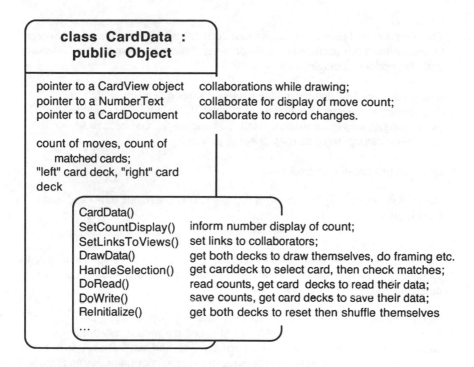

Figure 8.20   Design for class CardData.

- `ReInitialize()`
  Initialize both left and right card decks (`CardDeck::Initialize()`), passing as arguments data defining the position where a deck is to be displayed, and information defining the set of "pictures" that it is to use for its cards. The pictures would be "bit-maps" – possibly loaded from a named file, or held as a program "resource" data structure, or maybe defined in static arrays.

  Get decks to shuffle their cards.

- `SetLinksToViews()`
  Store pointers to collaborating `Views` and `CardDocument` in own data members; get each `CardDeck` to make its own copy of the pointer to the `CardView`.

- `SetCountDisplay()`
  Ask the `NumberTextView` to change the displayed value to match the current move count. (This will be called both from `CardData`'s own `Handle-Selection()` member function, and from member functions of `CardDocument`).

- `DoRead()` and `DoWrite()`
  Transfer from/to a file the move and matched card counts, then ask each `CardDeck` to transfer its data.

- DrawData()
  This function will take an argument that defines an "update rectangle". Screen updates should be optimized by only drawing those data elements that intersect with this update rectangle.

  Using Rectangles that define the deck positions determine which CardDeck(s) intersect the update rectangle; if a CardDeck does intersect with this rectangle, then get it to draw itself (passing on the update area so that the CardDeck can optimize its own drawing process).

  Draw frames around the decks.

  (Note, don't need to update the NumberText; it is a separate View and looks after itself.)

- HandleSelection()
  Called when a mouse down event occurred in main CardView; all such events count as "moves" so increment the move count, and get the NumberText to update itself to reflect the new value.

  Determine which CardDeck includes the mouse position (using the Rectangles that define the deck positions, and the mouse coordinates that are passed as an argument to HandleSelection()), and ask that deck to execute its HandleSelection function (passing the mouse coordinates as an argument so that the deck can identify the card selected).

  If both decks then have "open" cards (CardDeck::HasSelection() routine), find the selected cards (CardDeck::SelectedCard()), and see if they match.

  If the open cards in the two decks match, get each deck to record its current card as "matched" (CardDeck::TakeCard()) and update the count of matched cards. Tell the CardDocument that it has been changed.

  If all cards are matched, terminate the current game and reinitialize everything so allowing the user to start a new game.

Class CardDeck

A design for class CardDeck is shown in Figure 8.21. A CardDeck needs a pointer member to reference its collaborating CardView, some coordinate data defining its position, and an array of Cards. Further, a CardDeck has to keep track of any "open card"; so it will require a number of integer data members, e.g. integers for the row and column of the open card, and its "value". Since all Cards are the same size, and the layout is a simple rectangular array, the position of the CardDeck implicitly defines the positions of all of its Cards.

Private member functions of CardDeck   The main responsibilites for class CardDeck have already been introduced. A CardDeck has to be able to initialize itself and its own cards (setting their picture bit-maps etc.); it must be able to get its cards drawn; it has to shuffle them; it has to report whether it has a card open (and be able to report the "value" of any open card); and, it has to read and write its data. Most of the member functions are part of the public interface, but here there are some private implementation functions. When dealing with Draw() and HandleSelection() requests, a CardDeck will need to work out the area occupied by each card and do things like looping through its array finding the card located under a specific point. These processes are relatively elaborate, and are needed in more than one place in the code of CardDeck, so they are worth promoting into functions. Naturally, they are member functions of CardDeck and, since they are only used to facilitate other operations, they are private. As with all the other classes considered, there is no need to make provision for protected interfaces – class CardDeck isn't going to be reused or extended.

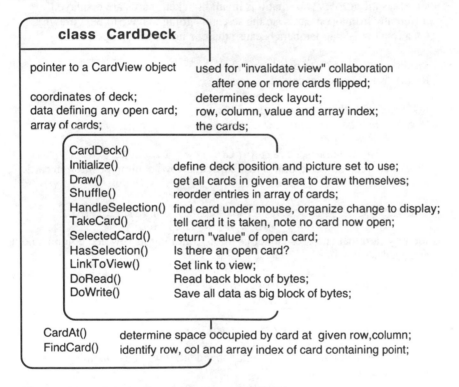

Figure 8.21   Design diagram for class CardDeck.

Class CardDeck does not depend much on the framework and so a more detailed specification can be given here. As would be usual in a detailed design outline, complete signatures can be given for most member functions:

- CardDeck()
  Constructor. Set the link to the collaborating CardView to NULL.

- void CardDeck::Initialize(Point topleft, Pictures ...)
  The Point argument (or maybe a Rectangle argument) defines the location of this CardDeck in terms of the coordinate system of its CardView; record this coordinate information in data members.

The other argument(s), `Pictures...`, will identify the bit-map pictures to be used for the 36 cards in this deck. (It is left to the implementors to decide whether this information should take the form of a filename, an open stream, a pointer into a static array, or a "resource" number.)

- `void CardDeck::Shuffle()`
  Randomize the array of `Cards`.

- `void CardDeck::TakeCard()`
  Inform "open" card that it has been taken (allowing it to update its internal state). Notify `CardView` that it is invalid ("taken" cards are displayed differently from open cards, so the screen is going to have to be redrawn). Clear out own data members because there is no longer an open card.

- `int CardDeck::SelectedCard()`
  Return "value" of open card.

- `Boolean CardDeck::HasSelection()`
  Return *true* if have an open card.

- `Rectangle CardDeck::CardAt(int row, int col)`
  Work out area that corresponds to a particular row, column position in card-array.

- `int CardDeck::FindCard(const Point& containedpoint,`
  `                        int& arow,int& acol)`
  Identify card that encloses cursor position, returning its row and column and its array index.

- `void CardDeck::LinkToView(CardView*)`
  Set link to collaborating `CardView`.

- `void CardDeck::DoRead(File* afile);`
  `void CardDeck::DoWrite(File* afile)`
  `File*` argument will be framework dependent; it will be something like a stream to be used for i/o.

  Transfer data defining current open card, then do a block transfer of all the data in the Cards array.

- `void  CardDeck::Draw(const Rectangle& area)`
  Loop through array of cards, getting each to draw itself in its appropriate area (if coordinate remapping is required, this process may require collaboration with the `CardView`, e.g. with the Macintosh might need to map from 32-bit `View` coordinates to 16-bit Quickdraw coordinates). Optimize the drawing process by only dealing with `Cards` that intersect with the specified redraw area.

- `void  CardDeck::HandleSelection(const Point& where)`
  Find card at selection point; if this card is "taken", do nothing.

Invert any previously open card, and notify the `CardView` of an invalid area.

Copy data defining selected card into data members used to record "open" card.

Invert newly opened card, and notify the `CardView` of an invalid area.

Class Card

The simplicity of class `Card` means that there is little detailed design work to do; most remaining decisions on the representation of this class can be left to the implementation phase. Some aspects are framework dependent; e.g. the bit-mapped picture might be represented as an integer "resource number", or as a byte array, or as an instance of some `ImageItem` class. The state of the card – face-up, face-down, or taken – would probably be implemented using some enumerated type defined for the class; again, this is really an implementation decision.

Its functions are:

- `Card();`
  Constructor; needed when creating array of cards. Record state as "face-down".

- `void Initialize(int id,...);`
  Gives a card an individual identity; the arguments will specify its "value" (for matching purposes) and its bit-map picture.

- `Boolean IsTaken();`
  Return *true* if state is taken.

- `int Value();`
  Return value used for matching.

- `void Draw(const Rectangle& box);`
  Draw in the specified rectangular region. If the card is face-down, fill the region with a uniform pattern. If the card is face-up, draw the bit-map picture scaled to fit in the specified area. If the card is taken, draw as for face up, then draw diagonals across the card-face to show that it has been matched and is out of play.

- `void Invert();`
  Change state back and forth between face-up and face-down.

- `void CloseCard();`
  Force card into face-down state (e.g. at start of new game).

- `void Take();`
  Mark state as taken.

Decisions as to which of these functions is made "inline" would be left to the implementors.

8.6   IMPLEMENTATIONS?

8.6.1 Implementing the classes

Coding from detailed class design diagrams and function specifications should be relatively straightforward. The design diagrams translate fairly directly into class declarations; for example from Figure 8.21, one can get the declaration of class `CardDeck`:

```
class CardDeck {
public:
    CardDeck();
    void        Initialize(Point …);
    void        Draw(const Rectangle&);
    void        Shuffle();
    void        HandleSelection(VPoint where);
    void        TakeCard();
    int         SelectedCard();
    Boolean     HasSelection();
    void        LinkToView(CardView *f);
    void        DoRead(File* afile);
    void        DoWrite(File* afile);

private:
    Rectangle   CardAt(int row,int col);
    int         FindCard(const Point& containedpoint,
                         int& arow,int& acol);

    Point       ftopleft;
    Card        fCards[kDECKSIZE];
    int         fopenCard;
    int         fopenRow;
    int         fopenCol;
    int         fopenVal;
    CardView*   fdisplay;
};
```

Many of the function implementations are almost as easy to obtain. However, there will be some problem areas. The definition of a program's display structure tends to be a problem with most frameworks. Drawing code naturally depends on the system software (so, programs on the Macintosh will use Quickdraw, X-Windows routines will be needed in Unix systems). Developers have to be familiar with the graphics facilities available on their chosen platform. Other difficulties are more framework dependent and relate to idiosyncrasies of the individual frameworks/development platforms.

For example, although written in C++, the MacApp framework was required to be capable of being used by development programmers working in Apple's Object Pascal language or those using a Macintosh-specific dialect of Modula2. Object Pascal doesn't understand constructors – so MacApp's classes can't make use of constructors. Function signatures must specify Pascal compatibility (so instead of something like `class View : public CommandHandler { ... void Draw(const Rectangle&); ... };` one get things like `class TView : public TCommandHandler { ... pascal void Draw(const VRect& ); ... };`). Further, the Macintosh's OS likes data structures in its heap to be moveable – and this requirement has a few interesting side effects on programming style. Other frameworks and development environments present their own challenges.

On Macintosh and PC platforms, programs usually use "resources". These are *"Resources"* simply preinitialized data structures that can be loaded by a running program. They are used to define things like the entries in a menu bar, the size and location of a program's main window, the text strings that are to be displayed when errors occur, sounds to be played, movies to be shown, and 101 other things. Resources are created using special resource editor programs. Developers must learn to use their system's resource editor and get to understand the resource-management facilities of their OS. A Macintosh or PC implementation of the Cards program would almost certainly use picture-resources for the images shown on cards.

8.6.2 Where does it all start? With "main()" of course

It is all very well having the classes, but you do need a program. It will be a typical OO main program of about three statements: initialize the environment (well, in some environments, that might take more than three statements), create an instance of the specialized application class, tell the application object to run. For example, the ET++ implementation of program Cards is:

```
int main (int argc,char** argv)
{
    Symbol preferredfile = cCardDocType;// a file "signature"
    Application* anApplication  =
            new CardApplication(argc,argv, preferredfile);

    InitUdata(); // Initialize random number generator etc

    return anApplication->Run();

}
```

Files available in association with this text include implementations of Cards using MacApp, ET++, Borland's OWL, and other frameworks. Figure 8.22 illustrates the display from the ET++ version of Cards.

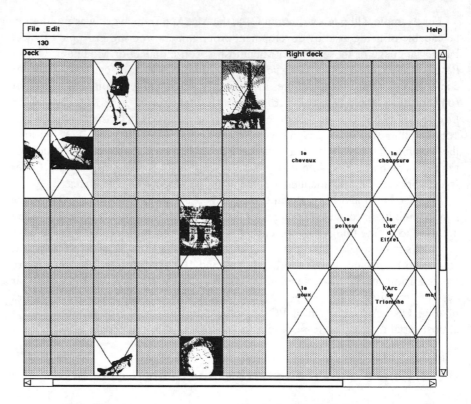

Figure 8.22   Screen Image from the ET++ version of Cards.

REFERENCES

M. Andrews, *Visual C++ Object oriented Programming*, SAMS, 1993.

T. Faison, *Borland C++ 3 Object oriented Programming*, SAMS, 1992.

S. Holzner, *Miccrosoft Foundation Class Library Programming*, Brady, 1993.

S. Holzner, *Borland C++ Windows Programming*, 3e, Brady, 1994.

N. Rhodes and J. McKechan, *Symantec C++ Programming for the Macintosh*, Brady, 1993.

A. Weinand, E. *Gamma, and R. Marty, Design and Implementation of ET++, a Seamless Object-Oriented Application Framework*, Structured Programming, **10**, 1989.

D.A. Wilson, L.S. Rosenstein, and D. Shafer, *C++ Programming with MacApp*, Addison Wesley 1991.

Additional Features of Framework Class Libraries

9

The classes `Application`, `Document`, and `View` (or their equivalents) form the core of any framework class library. But the framework libraries offer much more than just these core classes. Frameworks include specialized classes (or small clusters of interacting classes) that implement additional commonly used program abstractions. Further, program-specific classes can acquire extensive functionality either by inheritance from the framework's base class, or through some multiple inheritance scheme.

For example, most frameworks provide a "Command" (or "Task") class that can be used to implement a scheme for reversible ("undoable") user actions. There may be support for "object i/o" (i.e. a mechanism for making objects "persistent" so that they can retain their identity when saved in a file). Another common abstraction is some form of "dependency" mechanism; this will provide a standardized way of arranging for changes to particular objects to update, automatically, other dependent objects. Dependency mechanisms are realized through inherited functionality and support classes that keep track of the interdependencies that exist among objects. The sections in this chapter introduce some of these more advanced aspects of frameworks.

9.1 COMMAND OBJECTS

9.1.1 The role of command objects

The primary role of a "command object" is to record information that allows a user to reverse any changes that have just been made to the data associated with a document. For example, a user of some "Draw" program may select a shape and reshade it; review the overall diagram, consider that original shade was better and so reverse (Undo) the shading operation. On further review of the diagram, the user may again decide that the shape should be reshaded, and "Redo" the change.

Reversible ("undoable") commands

These processes are illustrated in Figure 9.1. The "Undo/ Redo" processing can be continued until the editing action is finally committed one way or the other. Typically, an editing action is committed when the user starts to make some other change or requests a process such as File/Save or File/Print.

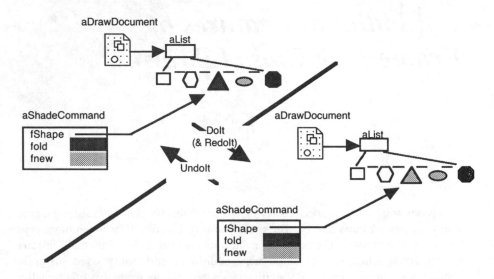

Figure 9.1   A "Shade Changer" Command Object responding to Undo/Redo requests.

Single level or multi-level undo

Some frameworks support only "one-level Undo" – a command object gets disposed of once it has committed and the associated editing action cannot subsequently be reversed. Other frameworks support multiple-levels of Undo. These frameworks preserve command objects for extended periods of time.

Command::Doit()
Command::Undoit()
Command::Redoit()

A framework's class Command will define a number of virtual functions including "DoIt()", "UndoIt()", and "RedoIt()" (or similar functions having slightly different names). When its DoIt() member function is invoked, a command object should change the data it affects to the "new" state. A call to UndoIt() should result in the command object restoring the data to its old state. The implementation for RedoIt() might default to a call to DoIt(). (DoIt() is only called directly when the command object is created and has to change its associated data for the first time; sometimes it is convenient for the DoIt() function to create temporary data structures that simplify subsequent undo/redo operations.)

Calls to methods of a Command object are built into the framework code

If command objects have been properly integrated into a framework, then at "appropriate points" the framework code will check for the existence of a command object and, if one exists, call its DoIt(), UndoIt(), or RedoIt() method as required. Again, this is one of the advantages of using a fully developed framework. The development programmer does not have to work out where these "appropriate points" might be. The authors of the framework will have determined where such calls are necessary, and consequently one doesn't get programs with glitches (e.g. a failure to commit an edit operation before saving data to a file).

However, there are frameworks where the built-in support for command objects is weak; the development programmer has to take on more of the responsibility of deciding when to perform editing actions and when to commit the process and release a command object.

9.1.2 Simple command objects that handle menu commands

The simplest command objects tend to be those created for menu selections (or any equivalent command-key selections) that just change some property of a selected data element. Typical examples would be those commands that change the font-face, font-style, or font-size of some selected text, or those that change the colour or shading of some graphic element. For instance, if one has a graphics program with selectable, shaded graphic shapes:

Simple commands

```
class Shape {
public:
    ...
    void    Reshade(int newshade)
            { fshade = newshade;
                ...    // probably additional code.
                // The view that displays this shape
                // might need to be told that a display
                // region is now invalid.
            }
    int     Shade() { return fshade; }
    ...
private:
    ...
    int     fshade;
    ...
};
```

(Methods are included in these class declarations to make these illustrations more concise; generally it is preferable to define them separately.) The program could have a menu selection Shade with options Shade/Dark Grey, Shade/Grey, ..., etc. If a user selects one of the shapes in the document and then picks one of these Shade menu options, a suitable command object will be created:

```
class ShadeCommand : public Command {
public:
    ShadeCommand (Shape* theshape, int newshade, ...)
            {
                fShape = theshape;
                fnew = newshade;
                fold = fShape->Shade();
                ... // other steps e.g. link to "document"
            }
    void    DoIt()      { fShape->Reshade(fnew); }
    void    UndoIt()    { fShape->Reshade(fold); }
private:
```

```
    Shape*          fShape;
    int             fold;
    int             fnew;
};
```

The command object would be created with a pointer to the data element that it affects (the chosen shape), details of the new shade (known on the basis of the particular Shade menu option that was selected), and probably a pointer to the document that "owns" the data element. The newly created command object would interrogate the shape as to its old shade, and store both old and new shades. Its `DoIt()` and `UndoIt()` methods simply reapply the appropriate shades to the shape.

Informing collaborators

Member functions inherited from class `Command` will also be called as needed. These would perform tasks such as keeping any related document informed as to changes (so that the document object can handle File/Revert and File/Save options appropriately). Other objects might also have to be informed of changes; any view displaying the shape should be told of an invalid region when a shape changes its shade. This dependency is not the concern of the command object that changes the shape; it is a responsibility of the shape itself. One way to handle the dependency is suggested in the code above where the `Shape::Reshade()` method notifies any associated view.

Representing the old and new states

Class `ShadeCommand` is possibly atypical because it allows a very simple, direct representation of the old and new states of the data. Just two data members in the command object suffice – the old shade, and the new shade. In other cases it is more difficult to devise a suitable representation of old and new states. Consider for example a word processor program that needs a command object to represent old and new states for a "delete paragraph" operation. It would not be appropriate to duplicate a data structure representing a complete chapter of a textbook just to get two versions, one omitting a particular paragraph. Instead, some more complex state representation would have to be used that held the text of the paragraph and some pointer to its original position. The ability to choose good representations for command objects can only really come with experience.

DoMenuCommand()

As illustrated in Figure 9.2, menu-based command objects are created in a command-handler's `DoMenuCommand()` member function. This command-handler may be the application, the document, or a view (the most common case is the document). If the command-handler is to be a view, the programmer must remember to get that view into the "target-chain" of command-handlers that will be given the chance to handle a command. Figure 9.2 shows a `ShadeCommand` being created by the `DrawView` of some drawing program.

A `DoMenuCommand()` function will have a form like the following:

```
Command* DrawView::DoMenuCommand(int cmd)
{
    switch(cmd) {
case cShadeGrey:
        ShadeChanger* aShadeChanger;
        aShadeChanger= new ShadeChanger(fData,
                            cShadeGrey, ...);
        return aShadeChanger;
case …:     // Deal with other commands
        …
```

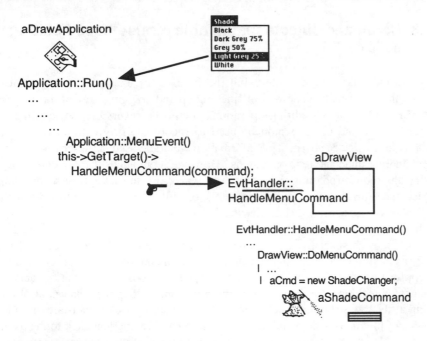

Figure 9.2   Creating a command object in response to a menu selection.

```
            ...
            break;
default:
            return View::DoMenuCommand(cmd);
    }
}
```

The user's menu-selection action gets translated into an integer command number which is given to the "target" object at the front of the chain of command-handlers. Each handler in the chain gets a turn to claim the command, or to pass the command on to the next command handler. In the example code above, a DrawView handles a cShadeGrey command by creating a ShadeChanger command object. If the command is not one of those that a DrawView can deal with, the inherited View::DoMenuCommand() function is invoked; there, the code will pass an unclaimed command on to the next command-handler in the chain.

As noted in Chapter 8, menu options are not always active. The options in a *DoSetupMenu()* "Shade" menu would only be active if there were a selected shape that could be reshaded. Objects that are responsible for dealing with menu-based commands are also responsible for activating and deactivating the entries in the menus. So, each command-handler class can have DoSetupMenu() member function; the code in this function will enable or disable menu items associated with particular command numbers.

Like display structures, menus may be created procedurally or by the *Creating menus* interpretation of some "MENU" resource data structure that is loaded at run-time.

9.1.3 Command objects that handle mouse-based editing actions

"Mouse-commands"

Some editing changes are made using the mouse; for example, in a graphics program there is usually some way of "picking up and dragging" any of the shapes shown in a view. Such editing operations should be reversible; so, a "Shape-Dragger" command object would be used to record the original and modified positions for a selected shape. This information would be used, as described above, when handling undo and redo operations. However, commands that are entered by moving the mouse (possibly referred to as `Trackers` to distinguish them from simpler `Commands`) have a few additional responsibilities. They have to work with the display views to provide visual feedback.

DoMouse-Command() method

The development programmer has to implement a `MyView::DoMouseCommand()` function for each specialized view that is to support mouse-based editing. This member function creates an appropriate mouse-command object that is returned to the application object. The application object (or, possibly, some "Tracker-Manager" object hired for the occasion) will then enter a loop that takes control for as long as the mouse button is held down. In this loop, a series of calls are made to methods of the mouse-command object so that it can provide feedback to the user. Calls may also be made to the view to cause it to scroll if the user performs some mouse action that drags something out of the area visible in the window.

TrackMouse(), TrackConstrain(), TrackFeedback()

Among the additional methods of a mouse-command object there will be the member functions `TrackMouse()`, `TrackConstrain()` and `TrackFeedback()` (or their equivalents, the names of these functions will vary). Their roles are summarized in Figure 9.3 which shows a "ShapeDragger" mouse-command object being used to move a shape.

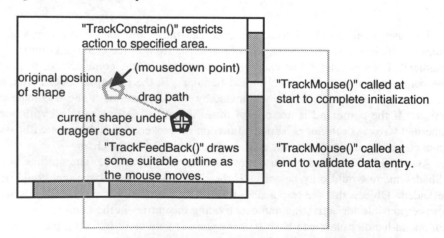

Figure 9.3   Mouse-command objects must provide visual feedback.

TrackConstrain()

Sometimes one gets situations where movements must be constrained. For example, a shape in some graphic design may be "pinned" to a particular region; although it can be moved around within that region, it must not be moved outside of the region. A `TrackConstrain()` method will be included to permit the

imposition of such constraints. If the mouse is moved, its new position is passed to
`TrackConstrain()` before being used in the other routines. `TrackConstrain()`
can "adjust" the mouse coordinates so that they are restricted to fit within any
chosen region.

A `TrackFeedback()` member function provides the visual effects for the   *TrackFeedback()*
mouse-based editing action. So, if a `ShapeDragger` command object was being
used to move a selected shape, an outline of that shape might get drawn repeatedly
along the drag path.

The framework code typically arranges to call the `TrackFeedback()` function
twice during each processing cycle while the mouse is held down. This simplifies
the processes needed to draw and then remove the outlines of moving shapes. The
display's "pen" can be set in XOR-mode, so an initial drawing action can be
completely cancelled by a second call.

The `TrackMouse()` method, or its equivalent(s), is called at the start (just after   *TrackMouse()*
the mouse button is first pressed), at the end (when the mouse button is released),
and possibly once during each processing cycle. An argument passed in the call
can be used to distinguish among these cases. (Some frameworks use separate
functions rather than a multipurpose `TrackMouse()` function.) The first call may
result in some additional initialization processes. Normally, some "validation"
checks are made at the time of the final call.

For example, a program might allow a user to place nodes and join these nodes   *Validating a mouse*
with edges (e.g. to build a model of some kind of network or circuit). An   *action*
`EdgeSketcher` command object would be used to draw edges; it would have a
"validation" requirement that the two ends of the edge should be located within
distinct nodes. If an edge is drawn but left with an end outside of any existing
node, it will fail its validation test. In this case, the `TrackMouse()` method would
return some indicator to show that editing action was void; the framework code
would dispose of the command object and never set up all the Undo/Redo
mechanism.

9.1.4 COMMANDS THAT ADD DATA

The `Shading` command and `ShapeDragger` command described above alter
existing information but do not add new data to a document. Commands that add
data, like an `EdgeSketcher`, require some additional code.

Figure 9.4 illustrates the state of various objects after an `EdgeSketcher`
command object has been created and validated. The `EdgeSketcher` command
object will have a pointer to the edge that has been created. If this edge is added to
the document (i.e. either the method `EdgeSketcher::DoIt()` or the method `Edge-`
`Sketcher::RedoIt()` has just been executed), then a second pointer to this same
edge should be inserted in the document's data structure (shown in Figure 9.4 as a
list of edges). If the user decides to cancel the editing action (i.e. `Edge-`
`Sketcher::UndoIt()` is executed), the link from the document's list to the edge
should be removed because the edge is not part of the document's data. Of course,
the `EdgeSketcher` command object needs to maintain its own link to the new edge
throughout any sequence of Undo/Redo actions.

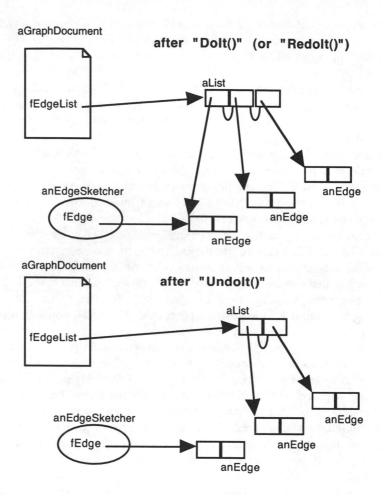

aGraphDocument

after "DoIt()" (or "RedoIt()")

fEdgeList

aList

anEdge

anEdge

anEdgeSketcher

fEdge

anEdge

aGraphDocument

after "UndoIt()"

fEdgeList

aList

anEdge

anEdge

anEdgeSketcher

fEdge

anEdge

Figure 9.4   Situation prior to the "commit" of a command object that adds data
to a document.

Eventually, the user will move on to other things and the EdgeSketcher
command object should be deleted. But there is a potential problem. When the
EdgeSketcher gets deleted, should it arrange to delete the edge, or should the
edge be left alone?

Death of a program.   If an EdgeSketcher object tries to delete an edge that now belongs among the
Hung from a   document's data elements, it will leave a dangling pointer and the program will die,
dangling pointer?   hung from a dangling pointer, as soon as an attempt is made to access the edge.
Bled by memory   But if the EdgeSketcher does not delete an edge whose addition was cancelled,
leakage?   then that edge object will remain forever in the heap. Slowly, heap memory will
leak away and the program will eventually die bleeding to death by memory loss.

Command objects can maintain a status flag (signifying whether the last
operation was an Undo or Redo). When a command object gets deleted, it should
check this flag and if this shows that the last operation was Undo then any
associated data elements should also be deleted. Otherwise, they must belong to
the document or some other data structure and should be left alone.

9.1.5 More advanced uses of command objects

Command objects have uses other than their primary role as the support mechanism *Filtered commands*
for Undo/Redo processing. They can be used to help optimize the handling of
changes to a document's data. A change to a document's primary data storage
structure may be costly in time (or space as when a change necessitates the creation
of large temporary data structures). If changes are costly, it is best to make the
changes only when an editing command is actually committed. However, in the
mean time the program must make it appear that requested changes have been
made. Various schemes are used where the display of a document's data will be
"filtered" with respect to any uncommitted commands. If data elements are being
removed by an uncommitted command, they will not appear in the display. Data
elements, that have not yet been added to the main data structure, will appear in
displays if they are being added by an as yet uncommitted editing command. Most
framework libraries include example programs illustrating this kind of processing.
However, it is a relatively advanced feature and shouldn't normally concern
beginners.

9.2   OBJECT I/O

While some of the objects present in a running application program are transient
and are only meaningful during the duration of program execution, most of the data
that are owned directly (or indirectly) by a document object should be capable of
being saved to disk files. After all, the application programs that are built with
these class libraries are basically "editors" that get run many times as their users
build and refine each document. They have to save partially defined data to files.
For example, a graphics program will have to save data defining its current list of
shapes, with each shape identified by type together with details of its shading,
position, and size. A word-processor program will have to save details of its list of
paragraphs, pictures and tables, with each item characterized by formatting
information and content bytes.

Such applications do not require an "Object Database". All the data in a file   *Data files NOT*
will be reloaded at the same time, and the file will be completely rewritten when   *Databases*
the File/Save menu option is next invoked. Individual objects (paragraphs, shapes
etc.) are not fetched on demand. There are no searches for objects that satisfy
particular predicates. Consequently, there is no need for any indexing of these
files nor indeed for any of the more elaborate structuring such as is necessary in a
database. Even so, there are some problems associated with saving and restoring
data to and from files.

Subsections 9.2.1 and 9.2.2 illustrate some of these problems together with
simple, ad hoc solutions that are generally usable. Subsection 9.2.3 contains a
short discussion of more general mechanisms provided by frameworks.

9.2.1 Pointers: the representation of objects in memory and in files

The data members of an object typically comprise both state information and links to collaborating objects. Thus, class CardDeck in the Cards program has a link to a cooperating view (the CardView* data member) and the state information as represented by the data fields that define the array of cards and specify which card is "open" etc:

```
class CardDeck {
public:
    ...
private:
    ...
    Card        fCards[kDECKSIZE];   // "State"
    int         fopenCard;          // "State"
    ...
    int         fopenVal;           // "State"
    CardView*   fdisplay;           // Link to Collaborator
};
```

The fdisplay link is transient; it only has meaning while the program is running. When a CardDeck object is restored from a file, the link gets set to a newly created view. The other data that define the state of a CardDeck object (its fCardDeck array, its fopenCard … fopenVal data members) can be transferred as a block of bytes:

```
void CardDeck::DoWrite(File* afile)
{   long datasize;
    void *Ptr;
    Ptr = &fopenCard;
    datasize = sizeof(int);
    afile->WriteData(Ptr,datasize);
    ...
    Ptr = &fCards;
    datasize = sizeof(Card) * kDECKSIZE;
    afile->WriteData(Ptr,datasize);
}
```

Beware of "pointer types" in state data   However, it is quite common for an object's "state data" to include pointers as data members. Consider a program used to build a specialized graph of nodes and edges. The edges could be defined by pointers to the nodes that they join:

```
class Edge : {
public:
    Edge();
    ...
private:
    Node*       fNode1;
    Node*       fNode2;
};
```

Naive programmers have been known to try to save an edge to a disk file using code such as:

```
void Edge::DoWrite(File* afile)
{    // The naive programmer's DoWrite()
    long datasize;
    void *Ptr;
    Ptr = &fNode1;
    datasize = sizeof(fNode1);
    afile->WriteData(Ptr,datasize); //NO, stupid!
    Ptr = &fNode2;
    datasize = sizeof(fNode2);
    afile->WriteData(Ptr,datasize); // NO!
}
```

Of course, such code saves the *addresses* of the nodes. Such data are useless when read back from a file. The code restoring the data structures will have created a new set of nodes at addresses quite different from those being used when the file was last saved.

Address-pointer data members have to be converted into some machine independent form when data structures are written to files. The typical way that this is done is to have a long-integer unique identifier associated with each object in those classes that will be referenced by pointer data members. Thus, in the graph-editing program, class Node could have a static (i.e. class) data member to record the number of nodes created, and each individual node could have its own fId unique identifier number:

Convert address-pointers to unique identifiers when saving

```
class Node : public Datum {
public:
    Node(Point,View* v);
    long        GetId() { return fId; }
    ...
    static void   InitializeCNodeCounter(long);
private:
    static long   sNodeCounter;
    long        fId;
    ...
};
```

As each Node is created, the class's counter is incremented and the new node is assigned the current value as its unique identifier; this would typically be part of the work of a Node::Node() constructor.

Providing the unique identifiers

```
Node::Node(Point p,View* v)
{
    ...
    fId = ++sNodeCounter;
    ...
}
```

The program has to be able to set an initial value in the class variable sNodeCounter. When a file is read back, the counter must be initialized to the identifier of the highest existing node so that any nodes that are added subsequently will acquire distinct identifiers:

```
long Node::sNodeCounter = 0;

voidTNode::InitializeCNodeCounter(long startval)
{
    sNodeCounter = startval;
}
```

(It might be more appropriate to assign node numbers based on the value of a data member that belongs to the document rather than a class variable like sNodeCounter. The use of a data member in the document will result in all the nodes in that document being numbered sequentially. A class variable like sNodeCounter is shared by all documents that are open simultaneously. Consequently, if nodes are added in turn to each of several open documents there will be gaps in the numbering of nodes in any individual document. Often, such gaps would not matter. In those applications that require that data items be numbered sequentially as well as uniquely; the numbering scheme has to be document-based rather than class-based.)

Using the identifiers in files   When edges are written to a file, they should save the identifiers of the nodes that they join:

```
void Edge::Write(File* aFile)
{
    ...
    long nodeID;
    nodeID = fNode1->GetId();
    aFile->WriteLong(nodeID);
    nodeID = fNode2->GetId();
    aFile->WriteLong(nodeID);
}
```

Convert unique identifiers to address pointers when restoring   Of course, when an edge reads its data from a file it is going to get two long integer node identifiers when it requires Node* pointers. The programmer implementing the cluster of nodes, edges and related classes has to provide a mechanism for converting from identifiers to address pointers.

One way might be to have nodes and edges are stored in separate lists that are both data members of a more elaborate Graph class. When a Graph is saved it could writes the list of nodes before the list of edges. So, when a Graph object is being read back from file, all the data defining nodes are input first. As the edges are read, they can ask their associated Graph object to convert a node identifier to a Node* pointer:

Start by reading data that defines the nodes
```
pascal void Graph ::ReadFrom(File* aFile)
{
    long          countersetting = 0;
    long          icount;
    {
    icount = aFile->ReadLong();
    for(int i=0;i<icount;i++) {
```

```
                Node* aNode;
                aNode = new Node;                              Each node reads
                aNode->Read(aFile);                            itself
                if(aNode->GetId() > countersetting)
                        countersetting = aNode->GetId();
                fNodeList->InsertLast(aNode);
                }
        Node::InitializeCNodeCounter(countersetting);
        }

        {                                                      Next read data
        icount = aFile->ReadLong();                            defining the edges
        for(int i=0;i<icount;i++) {
                Edge* anEdge;
                anEdge = new Edge;
                anEdge->Read(aFile,this);                      With each edge
                fEdgeList->InsertLast(anEdge);                 reading itself
                }
        }
        ...
}

void Edge::Read(File* aFile,Graph* g)                          Edges get their
{                                                              Graph to convert
        ...                                                    node identifiers to
        long ID1,ID2;                                          Node* pointers
        ID1 = aFile->ReadLong();
        ID2 = aFile->ReadLong();
        fNode1 = g->NodeWithId(ID1);
        fNode2 = g->NodeWithId(ID2);
        ...
}
```

The `Graph` object could identify the appropriate node by searching down its list of
nodes checking for one with the required identification number.

9.2.2 Transferring heterogeneous collections to/from files

The `Graph::ReadFrom()` function in section 9.2.1 was simplified by the fact that
the nodes and edges were kept separately. When a file was saved, the nodes were
saved first, then the edges were saved etc. So, when reading back a set of data
items, the code could specify the type of object to be created to read the next input
data:

```
for(int i=0;i<icount;i++) {
        Node* aNode;
        aNode = new Node; // I know its a node
        aNode->Read(aFile);
        ...
        }
```

More typically, one would have a single list of data items. For example, a graphics program might have a single list of `Shapes` – all the entries on this list would be instances of more specialized subclasses such as `Triangle`, `Pentagon`, `Rectangle` etc. (see Figure 9.5). The read routine that restores such data from file is going to have to be able to determine the type of the next shape so that it can create an instance of the appropriate class and then tell it to read itself.

A simple approach using a code number to identify classes

Figure 9.5 illustrates a simple approach to dealing with this problem. When `Shapes` write themselves to the file, they first output an identification code, e.g. code 0 for a circle, code 1 for a triangle, code 2 for a rectangle etc.; then they write the data defining their boundary points and their shade. The input routine reads the type code and uses this to select the appropriate type of shape to create:

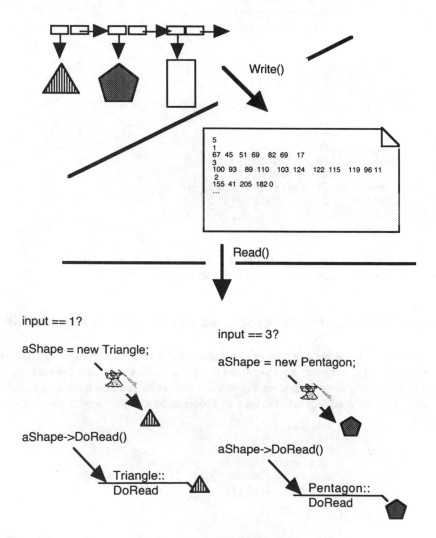

Figure 9.5   Simple i/o for a heterogeneous list of objects.

```
void ShapeDocument::ReadFrom(File* aFile)
{
    long          itemcount;
    itemcount= aStream->ReadLong();
    for(int i=0;i<itemcount;i++) {
            Shape*        aShape;
            long          itemCode;
            itemCode = aFile->ReadLong();

            switch(itemCode) {
case 0:
                    aShape = new Circle;
                    aShape->Read(aFile);
                    break;
case 1:
                    aShape = new Triangle;
                    aShape->Read(aFile);
                    break;
                    ...
                    ...
                    }

            fShapeList->Append(aShape);
    }
}
```

This style of input routine has the disadvantage of building knowledge about the range of shape types into the ReadFrom() routine of class ShapeDocument. Such knowledge can be factored out in a number of ways.

Who needs to know what types of shapes are possible?

One way is to use the approach illustrated, in section 6.6, for the Space Invaders example. The base class, Shape, could have a static member function for generating instances of its various specialized subclasses:

```
Shape*  Shape::MakeMEAShape(long id)
{
    switch(id) {
case 0:
            return new Circle;
case 1:
            return new Triangle;
    ...
    }
}
```

This allows the ShapeDocument::ReadFrom() function to be simplified to:

```
void ShapeDocument::ReadFrom(File* aFile)
{
    long          itemcount;
    itemcount= aFile->ReadLong();
    for(int i=0;i<itemcount;i++) {
            Shape*        aShape;
            long          itemCode;
            itemCode = aFile->ReadLong();
            aShape = Shape::MakeMEAShape(itemCode);
```

```
                              aShape->Read(aFile);
                              fShapeList->Append(aShape);
                   }
        }
```

This approach transfers knowledge about subclasses of Shape from class Shape-Document to class Shape itself. This still creates interdependencies among classes, but it is more appropriate for Shape to know about its descendants than have this knowledge held by some unrelated class.

Using prototype
objects
Another approach relies on the use of some global data that define the types of shape that exist. For example, one could have a global array of Shape* pointers that are filled with prototype instances in some initialization routine:

```
Shape*      gProtoShapes[10];
...
void ShapeApplication::IShapeApplication(...)
{
    ...
    // Now, set up the prototype shapes.
    gProtoShapes[0] = new Circle;
    gProtoShapes[1] = new Triangle;
    gProtoShapes[2] = new Rectangle;
    ...
}
```

This ShapeDocument::ReadFrom() function creates new objects by *cloning* the appropriate shape from the gProtoShapes array.

```
void ShapeDocument::ReadFrom(File* aFile)
{
    long           itemcount;
    itemcount= aFile->ReadLong();
    for(int i=0;i<itemcount;i++) {
        Shape*         aShape;
        long           itemCode;
        itemCode = aFile->ReadLong();
        aShape = gProtoShapes[itemCode]->Clone();
        aShape->Read(aFile);
        fShapeList->Append(aShape);
    }
}
```

Cloning objects
Cloning creates an identical copy of an object. A class can define a simple member function that creates clones using that class's copy constructor e.g.:

```
class Circle : public Shape {
public:
    Circle(const Circle& ref);   // copy constructor
    Circle* Clone();
    ...
private:
    Point  fCentre;
}
```

```
Circle::Circle(const Circle& ref) : Shape(ref),
                    fCentre(ref.fCentre)
{}

Circle* Circle::Clone() { return new Circle(*this); }
```

9.2.3 Framework support for input and output of objects

Figure 9.6 illustrates the type of data that begin to justify extensive support for object i/o. Here, a graph structure is used to represent a finite state machine, with nodes corresponding to states and (directed) edges representing transitions (the example is based on the program developed in Chapter 10). Both nodes and edges in this graph are labelled. The example graph represents a partially composed finite state machine that is to recognize character sequences that represent numbers. The partially constructed graph has just two states and one transition defined. Even in this simple case, the program's representation, as shown in the lower part of Figure 9.6, might be quite complex involving instances of class Node, class Edge, class List, class Link, class TextItem, class Font, etc.

The problems touched on in the previous two subsections are greatly magnified here.

The "structure" that is to be saved is composed of many separate interlinked objects. In memory each of these links will be represented as a pointer data member that holds a physical address. All such pointer data members require translation to some alternative representation during transfers to (from) disk files.

Pointer data members

Normally, in such structures there are multiple references to individual objects – in the example, the same font object is referenced by each of the label objects (because all the labels are displayed using the same font). This must be correctly handled during transfers to file; the file representation of the structure should have just a single font object. When the structure is restored from file, all the newly created label objects should again possess pointers to the same newly created font object.

Multiple references to objects

As well as multiple references to individual objects, structures often have circular chains of pointers. For example, in Figure 9.6, the nodes have lists of edges – so there is a chain of pointers node → list → link → edge. The edges contain pointers to their nodes, so there is a final link edge → node that results in a circular chain. Again, the structure as represented in a file should have just single copies of each of the objects involved in such a circular chain. The data saved for these objects should allow the chain to be rebuilt in terms of pointers when the structure is read back from file.

Circular structures

In addition to problems with pointer data members, one usually has the problem of heterogeneous collections that must be saved to disk – like the list of Triangles, Pentagons, etc. in the example in section 9.2.2. The file representation of such a collection is going to have to contain data that specify the type of the next object in the file. The input routines must be capable of using this type data so that an object of the appropriate class can be created.

File representation needs data specifying objects' classes

Some data:

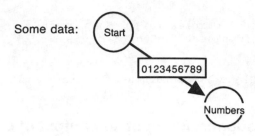

and its representation in a program:

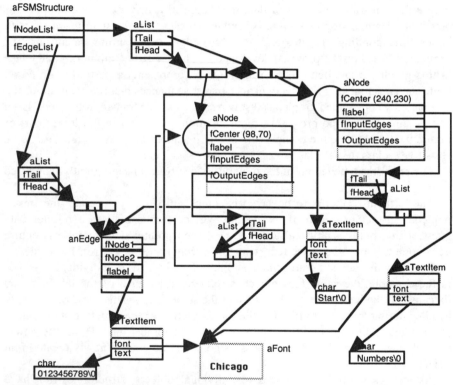

Figure 9.6   Some data and a simplified(!) view of the corresponding memory representation.

Framework support for "streamability" or "persistence"

Many frameworks now incorporate general purpose mechanisms for handling object i/o. The framework code handles the difficult parts. So, if one did have:

```
class Edge : public … {
public:
    …
    void    PrintOn(OStream&);
private:
    …
```

```
     Rectangle      fBounds;
     ...
     Node*          fNode1;
     Node*          fNode2;
};
```

it would be possible to code the output function as:

```
void Edge::PrintOn(OStream& os)
{
    ...
    os << fBounds;// save bounding rectangle
    ...
    os << fNode1;// in effect, save "link" to a Node object
    os << fNode2;
};
```

The framework code invoked for statements like os << fNode1 etc. will recognize that things like fNode1 and fNode2 are pointer data members and arrange either to get the referenced Nodes to write themselves to the output stream, or would write a "duplicate entry" record for any Nodes that had already been written to that output stream.

The framework code will also address the problem of heterogeneous collections, like the list of shapes in section 9.2.2. The data written to a file for each object includes information describing the class of that object. These data are used to help restore objects from file.

Specialized input/output "Streams"

Generally, frameworks use specialized "Stream" classes, e.g. a class OStream and a class IStream, to provide such support for object i/o. These stream classes would possess the same functionality (and, possibly, the same operator function interface) as the standard ostream and istream classes in the iostream library. But, in addition they help sort out pointers, and record and use data defining the classes of objects transferred.

Streams may present an interface using >> and << operator functions

Actual stream i/o can be achieved using the same general approach as was illustrated in a simplified form at the end of section 6.5.1 As shown there, a global output operator function can invoke a public virtual PrintOn() method of the object for which it was invoked; similarly, an input operator function can invoke some virtual ReadFrom() member. These global operator functions would be defined by the framework; e.g. for a system defining "streamability" as a property of its base class Object, one might have the basic functions:

```
OStream& operator<<(OStream& os, Object& op)
{
    op.PrintOn(os);
    return os;
}

IStream& operator>>(IStream& is, Object& op)
{
    op.ReadFrom(is);
    return is;
}
```

*Alternatively, streams
may use distinct
functions*
A framework might not use operator functions with its streams. As an alternative, it might provide stream classes with interfaces like:

```
class OutputStream : public Stream {
public:
    void    WriteData(Byte*,int length);
    void    WriteRectangle(const Rectangle&);
    ...
    void    WriteLong(long);
    void    WriteObject(Object& op) { op.PrintOn(this); }
    void    WriteObject(Object*);
    ...
};
```

with output routines being along the following lines:

```
void Edge::PrintOn(OutputStream* os)
{
    ...
    os->WriteRectangle(fBounds);
    ...
    os->WriteObject(fNode1);
    ...
};
```

*Inherit
"streamability" or
"persistent"
capability*
In order to use a framework's i/o facilities, the development programmer must make "persistent" or "streamable" all those classes whose instances are to be saved to file. A framework may define "streamability" as a property of its base class (e.g. ET++, MacApp, etc.), or it may utilize multiple inheritance and provide a separate class Streamable (e.g. as in Borland's ObjectWindows class library).

Often, as well as acquiring streamability properties by inheritance, a developer's class will have to define a few additional class specific functions. If needed, these would be used by the framework to provide things such as run-time class identification of objects, and some auxiliary i/o facilities. Although class specific, these functions have standard forms and their declaration and definition often require just the inclusion, in each class, of a framework supplied macro declaration.

*Streamable
(persistent) classes
must provide effective
"PrintOn" and
"ReadFrom"
functions*
Each class defined for a program will have to provide effective definitions for the virtual PrintOn() and ReadFrom() (or equivalent) functions. So for class Node derived from class Object:

```
class Node : public Object{
public:
    ...
    OStream&      PrintOn(OStream&);
    IStream&      ReadFrom(IStream&)
    ...
private:
    Rectangle     fBounds;
    TextItem*     fLabel;
    ...
};
```

there would be `Node::PrintOn()` and `Node::ReadFrom()` functions,:

```
OStream& Node::PrintOn(OStream& os)
{
    Object::PrintOn(os);// Must call inherited function
    // Output bounds data,
    os << fBounds;
    …
    os << fLabel;        // Save referenced TextItem
    …
    return os;
}

IStream& Node::ReadFrom(IStream& is)
{
    Object::ReadFrom(is);
    is >> fBounds;
    …
    // Get rid of any existing label
    if(fLabel != NULL) {
            delete fLabel;
            fLabel = NULL;
            }
    is >> fLabel;// Create new TextItem and load it with data
    …
    return is;
}
```

Apart from the handling of the pointer data members, which will be explained in the next section, this looks all pretty much the same as the simpler situation illustrated in section 6.5.1. One obvious difference is the call to the inherited `PrintOn()` and `ReadFrom()` functions. These calls are essential; the functions defined for class `Object` perform housekeeping tasks related to the ways in which pointer data members, and type data, are handled.

Call to inherited function allows streaming system to perform "housekeeping" tasks

In addition to performing the standard i/o roles, class `OStream` and `IStream` maintain records of those objects transferred to/from file. These records are held in a list or "dictionary" object associated with each stream. When an object invokes the inherited function `Object::PrintOn()`, an extra entry is added to the output stream's list. The entry might consist of a pointer to the object in memory and a sequence number recording its position in the file. Similarly, an entry is made in a list associated with an input stream when an input function calls `Object::ReadFrom()`.

Recording details of objects as they are transferred

The framework deals with pointer data members

The "smarts" provided by the framework come into play when transferring objects referenced by pointers. A statement like `os << fLabel` in `Node::PrintOn()` does *not* result in the output of a hexadecimal number representing the address of the `TextItem`. Instead, either the `TextItem` itself is saved, or a "duplicate reference indicator" gets written to the file.

Transferring objects referenced by pointers

The framework will provide something equivalent to a global function:

```
OStream& operator<<(OStream& os, Object* op);
```

The statement os << fLabel would be compiled into a call to this output function. The code for this function will be something like

Code to output an
object referenced by a
pointer

Find whether object
already saved by this
stream
Code to save just the
"identifier" for any
duplicated items

Get any new object to
save itself and add it
to the list associated
with the stream

```
OStream &operator<< (OStream &os, Object *op)
{
    int     index;
    // Check for object in a list associated with the OStream
    index = os.CheckFor(op);
    if (index > 0 ) {
            // It was in the stream's list, i.e. already saved
            // so just save data identifying a duplicate
            os.put('@'); // flag character,
                         //        indicating a duplicate
            os << index; // write index of original
            }
    else {
            ...

            ...
            // Now call object's own PrintOn function
            // so that it can save its own data
            op->PrintOn(os);
            ...
            }
        ...
    return os;
}
```

When called to transfer an object, the function first checks whether that object is present in the list of those already written to file. If the object has already been written, it is sufficient to output its index number. So for example, if in a set of nodes there were some with the same label (i.e. pointers to the same TextItem) then this TextItem would get written to file while the first of these nodes was executing its PrintOn() routine. When the other nodes sharing the label executed their PrintOn() routines, the statement os << fLabel would result in output of "duplicate entry" flags that referenced the original copy.

Input functions

The framework provides corresponding input functions. For each program-specific class there will be an associated global function (possibly generated by macro), e.g. for class TextItem one would get:

```
IStream& operator>>(IStream& s, TextItem*& op);
```

(the second argument for this function is a "reference to a pointer", i.e. a pointer to a TextItem is being passed by reference so that it can be set inside the function). This function would be called for the statement is >> fLabel in class Node's input routine.

Read a new object or
a duplicate reference
indicator

The input routine will expect to find in its stream either the data characterizing some new object or a "duplicate entry" flag with index number. If it finds the duplicate entry flag, the routine can use the index number to "look up" the required object in the list of previously read objects that will be associated with the input stream. The pointer op, that is passed as an argument to the routine, will be set to reference this previously read object (resulting in multiple references to the same object). However if the input routine finds that its stream contains data that define a new object, it will arrange for an object to be created in the heap and set the op parameter to reference this new object. The newly created object will be asked to execute its own ReadFrom() routine.

Using mechanisms such as those just outlined, framework code can sort out all the pointer links that have to be saved and restored.

Frameworks handle any necessary type information

As explained above, the framework routine that reads the information for an object referenced by a pointer expects to find either a "duplicate entry" flag or data defining a new object. The duplicate entry is easy; there will be some readily recognized "escape character sequence" (e.g. the '@' character used in the code shown above) and then an integer index number. It is the new objects that are more difficult.

The data file is going to have to contain information specifying the object type (i.e. its class) and then its content data. The input routine has to check that the specified class of the object is appropriate, create an instance of that class, and get the newly created object to read its own data. The mechanism used will be a generalization of the simple scheme, illustrated in section 9.2.2, where it was sufficient for the output routine to use integers to identify the classes of the Triangles, Pentagons etc. that could be in the file, and where the input routine worked with a simple array of prototype shapes.

It is fairly easy to put more general type information (e.g. a class name) into a file as it is written. If objects "know" what class they are, the output routine (OStream& operator<<(OStream& os, Object* op)) can ask an object to identify its class and write this information to the file before getting the object to write its own data. The forthcoming standard for C++ will specify a mechanism for run-time type identification. Currently, most frameworks provide some means of run-time type identification; their mechanisms deviate to varying degrees from the proposed standard.

Run-time type information (RTTI)

One way of providing run-time identification information relies on an extra data member in each class. The extra data member is an instance of a framework-provided class (e.g. class "ClassRecord") that can hold information such as a class name, and some details of a class's ancestry (see Figure 9.7). This data member would be public (or, better, there will be a public access function that provides "read" access to it). The data member would be static (so all instances of a class would actually share a single pointer to their corresponding ClassRecord object). A framework may provide macros that can add such an extra data member (and any access function) to a class declaration.

As suggested in Figure 9.7, each object's type would be written to file followed by its data. The example shown is for a "shape list" (a list that contains objects of classes derived from class Shape). When written to file, this might save the number of elements followed by each element. The type of each data element would be written before its data.

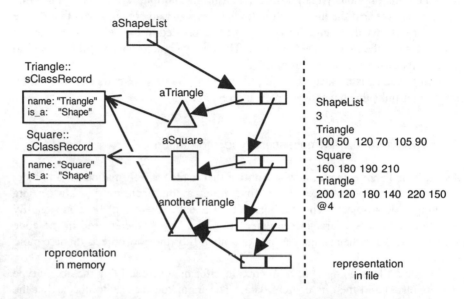

Figure 9.7   Provision and use of run-time type information.

The input routine has to be able to use these class names. Given a class name, it must be able to create an instance of that class. Figure 9.8 illustrates the basis of one approach (not the only approach, e.g. MacApp uses something quite different based on mechanisms devised for Object Pascal). In the approach illustrated in Figure 9.8, the framework's run-time support environment contains some form of "ClassTable" object. This table would have entries for all known types of persistent object. Each entry would consist of a pointer to a ClassRecord object and a pointer to a prototype instance of that class. (The ClassTable is really just a slightly more sophisticated version of the prototype table used in section 9.2.2.)

When the input routine encounters a class name in its input stream, it asks the ClassTable object to return pointers to the ClassRecord and prototype instance for that class.

Clone the prototype

A new object of the required class can be obtained by getting the prototype to *clone* itself. The new object can then read in its data. (Rather than use a prototype and a cloning process, the table could hold a pointer to a specially generated function that can be called to create a new instance of the required class.)

Checking whether the object read is of the required class

As noted earlier, for each class there would be a distinct input function; e.g. for a Shape* pointer there would be the function IStream& operator>>(IStream& s, Shape*& op). The function sets the pointer argument op to reference the newly created object – which had better be some form of Shape!

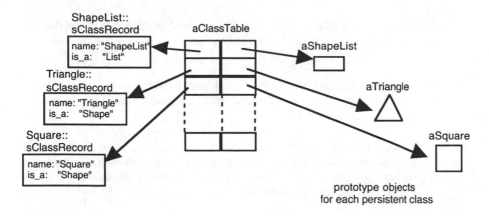

Figure 9.8   Prototype object table

At some stage, the input function will check the class of the object whose address is about to be saved in the pointer op. The code generated for each of these class-specific functions "knows" the class that it deals with; so, the code can include a check that asks the newly created object for its class and compares this with the class expected.

This process must allow for polymorphic pointers. A Shape* pointer can reference a Triangle, a Pentagon, a Circle, or an instance of any other subclass of class Shape. When checking the acceptability of an input object, the code may have to use the data in ClassRecord entries that define hierarchical relations.

For such a scheme to work, the run-time environment must include the ClassTable object, and this must contain details of all persistent classes. Somehow, these structures must be created before a program starts executing the main Application::Run() function. The approaches used vary widely.

Setting up the ClassTable etc.

The ET++ framework has a rather complex, but completely automatic mechanism. There is in effect a static instance of ClassTable (which gets initialized prior to entry to main()). The required ClassRecord objects and prototype objects are created by the macros that are inserted into those classes that are to be persistent. The constructor for class ClassRecord arranges for its instances to add themselves (and the associated prototype class instances) to the ClassTable.

In other frameworks, things equivalent to the ClassTable, class prototypes, and the ClassRecords are set up in an explicit initialization phase. The setup procedure would be called from main(), prior to the call to Application::Run(). The setup code, as written by the developer, would include a series of calls that "register" those classes for which "streamability" support is required.

Using a framework's facilities

Naturally, the things that a developer must do to exploit "persistence/streamability" vary with the framework. The basics are:

- Make those classes whose instances are to be saved to file inherit from the framework class that defines "streamability" (class `Streamable`, or class `Object`, or class `TObject`, ...).

- Provide effective definitions for the virtual `ReadFrom(InputStream)` and `PrintOn(OutputStream)` (or equivalent) functions that are defined in the base "streamable" class.

- In the code written for `ReadFrom()` and `PrintOn()` functions, leave worrying about pointer data members etc. to the framework. To save a pointer data member, e.g. an `Edge`'s `fNode1 Node*` pointer, simply write code like `os << fNode1` (or `outstream->WriteObject(fNode1)`).

   When coding the `ReadFrom()` routine, remember to deal with any data members that are not saved to file. Some will correspond to data that are simple to recompute, others will be pointers to collaborators. Set them in the `ReadFrom()` routine! When restoring pointer data members, watch out for possible memory leaks. Usually, it will be necessary to delete an object currently referenced by a pointer data member before resetting that pointer to reference a newly read object.

- Make certain that the framework can set up whatever data structures it needs for run-time support of streamability. This may involve the use of macros in class declarations, or may require some initialization routine.

   Ideally, framework classes will be streamable and have effective definitions of the input/output functions. For example, a `List` would have a `PrintOn()` function that writes its length and then gets each element to write itself. Then, if class `Graph` has a `List* fNodeList` data member that references a list holding its `Nodes`, the list and its contents can be saved to a stream with a single statement like `os << fNodeList`. Ideals are not always achieved; e.g. "streamability" came late to MacApp and the majority of its classes lack effective definitions for their `ReadFrom()` and `PrintOn()` functions.

9.3   THE CLONING OF OBJECTS

Generalized "shallow clone" functions

The object i/o mechanisms described in sections 9.2.2 and 9.2.3 both made use of arrays of prototype objects and cloning. When files were read, new objects were created by cloning from prototypes. Many class libraries provide some general implementation of cloning as part of the functionality defined in their base class (class `Object`, or `TObject` etc.). Thus, ET++ provides `Clone()`:

```
Object *Object::Clone()
{
    // Find out size of object (IsA()->Size())
    // Then use a specialized memory allocator to allocate
    // a block of bytes of the required size
    Object *op= (Object*)
        Storage::ObjectAlloc(IsA()->Size());
```

```
    // duplicate the bytes from original to copy
    // (hopefully including things like 'vtable' pointers!)
    memcpy(op, this, IsA()->Size());
    ...
    return op;
}
```

These general methods generally work fine for simple classes whose data members
don't include pointers. (Actually, it depends on the C++ compiler and how it
chooses to represent objects; the allocation of a block of bytes and the use of
memcpy() might not work with all C++ compilers.)

However, if there are pointers among the data members, then one can get the
same sort of problems from shared structures as were illustrated in section 5.2.2
where copy constructors and class-specific assignment operations were introduced.
The problems are illustrated in the next few figures for an object that is a simple
graph of nodes and edge (Figure 9.9 illustrates an example graph with three nodes
and four edges).

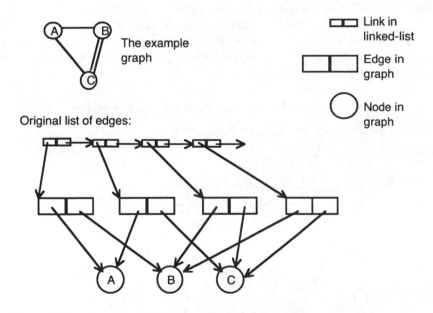

Figure 9.9   A graph as a list of edges that link nodes.

```
class Graph {
public:
    ...
private:
    LinkedList    fEdges;
    ...
};
```

Shallow cloning leads to structure sharing   The simple clone mechanism provided by class `Object` would be unsatisfactory. It would merely duplicate the bits in the primary record; this produces what is called a "shallow clone". A shallow clone mechanism can result in an original and "cloned" copy of an object having pointers to the same structures; e.g., a `Graph` and its clone will have pointers to the same list of edges. The edges would be shared and so a change to the "copy" would alter the original. If the copy was deleted, the original would be left with dangling pointers. This is exactly the same problem as occurred with class `Player` and its `char[]` in section 5.2.2.

Deep clones (copies)   If a class has pointer data members that will reference data that cannot be shared, it is expected to provide a "Deep Clone" (or "Deep Copy") mechanism that builds on the shallow clone. The programmer implementing the class may have to write a deep clone function. This will first invoke the shallow clone method to produce a new object, and then will "deep-clone" every object that is referenced by a pointer in a newly created object. The process is recursive; every object that is created must itself deep-clone those other objects that it references by pointers.

Unfortunately, graph structures that may contain multiple pointers to the *same* object cause problems (e.g. two edges both point to the same node "A"). A simple recursive deep-clone process applied to the original graph would result in the graph shown in Figure 9.10. This is an invalid copy because it contains eight nodes instead of the original three. Some of these nodes contain identical data, such as their labels, but they are distinct nodes. The copy does not represent the same graph as the original.

Copy by making new object for everything accessed by pointer:

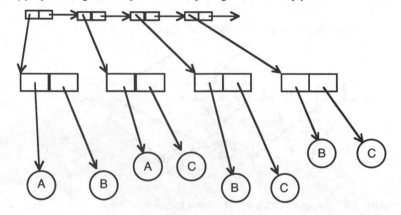

Figure 9.10 An invalid copy of the graph that would result from a naive approach of duplicating all objects referenced by pointers.

If it is to result in a valid copy, the copying process must keep track of those objects that have been duplicated. If the copying process finds that it has made a duplicate of a particular object, it should not make a further duplicate but should simply use a reference to that copy already made. This is just a variation on the approach to solving the problem of getting a valid copy written to file, as discussed in section 9.2.3. (The ET++ framework actually implements a form of deep clone that uses its i/o routines writing/reading an in-memory buffer rather than a file.)

In their text book, Gorlen et al. outlined a general approach to the deep-cloning of objects. In the NIH class library, class `Object` provides a method `deepCopy()` that keeps track of those objects that are already involved in the process of making a copy, and arranges for copies as needed:

Gorlen's NIH class library offers a model for a deepCopy operation

```
Object* Object::deepCopy() const
{
    bool firstObject = NO;
    // If this is the first step in copying a complex object
    // then create a storage structure that will keep
    // pointers to all those objects involved in current
    // copying operation.
    if (deepCopyDict == 0) {
        deepCopyDict = new IdentDict;
        deepCopyDict->add(*new Assoc(*nil,*nil));
        firstObject = YES;
    }
    // Lookup current object in table of those copied
    Assoc* asc = (Assoc*)deepCopyDict->assocAt(*this);
    if (asc == 0) {
        // if object was not found in the table,
        // then it has not yet been copied.
        // Make a shallow copy, note object and
        // copy in storage structure
        Object* copy = shallowCopy();
        deepCopyDict->add(*new Assoc(
                        *(Object*)this,*copy));
        // Then "deepen" the shallow copy by duplicating
        // non-sharable parts.
        copy->deepenShallowCopy();
        // When copying operation completed, tidy up
        if (firstObject) {
            …
            …
        }
        return copy;
    }
    // but if object had already been copied,
    // it is sufficient to return reference to the
    // existing copy
    else return asc->value();
}
```

Make a storage structure to keep track of parts already copied

Create copy of any new parts

But just use previously made copy if get a duplicate reference

Each class derived from class `Object` must provide its own `deepenShallowCopy()` member function to organize the duplication of those of its data members that relate to data that cannot be shared. So if class `Graph` had been built on top of the NIH library (using NIH's class `Object` as a base and NIH's `LinkedList` for a data member) it would be:

```
class Graph : public Object {
public:
    …
    void deepenShallowCopy();
    …
```

```
private:
    LinkedList    fEdges;
    ...
};

void Graph::deepenShallowCopy()
{
    // Usually would invoke base class's deepenShallowCopy
    // but Object::deepenShallowCopy(); doesn't do anything.

    // duplicate non-sharable edge data
    fEdges.deepenShallowCopy();
    ...
}
```

and a Graph would be copied as follows:

```
Graph* mastergraph = new Graph();
...
...; ...; master->AddEdge(); ...;
...
Graph* copygraph - (Graph*) mastergraph->deepCopy();
```

The Graph object uses the deepCopy method inherited from class Object, so it gets shallow copied and then told to deepen itself. Its deepen routine gets its edge list copied by invoking LinkedList::deepenShallowCopy() which iterates along the list deep-copying each item.

Duplicating a list

```
void LinkedList::deepenShallowCopy()
{
    SeqCtln::deepenShallowCopy();
    register Link* p = firstLink;
    // List head and tail pointers currently point to links
    // belonging to original list!  Reset them. This
    // list starts empty!
    firstLink = lastLink = Link::nilLink;
    // count of elements is count for original list, reset it
    count = 0;
    // Now, march down original list deep copying each
    // data item, and then add the copy to this list.
    while(!p->isListEnd()) {
```

By duplicating each element

```
            add(*(p->deepCopy()));
            p = p->nextLink();
            }
}
```

If class Edge provides a suitable deepenShallowCopy() routine, i.e. something like:

```
class Edge : public Object {
public:
    ...
    void    deepenShallowCopy();
    ...
```

```
    private:
        Node*   fNode1;
        Node*   fNode2;
        …
    };


    void Edge::deepenShallowCopy()
    {
        …
        // Can't share nodes, replace with appropriate copies
        fNode1 = (Node*) fNode1->deepCopy();
        fNode2 = (Node*) fNode2->deepCopy();
        …
    }
```

then the entire graph would be cloned correctly.

9.4 "DEPENDENCIES"

It is unusual for a change to a data element not to have side effects that ramify through a system. For example, in section 9.1.2 it was noted that when a "shape" had its shade changed it would have to inform the view in which it was displayed; the document that owned the shape would also need to be informed (this would usually be done as part of the framework's standard command-handling code).

A change to an object may propagate to "dependents"

A single simple dependency relationship, such as a `DrawView`'s dependence on the shade of a `Shape`, is easy enough to accommodate. The programmer implementing `Shape::Reshade()` must just remember to include a call that sends an update message to the view. But there would need to be similar code in `Shape::Resize()`, `Shape::Recolor()`, `Shape::Rotate()` etc. If shapes are to be displayed in more than one view, all such statements notifying views of changes in shapes would need to be duplicated. For shapes to collaborate with views, class `Shape` will have to declare a `View*` data member (or maybe several such data members) and program code will have to arrange to set this member (or these members) in each newly created shape.

Problems of ad hoc methods of handling dependencies

In a more complex program, there are usually many different kinds of objects that would have to be notified so that they too can adjust to a change in any particular data element. Consider for example a spreadsheet program; a change to a value in one data cell would lead to i) the updating of all those cells whose values depended on the changed cell, ii) the redrawing of the view displaying these numbers, iii) possibly the redrawing of some other view such as a histogram or pie chart displaying part of the information in the spreadsheet, and iv) the updating of the change count for the spreadsheet document.

When there are multiple instances of simple dependency relationships (e.g. several `Views` dependent on a `Shape`'s shade), or when there are complex arrangements of interrelated data items (as in a spreadsheet), it is no longer satisfactory to use individual data members to hold links to dependent objects and have explicitly coded statements notifying each dependent. Instead, some more general mechanism is required.

A list of dependents

The simplest general approach is to define an `fDependents` data member (of type `List*`) in class `Object` (or whatever class serves as the root for the main class hierarchy), along with `AddDependent(Object*)`, `RemoveDependent(Object*)`, `Changed()`, and `Update(…)` member functions. The `AddDependent(…)` method appends its argument to the list of dependents (creating the list if necessary). The `Changed()` function iterates along the list of dependents telling each to update itself (passing as arguments some details of why an update is necessary). An `Update()` function performs whatever work is necessary to get an object updated; this function has to be overridden in any class whose instances are expected to react to changes. The entire mechanism is recursive in nature. As part of the handling their own updates, dependents might call their own `Changed()` functions and so propagate the effects of an initial change through a complete network of interrelated objects.

Change-manager objects

Practical implementations of a dependency mechanism are usually a bit more complex than the naive approach just outlined. A naive implementation can result in a dependent object receiving multiple requests to update itself. For example, a view object displaying spreadsheet data would be told to update itself for the cell that was changed, and again for each other cell whose value depended in some way on the changed cell. Since handling the update of a view might well involve some graphics operations, this process would be extremely inefficient. Further, serious problems can arise if a simple recursive updating mechanism is applied to a network of interrelated objects where there are dependency cycles (the program would blow its stack). Rather than using a simple list structure to store dependents and a `Changed()` function that iterates along this list, the more sophisticated systems will use an instance of some class that provides specialized facilities for dependency management and change propagation. This change-manager class would be provided as part of the framework library and would incorporate algorithms for linearizing update sequences so as to minimize or eliminate multiple updates.

Smalltalk provided the original model for the "dependencies" concept, a concept that is now beginning to be adopted by the more sophisticated of the C++ libraries. For example, ET++'s class `Object` has methods `AddObserver(…)`, `Remove-Observer(…)`, `Send(…)` (or `Changed()`), and `DoObserve(…)` that provide the programmer's interface to its dependency mechanisms. Details of dependencies are handled centrally by a form of dictionary object that is one of the server objects in the run-time system (consequently, individual objects do not require a data member that points to a list of dependents). The dependency mechanisms are fairly complex and sophisticated; they provide a variety of special options, e.g. a particular object can request that its updates don't immediately propagate to dependents (which might be a useful optimization if several related changes have to be made to a particular object). The dependency mechanism is fairly consistently integrated into the framework code, e.g. the various collection classes have calls to `Send(…)` (or `Changed()`) in all those member functions that alter their contents.

9.5 USING THE COMMAND-HANDLER CHAIN TO AVOID INTRODUCTION OF PROGRAM-SPECIFIC CONTROL CLASSES.

Frequently, a program will appear to require specialized subclasses of standard user interaction elements such as `ActionButtons` etc. Figure 9.11 illustrates a situation where one might be tempted to introduce a specialized control class. The program displays a three-dimensional figure in a specialized view, and there are to be three button controls that allow a user to rotate the figure about different possible axes. One approach would be to define a specialized class, e.g. `class MyButton : public ActionButton`. Instances of this class would each own a link to the corresponding data object so that rotation requests could be forwarded; the class would have to have an extra member function to set the link, and would have to override one of the normal functions for the control. When the mouse is clicked inside a button, the framework invokes that button's `Control()` function (or its `DoEvent()` function, or other equivalent function). A replacement `MyButton::Control()` function could forward a request for rotation to the data object. This scheme is simple, but does involve introduction of extra classes that are not really essential.

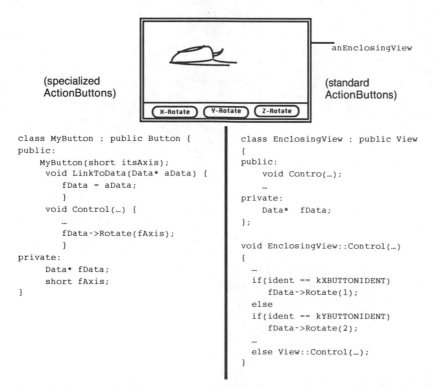

```
class MyButton : public Button {
public:
    MyButton(short itsAxis);
    void LinkToData(Data* aData) {
       fData = aData;
       }
    void Control(…) {
       …
       fData->Rotate(fAxis);
       }
private:
    Data* fData;
    short fAxis;
}
```

```
class EnclosingView : public View
{
public:
    void Contro(…);
    …
private:
    Data*  fData;
};

void EnclosingView::Control(…)
{
  …
  if(ident == kXBUTTONIDENT)
     fData->Rotate(1);
  else
  if(ident == kYBUTTONIDENT)
     fData->Rotate(2);
  …
  else View::Control(…);
}
```

Figure 9.11   Alternative approaches for user interaction – specialized classes or unified control.

Centralizing control   An alternative approach, which avoids creating specialized versions of simple interaction components, takes advantage of the way signals get passed along the command-handler chains. The normal `Control()` member function of an interaction element may have some code for a specific response (e.g. `Control()` code for a button might cause the button to flash) but it will also arrange to pass on details of the request to the next handler. The details would include an integer that identifies the kind of event and a pointer to the object that received the event (or some other identifier associated with that object). Instead of having individual specialized control elements each communicating with the data, the control functions can be centralized at some higher level (e.g. the document or the program's main view) whose `Control()` (or `DoEvent()`) method sorts out the various special cases as suggested by the alternative code fragment in Figure 9.11.

An Example "Editor" Program 10

The "Cards" application illustrates only the three most important classes in a framework library – class Application, class Document, and class View. Its simplicity is due to the fact that Cards does not require any real data entry, and user-actions are not reversible.

The program illustrated in this chapter adds a few more of the features of a typical application. It is yet another editing program with a "palette" of editing tools that can be selected and used to add or manipulate data elements.

In addition to the features that were present in the Cards program, the example program in this chapter supports both menu-based, and mouse-based editing actions. Some of these editing actions are reversible; so, the Edit menu supports Undo/ Redo options. The program also makes some limited use of the "collection class" data storage structures provided with the class libraries

10.1 THE "FINITE STATE MACHINE (FSM)" PROGRAM

Like "Cards", this is an "educational" program, but it is aimed at a slightly more sophisticated audience – first year students enrolled in a Computer Science degree. The program is to be used to help introduce ideas relating to finite state machines and lexical analyzers.

Figure 10.1 illustrates a very simple "lexical analyzer" based on a finite state machine. This lexical analyzer is supposed to accept character sequences that represent decimal numbers. Its finite state machine has four states, each represented by a node in Figure 10.1 In its "Start" state, it is waiting for the first digit. Any leading whitespace characters (space, tab, newline, etc.) will be consumed. This is indicated by the loop, labelled "whitespace characters", that is attached to the Start node. If the machine receives a digit as input, it makes a transition along the edge to the "Number" state. Any other input character should cause a transition to the "Error" state.

Subsequent digit characters should leave the machine in its Number state. This is indicated in Figure 10.1 by the loop, labelled with the allowed digit characters, that is attached to the Number node. If a whitespace character is received when the machine is in its Number state, it should make a transition to its "Finish" state and terminate. Any other characters received as input are to cause a transition to the error state.

The program is to let students define a network of named nodes, labelled directed-edges, and labelled loops. Naturally, the program should allow networks to be saved to, and restored from files on disk. If possible, the program is to "run" the finite state machine. This will require some way of specifying the input characters and the starting state. The "run" process should highlight the current state, show a character being consumed from the input sequence, highlight the edge (or loop) taken, and finally highlight the new state. This process should be repeated until the input is exhausted, or there is no transition from the current state that can accept the next input symbol. The run process should proceed at a sufficiently slow rate that the transitions can be observed.

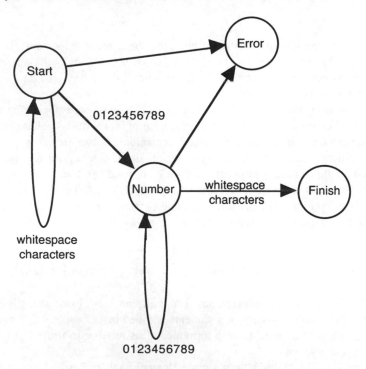

Figure 10.1   A simple finite state machine.

10.2 FSM: ANALYSIS

The required program is simply another specialized variant on the graphics-editor theme. Its obvious implementation is as a palette-based editor. The palette will provide "tools" for placing nodes, drawing edges, adding loops, and moving nodes. With these tools, a student using the program will be able to build a network that will be displayed in a scrollable main view. Given the standardized nature of the program, and the constraints of framework libraries, many of the objects required are immediately apparent, see Figure 10.2. Of course, some scenarios illustrating interactions among these prototypical objects will be needed to refine any initial rough descriptions of their classes.

10.2.1 The principal objects

There will have to be a FSMApplication class, a FSMDocument class, and a couple of specialized views – one showing the palette (PaletteView) and the other displaying the main network data structure (FSMView). These classes are obvious given that the program is to be constructed using a standard framework library. These classes are shown in Figure 10.2 along with classes for some of the other objects whose presence is relatively easy to infer.

FSMApplication, FSMDocument, FSMView, PaletteView

As in the Cards program, it will be sensible to define a FSMStructure class that is quite separate from the FSMDocument. A FSMDocument will create and own a FSMStructure object, and will pass some requests (e.g. read and write requests) on to that object. The alternative would be for the document object itself to hold lists of edges, nodes, loops etc. and for it to deal with requests to add a node, join two nodes with an edge, or place a loop. This second approach seems less satisfactory; the document class has more than enough responsibilities without this additional work. A FSMStructure seems a useful abstraction. There are data that obviously could be owned by a FSMStructure (the nodes, edges, and loops). There are well defined processes that involve these data as a whole (e.g. Run() and PrintOn()). Besides, it is generally wise to separate, as far as practical, the program-specific aspects from framework code; such a separation increases the chance of one being able to port an program to a different framework running on another platform. So being a FSMStructure is an obvious role for an object.

FSMStructure

Of course, the nodes, edges, and loops will be objects. They may belong to unrelated classes; or their classes may represent specializations of some more general "data item" class. The analyst, or maybe the designer, will resolve this once more is known about the behaviours of such objects and it is possible to determine whether they really have anything in common.

Node, Edge, Loop

Nodes, edges, and loops are all to be added using mouse actions. There is also a requirement that nodes be capable of being dragged around (presumably chased by their attached edges and loops). All of these operations will involve mouse-command objects (or "trackers"). Thus, there will be NodeMover, NodeSketcher, EdgeSketcher, and LoopSketcher command objects. The placement of names on nodes, or of labels on edges and loops, is also likely to be a process that should be reversible and therefore the program will need some form of LabelChanger command object.

Command objects

The classes noted above would all have to be specially defined for the program. It would also use a few instances of standard classes. It is going to need instances of class TextItem that can be used to enter the names of nodes and the labels for loops and edges. If a "run" option was provided, there would also have to be another editable text item that could be used to enter the input string that was to be processed by the running transition state machine.

Views for data entry

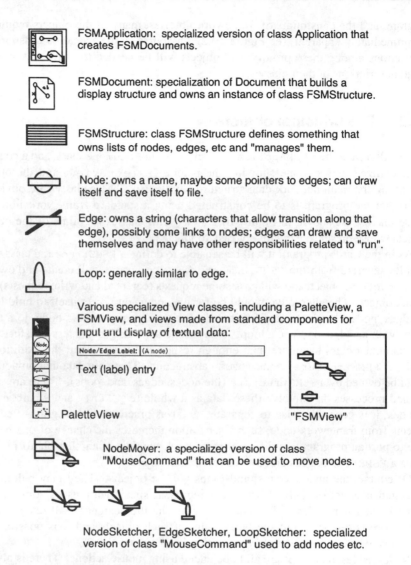

FSMApplication: specialized version of class Application that creates FSMDocuments.

FSMDocument: specialization of Document that builds a display structure and owns an instance of class FSMStructure.

FSMStructure: class FSMStructure defines something that owns lists of nodes, edges, etc and "manages" them.

Node: owns a name, maybe some pointers to edges; can draw itself and save itself to file.

Edge: owns a string (characters that allow transition along that edge), possibly some links to nodes; edges can draw and save themselves and may have other responsibilities related to "run".

Loop: generally similar to edge.

Various specialized View classes, including a PaletteView, a FSMView, and views made from standard components for input and display of textual data:

Text (label) entry

PaletteView   "FSMView"

NodeMover: a specialized version of class "MouseCommand" that can be used to move nodes.

NodeSketcher, EdgeSketcher, LoopSketcher: specialized version of class "MouseCommand" used to add nodes etc.

Figure 10.2   The "obvious" classes – obvious from the nature of the problem and the general design dictated by the use of a framework.

Scrollers and others   There will be a `Scroller` object (with its `Scrollbar` objects) associated with the main view, a `Window` object, and assorted `List` objects owned by the `FSMStructure` and used to hold nodes, edges, and loops. These objects can be ignored during the rest of the analysis process and throughout much of the design, and they are not illustrated in Figure 10.2. They are just standard components whose use can be examined in the detailed design phase or maybe left to the implementors.

All the classes outlined so far are "obvious". Some are obvious from the nature of the problem – the task is building a "FSMStructure" that consists of "nodes", "edges", and "loops". Others arise from the use of a framework library which inherently dictates a particular design style and use of its "application", "document", "view", and "user-interaction" classes. Other classes will also be required, e.g. this program was found to need "PaletteItems" and a "Palette-Manager" class. These were identified during the process of analyzing interactions among objects from the "obvious" classes as described in the next section. Some of the interactions were overly complex, and so were simplified by introducing extra classes that acted as intermediaries and took on some of the responsibilities.

10.2.2 Analyzing interactions among objects

Many aspects of the program are standard and do not require further detailed analysis. For example, the process of opening an old document, creating a FSMStructure and some views, and linking of standard components will be very similar to that illustrated for the Cards application. Aspects that do require further consideration include i) the handling of mouse commands by the main view, ii) the mechanisms that will be used to associate editable text labels with nodes, edges and loops, and iii) the "run" feature.

Responding to a mouse-click in the main view

In the Cards program, the handling of a mouse click by a CardView was extremely simple. There was only one possible interpretation – the user wanted to turn over a card. Consequently, it was easy to code a response; the CardView object simply forwarded the request to the CardData object.

Here, the interactions involving the main view and the mouse are more complex. A click on the mouse button will, via various intermediaries, get translated into a call to the FSMView::DoMouseCommand() member function. This routine will probably have to create some form of "command object" that will look after a reversible editing operation. As illustrated in Figure 10.3, there are several possibilities for the command object – the user may want to move existing components, or add new components. The command object created must match the user's need.

The tool selection defined in the tool's palette will identify the kind of command object that is required. But each different kind of command tends to have distinct specialized requirements. A "node sketcher" can start anywhere (the problem specification didn't include any constraints on overlapping nodes). But it is only meaningful to create an "edge sketcher" if the initial mouse point lies within an existing node. A "loop sketcher" can only be created if the mouse point lies within an existing node and that node doesn't already have a loop (there is a limit of one loop per node). Thus, the processing of a mouse click involves knowledge of specific constraints appropriate to the different tools. If the selection of command object is left to the view, the code will become complex. A possible implementation would be along the following lines:

```
Command* FSMView::DoLeftButtonCommand(Point p, ...)
{   Command*     aCommand;
```

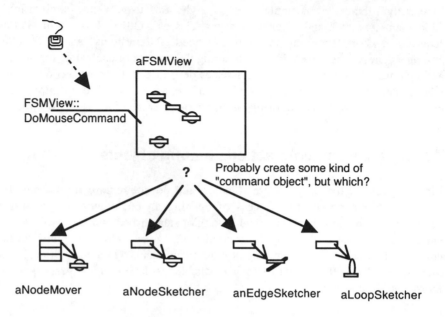

Figure 10.3   A mouse click in a FSMView should result in the creation of a command object that will look after an editing action.

Simple, but unsatisfactory, approach to command object creation

```
...
// Use link to PaletteView to ask what tool is currently
// selected
int    aTool = fPaletteView->SelectedTool();
switch(aTool) {
case kNODETOOL:
         aCommand = new NodeSketcher(fData,this);
         break;
case kEDGETOOL: // Click must be in a node
         Node* aNode = fData->NodeAt(p);
         if(!aNode) aCommand = gNoChanges;
         else aCommand = new EdgeSketcher(fData,aNode,this);
         break;
...
         }
     return aCommand;
}
```

The trouble with this code is that it distributes too much knowledge about the different possible commands and their individual requirements. The code becomes difficult to maintain and extend (e.g. the addition of other tools to the palette will necessitate reimplementation of this routine). It is more appropriate for the view object to communicate with some other entity that can create a suitable command object.

A general approach: PaletteItems

A possible approach is illustrated in Figure 10.4. The program could use a set of PaletteItem classes – one class for each different type of tool in the tools palette. These PaletteItems perform tasks for the PaletteView and for the main FSMView. A PaletteItem "knows" the requirements of the kind of command object with which it is associated and, by cooperating with the main FSMStructure object, can check whether constraint requirements are satisfied. A PaletteItem can also draw some representation of the tool that it represents and so can represent itself within the PaletteView. A PaletteItem can be accorded some region of space from the PaletteView and be asked to deal with any mouse clicks events in that region.

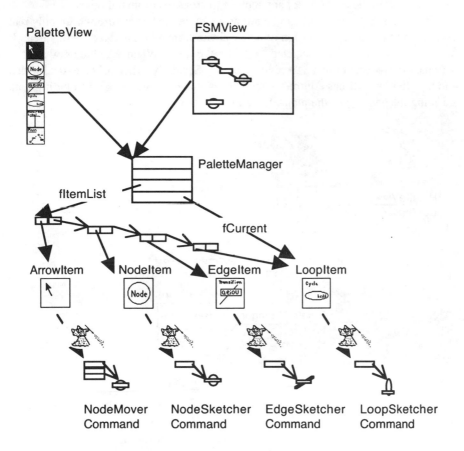

Figure 10.4   PaletteManager and PaletteItems.

Of course, there are several PaletteItems and it would be inconvenient for the *PaletteManager* PaletteView and the main FSMView to have to know about all the items. (Direct collaborations between views and palette items would necessitate too many links, and would be inconvenient because there is really an indeterminate number of palette items – extra items can be added as needed but such additions should not affect the operations of the views.) Instead of allowing direct collaborations between views and palette items, a "palette manager" acts as an intermediary.

The PaletteManager can be set up to work with a list of PaletteItems. This facilitates extension (extra tools can be added to the palette relatively easily) and also reuse. The PaletteManager/PaletteItem arrangement is something that is likely to be needed in other programs (it is a kind of programming "cliche"); this tiny cluster of classes doesn't offer much in terms of reusable code, it is more a reusable design. The `PaletteManager` can be asked to get its `PaletteItems` to draw their icons in the `PaletteView`; it can deal with a mouse click for tool selection (by asking each `PaletteItem` in turn whether they contain the mouse's coordinates and then setting a private pointer to the "current" tool); and, it can be asked to pass to the current tool any request to create a command object.

Interactions between the PaletteView and the PaletteManager

The `PaletteView` itself can be left as a simple view structure whose only real responsibilities are the provision of a coordinate framework and a drawing area. It will own a pointer to its associated `PaletteManager`. When asked to draw itself, it will pass the request to its `PaletteManager`. Similarly, when asked to deal with a mouse click, it will pass on the request to its `PaletteManager` (including data defining the position of the mouse), see Figure 10.5.

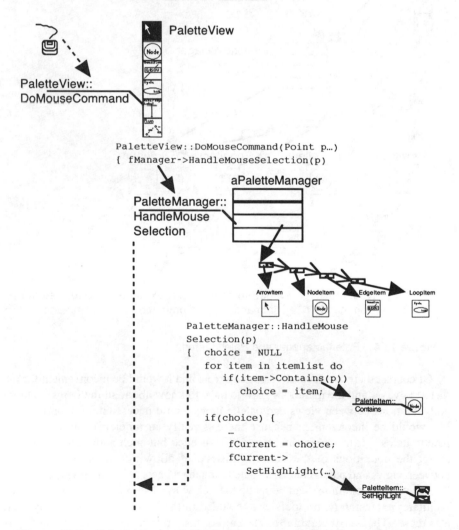

Figure 10.5 PaletteView and PaletteManager cooperate to select the current tool.

The main `FSMView` can also be kept as a simple view. It will pass on "draw" requests to the `FSMStructure` object that owns the data. When asked to deal with a mouse down, it can pass the request on to the `PaletteManager` and request return of a suitable command object that can handle the user's intended drawing action.

As illustrated in Figure 10.6, the process of creating a new command object involves collaborations among many different objects. The `FSMView`'s response to `DoMouseCommand` is to request that the `PaletteManager` create a command object. The `PaletteManager` passes this request to the currently selected `PaletteItem` which, for example, could be a `LoopItem` (the one responsible for creating `LoopSketcher` command objects).

Interactions among objects to create a new command object

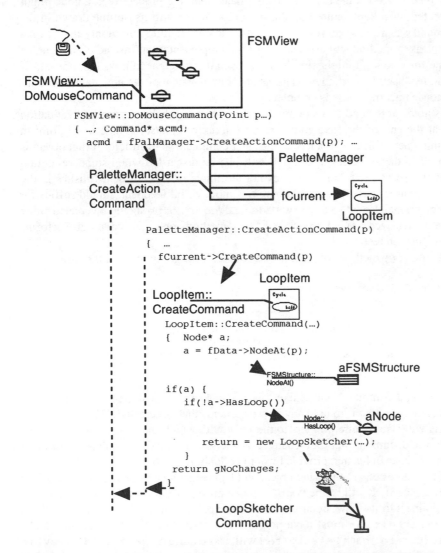

Figure 10.6   A FSMView, a PaletteManager, a PaletteItem, the FSMStructure and a Node all collaborate in the process needed to create a new command object.

The LoopItem will need to communicate with the main FSMStructure which reports whether there is a node under the mouse. If there is a node, the LoopItem must ask that node whether it already has an attached loop. Only if there is no existing loop can a LoopSketcher command object be created.

Creating the actual Nodes, Edges, and Loops

At some stage during their operation, the NodeSketcher, EdgeSketcher, and LoopSketcher command objects have to create either a new Node, or a new Edge, or new Loop object. This may be done in the command object's constructor (or some other initialization routine) but frequently it can be deferred. If a data item is actually used to help provide feedback during the sketching process, then it does have to be created at the start of the command handling process (e.g. a node might be created, then kept centred on the mouse position, with its outline drawn in the command's TrackFeedBack() method). However, quite commonly the Track-FeedBack() method will use a simpler visual representation (instead of drawing an outline for a node, it might simply draw a small circle that follows the mouse). If simple feedback is used, there is no need for an instance of the new data item until the mouse tracking phase is complete.

As noted at the end of section 9.1.3, many commands have to satisfy validation tests at the end of the "sketching" (mouse tracking) phase. There is no point in creating a new object, e.g. a new Edge, in a command's constructor if that object is not used in the sketching phase and will just be discarded when some validation test is not passed (e.g. an edge does not terminate in a node, or it finishes in the same node as it started). In this program, simple visual feedback should suffice for editing commands and so any new Nodes, Edges, or Loops should be created either in the sketcher-command objects' DoIt() methods or in the code that performs their validation tests.

The DoIt() methods of NodeSketcher etc. are going to be of the form:

```
void NodeSketcher::DoIt()
{   // create a node at the current mouse position
    // as recorded in fFinalPosition
    fNode = new Node(fFinalPosition, …);
    // and add it to the FSMStructure
    fData->AddNode(fNode);
}
```

Obviously, a command object like this NodeSketcher will need a pointer to the FSMStructure that is to hold any new data items that get created. These command objects will also require pointers to the view with which they work (the FSMView) and to the document that they change. Somehow, it must be possible to set these various pointer fields that provide links to collaborating objects.

Linking up collaborating objects

A NodeSketcher command object will be created by a NodeItem (a specialized PaletteItem). It will be the NodeItem that must provide the new command object with pointers to its view, document, and other collaborating objects. So it follows that PaletteItems must have pointers to a FSMView, a FSMDocument, and a FSMStructure. Each PaletteItem will also require a link to the PaletteView. When a PaletteItem is selected as the current item, it can change its record of its highlight state but it shouldn't simply draw itself. Instead, it should send a message to its PaletteView identifying an area of the view as being invalid thus allowing the view, and the underlying framework code, to arrange an update at some convenient time in the future.

`PaletteItems` can only hold pointers to views etc. if they are given these pointers by something else. The only thing that knows about the existence of `PaletteItems` is the `PaletteManager`. Consequently, the `PaletteManager` must also own such pointers.

Each `FSMDocument` will have its own `FSMStructure`, its own window containing a `FSMView` and a `PaletteView`, and a distinct instance of class `PaletteManager` to work with its `PaletteView`. The objects that must be linked together will all exist once a `FSMDocument` has completed the construction of its display structure. However, some may get changed as the program runs (e.g. the "revert" process applied to a document may throw away an existing `FSMStructure` object and construct a new one to hold data from a file). Methods for setting links are going to have to be provided in most of the classes that get to be defined for the program and calls to these methods will have to be added at "appropriate" places.

While an analyst can identify the need for links among objects that must collaborate, decisions as to how these links are set up are best left to the designers. Other design decisions, and conventions imposed by the use of a particular framework, will determine the times at which the necessary data are available. For example, one framework might load an existing file by creating a document, making the display structure and then reading the data (so links to views can be set at the same time as other volatile information, not stored in the file, are recomputed); another framework might read the data before the views are created, in which case the establishment of links to views cannot be part of the input routine. Such issues are the concern of the designers and not the analyst. The analyst must simply highlight the need for links among those objects that collaborate.

The analyst defers details of links to the designers

Labelling nodes, edges, and loops

Loops, edges, and nodes all have labels that have to be editable. The simplest approach is to use a separate editing area for entry of a character string and have some mechanism (probably a menu command) that causes a selected node/edge/loop data element to accept that string as its label.

The separate text-editing element will be simple to implement. One part of the display structure used in the program will be an instance of a framework class that deals with text entry. This standard component will have associated with it text editing commands that support undo/redo editing operations.

There would have to be some means for selecting the node, or the edge, or the loop that was to be labelled. The fact that all three kinds of object are here going to have to be treated similarly suggests that there really should be some more general class, class `Datum`, that is their base class. If all elements can be treated as instances of some class `Datum`, then it will be possible to have a pointer to the `Datum` element whose name can be changed and a `LabelChanger` command object that changes the label of a `Datum`. If the classes do not have a common base class,

Need for class Datum: base class for Node, Edge, and Loop

there will have to be similar but distinct `NodeLabelChanger`, `EdgeLabelChanger`, and `LoopLabelChanger` commands.

Obviously, some kind of mouse interaction would be involved in the selection of the "datum" that was to be labelled. There would be no need to support an undo/redo mechanism for datum selection. As selection involves simply a click of the mouse button, there is no need to protect the user from loss of work if an error is made; if the user clicked on the wrong datum, they only need to move the mouse to the correct datum and again click the button. Once a data element has been selected for re-labelling, "something" has to remember which element was selected. The obvious candidate is the `FSMStructure`; so a `FSMStructure` will have to have a `Datum*` pointer field that references the data item most recently selected for labelling.

The tool palette will have to provide a tool that can be used to select a datum that is to be re-labelled. The "arrow" tool (which really consists of the `ArrowItem` and the `NodeMover` command) is intended for dragging nodes; it would be unreasonable to try to incorporate a secondary use like data element selection. There will have to be a distinct `PaletteItem` – a `LabelItem`.

A `LabelItem` will simply be another `PaletteItem` in the list belonging to the `PaletteManager`. A mouse click in the `PaletteView` can lead to its selection as the current tool. Subsequently, a mouse click in the main `FSMView` will result in a request for action going first to the `PaletteManager` and then to the `LabelItem`. Unlike the other `PaletteItems`, a `LabelItem` will never create a command object. It will simply get the `FSMStructure` to check whether a datum was located at the mouse position, and if so make that datum the current selection for labelling. The process is illustrated in Figure 10.7. (Each framework has some convention for dealing with situations where there is no need to create a `Command` object although one was expected. The object asked to create the `Command` object may return a `NULL` pointer, or some other flag – like a `gNoChanges` flag as illustrated in Figure 10.7.)

A menu-based command could be used to arrange the (reversible) changing of the label. Menu selection could be handled either by the `FSMView` or by the `FSMDocument`. The document seems slightly more appropriate. It would have to deal with DoSetupMenu requests for which it would enable or disable a "Change Text" menu item according to whether its associated `FSMStructure` had, or did not have, a currently selected data item. A DoMenuCommand request specifying the "Change Text" option would be handled by the creation of a new `LabelChanger` command object.

A `LabelChanger` command object would need to store two `char*` pointers to strings (the old and new labels) and a pointer to the data item that is to be labelled. Data items will have to provide `GetLabel()` and `SetLabel()` methods.

The "Run" option

The "run" option should provide some form of animated display showing state transitions made as input data are processed. Figure 10.8 illustrates a simple network processing a sequence of digits and trailing spaces.

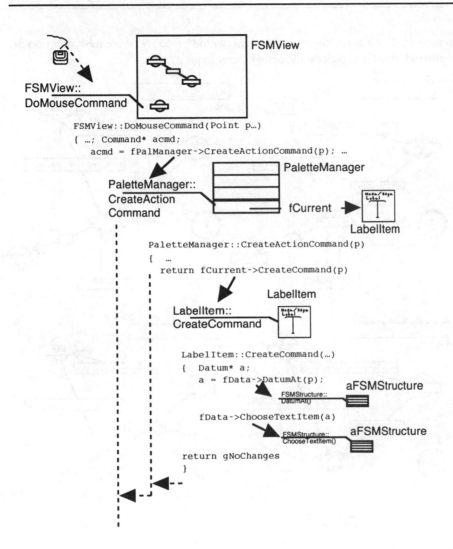

Figure 10.7 A FSMView, a PaletteManager, a LabelItem, and the FSMStructure collaborate to select a data element that is to be relabelled.

The run option will require a mechanism for selecting a starting node, some means of entering input data, and a way of starting the actual scan of the input. The selection of the starting node can be handled in a very similar manner to the selection of a data item for relabelling. There will have to be one more item associated with the tools palette – a RunItem. Like the LabelItem, a RunItem will not set up a reversible command, it will simply check that the mouse click was within a displayed node and then pass the identity of this node to the FSMStructure object. Input data could be entered prior to the start of the scan using a standard editable text field; the scan process would involve removing successive characters from the input string entered (changing the data displayed in the text field). The start of the scanning process could be activated by a menu

A tool for selecting a starting node, and menu command to "run"

command. Once again, the FSMDocument would probably be the most appropriate command handler to pick up the actual menu request.

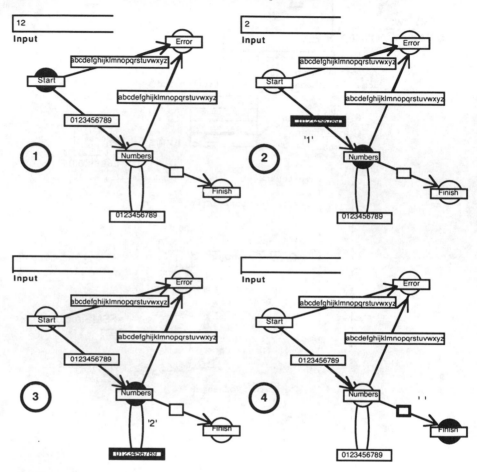

Figure 10.8   Form of display required by "run" option; successive frames show stages in the processing of input characters.

The "run" process

The code for running the scan would be encoded as a method of class FSMStructure. The code would involve a loop, with each iteration consuming one character from the input string. Since the FSMStructure object must be able to access the input string, it will have to own a pointer to the TextItem that owned the string. As characters were removed from the string, the TextItem would have to be updated and redisplayed. The current node would have to redraw itself with some suitable highlighting. The edges leading out from the current node would need to be identified and each in turn asked whether it could "accept" the next character. An edge, or a loop, will "accept" any character that occurs in its label. (The code could be written to detect and report as errors any situations where more than one edge would accept a character; but it would be sufficient just to use the first edge that accepted a character.) Since input edges and output edges behave differently, they will have to be distinguished within a node's internal structure. If no output edge could accept the next character, the program should check any loop attached to the current node. If the input character was not acceptable by either an edge or a loop, the run process would terminate. Otherwise, the accepting edge/loop should be briefly highlighted. If the character was accepted by a loop, the current node would not change; but if an edge was used, the `FSMStructure's` record of current node would be updated to the node at the terminus of that edge. The iterative process would continue by consuming the next input character and by highlighting the new current node. The process would terminate either when all input was exhausted or when the next input character could not be consumed.

An "integrated run" process?

The actual "run" process might be integrated with the operation of the underlying framework. If the "run" process is integrated with framework operations, menus and other controls will operate normally. Consequently the program could have a "Stop" menu option, and would allow the user to scroll the view or quit the application while the simulated transition state machine was running. A "`DoIdle`" mechanism (often available with event handlers) would provide a basis for such integration. Selection of the "Run" menu option would set a "running" flag in the `FSMStructure` and schedule regular ($\approx 1/2$ second) processing for the `FSMDocument` event-handler. Each time it was scheduled to perform idle time processing, the `FSMDocument` would request that the `FSMStructure` perform one more step. When the simulation terminates (input exhausted or invalid input character), the running flag would be cleared and the scheduling of idle time processing would be cancelled.

A separate "run" process?

Alternatively, program-specific code could grab control. System dependent timing functions could be used to control the speed of execution. The highlighting of nodes, edges, loops etc. would be achieved by direct drawing operations. The implementation would probably be simpler than the integrated approach. The locking out of normal controls (such as the File/Quit menu option) might be unattractive, but would probably not matter much for processes that only take a short time.

Either approach might prove suitable for the program. The implementors would probably have to try both and then select the mechanism that worked more effectively.

10.2.3 Classes and responsibilities resulting from the analysis

The objects, and their classes, that emerge from the problem analysis are summarized in this subsection.

Class FSMApplication

The program will use one instance of a class `FSMApplication` – a minor specialization of the framework's `Application` class. The only change will be the provision of an effective `DoMakeDocuments` function.

View classes: FSMView ,PaletteView, and others

The general form of the program's display structure is shown in Figure 10.9. The top portion of the display will be constructed using standard framework components for the display of text strings and the entry of editable text; these components need not be specialized in any way.

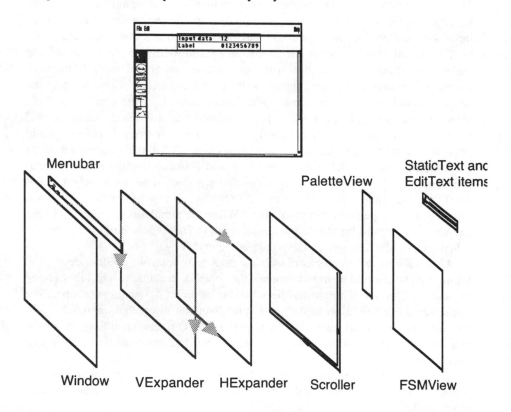

Figure 10.9   A display structure for the FSM program.

The two specialized view classes – FSMView and PaletteView – will be derived from the framework's View base class. The rest of the display structure will be assembled from standard framework components such as a Scroller (with its associated ScrollBars) and other components as required.

Collaborations and responsibilities of class FSMView
An instance of class FSMView will collaborate with a PaletteManager and a FSMStructure. Really, the only responsibilities of a FSMView are the provision of a coordinate framework and the forwarding of framework requests Draw and DoMouseCommand to appropriate program-specific entities. A Draw request to a FSMView will be forwarded to the instance of class FSMStructure that is displayed in that view. A DoMouseCommand (or the equivalent DoLeftButtonDownCommand) request will be forwarded to the PaletteManager which will arrange for the creation of the appropriate command object. A FSMView will have to own pointers to these collaborating objects. These pointers will have to be initialized when the view is created and, possibly, will have to be updated if a collaborator is changed

(e.g. a File/Revert operation may destroy the existing `FSMStructure` and create a new one; if so, the corresponding `FSMView` will have to be informed).

Class `PaletteView` is similar to, but even more limited than, class `FSMView`. A `PaletteView` will collaborate with a specific `PaletteManager` object; they will both have been created during the process of opening a document. A `PaletteView` forwards `Draw` and `DoMouseCommand` requests on to its collaborating `Palette-Manager`.

Collaborations and responsibilities of class PaletteView

Class FSMDocument

Class `FSMDocument` will be a specialization of the framework's `Document` class. Its responsibilities will include the creation of the program's display structure and the creation of instances of class `PaletteManager` and class `FSMStructure`. Because a `PaletteManager` object records details of the editing tool selected for its document, each document must have a separate `PaletteManager` (it is not possible to have a single instance owned by the `Application` object). A `FSMDocument` will also be responsible for linking up views and data objects after these have been created.

Creation of views and other objects

Instances of class `FSMDocument` will handle `DoRead` and `DoWrite` requests from the framework by invoking the inherited method from the base class and then forwarding the read/write request to their associated `FSMStructure` objects. The class may also have to provide methods for dealing with File/Revert requests; the code in these methods may have to reestablish links between collaborating views and data objects.

Management of file transfers

Class `FSMDocument` will have to provide implementations for methods called during a File/Close operation. The code written for these methods will have to dispose of other objects, such as the associated `FSMStructure` and `Palette-Manager` objects.

Deletion of objects

A `FSMDocument` object will be responsible for a specialized "FSM" menu that offers options allowing the relabelling of a data item or the initiation of a run of the modelled finite state machine/lexical analyzer. Class `FSMDocument` will provide the two menu handling methods; its `DoSetupMenus` method will entail collaboration with the `FSMStructure` object. The FSM/Run menu option should only be enabled if a node has been selected as the starting state for the machine, and the FSM/Change-Text menu item should only be enabled if a data item has been selected for relabelling. Class `FSMStructure` must provide boolean functions that return such status information when requested.

Menu handling

Selection of the FSM/Run menu option should result in a `Run` request to the `FSMStructure` which will either run the simulation or initiate the system so that the machine will be run during subsequent "idle time" processing. The FSM/Run option is not associated with a reversible command object.

Selection of the FSM/Change-Text option should result in the creation of a `LabelChanger` command object.

Creation of a command object

If the running of the finite state machine simulation is to be integrated with the framework event handling, a `DoIdle` method must be provided in class

DoIdle method?

Data owned
FSMDocument. The implementation will consist simply of a request to the associated FSMStructure object asking that it advance the simulation by one step.

An instance of class FSMDocument will own pointers to collaborating objects
such as the FSMStructure, the PaletteManager, and the various views.

Class FSMStructure

Data resources owned
Figure 10.10 shows a "fuzzy" analysis diagram for class FSMStructure. The
diagram is a little more elaborate than those shown previously (e.g. Figure 8.12)
because of the greater complexity of the problem. Class FSMStructure must
define the form of objects used to store the graphs of nodes and edges and provide
all the functionality associated with these graphs. Details of representation of the
graph can be left to the design phase; basically a FSMStructure will maintain
"lists" of nodes, edges, and loops. A FSM-Structure must hold pointers to all
those views with which it will collaborate (the main FSMView and the edit-text
fields used to enter labels etc.). In addition, it will have to store data defining any
node designated as a starting node for the simulation or any data item selected for
labelling.

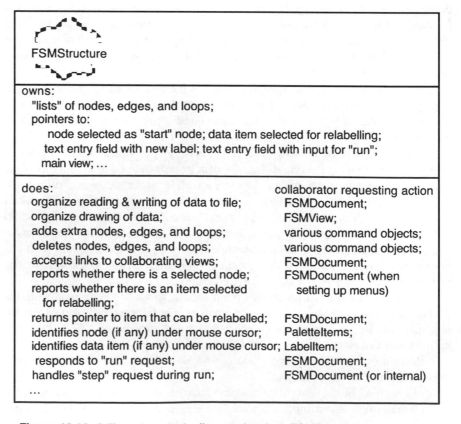

Figure 10.10 A "fuzzy" analysis diagram for class FSMStructure.

The list of responsibilities identifies most of the member functions that are likely to be required in class FSMStructure. With more elaborate examples such as this, it is helpful if the analyst tabulates the responsibilities (things that a FSMStructure does) along with details identifying those other classes whose instances might request such actions. *Responsibilities and collaborators*

As shown in Figure 10.10, a FSMDocument can ask a FSMStructure to organize the reading or writing of data to a file. A FSMView can ask a FSMStructure to organize the drawing of data.

Command objects, such as a NodeSketcher or an EdgeSketcher, are going to ask a FSMStructure to add or delete nodes or edges. Even though the program is not required to support an Edit/Cut menu option, a FSMStructure still has to be able to delete items. The "DoIt/RedoIt" operations of command objects add data items; the "UndoIt" operations will request deletions.

A FSMStructure has to respond to queries regarding the existence of a node selected as starting node, or a data item selected for relabelling. These methods will be invoked when the FSMDocument is administering its menus. A FSMStructure will have to provide a pointer to the data item when a FSMDocument is constructing a LabelChanger command object.

Class FSMStructure will define a public Run() member function that can be invoked by a FSMDocument to start the simulated run of the finite state machine. The run process might be fairly complex and involve various private member functions; these would be identified in the subsequent design phase. The Step() function, that advances the simulation by one step, would be part of the public interface if the simulation was effected using the DoIdle() mechanism of class FSMDocument. But, if the implementation of Run() was not integrated with the framework, the Step() function would be private.

Classes Datum, Node, Edge, and Loop

As described above (in the section considering how the user might change labels on nodes etc.), it seems advantageous for classes Node, Edge, and Loop to form part of a hierarchy involving the common abstraction of a "data item" that has a label, a link to a view, and various position data including at least a centre-point and a bounding rectangle. A possible hierarchy is shown in Figure 10.11. Classes Edge and Loop have more in common – such as behaviours involved in the run process. It is possible that later design decisions might expand this hierarchy to include an additional intermediate class between class Datum and classes Edge and Loop.

Class Datum will have to have data member to hold a label (possibly as a char* string) and member functions that can be used to get the current label and to set a new label. (A LabelChanger command object will have to ask for the current label so that if told to "Undo" it can restore a data item's original state.) *A label string that can be read and set*

Data items are odd shapes, and their shapes can change when their labels change (the label boxes should adjust to the size of the included text). They will have to be characterized by various coordinate data. The exact nature of this data can be left to the design phase but will include at least information defining the *Coordinate data and bounding rectangles* centre of the item, the bounds for the rectangle that will include the label, and an overall bounding rectangle.

Collaborations with a FSMView

Whenever a data item is moved or changed in anyway, it will have to inform the FSMView in which it is drawn that the areas under the old and new versions of its bounding rectangle will have to be redrawn. The Datum object will actually send "Invalidate" requests to its associated view. Thus, each instance of class Datum will have to own a pointer to a FSMView, and this pointer has to be set when a Datum is created by a command object or has been read in from a file. It may be necessary for class Datum to define a member function that allows this view pointer to be reset.

Redefinable methods to Draw, Read, Write and "Flash"

All variants of class Datum will be required to draw themselves, read their data from file, and write their data to file. They will also have to be able to "flash" – that is, provide some form of visual highlighting during the run process. The implementations of these methods would have to be deferred to the specific subclasses.

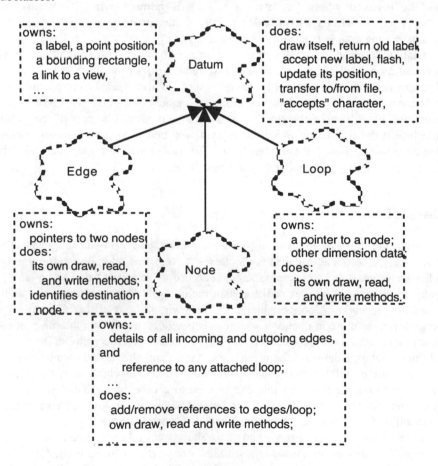

Figure 10.11 Analysis hierarchy for classes Datum, Node, Edge, and Loop.

The FSMStructure object is going to have to ask each data item whether it contains the mouse position. Such a containment check is needed at number of different processing stages, e.g. when palette items are creating node mover command objects or label changer command objects etc., or when an edge sketcher is validating a new edge. It would be inappropriate to use the overall bounding rectangle of the data item (the bounding rectangle of a diagonal edge encloses an awful lot of points that are quite remote from the edge). There will have to be some alternative definition of some smaller area that should be used in such Contains(Point&) containment checks. This could be defined as the area occupied by the data item's label, but it is possible that the various specialized types of data item would need different definitions. Again, this is a decision that can reasonably be left to the design phase.

ContainsPoint method

Edges and Loops must both provide a method that can be used to check whether the next input character corresponds to one of those that can cause a transition along the edge or round the loop. The test entails simply a check whether the data item's label contains the specific character. The test is simple to implement at the level of class Datum. The method will not be needed for Nodes but that doesn't really matter much. If the designers create a more complex class hierarchy with an additional class between class Datum and classes Edge and Loop, then this responsibility can be relocated.

NameContains-Character method?

Each of the subclasses of class Datum will have its own special responsibilities and additional data fields. In addition, there are several methods where each subclass will either override the default implementation in class Datum, or provide an effective implementation for a deferred function.

Edges need to own pointers to the nodes that they join; as Edges are directed, the ends should be distinguished – starting node, and ending node. Class Edge must also provide a function that returns a pointer to the end node; this information will be required when the FSMStructure is simulating the running of the finite state machine.

Special responsibilities of class Edge

A Loop will have to own a pointer to the node to which it is attached and, possibly, some additional coordinate data that define the size of the Loop drawn by the user. Class Loop will override or define several methods of class Datum but does not appear to introduce any extra functionality.

Special responsibilities of class Loop

Nodes are going to have several extra responsibilities and additional data fields. Nodes need additional data members for incoming and outgoing Edges, and for any attached Loop (probably two lists of Edges and a pointer). Class Node will have to define methods for adding and deleting Edges and Loops. The methods for adding Edges will have to distinguishing incoming and outgoing Edges. Nodes are going to be asked whether they own Loops, and during the run-process they will be asked for their lists of output Edges or for a pointer to aLoop.

Special responsibilities of class Node

Rearrangments in the layout of a network of Nodes, Edges, and Loops are to be made by dragging a Node. If a Node is moved, it will have to organize the updating of the positions of its associated Edges and Loop. While a node is being moved by the user, it would probably be appropriate to hide its Edges. Details of these responsibilities can be left to the design phase.

Class PaletteManager

The `PaletteManager` is quite simple. It will own a list of `PaletteItems` that can be built in its constructor; it will also have a pointer to the currently selected item from this list and pointers to its collaborating `FSMDocument` and `FSMStructure`.

The `PaletteManager` will receive requests, from the `PaletteView`, to draw the palette and handle mouse actions. Draw requests would be handled by iterating along the list of items asking each to draw itself. A mouse event represents an attempt by the user to select a new tool; the `PaletteManager` would ask each item in turn whether it contained the mouse point and would update its record of the currently selected item. The individual `PaletteItems` have to have links to the main views. The `FSMDocument` would pass pointers to the views to the `PaletteManager` which would distribute copies to the individual `PaletteItems`. There might have to be a mechanism for resetting a pointer to the `FSMStructure` if this can be changed (e.g. during a File/Revert operation). The `PaletteManager` would have to update its own record of the associated `FSMStructure` and pass on this information to the individual `PaletteItems`.

The main `FSMView` will send requests for the creation of a command object to the `PaletteManager`. Such requests will be forwarded to the currently selected `PaletteItem`.

PaletteItem classes: ArrowItem, NodeItem, EdgeItem, LoopItem, LabelItem, and RunItem

The `PaletteItems` form another simple hierarchy. The abstract class `Palette-Item` defines the data members and specifies all the behaviours of these kinds of objects. `PaletteItems` own some image that they can draw in the `PaletteView`. Each has a rectangle that defines the particular part of the view to which they correspond. They may need to identify a cursor that is to be used after tool selection and so may need to have an integer cursor-identifier. Another data member might be a flag to show whether they should be drawn with some form of highlighting. They will require pointers to the views and other objects with which they collaborate.

`PaletteItems` can draw themselves, check whether they contain a point, reset links to views and other collaborators, possibly set a cursor, set their highlight state, and most important can be asked to create a command object. Most of these behaviours are common and can be fully specified in class `PaletteItem`.

Specialized `PaletteItems` differ only in the command objects that they create and the conditions that they have to apply to this creation process. Each specialized `PaletteItem` subclass will provide a distinct implementation of `CreateCommand`.

A `RunItem` never actually creates a command object, it just checks whether a node lies at the mouse location as defined by a `Point` argument (this requires collaboration with the associated `FSMStructure`). Similarly, a `LabelItem` just checks whether any data item contains the mouse point and, if so, makes that item the target for labelling. The `ArrowItem` will create a `NodeMover` provided that the mouse is located over a `Node`. The others have been described previously in the subsection that introduced the `PaletteManager` and its `PaletteItems`.

The Command classes: NodeMover, NodeSketcher, EdgeSketcher, LoopSketcher, and LabelChanger commands

The command objects are all quite simple because there are no problems in representing "old" and "new" states of the data.

The `NodeMover` command object will need to record the old and new positions *NodeMover* of the `Node` with which it is associated (along with a pointer to that `Node`). It has to collaborate with the `FSMView` and the `FSMStructure` and so will need pointers to each. As a mouse-based ("tracker") command, it has to provide visual feedback that somehow shows the selected `Node` following the cursor as the mouse is moved. Details can be left to the design phase; there may be some complexity in the code as it will be necessary to hide `Edges` etc. while a `Node` is moving and then show them again when mouse dragging terminates. The `NodeMover`'s `DoIt` and `UndoIt` methods will operate directly on the selected `Node` telling it to change its position to the new or old value respectively.

There are no specified constraints on a `NodeSketcher`. Visual feedback will *NodeSketcher* again be necessary; it will probably be sufficient to have a circle following the cursor indicating the position that would be occupied by a new `Node`. When tracking is terminated, the `Node` can be created. The `DoIt` and `UndoIt` methods of the `NodeSketcher` would communicate with the `FSMStructure` asking it to add/delete the node. Like all command objects that create new data items, a `NodeSketcher` will have to check some "undone" status flag to determine whether it should or should not dispose of its `Node` when it is deleted.

The `EdgeSketcher` command class will be generally similar to `NodeSketcher` *EdgeSketcher* class. An `EdgeSketcher` will only be created if its starting point is located within an existing `Node`. Its visual feedback mechanism could simply show a line leading from that `Node` to the current cursor position. `EdgeSketchers` do require some validation tests at the end of the mouse tracking phase. The end point of the line must lie within the bounds of some second `Node`. This constraint would have to be validated before an actual `Edge` object was created; if the line drawn failed to satisfy the constraint then the editing action should be abandoned. The `DoIt`/`UndoIt` operations of an `EdgeSketcher` will have to notify the nodes involved as well as the `FSMStructure`. The starting node would have to be told to add or delete an outgoing `Edge`, while the ending `Node` would have to add or delete an incoming `Edge`.

The `LoopSketcher` command class is another standard mouse-based command *LoopSketcher* object. Its visual feedback mechanism would show an oval connecting the current cursor position and the centre of the starting `Node`. It would probably be helpful to constrain `Loops` so that they lie along either a horizontal or vertical axis (otherwise they get to be difficult to draw with the graphics primitives available on most systems). Thus class `LoopSketcher` might implement some form of `Track-Constrain` function.

LabelChanger   LabelChanger command objects are slightly different from the others being menu-based rather than mouse-based. Consequently, class LabelChanger does not have to provide any visual feedback or tracking mechanisms. A LabelChanger object will own copies of old and new labels and a pointer to the data item that is to be changed.

The "main program"

Naturally, there is a main program. It will have the usual form for an OO program – create the principal object and tell it to run. Thus, it can be sketched out by the analyst; for the ET++ framework it would be:

```
int main(int argc, char** argv)
{
    Symbol preferredfile = "FSMDemoDocument";
    Application* myApplication = new
            FSMApplication(argc,argv,preferredfile);

    return myApplication->Run();
}
```

10.3 SOME DESIGN DECISIONS

The design phase in the development of the program has to focus mainly on issues such as integration with the framework, the representation of Nodes etc., and the organization of some of the more complex interactions like those occurring when Nodes are moved. The design phase must also complete the characterization of inheritance and other relationships that exist between the classes. Other aspects arising in this phase will include the form of files and the mechanisms used for object i/o and the approach that should be used for handling the simulated run of a finite state machine.

Integration with a framework   Integration with the framework should be fairly straightforward. Figures 10.12-10.15 illustrate how the program specific-parts might be integrated with a framework (using the ET++ framework as the example).

Figure 10.12 shows the main "command handler" classes – Application, Document, and Views. The FSM classes involve small extensions to the base classes. Thus, the methods of class Document that must be overridden (DoRead, DoWrite etc.) devolve all responsibility onto the associated FSMStructure. The setting up of the display, and the handling of the extra FSM menu, are the only real additional responsibilities of a FSMDocument. Similarly, the FSMView and PaletteView pass on to associated FSMStructure and PaletteManager objects all the real work involved in drawing or responding to mouse actions.

The program-specific classes, FSMStructure, Node, ..., ArrowItem (shown in Figure 10.13) could be based on the framework's base class. This would allow use of any framework facilities for object i/o.

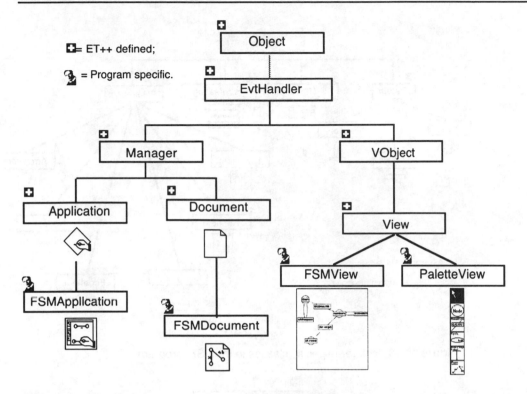

Figure 10.12 The main command handlers for the FSM program.

Figure 10.14 illustrates the various Command classes defined to handle reversible commands, while Figure 10.15 summarizes some of the other classes from the library that were used in the construction of the program. In addition to obvious classes such as Window, Scroller, and ScrollBar, the program uses TextField objects for entry of labels and other strings and ImageItems (which incorporate bit maps) as the basis for representing tools in the palette. A framework collection class, e.g. a list, would probably be used for the dynamic storage structures that might be required.

Classes Datum, and FSMStructure are explored further in the rest of this section; they help illustrate some of the things that the designer would have to consider.

Class Datum

Class Datum serves as an abstract base class for the specialized Node, Edge, and Loop classes. In this case, the abstract class was identified during the analysis; but quite often, the need for an abstraction will only appear during the design. The designer will have to refine any initial abstraction. The design should try to maximize effective use of the abstraction – data members that are needed by all variant subclasses should be moved to the abstract class, its interface should express their common functionality.

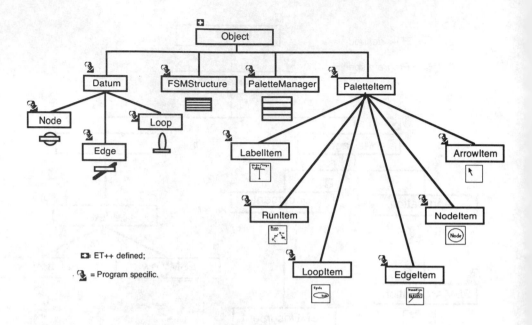

Figure 10.13 Program-specific classes in the FSM program.

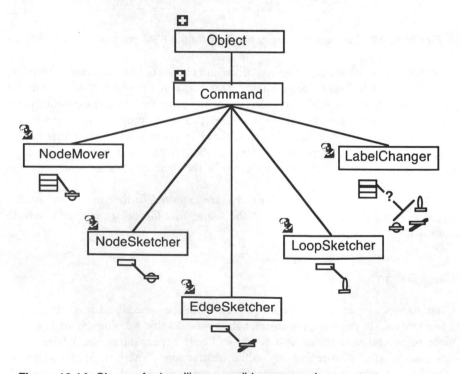

Figure 10.14 Classes for handling reversible commands.

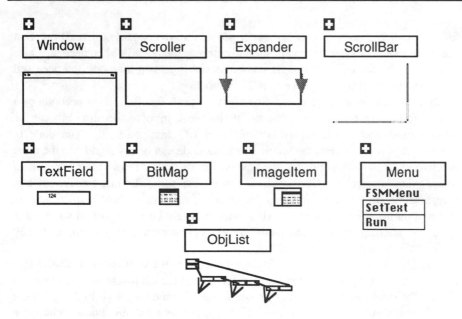

Figure 10.15 Standard framework classes used in the program.

When planning for a C++ implementation, the designer will have to consider the accessibility of data members. If special functions written for subclasses are likely to use data members defined in the base class, then protected access must be provided. Decisions have to be made with regard to the member functions that can be defined at the base class level, and those that must be left open for redefinition by subclasses (the virtual functions). Among the virtual functions, the designer must distinguish between those that can have a default implementation and those that can only be left as pure virtual (deferred) functions. All these considerations are in addition to the normal design work involved in choosing an appropriate representation for the classes' data, and in refining ideas regarding the classes' functionality. Consequently, there can be quite a lot of design work for a class like class Datum.

The analysis, Figure 10.11, identified class Datum having a point position, a label, and a bounding rectangle. The label (character string) would be centred at the Datum's point position. Various "areas" are going to have to be associated with instances of class Datum. The label would be drawn with a box; the area of this box could define the Datum for operations such as using the labelling tool to pick a Datum that is to be relabelled. A Node has a circular display area in addition to its label box; this area, or for simplicity its surrounding square, should also probably count as being a part of the Node during operations such as checking whether a newly drawn Edge terminates within an existing Node. In addition to its label and any additional areas, a data item will have to define an overall bounding rectangle. This bounding rectangle should include any area that might be affected if a Datum is changed. A Node's bounding rectangle would have to include its label and circle shape; an edge's bounding rectangle had probably better include the Nodes at each end and its label; while a Loop would include its label and the Node to which it is attached. Since each type of Datum has a slightly different definition

Bounding areas etc. for displayed data items; and a "virtual" ReComputeBounds() function

for its "bounds", class `Datum` had better provide an overridable (virtual) function (`ReComputeBounds()`) that can be used to compute the bounds when needed. Once calculated, the bounding rectangle can be stored in a data member and accessed using a non-virtual `Datum::GetBounds()` function.

An extendible Draw function

Class `Datum` obviously has to define a virtual `Draw` function. This need not be a "pure virtual function" because some of the work involved in drawing can be standardized and implemented at the level of class `Datum`. The code in `Datum::Draw()` can draw the boxed label; this code can be invoked from the more specialized `Draw()` functions of subclasses after they have drawn their appropriate lines, ovals, or circles. Class `Datum` can define some default behaviour, e.g. inverting the label briefly, for the other drawing operation (the "flash" operation required during the run process). This function might be made virtual so that it is open for redefinition in subclasses if necessary; however, the proposed default behaviour would probably suffice.

Target for a relabelling operation

The style in which labels get displayed should make it possible to distinguish the `Datum` (if any) that has been selected for relabelling. It might be appropriate to change the label's font style, though some simpler form of highlighting, e.g. a more heavily outlined box, could suffice. This requires that a `Datum` "know" whether it has been selected.

Consequently class `Datum` had better provide a boolean state variable (`fLabel-Target`) that indicates whether it is the target for labelling operations; there will also have to be a public member function that can set the state of this flag (`MakeTarget(…)`). This function would be called by a `FSMStructure` while it was updating its record of the item selected using the labeller palette tool. Other member functions involved in the labelling process would include `GetLabel()` (which returns the current label as a `char*` pointer) and `SetLabel(…)` (which would accept a `char*` pointer to a replacement label string).

Collaboration with FSMView

Whenever an item is changed, it has to advise the associated `FSMView` to invalidate those areas under the original bounding area and its new bounding area. Consequently, class `Datum` will have to define a data member that can hold a pointer to its collaborating `FSMView`. This pointer can be set when a `Datum` is constructed (as part of the process of constructing a `Node`, or `Edge`, or `Loop`) by a sketcher command; a `View*` pointer can be included in the arguments to the constructor or other initialization routine. A `SetView(View*)` member function, that changes the pointer, might also be needed.

Datum::SetLabel() and view invalidation

A `Datum` is changed when its label is altered. The code for changing a label will have to get the view to invalidate the existing area, make the actual change to the string, recompute bounds, and then again ask the view to invalidate the new area.

Datum::Move() and virtual Datum::UpDate-Position

Similarly, when a `Datum` is moved, it will have to get the view to invalidate its area, update its coordinates and compute the new bounds, and again get the view to invalidate an area. `Datum::Move()` can define these operations. However, the mechanism for updating coordinates of a `Datum` will have to be virtual. A `Datum`'s position will be updated by shifting its centre, and other components, by some point-offset; but each different type of datum has distinct components. Also, when a `Node` has its position updated it has to move any attached `Loop` and `Edges`.

"Hiding" data items during a drag operation

Data items are moved after a `Node` has been repositioned using a `NodeMover` command object. During the actual dragging operation, an outline of a `Node` should follow the mouse cursor. The Node, together with any associated Loop and Edges, should be hidden until the mouse button is released and the Node's new position is defined. Class Datum should have a boolean state variable (fHidden); the Draw() method of each variant of class Datum should start by checking this flag and omitting any drawing actions if the variable is set. Class Datum will have to provide a Hide(…) method that can set this variable. This should probably be a virtual function. An Edge or Loop will just reset its fHidden flag; but when told to "hide" a Node will have to notify its associated Loop and Edges.

Protected access to Datum's data members

The code for member functions in derived classes like class Node, and class Edge, will need access to class Datum's data members such as the centre point, the hidden flag, and the bounds rectangle etc. These data members will either have to have "protected" status, or a set of protected access functions will have to be provided for private data. In this program, protected data members would probably suffice.

Figure 10.16 summarizes the design ideas for class Datum.

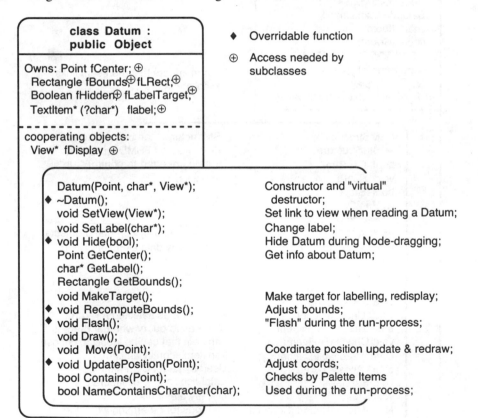

Figure 10.16 A design for class Datum.

FSMStructure

Design compromises Figure 10.17 illustrates a preliminary design for class FSMStructure. Most design decisions involve compromises. Here a problem arises with the structures used to hold the nodes, edges, and loops. One approach would be to have a single list of Datum* pointers; the alternative approach, shown in Figure 10.17, uses three distinct lists – one each for Node* pointers, Edge* pointers and Loop* pointers.

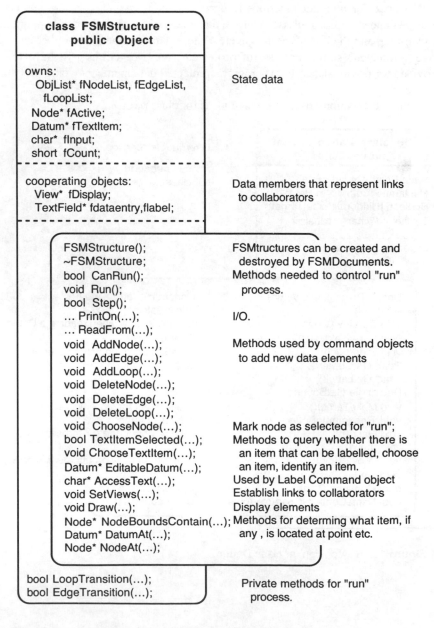

Figure 10.17 FSMStructure preliminary design diagram.

A three list structure simplifies the implementation of some of `FSMStructure's` responsibilities; for example, one place where a three list representation helps is in the process of selecting a starting point for a "run" of the finite state machine. This process begins with a mouse click in the main view being passed to the `RunItem` object associated with the palette. The `RunItem` object wants a `Node`, so it will be asking the `FSMStructure` for a `Node` at a particular point. With the representation using separate lists, it is easy to implement a `FMStructure::NodeAt(Point)` member function that checks the entries in the node-list.

If the `FSMStructure` held `Nodes`, `Edges`, and `Loops` in a single list of data items, then it would not know whether the specific `Datum` at the mouse point represented a valid starting point. After finding a `Datum` at the mouse position, the `FSMStructure` would have to query it as to whether it was a valid starting point. Class `Datum` would have to be extended; it would have to provide a method such as `Boolean CanBeStartingPoint()` that returns `true` for `Nodes` and `false` for other specialized `Datum` variants.

However, a disadvantage of a "three list" structure is that class `FSMStructure` has to provide distinct `AddNode`, `AddEdge`, `AddLoop`, and `DeleteNode`, `DeleteEdge`, and `DeleteLoop` methods. This complicates the class's public interface.

Distinct "AddNode", "AddEdge" and "AddLoop" methods

The designer would have to choose between two alternative approaches to arranging the lists within a `FSMStructure` object. The lists can be "expanded" – so that their head and tail pointers and other data form a part of the `FSMStructure` itself (of course, the links and other parts of the list will be separately allocated). Alternatively, the `FSMStructure` can have `List*` data members with the list structures being separately allocated. Either arrangement would be acceptable, the "expanded" representation would probably be more appropriate because there is no reason for the lists of components to be passed as arguments to member functions of other objects.

Composition structure: "expanded" or separately allocated components

The responsibilities for "adds extra nodes, edges, and loops" identified during analysis (Figure 10.10) have been expanded to the set of methods `AddNode` ... `DeleteLoop`. The other responsibilities identified during analysis have also to be defined in more detail. While some are straightforward, others such as managing the "run" process and helping with the labelling commands, require further consideration.

An instance of class `FSMDocument` will trigger the run process in response to a menu selection. The FSM/Run menu option should only be enabled when the `FSMStructure` has a record of a starting node in its `fActive` data member; there will have to be a member function `CanRun()` that can be used to check this. `FSM-Structure::CanRun()` would be called by the `FSMDocument` when it is enabling the menus that it controls. Member function `Run()` would be called from `FSMDocument::DoMenuCommand`. The run process will entail transitions along edges or around loops; the implementation should probably use two distinct private member functions that cater for these situations. The run process needs to access the input characters in the data entry text field; consequently, one of its data members must be a pointer to this field (the pointer will have to be set in the `SetLinksToViews` method). While running, a `FSMStructure` object can

Member functions for the "run" process

communicate with the text field object, changing the character string that it displays.

*Interactions with
RunItem and
LabelItem*
When a `RunItem` is used to select a node as the starting node, or the `LabelItem` is used to select an item for labelling, a message will get sent to the `FSMStructure` notifying it of the selected item. Hence, class `FSMStructure` must provide functions such as `ChooseNode(Node*)` and `ChooseTextItem(Datum*)`. The implementation of these functions will involve recording the identity of the item in the `FSMStructure`'s `fActive` or `fTextItem` data members and further notifying the selected object (it may have to do something like highlight itself).

*Responsibilities on
behalf of
LabelCommand*
When a `LabelCommand` object is created it is going to have to get a reference to the selected `Datum`. It will also need a copy of the character string in the label text field. The `FSMStructure` object can act as an intermediary; it will have links to the other views already and can be made to interact with the label text field to get its character string data that can then be passed to the `LabelCommand`.

*Design
documentation*
As well as producing a sketch for the design of a class, such as that shown in Figure 10.17, the designers would also be responsible for defining the processing effected by each member function. These initial function definitions would be brief natural language descriptions along the following lines:

- `void FSMStructure::AddNode(Node*), ::DeleteNode(Node*)`: Adds (deletes) the `Node` to the list (`fNodeList`); also, inform the `FSMView` that an area surrounding the node is "invalid" (and, possibly, force an immediate redrawing of the view).

- `bool FSMStructure::TextItemIsSelected()`: Returns `true` if the `fTextItem` data member is not `NULL`.

- `Node* FSMStructure::NodeAt(Point)`: Checks through the `Nodes` in the list `fNodeList` asking each in turn whether they contain the specified point. Returns `NULL` if none contain the specified point. Otherwise, returns a pointer to the `Node` containing the point (or the last such `Node` if the point is contained within more than one `Node`).

Here it is possible to define (at least approximately) the signatures for the member functions. In more complex problems, the definition of signatures might be deferred to a subsequent more detailed design phase. In any case, the signatures of the member functions are likely to be revised during implementation (e.g. the signature of `FSMStructure::NodeAt` might be refined to `const Point&` rather than `Point`).

*Summarizing
collaborations*
Class `FSMStructure` is involved in rather more collaborations than the simpler classes such as those in the Cards program or SpaceInvaders program. The designers of a class that has this level of complexity would need to provide some form of summary documentation illustrating all its collaborations. Figure 10.18 illustrates one possible style.

*Collaboration
diagrams*
Each class involved in numerous collaborations is used as the focus for one set of diagrams. These diagrams show client classes that request actions by instances of the focus class, and server classes to which instances of the class make requests for actions. Often, collaborating classes appear as both clients and servers. For

example, in Figure 10.18 class FSMView is a client of FSMStructure (a FSMView will ask a FSMStructure to organize drawing of the data), and a server (a FSMStructure will ask its associated FSMView object to InvalidateRect when items are added or removed from its lists).

The "design documents" that should be produced for a moderately complex class such as FSMStructure would include:

- some version of the class hierarchy diagram (part of Figure 10.13);

- a class design diagram (Figure 10.17);

- natural language descriptions of all the member functions;

- some form of class composition diagram (this would show whether components such as FSMStructure's lists are "expanded" or separately allocated);

and

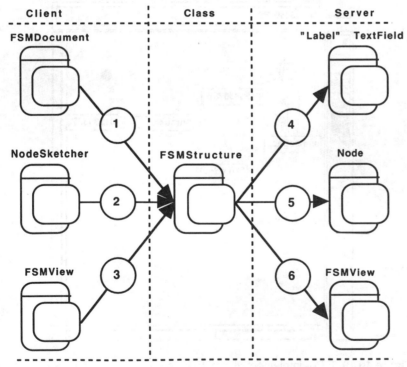

1. SetViews, PrintOn, ReadFrom, Run, TextItemSelected, ...
2. AddNode, DeleteNode.
3. Draw
4. GetString.
5. SetSelectionStatus, ...
6. InvalidateRect.

Figure 10.18 Some of the collaborations involving class FSMStructure.

• diagrams similar to Figure 10.18 that summarize the various collaborations that involve instances of the class.

In addition, the documentation should specify how instances of the class get created and destroyed and whether instances of the class create or destroy other objects.

10.4   IMPLEMENTATIONS

Some example implementations of the complete program are provided in the files associated with this text. Figure 10.19 illustrates the display from an ET++ implementation of the program.

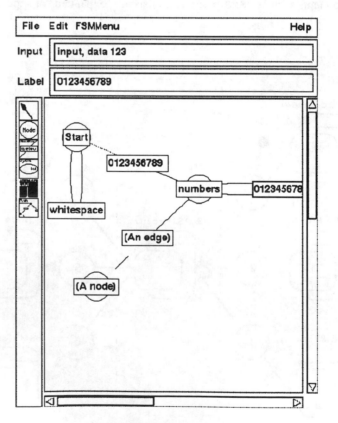

Figure 10.19 ET++ implementation of the program.

EXERCISE

Implement a version of the Map editor program described below.

The Map program allows a user to create simple city maps. These maps can be printed and stored in files. A "map" can contain an arbitrary number of elements. Three main kinds of element may be present in a map:

- linear features (roads, railways, power lines, etc.);
- area features (major building complexes, parks, lakes, etc. – for simplicity, these are restricted to being rectangular in shape);
- "picture" features (small graphic elements that can be used to represent features that have particular importance).

Every feature has a name. Large area features have their names displayed on the map.

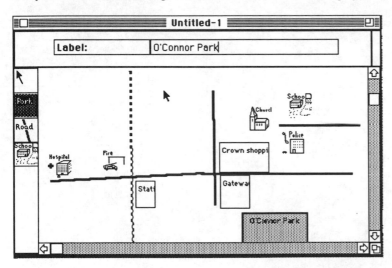

The program will provide a palette of tools for manipulating map elements. There will be an "arrow" tool that can be used to select an existing element or to drag an element to a new location. An element is selected by clicking the mouse within it while the arrow tool is in use (any previously selected element should be deselected). If the mouse button is held down after selection, the chosen element can be dragged. If the mouse is simply clicked in a background region of the map, any previously selected element should be deselected. When an element is selected, its name will be displayed in a separate view region. It should be possible to rename a selected element. A selected element can also be "cut" from the map.

The other components in the palette will represent tools for adding elements to the map. There are three tools – one for adding linear features, the second for area features, and the third for graphic features. Menu options should allow the user to select the particular type of element that will be added using the tool (thus, the linear feature tool can be made a "road sketcher", or a "train track sketcher", or a "power line sketcher").

A minimal implementation should be able to:

- Create an empty map document, a map view for displaying that document and the associated scrollers and windows. (The map view can be fixed in size – so there is a one page limit on the size of the map.)
- Support multiple documents open simultaneously (but no "cut-and-paste" between documents).

- Interact with the user accepting editing commands that add a variety of data elements. The program must provide support for linear, area, and graphic elements; it must have some means for defining different variants (i.e. graphic elements that are associated with different pictures, area features with different fill patterns, and a variety of line styles for linear features).
- Allow a single existing element to be selected and, subsequently, be moved, be cut, or be renamed.
- Save data to files and correctly restores data.
- Handle file menus appropriately (so that the act of changing a document that was restored from a file will enable the File/Revert menu option).

"Intermediate Topics"

Intermediate Topics

The three chapters in this section introduce slightly more advanced topics in object-oriented programming.

Chapter 11 looks at the development process from analysis, through design to implementation. The examples in chapters 6, 8, and 10 have already illustrated some ideas. These are summarized and illustrated with further examples. The approaches illustrated will suffice for simple examples, references are given to a number of texts that present more sophisticated methodologies.

OO Analysis, Design, and Implementation

Chapter 12 introduces a few more features of the C++ language including nested classes, parameterized classes, and exception handling mechanisms. Some small examples are also included that illustrate somewhat simplified applications of multiple inheritance.

Intermediate C++

Chapter 13 is a brief overview of Eiffel 2.3. Eiffel is an innovative OO language that in the late 1980s was offering multiple inheritance, parameterized classes, exception handling and other features not then available in C++. Further, Eiffel offered a complete solution to OO development as the compilers came with extensive class libraries for Unix systems, and with a graphics based development environment that could to some degree be used as a design tool. The newly released Eiffel 3 again seems to be on the leading edge with an integrated development environment that will facilitate the incremental extension of an OO program. Eiffel is possibly too revolutionary. Software houses have tended to be reluctant to adopt Eiffel as it represented too big a break from previous practice and the limited range of suppliers of compilers etc was seen as a risk. Consequently, Eiffel has tended to be regarded more an interesting essay rather than a practical tool.

Eiffel

OO Developments from Analysis, through Design, to Implementation

Alger: *What is wrong with software development today? It doesn't work: 75% of all projects either not completed, or not used when complete. Myths of software engineering: 1) the project team understands the problem, 2) there are fixed specifications, 3) …. Communications breakdowns. ….*

Brooks: *We hear desperate cries for a silver bullet – something to make software costs drop as rapidly as computer hardware costs do. But, as we look to the horizon of a decade hence, we see no silver bullet. There is no single development, in either technology or in management technique that promises even one order-of-magnitude improvement in productivity, in reliability, in simplicity.*

Software is difficult to produce. Some difficulties are inherent. Software systems are complex and, as argued by Brooks, this complexity is a part of the essence of a software system and cannot be avoided through any process of abstraction. Other difficulties are "accidental"; they are consequent upon the ways that software is normally constructed. Changing the approach to software construction may, possibly, ameliorate these difficulties.

Going "object-oriented" does help; but it does not eliminate all difficulties. Further, going object-oriented does introduce new problems – the problems of finding the objects and identifying their classes. It is easy to find the objects and their classes in problems like Space Invaders or Cards. A program such as FSM is a bit more realistic and more work was needed to identify its classes; but the analysis for FSM was greatly simplified from the start by the knowledge that the program was intended to be a standard palette-based editor built on top of a framework library. With more complex applications, it can be quite difficult to identify the "correct" classes. In extreme cases, one analyst's object can be another

Modular encapsulated classes help control complexity! But what are the classes?

An OO methodology guaranteed to find the classes?

analyst's method (just as in US-English dialect, where one can "nounify" every verb and "verbify" every noun).

There are frequent calls for an OO methodology – some definitive prescription of a reliable way of developing OO programs. The dictionary definition of methodology is *"a body of methods used in a particular branch of activity"* and a method is *"a special form of procedure"*. If there was a methodology we'd all be out of jobs. If there really was a set of procedures that could be executed to create a new program, someone would code them up, add a Windows-3.1 interface and sell the ultimate CASE tool. There is no methodology.

> Stroustrup: *Design and programming are human activities; forget that and all is lost.*

> Brooks: *Software construction is a creative process. Sound methodology can empower and liberate the creative mind; it cannot inflame or inspire the drudge.*

> Booch: *The amateur software engineer is always in search of some sort of magic, some earthshaking technological innovation whose application will immediately make the process of software development easy. ... Professionals know that rigid approaches to design lead only to largely useless design products that resemble a progression of lies, behind which developers shield themselves because no one is willing to admit that poor design decisions should be changed early, rather than late.*

Suggestions for (simple) projects

Some guidelines for OO development are provided in the following sections; but, really, they are just hints. They should be sufficient to enable beginners to complete individual projects. These hints are presented in the context of the Chemistry Editor program originally used as an example in Chapter 1. They concern the kinds of issue that should be considered at different stages in project development and identify some approaches to documenting design decisions.

However, there is no list given of activities that one can work through mechanically (and have ticked off on some managerial chart) and be guaranteed of arrival at a working OO system. Moreover, real projects involve work by development teams and such team projects entail a whole new series of problems.

Advice on complex projects

There are numerous (somewhat conflicting) proposals as to how to carry out OO analysis and design of more complex projects. The text books listed at the end of this chapter present a few of the proposed approaches. Some of the texts present relatively simple approaches that may be relevant even to small individual projects (e.g. the book by Wirfs-Brock *et al.*, or the text by Henderson-Sellers). The other texts should be left until beginners have gained some familiarity with an OO language and OO approaches on small projects, and are starting to work with a group on some larger development.

Repeat Analysis, Design, Implementation, until ...

The following sections focus on the usual phases of a project: Analysis, Design, Detailed Design, and Implementation. However, it must be remembered that these phases are not as clearly distinct in an OO development as they were in conventional projects. The individual phases tend to merge and the entire process often involves some degree of iteration. For example, the initial analysis process will often identify clusters of "proto classes" (things that look like candidates for being classes, but may not correspond to classes in the developed system). These clusters are formed around proto classes that apparently should have strong mutual interactions, but only limited interactions with the rest of the components of the proposed system. One such cluster might have to be explored first; the analysis process being extended into design, and possibly even prototype implementation, in order to refine ideas as to the kinds of service that the cluster might reasonably offer. Once these services are better defined, the development group can return to the analysis of the overall problem. Subsequently, the initial design of the classes in the cluster may be revised; certainly any prototype implementation will be reworked. If changes at these levels alter the services provided by the cluster, then the developers would have to reiterate through the overall systems analysis stage.

Some limited amount of prototyping of code is likely to be necessary in OO developments. Smalltalk environments encourage an approach where one analyses a bit, designs a bit, implements a bit and starts over; eventually, the entire proposed system will get to be "prototyped" (sometimes with the unfortunate result of the prototype implementation being put into production use). Prototyping of a complete system is not quite so practical with C++ developments, but typically some parts of a project are prototyped in order to refine the analysts' understanding of the problem. This is not in the Smalltalk style of building a complete system; the prototyping is done to resolve specific problems that arise during the analysis. Preliminary implementation work may be done on a small part of the system to resolve whether some proposed algorithm is practical, or to allow potential users to evaluate a proposed mechanism for interacting with the program. Prototyping of an algorithm should be done in C++, but prototyping of interfaces is usually better done using specialized tools that allow "working" mock-ups of interfaces to be constructed quickly.

Prototyping

11.1 ANALYSIS

The primary objective of the analysis phase is the identification of the major components required in a system, and the characterization of the ways in which these will interact.

One does not typically start by trying to identify "classes". The classes come later, and class hierarchies come later still. The place to start is the "objects". (The "objects" identified during this first preliminary analysis may not be quite the right ones, they may turn out not even to be objects. They'll be referred to below as "candidate objects" or "proto-objects".)

"Objects" first, "classes" later

The objects of interest are going to be those that exist in the computer's memory when the program is executing. Each object has (should have!) a specific, well defined role – it will own some resources and provide services based on the resources that it owns. A program's objects are supposed to be analogous to real world physical objects that obviously do exist, do own resources and do perform tasks. Identifying real-world objects is sometimes a useful point to start a search for the objects needed in a computer program; but only a starting point.

Simulating the real world?

While an OO program is often a "simulation of the real world", it is rarely an exact simulation. The objects that exist in an OO program may well be fictions. For example, an OO payroll program is more likely to have intelligent "paycheck" objects (that can work out their amount and then print themselves) rather than a collection of objects that really do model the actual time-sheets, finance clerks, and accounting officers that are involved in the real world situation. Such intelligent paychecks are a fiction; but they constitute a more useful basis for an effective computer implementation than would classes that more accurately model the real world.

The analyst's first task is to identify the useful objects such as intelligent paychecks, along with display view objects and all the rest. Given the current fashion for "user-friendly" WIMPy (Window, Icon, Menu, Pointer) programs, it is inevitable that similar subsystems, groups of objects, are found in most applications (e.g. a subsystem with View objects etc. for the user interface; a "document" object to organize the storage of other objects in working memory and the transfer of persistent objects to disk storage). Such standard components can be dealt with fairly quickly and superficially during the preliminary analysis which should focus primarily on the new, application-specific components.

11.1.1 Finding the objects in the specification documents

Underline nouns and verbs?

One approach to finding likely candidate objects starts with a conventional formal specification of the required application. In textual descriptions, nouns represent things, verbs describe actions. So at least in principle, one might proceed by underlining the nouns to identify the things involved (i.e. candidate objects) and marking the verbs (i.e. the responsibilities of, or methods of, these objects).

Of course, nothing is that simple. The typical application specification will refer to things that are outside the scope of the actual implemented program (e.g. "the user"); the initial list of nouns must be filtered to exclude these external components. Usually, several different nouns, or noun-phrases, are used to describe the same thing (e.g. the user, the customer, he/she); the list of nouns must again be filtered to eliminate such alternatives and retain only a single standard term (a noun or noun phrase) for each "thing" involved.

Real object, or just an attribute?

Some of the remaining nouns will indeed identify objects that will be present in the program; others don't merit the role of being separate objects (except in Smalltalk where everything is inherently an object), they will exist simply as attributes (data members) of "real" objects. To be worth considering as a real object in this analysis phase, the "thing" has got to own a reasonable number of resources and provide a range of services relating to those resources. So in a banking application, an Account is likely to be a reasonable analysis object. In some way or other an Account "owns" details of an account-holder, a balance, and a transaction record, and it can accept a deposit, process a debit, print a warning letter when it goes into deficit etc. A Balance is probably not worth considering as a separate object during analysis. A Balance can only store a value and, at most, respond to `Value()`, `+(increment)`, and `-(decrement)` "messages". A `Balance` should be treated just as a data attribute of an `Account`, probably being represented as a fixed-point number. As far as the analysis is concerned, details of the "account-holder" could also be treated as a simple `struct` data attribute – even though the implementation might well utilize a class `AccountHolder` rather than a simple `struct` for this data attribute.

After filtering the noun list to remove external entities, alternative names, and "data attributes", the analyst will be left with a list of candidate objects that would need to be evaluated further. The responsibilities of different candidate objects must then be determined using action verbs as an initial guide to the range of possible responsibilities.

11.1.2 Finding the objects by exploring "scenarios"

An approach based on textual is feasible. It is quite well illustrated in the text by Wirfs-Brock *et al*. However, working from a standard specification does not seem to be an ideal approach. It suffers from the problem of separating the developers from the clients. Even if the developers successfully implement a system that matches the specification, there is a good chance that it won't match the clients' real requirements. Furthermore, because they lack detailed domain knowledge, developers who must guess at objects from nouns in a specification may well misinterpret the requirements and fail to produce a workable product. Whenever possible, the domain expert(s) should become part of the analysis team.

Interact with your tame domain expert(s)

Thus, an analyst working on the chemistry editor problem, Figure 11.1, should start by recruiting Sue (the chemist client) into the analysis team.

I require a system that allows me to enter a chemical structure graphically, converts it to a standard representation, transmits this to the Chem. Abstracts search (CAS) service in Colombus Ohio, retrieves details about its properties, files either or both structure and retrieved data and prints selections.

Sue

Figure 11.1   The example problem – Sue's chemistry editor.

The analysis group should then proceed by developing numerous "scenarios" that each represent some part of the work carried out by the planned program. One such scenario is illustrated in Figure 11.2. This scenario is one of those created when considering how the actual editing of the chemical structure might be performed. The editing process would almost certainly involve some form of palette-based editor, with a palette of "tools" that the chemist can use to add substructural components (possibly overlapping with existing parts of a structure), change atoms, alter bond orders etc. The client and the analyst have to resolve how the editing is done, determine the range of "tools" required, and identify additional requirements (such as the possibility that a subsequently added component may have to overlap with and merge into existing structural features).

Scenarios

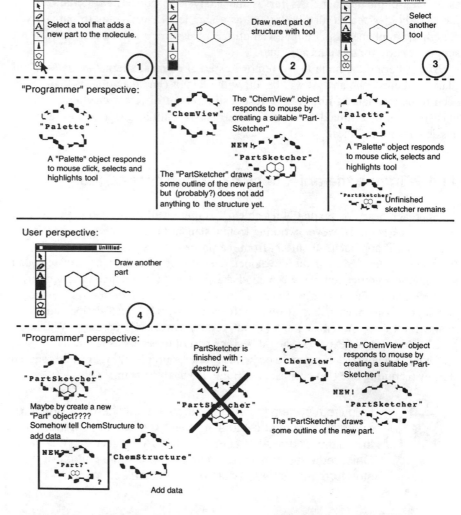

Figure 11.2 Exploring the problem domain from the user's perspective – and identifying possible "objects" in the system perspective..

Client perspective   As shown in Figure 11.2, each scenario involves a number of perspectives (just two in Figure 11.2). Working with the client, the analyst develops the "user perspective". This details the input actions of the user and suggests the form of the program's responses. In the various panes shown in Figure 11.2, the user is shown selecting a tool from the palette and using this to make a substructure (a predefined standard assembly of atoms and bonds) the basis of the new structure.

System perspective   Along with each pane in the client's perspective, the analyst sketches out a first guess as to the "objects" that might be involved in carrying out the tasks proposed. If the program is to have a tools-palette, then there must be some kind of `Palette` object present at run-time that will look after these "tools".

Similarly, handling the user's drawing action is going to have to involve some objects. The analyst would probably be able to identify the need for a ChemView object (with responsibilities that include arranging the display of the structure and responding to mouse-clicks). The ChemView object must be in existence prior to the start of this scenario (note to self: must determine when the ChemView was created and what created it). Its response to the mouse click will be the creation of some form of PartSketcher object that will actually draw the additional structural features and allow them to be positioned.

The fourth pane in Figure 11.2 shows, in the user perspective, the start of a subsequent editing operation. In the system perspective, the analyst must diagram the interactions among objects that are necessary to complete the first editing step. If it had not already done so, the PartSketcher would have to create a Part object that represents the appropriate substructure with its atoms and bonds, and then arrange for the Part object to be handed over to a ChemStructure object with the request that it be merged with any existing structural components.

Every processing step in the program will be initiated by some user action. So, by exploring scenarios for all the different options required by the user, the analyst can identify candidates for all the main objects that will be needed in the system. Figure 11.3 contains a fragment from another scenario, which explored the program's response to a user request to transmit the structure as a query to a database server.

User perspective:

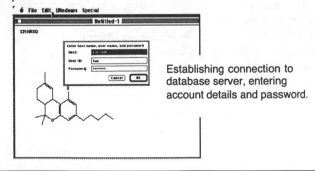

Establishing connection to database server, entering account details and password.

System perspective:

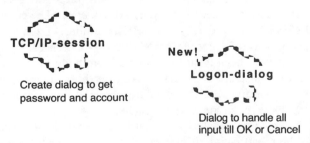

Figure 11.3 Fragment from another scenario from the analysis for the chemistry editor program.

The analyst may work with several domain specialists while developing such scenarios. For the scenario relating to Figure 11.3, the analyst had to work with a programmer familiar with interprocess communication on the Internet. Discussions with the programmer helped identify the "objects" that would be required during an Internet session.

Uses of scenario documents

The scenarios that are developed by the analyst and domain experts have value quite apart from their role of assisting the analyst in the identification of proto-objects. The information in the "User perspective" becomes the basis for the user manuals and tutorials that will eventually have to be created for the program. The "system perspective" information has a similar use as a "tutorial" for the eventual maintenance programmers; these diagrams represent the program's principal objects and interactions. Of course, both parts of these scenario diagrams are going to have to be revised as the initial analysis is refined and again later when design glitches necessitate some reworking of these initial ideas.

By exploring all the ways that the user can interact with the program, the analyst obtains a list of proto-objects and ideas of their responsibilities. Some, but not all of these proto-objects will correspond to objects in the implemented program. The next step in the analysis is to begin to identify classes.

11.1.3 Identifying and documenting possible classes

Starting to find the classes

Each of the proto-objects thus identified is an instance of some class. Consequently, the analyst can derive a list of likely candidate classes (or proto-classes). For the chemistry editor problem, this would include class `Chemstructure`, class `TCP_IPSession`, class `Part`, class `PartSketcher`, class `Table` (something to hold the data retrieved from the database server), etc. The responsibilities of each of these proto-classes can be identified by examining the behaviours of the object, for which the class is being defined, in each of the scenarios where it appears.

These initial ideas for proto-classes have to be documented. For very simple cases, one can use the "fuzzy analysis" diagrams that have been employed in earlier chapters (e.g. Figure 8.12). Often, some form of "Class-Responsibility-Collaboration" card record, e.g. Figure 11.4, is useful.

"Class-Responsibility-Collaboration" (CRC) cards (originally proposed by Beck and Cunningham) are recommended by Wirfs-Brock *et al*. The cards are typically real physical (5"x3") record cards or something similar. Each card identifies a proto-class, and provides for later insertion of cross-references to super- and sub-classes once these have been identified. In the main body of a card, one lists the responsibilities of an instance of the class, and notes those other classes with which there will be collaborations. These data are obtained by examining the various scenarios that have been developed earlier. (Alternatively, in the more linguistically-oriented approaches, such cards are created for each of the nouns selected from the specification document, with the responsibilities being derived from the verbs.)

Class	ChemStructure	
SuperClass		
Subclasses		
Collection of Atoms and Bonds (data)	List (for data elements)	
AddPart, ChangeAtom, ...	Document	
	View	
Save to file	...	
Restore from file		
Draw ....(Ideas as to responsibilities of an instance of this class)	(Ideas regarding those other classes with which a ChemStructure object might interact)	
...		

Figure 11.4   A simple "Class-Responsibility-Collaboration" card.

The use of real physical cards has been justified by empirical results from observations of developers trying to create OO programs. The actual physical cards help make a problem more concrete. Developers can move cards around, join cards with threads that represent interactions ("my document will need to tell your view to do an update, so let's join them with this blue thread to represent that interaction"). Use of the physical cards enables designers to play out scenarios and explore different organizations. Even if such playacting doesn't improve the analysis and design, it may well help to develop cohesion in a project group that is just commencing a major development.

Often, more detail is required than would be included in a simple CRC card. An alternative version is illustrated in Figure 11.5. The form shown is intended primarily for a hypertext environment where there are active links among different records. The class is identified, with a Booch-style fuzzy class outline (identifying this as an analysis level diagram). The responsibilities of the class are separated into "resources owned" and "services provided". Similarly, collaborations are made more explicit, with clients and servers separately identified. The class of a client that would use each service is indicated along with that service. Instances of the displayed class, e.g. `ChemStructure`, will collaborate with instances of other "server" classes to get particular work done; thus, a `ChemStructure` object will require the services of some collaborating `Collection` object to store its atoms and bonds. There should be cross references to each of the scenarios where instances of the class appear.

If a hypertext system is used, the fields in such records can all be active. Selection of a data resource will automatically link to a more detailed description of that data member; this record would suggest the data type, initial values etc. The references to client and server collaborating classes would link to the appropriate class descriptions. If the scenario data are also held in hypertext form, these can be reached via the scenario references. Other links would connect an analysis level diagram to the corresponding record(s) that detail the classes as revised during the design phase. Another link could connect to any "transition state diagram" (see below) that might be used to characterize a class.

Initial "analysis" level classes may get replaced in the design phase

The classes identified during these early analysis stages are not necessarily going to appear in the later design. The analyst has to describe what the program is to do, not how its work is accomplished. This allows the analyst to use more abstract, higher level descriptions. For example, in the scenario in Figure 11.2, the analyst identified a `Palette` object (instance of class `Palette`) that represented the tools palette. Class `Palette` is a useful fiction for understanding the overall behaviour of the program.

However, if a single class `Palette` were carried through into the design and implementation, it would almost certainly prove to be overly complex with too many and too varied responsibilities. A more likely implementation would be as in the FSM program, where the work ascribed here to `Palette` was divided among class `PaletteView`, class `PaletteManager`, and the various `PaletteItem` classes. Class `Palette` is not in any way an abstraction of these actual design/implementation classes; it exists only at the analysis level, and its responsibilities are completely remapped in the design into other quite distinct classes that come from quite separate parts of the final class hierarchy.

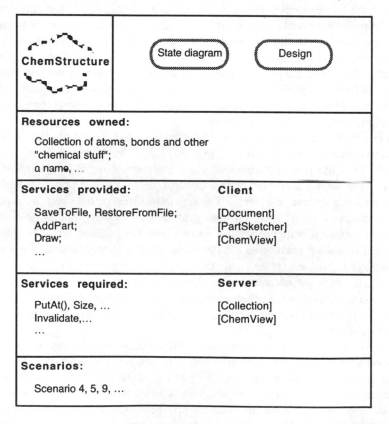

Figure 11.5 A more elaborate form of analysis level "Class, Responsibility, Collaboration" record.

A multifaceted analysis class, like class `Palette`, will often be replaced in the design by separate classes each of which captures only one aspect of the original behaviour. Another quite common situation is for the analysis to involve a "class" that seems to represent a useful abstraction but again in the design is replaced by a set of classes that are not hierarchically related.

For example, it may well be convenient for the analyst to work in terms of a `Tools` class – the chemistry editor might need one set of tools for adding substructures, changing bond orders etc. and another set for selecting data from the tables retrieved from the database server (e.g. a tool that selects a series of numerical values and draws them as a graph). As far as the analyst is concerned, such "tools" are very similar and belong to some overall `Tools` class. At the design level, it might well emerge that there was essentially nothing in common to among these different kinds of tools. They might all be related to a framework provided class `Command` (because most would be associated with reversible actions) but apart from this they might have no common data members and no common behaviours. The analysts class `Tools` would again disappear at the design level being replaced by a number of classes.

The analyst's concept of `Tools` is more like a category rather than a class: *Categories not classes*

categories : *a priori* conceptions applied by the mind to sense impressions, or
    relatively fundamental philosophical concepts.

class   : group of persons or things having some characteristic in common.

The concept of a "tool" is useful as a means to get to understand the overall system. But there should be no mandatory requirement that the later design and implementation actually include any abstract class `Tool`. If there are no common characteristics, the individual "tools" should be instances of distinct, essentially unrelated classes.

"Classes" that turn out to be categories are introduced quite often during the early stages in analysis. This is only natural. The analysts are trying to impose some order on the chaos of different things that emerge from the scenarios that are being created to characterize a proposed application. Consequently, they impose preconceived ideas ("this application is a specialized editor, it is bound to have a lot of editing 'tools', let's identify some"). The development team must simply be aware that some of the analysis level classes may turn out to be incoherent when, during the design process, an attempt is made to find common behaviours that can be associated with an abstract class. Then, another iteration can be made through the analysis process, or the developers can simply note the mapping from the analysis category to the various designed classes.

The analysts are not responsible for finding all the classes that exist in the   *Analysis need not*
implemented system. The analysts are concerned only with what the program must   *identify all the classes*
do. Many other classes are introduced as part of the design process when decisions are made as to how the program's tasks are to be accomplished. A few classes may be introduced during implementation.

For example, during the analysis phase for the chemistry editor, there is no need to consider how the atoms and bonds are to be represented. Obviously, they can be represented – maybe as integers (held in an array), or as a list of `struct`s, or as instances of specialized classes. The decisions regarding the representation of atoms etc. can be left to the designers who may choose to introduce a class `Atom`. Similarly, the designers could specify the storage structure to hold the atoms belonging to a `ChemStructure`, but, again, this decision can be deferred to the implementors. It might be sensible to implement some simple storage scheme (using an instance of a standard `TList` class or `ObjList` class from a framework class library) and then obtain some performance statistics through trial use in order to determine a more optimal storage structure. For example, if it emerged that the majority of the chemical structures created were natural products having a total of between 50 and 70 atoms, it would probably be sensible to change the storage structure to some form of dynamic array with an initial size of 60 elements. If the structures varied widely in size, the default list based implementation might be more satisfactory and so could be retained.

Class hierarchies identified during analysis?

Sometimes, the problem domain provides a natural classification hierarchy that can be used to interrelate different objects. Thus, a banking application might require "Account" objects with specializations such as "Checking Account", "Savings Account", "Mortgage Loan Account", "Personal Loan Account", etc. The analyst can obviously identify such class hierarchies and specify their existence as constraints on the design. With the exception of those hierarchic structures that are inherent in the problem domain, the analyst should not be overly concerned with hierarchic relations among classes.

State diagrams

In the discussion of Figure 11.5, it was noted that the analysts might need to include a state diagram in the document of a class. This is not that common (such diagrams were not needed in any of the examples in Chapters 6, 8, or 10), but does sometimes occur. An object that is an instance of some class may exist in only a finite number of well defined states; with each state accepting only a subset of the "messages" that can be sent to such an object, and with most "messages" causing a transition to a different state. For example, one of the communications objects used in the chemistry editor program might have conceptual "states" such as 'initializing', 'getting password', 'logging on', ..., 'logging off' etc. A simple diagram showing the states and allowed transitions is often an effective way of characterizing such classes. Where a state model exists, it should be documented as part of the analysis process.

A biased analyst is a good (well, effective) analyst!

Just as a final point on the analysis, when working with the client on scenarios, such as that in Figure 11.2, the analyst should *not* be thinking in terms of specific classes in some class library. It should not be a matter of "I can let you scroll that (note to self: specify a `TScroller`) and you'll be able to enter text directly on the item (note to self: specify a floating `TEditText`)". However, the analyst should be somewhat biased so that the chosen user interaction elements are easily realizable. This biasing is simply a matter of the analyst being familiar with the systems that the development team has successfully implemented. If a particular framework makes it a real pain to have a floating `TEditText`, then probably no delivered application would have included this feature; instead some data entry bar would always have been used. Knowing this, the analyst would propose a design with a data entry bar; the client will probably be quite content not to have to select between this and alternative mechanisms. It is the same kind of biasing towards the realizable that you'll find in other fields. Thus an architect, who has worked successfully with a subcontractor who can provide dry-climate indoor plants, would propose "a few Mexican cacti, or maybe a couple of Australian gum-trees" as the decoration for the atrium of a new office building. More difficult to achieve alternatives – "a nice bit of tropical rainforest, or a patch of subarctic tundra" – would simply not be offered to the clients.

11.2 DESIGN: FROM BLOBS TO BOXES

Figure 11.6 summarizes some of the more important tasks of the designer. First, the analysts' classes must be "firmed up". Rather than fuzzy blobs, one now wants classes that have well defined boundaries. For example, an analysts' class `Table` (something used to hold data retrieved from a remote database) must be replaced by "`abstract class Table`". The data resources owned by a class must be clarified, and the services provided by a class must be identified and split into private "housekeeping" services needed to maintain state data and public services offered to clients.

Clarify responsibilities of classes

Later, the designers start to explore hierarchical relations among classes. Often, as in pane 2 in Figure 11.6, this will be a matter of growing the tree downwards as specialized variants of classes are recognized. For example, the designers might identify two specialized kinds of table in the chemistry editor application – one for physical and spectral data (class `SpectrumTable`) and another for literature references (class `LitRefTable`). The designers will be hoping to reveal scope for shared implementations of routines at the abstract level, and "programming by differences" to clarify what is unique about the specialized subclasses.

Explore hierarchic relations among classes

At a still later stage, the designers will identify opportunities for the reuse of standard components. These opportunities may involve extending class hierarchies, as when "`abstract class Table`" is derived from a library class `DynamicArray` (pane 3 in Figure 11.6), or may be a matter of composition (e.g. class `LitRefTable` using an instance of class `TextItem`).

Reusing standard components

The early phases of the design process tend to be where one gets most iteration as ideas for classes and inter-class collaborations are proposed, evaluated, and revised. The analysts will have identified what the program is to do, but usually there will be many ways in which each task can be accomplished. The designers have to explore different approaches. The approaches will involve differing patterns of interaction among objects and, probably, slightly changed responsibilities for their classes.

11.2.1 Firming up the classes and identifying additional classes

The analysts' classes serve as starting points for the designers. All these initial classes will be characterized in terms of lists of resources owned and services provided – but both lists will be incomplete (and private housekeeping functions will normally have been glossed over).

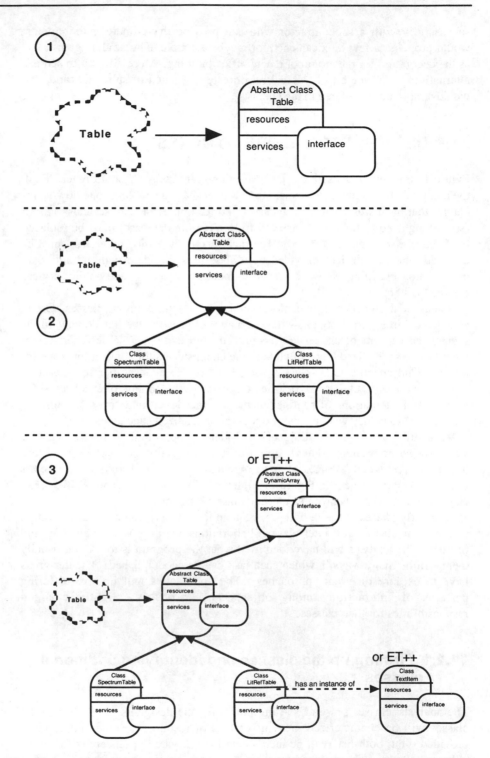

Figure 11.6 Tasks of the designer: firm up the classes, identify hierarchies, exploit reuse.

More detailed specifications have to be created for the classes. Again, one of the more effective approaches is to follow through the scenarios. The process is similar to that in the analysis phase, it is simply much more detailed.

Scenarios again

The level of detail should be at least that used when diagramming interactions among `Document`, `Views`, and other objects in the Cards and FSM example programs (e.g. Figures 8.15, 8.16 etc.). The diagrams should make explicit the sequence of method calls involved in each interaction. The designers should make a first attempt at identifying the data that must be transferred during these calls. (It is not worthwhile defining the method signatures at this stage. There will normally be some changes to the objects and classes during this initial design phase. Details of method signatures can be deferred to the detailed design phase that commences once the class definitions have become more stable.)

The interactions diagrammed in the analysis scenarios will omit details – details such as interactions with auxiliary objects and the handling of exceptional cases. For example, an analysis level scenario for showing the setting of "atom-type" in the chemistry editor would be similar to Figure 11.2 and simply show first the selection of the "Atom Namer" tool in the palette and then an `AtomNamer` command object being created after a mouse click in the main `ChemView`. Of course, the actual interactions are more complex. The mouse-click in the main `ChemView` would have to be shown as leading to it executing its `HandleMouse()` method. This method would involve a request to an associated `Palette` object to identify the current tool. The current tool identifier would be shown as being used in a switch statement that selects the processing option. The processing for an Atom Namer tool would involve a check that the mouse was located over an atom (one can't apply an Atom Namer tool to a bond or to a bit of empty space); only if the check were satisfied would an `AtomNamer` command object get created. The design scenario would have to make these details explicit.

Filling in missing details – like interactions with auxiliary objects

The more detailed characterization of the processing involved in each method frequently requires the invention of additional private member functions that perform "housekeeping" duties. For example, a `ChemView::HandleMouse()` method that attempted to deal with all possible command objects and the constraints on their creation would prove overly long and complex. The designers would typically split out a number of private member functions to deal individually with different types of tool selection and constraint.

Identifying private house-keeping functions

Information gained from each scenario should be transferred to more extended CRC cards defining each of the classes involved. Usually, analysis of design scenarios will result in additional data members that are pointers to instances of collaborating classes (and there may be a need for public interface functions that allow such pointers to be reset). As noted in the previous paragraph, the list of private housekeeping member functions may be expanded significantly. Additional public interface functions may be identified as extra responsibilities are accorded to the classes.

While scenarios can help identify the objects present at run-time and their mutual interactions, each scenario provides only a small glimpse as to the way in which an object from a particular class behaves. The designers should sketch out "life histories" for typical objects from each class. Such documentation helps implementors and maintenance programmers understand the role of the class of

Life history diagrams for typical objects in each class

which the objects are instances. Further, it provides a way of checking that links are established before the object tries to interact with collaborators. A stylized example of such a life history diagram is shown in Figure 11.7.

Example "Life histories" for instances of class Atom:

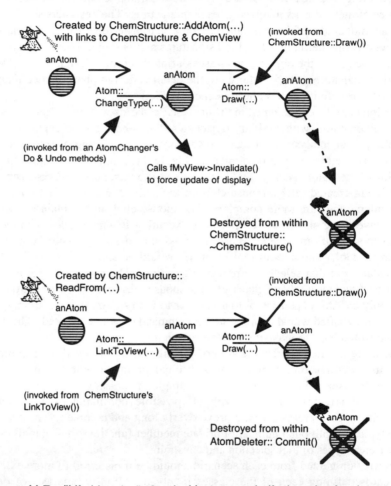

Figure 11.7 "Life histories" of typical instances clarify the role of a class.

Designing the algorithms for member functions   Of course, the designers must create the algorithms for any non-trivial member functions. At worst, this is no harder than it would be when designing normal procedural code; often, it is simpler. Much of the code in a normal procedural program consists of selection statements (switch()… or if()… else… statements). Each of these selections is a repeat of the original specification where variant data forms were defined. In a proper OO implementation, most of these selections should disappear – being replaced by calls to dynamically bound functions (as has been illustrated in many earlier examples). But, of course, if the algorithm to be implemented is some elaborate iterative or recursive process, there is really no change from the ordinary procedural implementation.

When a complex data processing task is required, the public member function that defines the service will typically be implemented in terms of a number of private member functions. For example, in the chemistry editor program, one of the member functions of class ChemStructure would have to be something like ConvertToStandardForm(). When a structure is sketched in using the editing tools, its atoms will be numbered according to the order in which they were added. The programs at the database servers can only handle structures whose atoms are numbered in a standard, or "canonical" manner. There are rules that define whether a structure's atoms are numbered canonically. The ConvertToStandardForm() function would implement a form of search through the different possible ways of numbering the atoms to find the "canonical numbering". This recursive search would probably involve ten to fifteen auxiliary routines, each of which would end up as being defined as a private member function of class ChemStructure.

Adding many private member functions

As details of the properties of classes are filled in, designers will encounter situations where extra classes are needed. Some of these extensions are simple and don't really change anything from the analysis model. For example, the designers creating class ChemStructure would almost certainly introduce class Atom and probably a class List to hold the atoms. Class Atom would be introduced because instances of this class can group together conceptually related simple data items (integer coordinates, an enum specifying the atom type etc.); such a class is little more than a struct record with a few methods for displaying itself, changing its data members, and performing file i/o etc. Class List would obviously be some simple collection class that would be expected to come from some standard class library.

Uncovering extra classes

Often, extra classes will be created either to isolate knowledge, or to reduce mutual interactions among existing classes, or to separate an overly complex class into a number of classes each capturing just one facet of the original class's behaviours.

As a design develops, it is not uncommon to find a class that needs to know too much about other classes. For example, if class ChemView really must handle creation of command objects for visual editing, it will need to know about all the different kinds of editing command and the various constraints that apply to their use. Such dissemination of knowledge makes a program more difficult to maintain and extend. Any change to the palette of editing tools is going to have effects on class Palette, class ChemView and maybe others. When a class appears to have too much knowledge of other classes, designers should consider introducing an intermediary. This kind situation arose in the FSM program and was one of the reasons behind the introduction of class PaletteManager and class PaletteItem. Class PaletteManager localized knowledge concerning the number of different tools. The individual PaletteItems localized knowledge concerning the constraints that applied to the use of the different tools.

Classes to localize knowledge

Sometimes, examination of the design scenarios will reveal cases where instances of different classes have too many interactions. An object of one class will be constantly interrogating another object as to the values of particular data members and requesting changes to those data members. Such interactions may be mutual; both objects may be "living in one another's pockets". One possible cause of such interactions is misassignment of responsibilities – ownership of the data

Reducing excessive interactions

resources in question may have to be reallocated. However, it is possible for such patterns of interaction to be evidence that a new class should be introduced. The new class would own those data members that seemed to need to be in common to the other classes, and would package all the services related to the management of these data resources. The original classes would be changed so that each had a pointer to an instance of the new class (and one would be made responsible for creating the instance at run-time and informing the other cooperating object of its existence).

Decomposing a complex class

As noted earlier, a class proposed by the analysts may turn out to have too many and too varied responsibilities. The designers may split such a class into several classes. Thus, class Palette may become class PaletteView (part of the View hierarchy in the final design) and class PaletteManager (from some totally unrelated branch of the final class hierarchy).

Revising the analysis to accommodate new design classes?

Addition of classes like class Atom and class List do not necessitate any change to the original analysis. Other additions to the set of classes planned for a program will constitute at least minor amendments to the original analysis model. Sometimes the changes can be described adequately in a simple concordance document that relates the design and analysis models (e.g. "Analysis class Palette was replaced by class PaletteView and class PaletteManager, PaletteView takes Palette's event-handling responsibilities with PaletteManager organizing data"). In other cases, the development group may need to make another quick iteration through the analysis phase to explore the impact of a proposed change.

Eliminating redundant classes

Very occasionally, the designers will be able to eliminate one of the analysts' classes. Detailed consideration will show that two apparently distinct proto-classes are really the same class viewed from different perspectives. For example, in Figure 11.2 the analysis scenarios show a PartSketcher command object creating an instance of class Part (that was to represent the extra atoms and bonds that were to be added to a structure), and then getting the ChemStructure to merge in the Part. At the design phase, it emerged that a Part was really just a ChemStructure. Instances of these classes held similar data (although a few of the ChemStructure's data members would always NULL in a "Part") and had essentially the same behaviours. Class Part could be discarded as redundant.

11.2.2 Identifying class hierarchies

Once the individual classes have begun to be "firmed up", the designers can start to explore class hierarchy relations other than those that are intrinsic to the problem (which would already have been specified by the analysts). The processes of "firming up" classes and discovering hierarchies do overlap significantly. Quite often, examination of design scenarios involving different instances of a class will reveal the need for specialized subclasses (because different data are needed, or different behaviours are exhibited in specific circumstances).

Identifying specialized subclasses

Some of the analysts' classes will represent abstractions, with the actual objects created at run-time being instances of more specialized subclasses. Thus, the analysts for the chemistry editor might specify that instances of class Table get used to display the various different kinds of information retrieved from the remote databases. The designers would find that the information retrieved for literature references (a series of items each in the form authors, journal, volume, issue, pages, ...) was quite different from that returned for spectral data (a series of number pairs, e.g. m/z: [41,78], [43,83], [44,6], ...). Both kinds of information need to be displayed in a tabular form with a surrounding boxed outline, labelled columns etc. Given such substantial differences in the data, the designers would find it necessary to create specialized subclasses, class SpectrumTable and class LitRefTable. (Somewhat less commonly, the analyst will have specified the use of different kinds of table and left it to the designers to capture commonalities in what would normally be a fairly obvious common base class.)

Generally, the abstract class defined as a base for the various specializations will be "partially implemented". Some of the behaviours will be common to all (or most) of the specialized variants. Thus, there might be common code in class Table to draw an outline box with titles and to arrange column widths etc. This common Draw() method would include a call to a private virtual member function, DrawContent(), that would be implemented differently in the various specialized subclasses. As much as possible of the code should be defined in the base class to maximize reuse and avoid duplication of essentially equivalent functionality in the individual specialized classes.

Partially implemented abstract base class

If class Table has two specializations in the initial design, it is likely that in the final program it will have several more. (Additional specialized tables will inevitably be needed to accommodate all the other requirements that the client identifies on first playing with a prototype of the program.) The designers are going to have to give careful thought to the definition of their base class, providing for extension (through overridable virtual functions) and for controlled access to common data (through protected access methods). These provisions must be clearly documented so that other programmers coming later can understand how they may create further specialized variants.

Designing, and documenting, the base class for future extension

It is at this stage of finding specialized versions of the analysts classes that the designers may discover that some of the original "classes" were merely "categories". An analyst might well describe editing actions on a ChemStructure and on Tables in terms of "tools" and have some class Tool that appears to serve as an abstract base class. The designers would find numerous specializations – AtomNamer tool, PartSketcher tool, ColumnSelector tool, etc. – as the individual scenarios were examined in detail. Apart from the fact that all these tools might appear to be related to command objects (as all will be involved in reversible editing actions), they may have nothing in common. Class Tool as an abstraction is eliminated. Either the individual specialized tools classes get to be derived directly from a Command class, or maybe separate intermediate abstract classes are identified (e.g. class StructureTool and class TableTool both derived from class Command).

Eliminating "categories"

Another important task of the designer is to discover new simplifying abstractions. For example, in the chemistry editor program, the analysts might have identified a Document object that owns an instance of class ChemStructure, two Tables (later differentiated by the designers into an instance of class SpectrumTable and an instance of class LitRefTable), an instance of class CASData (some other kind of retrieved information), etc. On the surface, there

Discovering a totally new abstractions

seems very little in common to these different classes. However, similarities are revealed if one looks at the pattern of communications between the `Document` object and these various other objects (Figure 11.8).

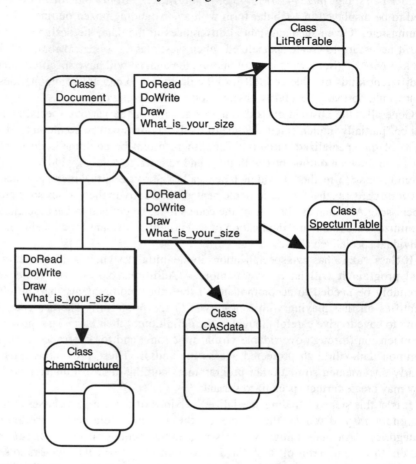

Figure 11.8 Similar patterns of interactions can suggest the creation of a new abstract class that subsumes existing classes.

The similarities in these communications patterns suggests a possible role for an abstract super-class, class `DataItem`, that subsumes the various specialized classes that arose naturally during the analysis of the problem. Class `DataItem` could define a common interface – with consistent definitions for the signatures of the similar member functions `DoRead()`, `DoWrite()` etc.

Pure abstract class with all methods "deferred"

Class `DataItem` would be a "pure abstract class". It would not implement any of the methods that it defines:

```
class DataItem {
public:
    DataItem() {}
    virtual ~DateItem() {}
    virtual      long  What_Is_Your_Size() = 0;
    virtual      void  DoRead(File*) = 0;
```

```
    ...
};
```

This type of abstract class does not bring any benefit from code sharing. There *No code sharing*
is no code at the abstract level to share. Each specialized class implements all of
the defined methods along with any specific methods.

The benefit from such abstractions is that they help factor out the original   *Factorizing out the*
specification. Instead of having a class `Document` that owns one each of a Chem-   *original specification*
Structure, CASData etc:

```
class Document {
public:
    ...
private:
    ChemStructure*       fChemStruc;
    CASData*             fCas;
    LitRefTable*         fLits;
    ...
};
```

a revised design can have a `Document` that owns a list of `DataItems`:

```
class Document {
public:
    ...
private:
    List                 fItemList;
    ...
};
```

Knowledge of the original specification for the program can thus be removed
from class `Document`. Rather than explicitly saving, restoring and displaying the
individual tables etc., its method can be written to iterate along the `fItemList`
telling each item in turn to save, or restore, or draw itself. Such changes to the
design make it practical to accommodate extensions requested later (such as some
of Sue's subsequent requests, Figure 1.3, for multiple `ChemStructures`, additional
data tables etc.).

The discovery of abstractions that allow for significant simplification of the
design can enhance a system's reliability and potential for extension. These
benefits outweigh the costs of the extra iterative cycles through the analysis and
design that will be necessary to revise the structure that was initially proposed for
the system

11.2.3 Using class libraries

As the design becomes more concrete, the designers will start to consider how to
exploit reusable components from class libraries. This will involve finding library
classes that can serve as base classes for classes proposed for the new system, and

the somewhat simpler use of concrete classes from the library as components in the new classes.

The use of instances of concrete library classes as components in new classes presents no problems. A bounding rectangle must be specified? Have a data member that is an instance of library class `Rect`. A header must be displayed above the data? Use an instance of library class `TextItem` (or its equivalent). Some items have to be stored in a list? Use an instance of library class `ObjList` etc. Quite a few of the decisions relating to the choice of concrete library classes can be deferred to the implementation phase.

The use of data members that are instances of concrete classes only comes up as a design issue if for some reason it becomes important to consider trade-offs between having an instance as an actual data member of a program specific class (expanded representation) or having a pointer data member with a separately allocated object in the heap. Small objects (such as `Rect`s) should almost always be incorporated as (expanded) data members (the overhead of having them as separately allocated heap-based objects is excessive). If the data represent optional information that is potentially large (e.g. an instance of class `TextStyle` that is needed only if a non standard font is used), then separate allocation is almost always favourable. Apart from these two heuristics, little guidance can be given.

If a data member represents something that might be shared with another object, then the data should *not* be regarded as a component (an owned resource). The data member becomes a pointer. The instance of the concrete class becomes a collaborator and not a component.

Inheriting from a
framework
The designers of programs that are to be built on top of framework class libraries will have been guided in their choice of classes by the concepts that are represented by the framework's principal classes. Consequently, the designers will have come up with a `ChemDocument` that owns data and talks to files, data will be displayed in `ChemViews` or `PaletteViews` or `TableViews` etc. There should be no difficulty in threading the program specific classes onto the framework classes. So, classes `ChemView`, `PaletteView`, and `TableView` become extensions of the framework's `View` hierarchy. Class `ChemDocument` is recognized as a specialization of `Document` etc.

When a program is being built using a framework class library, the designers must determine the extent to which specialized facilities such as object i/o and dependency mechanisms (see Chapter 9) will be used. Such decisions cannot be deferred to later detailed design and implementation stages because they have implications regarding the structure of the class hierarchies, the collaborators required by classes, and the patterns of interactions among instances of classes. For example, if the chemistry editor program were implemented in a simple fashion, any changes to an atom (e.g. change of type or position) will require that the `ChemView` be explicitly told to invalidate some subregion – so each atom would need a link to the collaborating `ChemView` and function `Atom::ChangeType(Atype n)` will need a call such as `'fMyChemView->Invalidate(aRect)'`. If a dependency mechanism were being used, a change to an atom would be more likely to result in a "changed" message being sent to its `ChemStructure` and then propagated to the `ChemStructure`'s dependents such as the `ChemView` (the message would specify that it was the atom that changed, so the view could ask the atom for its area and so do a minimal invalidate).

Inheritance can be overused. For example, Figure 11.6 suggests that class Table might be based on class DynamicArray. Such usage is dubious. It seems to be a case of "implementation inheritance". Class DynamicArray obviously provides some form of resizable storage and presumably has a public member function for adding another array entry; such facilities are likely to be quite useful in, for example, the communications code that builds a LitRefTable item by item as the literature references are returned by the database server. But in other ways, a LitRefTable doesn't behave much like a DynamicArray. It doesn't seem very likely that there would be much code like 'fLits-> At[index]' or code using any of the other member functions DynamicArray.

Overuse of inheritance

Rather than claim spurious inheritance from other classes, it is usually better design for a class to delegate work. Class Table should not be made a subclass of DynamicArray; instead, it should employ an instance of DyanamicArray.

11.2.4 Documenting the design

As soon as it begins to stabilize, the design should start being documented in detail. Class diagrams, CRC cards, or Unix-style man(ual) pages are a necessary part of this documentation, but they are not sufficient. Such documentation leaves the poor maintenance programmers wailing the refrain "I can understand the classes but I can't understand how the program works". Design scenarios and "life histories" of typical class instances are also essential.

A class design diagram, such as shown in Figure 11.9, provides a succinct overview of a class. Some notational schemes rely on subtle visual cues to distinguish between abstract and concrete classes etc. There is no need to be coy. A class should announce explicitly whether it is intended as an abstract class (in which case it should indicate provision for extensions), or a concrete class. The class diagram should identify data members, distinguishing between those that encode the state of an object and those that serve merely as links to independent collaborators. The member functions should be separated into those that form part of the public interface (the services provided by the class) and the "housekeeping" routines (which may be private or "protected"). If there are shared class data and/or functions, these should be indicated separately. Friendship relations, if any, can be shown as external entities that penetrate a class's boundary. (Class data, class methods, and friend relations are usually added in the later detailed design phase.)

Class diagram

Each class has a substantially extended Class-Responsibility-Collaborations "card" (more like a short essay than a "card"). Figures 11.10 and 11.11 illustrates a form that might be suitable for a hypertext based documentation system. In a hypertext system, each of the entries on a card would act as an active link to detailed documentation. Thus, each data resource would link to information defining the data type and role of that resource. In printed documentation, these links would have to be represented by page number cross references.

Extended CRC card

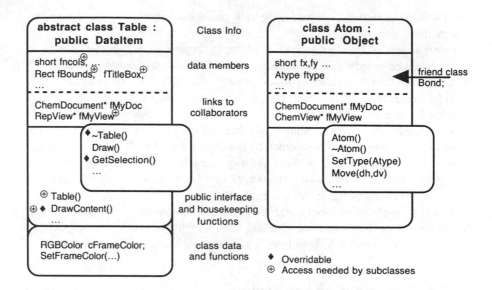

Figure 11.9   Design level class diagrams.

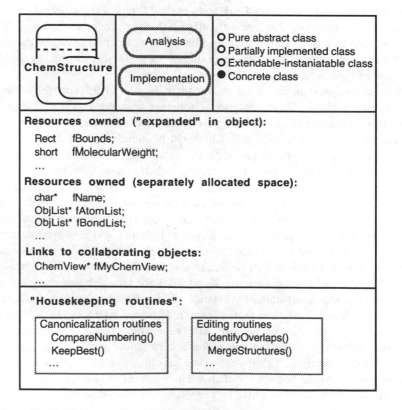

Figure 11.10   First part of a design level CRC card.

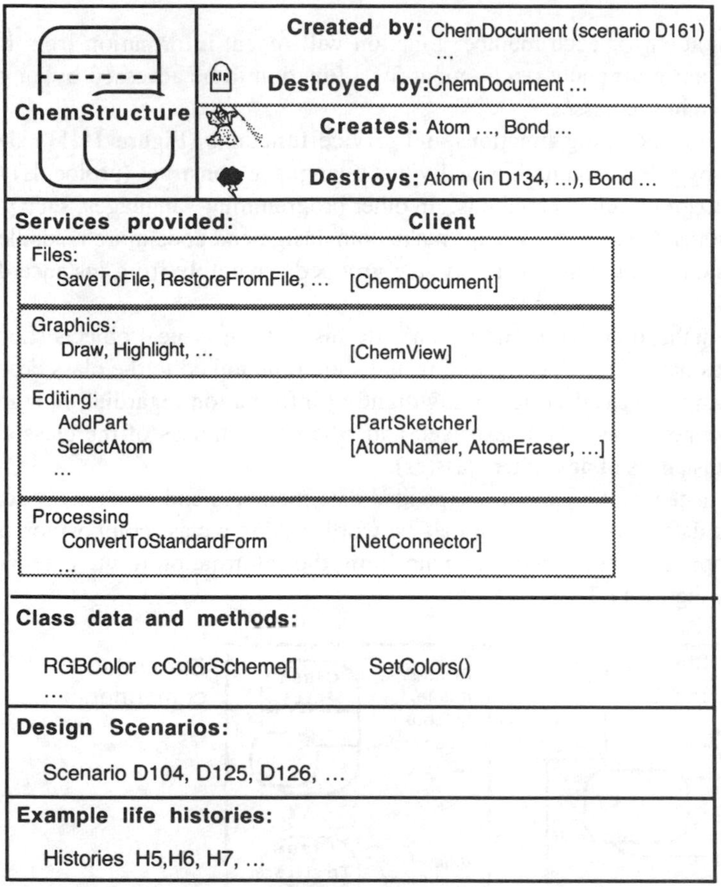

Figure 11.11 Second part of a design level CRC card.

The class should be identified as a design class (rather than an analysis class) and should indicate whether it is intended as a pure abstract class, a partially implemented abstract class, an instantiable class with provision for further extension, or a concrete class. Hypertext links (or cross references in printed documentation) should connect the design details to analysis and implementation details.

The resources owned would be grouped by general type – expanded data members that form part of the space allocated to the object, owned resources that are separately allocated (and for which an object of this class has only a pointer data member), and links to independent collaborators (again, a class instance will have pointer data members for these). The detailed documentation specifying the data type and role of each data resource would also identify those that might need to be accessed by subclasses (this information will also be available from the class diagram where tag marks are used to denote a need for such access).

Housekeeping routines would "link" to specifications of their algorithms. By the time the design is finalized, the details on each member function will include a specification of its signature. As with the data resources, the detailed

Documented algorithms of methods

documentation for each member function will repeat information from the class diagram concerning any requirement for a function to be accessed by, or possibly overridden in subclasses.

Both housekeeping functions and service functions (Figure 11.11) should be grouped by role. Smalltalk actually has a language construct (protocols) that can be used to group related methods. In other programming languages, such grouping has no semantics, it simply helps make both design and code more readable.

Class data and methods, if any, are grouped separately from instance data and methods.

Links to design scenarios and object life histories

The finalized design scenarios and life-histories of typical objects should both be kept as part of the documentation and should be linked to the class description. It is worth the class description highlighting information regarding how instances of a class get created and destroyed and whether instances of the class create or destroy instances of any other class(es).

Class composition diagram

If an instance of a class has expanded data members and owns many separately allocated data resources, then it may be worth having a class composition diagram. This simply repeats, in more graphic form, the information in the extended CRC card, see Figure 11.12.

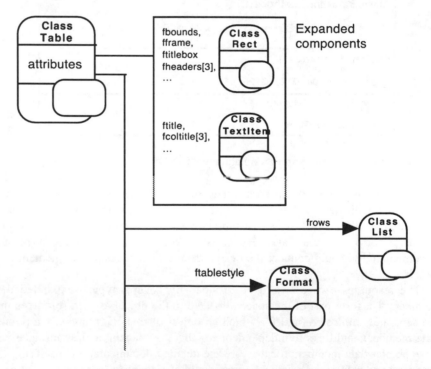

Figure 11.12 A composition diagram may be necessary for classes that have many components that are instances of other classes.

Figure 11.12 is a sketch for such a class composition diagram; it shows a class `Table` that has several (expanded) data members that are instances of class `Rect` and a few more that are instances of class `TextItem`. In addition, a `Table` owns a format record (a separately allocated object that is an instance of class `Format`) and a list of entries for its rows (as a separately allocated object that is an instance of class `List`). A `Table` also owns many other separately allocated `TableEntry` objects. These are not shown in Figure 11.12 (though they could be shown as attached via the `frows` list). Links to independent collaborating objects would be shown as pointer data members in the "Expanded components" part of the diagram.

The design scenarios provide the real data that describe class collaborations. However, some summary documentation is also required. Many books on OO analysis and design include diagrams purporting to show such collaborations. These diagrams show graphs with the nodes representing instances of different classes and the arcs representing edges. In a real program, there are so many nodes, many joined by multiple directed edges, that a single diagram becomes impossibly complex.

In a large system, many "class collaboration" summary diagrams will be required. One approach was illustrated for the FSM program in Figure 10.18. A set of class collaboration diagrams is used; each diagram focuses on a specific class and shows the classes from which it requests services and the classes to which it provides services. Each of the links shown on the diagram has to be annotated with a list of all the services; there should also be cross references to design scenario diagrams. Naturally, there is redundancy among the class collaboration diagrams that are produced (a class will appear as a client or as a server in the diagrams for all the other classes with which it interacts). Figure 11.13 provides a further rather similar illustration of the type of diagram proposed; this figure is a (highly simplified!) view of a collaboration diagram focussed on class `ChemStructure`. (Again, there are advantages in using a hypertext system. Popup menus can be associated with each arc that list all the requests one class might make on another; selection of an entry from a menu links to a description of that interaction.)

Class collaboration diagrams

Inheritance hierarchy diagrams will be required as part of the documentation. These will be similar to many shown before, such as Figures 10.12 – 10.15 that formed part of the documentation for the FSM program. It is not necessary to document the inheritance of concrete classes taken from a library (see Figure 10.15). There is no need to know such details for standard things like `NumberTexts`, `Scrollers` etc, as instances of these classes are simply employed to do a job. It is quite democratic; provided that you do you work right, no one will question your ancestry.

Inheritance hierarchies

Some of the schemes proposed for documenting OO programs allow diagrams that combine data defining class inheritance, class composition, and class collaborations. This is simply not worthwhile. In any real program, each of these aspects is individually complex. Overlaying different aspects in a single diagram simply obscures the relations that a diagram is supposed to highlight.

Combined diagrams?

Even if given inheritance hierarchies, class diagrams, composition diagrams and collaboration charts, a maintenance programmer may still be quite justified in complaining "I understand the classes, but not the program". The forms of documentation that focus on classes provide a limited, static view of a program's complex and dynamic behaviour.

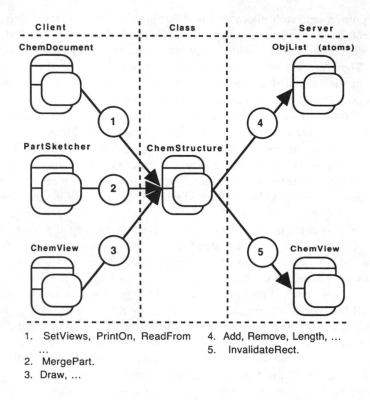

1. SetViews, PrintOn, ReadFrom   4. Add, Remove, Length, ...
    ...   5. InvalidateRect.
2. MergePart.
3. Draw, ...

Figure 11.13 Class Collaboration diagram.

Include the object life histories and the design scenarios   Some guide to a program's dynamic behaviour is provided through the object life histories and the design scenarios. It is essential that the designers prepare these carefully so that the system can be made intelligible to maintenance programmers.

11.3 DETAILED DESIGN: HAVE YOU COVERED ...?

There is never a distinct detailed design phase. One simply keeps iterating through the scenarios, revising (and hopefully improving) the way in which data resources are allocated and objects collaborate.

Partial implementations are normal   A class does not have to be completely defined before implementation begins because the responsibilities of a class can often be factored into distinct aspects. Some aspects can left unimplemented in partial implementations of a program. For example, with the chemistry editor, the development team could go ahead and implement the interactive parts of the structure editor long before the design of the algorithms for ChemStructure::ConvertToStandardForm() had even begun to be researched. A partial implementation keeps the client enthused (and even may help keep the development team enthused), and it can also serve as a framework for testing out ideas for code that will deal with the difficult bits of the problem.

However, before even a partial version of a class starts being coded up, the designers must resolve some issues. The following points are simply reminders of things that designers should be certain to cover before implementation work starts on a class. Most are already covered in the forms suggested for design documentation. Several points are C++ specific.

Creation and destruction of objects

The processes of instance creation and destruction tend to be somewhat neglected in most prescriptions of object-oriented design. They are worth highlighting. Classes should be documented with information specifying: i) how are instances created (in which methods of what other classes)? ii) how are instances destroyed? iii) can an instance of this class create instances of some other class (when, why, how)?, and iv) can an instance of this class destroy instances of another class?

Initialization

When instances of a class are created they must be put into an appropriate initial state. The implementation language determines whether the initialization process is invoked automatically as part of the act of creating an object or whether a separate call to an initialize method is required (although, the conventions used in a class library, e.g. MacApp, may differ from the standard for the language). The designer must determine, and document, the allowed initial states for an instance of a particular class.

Class invariants? Preconditions and postconditions for methods?

Can a class be characterized by any invariants? If it does have invariants, they should be documented so as to allow for possible formalized, automatic checking of correctness.

The designers may also be able to specify preconditions and postconditions for each method. These can be used to guide implementors and may also be used in automated systems for checking correctness.

Controlling access to data in inheritance hierarchies

The best rule is that all data fields are private to the class where they are defined. Methods can be provided that allow the values of data fields to be interrogated and updated; restricting access to these methods helps enforce encapsulation. The designer should identify those data-access and data modification methods that are public, really private, and "protected" (open to use in more specialized subclasses).

Provision for further specialization

In C++, the designer of the class must select and identify those methods to which changes are permitted in derived classes.

Method signatures

By the detailed design stage, the designer should be able to specify the message protocols – i.e. specify the data that are provided and returned in each method call.

Resource manager classes and assignment

If designing for a C++ implementation, the designers should address problems that can arise from assignment of class instances, either by the = assignment operator or by a "copy constructor". As explained in Chapter 5, assignment operations can lead to unintended sharing of data. These problems arise with "resource manager" classes whose data members include pointers to separately allocated structures in the heap, such as the character array referenced by instances of the class Player that was used to introduce the problem. (Implementation languages like Smalltalk, Eiffel, and Object Pascal, that use reference semantics for all objects, don't encounter this particular problem.)

An OO program will often have many such resource manager classes that are never expected to be used in assignments. Assignments involving things like Windows are rare! It may not be worthwhile implementing support for assignment. A class can simply be documented as not supporting assignment.; but, it may be wise to get the C++ compiler to enforce such a restriction. One can declare an assignment operator=() function, and a copy constructor, as private members of a class (while not providing any implementation) The compiler will then prevent any mistaken attempts to assign instances of such classes in code where they are used.

const

The designers should consider the support of const instances of classes. It is usually only things like Points and Rects that have const instances; there aren't that many occurrences of const Window, const ObjList, const Tracker etc. in the typical OO program. But class declarations do become a bit clearer if those member functions that don't change the state of instances of a class are declared as being const functions (e.g. Rect Window::GetExtent() const { …}).

Of course, if a member function returns a pointer to a data member, the returned type should be const (because otherwise the encapsulation on the object can easily be breached). For example, if class Player does have a char* name data member, the member function that gives access to this name should return a const char*:

```
class Player {
public:
    Player();
    ~Player();
    ...
    const char* Name() const { return fname; }
    ...
private:
    char*  fname;
    ...
};
```

The return of even a const pointer weakens encapsulation (the "const"-ness can be type cast away exposing the structure). So be cautious.

Views

If it is to be a WIMPy program, the designers must decide on the display structure(s). This may be a matter of creating a data resource (e.g. with ViewEdit for MacApp) that will be used in the implementation. Alternatively, the designers may simply provide a sketch of the nested view structures, something similar to the kind of sketch shown in Figure 10.9.

References or explicit pointer arguments

In most situations, it doesn't really matter whether a function signature is defined using reference arguments or explicit pointer arguments. However if the styles are intermixed, the code tends to look a bit messy. It is probably wise for the designers to specify a preferred style for the implementation.

Memory management

Memory management is easy in an environment that provides for automated garbage collection (Simula, Eiffel, Smalltalk, CLOS etc.). For C++ (also, for Objective C and the various object-oriented Pascal dialects) memory management becomes a significant design issue. The designers must specify how the program disposes of objects.

Memory leaks, resulting from a failure to release space associated with discarded objects, are common in implementations of OO programs. A program with a memory leak may seem to run correctly; but, sooner or later, the program will mysteriously run out of heap space and fail during full scale use.

Some debugging aids and environments can help the implementor find leaks. For example, the ET++ memory allocator (which substitutes special implementations for C++'s standard new and delete operators) effectively keeps track of all objects that were created and can generate a report identifying those objects that were not properly freed before a program terminates. However, it shouldn't really be the responsibility for the implementors to have to find and plug leaks; the designers should have come up with a no leak system.

11.4 IMPLEMENTATION

If the design phases have been done properly, implementation should be trivial. That statement should probably be qualified by one of the "smileys" that one sees in Internet messages :-). But, really, if the design has been carried through correctly, the worst problems facing the implementors should be errant semi-colons etc. leading to a few syntax errors at compile time.

The class designs should map easily into C++ class declarations. The algorithms for the methods should be complete so that implementation involves simply some straightforward coding. The most common run-time problems with OO programs seem to relate to missed connections – one object tries to communicate with another but its pointer is incorrect and it calls a non-existent collaborator. But, if the designers have followed through object life histories, they should have covered all the ways in which objects can be created and then get linked to their collaborators – there should be no missed connections.

Additional classes added during implementation   The implementors will utilize many simple, concrete classes taken from class libraries. These concrete classes serve simply to package standard data structure and provide a few useful routines for utilizing such structures. Favourite examples are a class `Date` that packages date-and-time services such as are commonly provided by an operating system, or a class `Point` that packages the concept of a point in 2-space. The use of such classes in the implementation should be noted in an appendix to the design level documentation.

11.5 GENERALIZING TO CREATE A NEW CLASS LIBRARY?

Unfortunately, generalization of code to form the basis of a specialized class library is a little harder than you may have been lead to believe. Firstly, the world has no need for another GUI class library, nor any need for the Smalltalk collection classes to be reimplemented yet again. Any new library created by a company is going to have to be targeted towards a specific vertical market – some area in which that company specializes. Secondly, even within a relatively narrow vertical domain, work on a single product will not provide a sufficiently broad view for developers to identify those abstractions that are most useful for that domain. Generally, it is necessary to accumulate experience over about three products before it is possible to identify the classes that really are reusable.

For example, the company that developed Sue's chemistry editor might then start on a project that displayed drug molecules and drug receptor sites. Such programs are used in pharmacology research where drug-"designers" try to imagine molecules that are likely "fit into enzymes" etc. and so have desirable therapeutic properties; if a molecule looks the right shape, it can be synthesised and tested. Obviously, such a program uses "ChemStructures" – but they aren't quite the same, because in the second program the focus is more on three dimensional shape and the options for displaying and viewing structures. Still another vision of the "ChemStructure" abstraction would emerge from a project that aimed to create a reaction library that could be used to document known chemical transforms (and which would be used by the synthetic chemists to plan the synthesis of a molecule that was expected to be useful as a drug). A reasonable view of what a reusable "ChemStructure" component would only emerge after such accumulated experience. Once this generalized "ChemStructure" and related components had been identified, the company would start to get a useful specialized class library for further developments in this narrow vertical domain.

Such developments can be worthwhile. For example, data reported by Love show that Hewlett-Packard was eventually able to obtain a very significant amount of code reuse. This came in a series of products in the biomedical area. The software products controlled and analyzed the results from scanners and other bio-analysis devices. The common classes that were abstracted included user interface elements (specialized "dials" and other displays that showed technicians the states of the devices, and associated visual control elements that provided control over the devices), together with various classes for storing and manipulating image data etc. In some of the later software products developed for this domain, only 8% of the code was really new – the remaining 92% consisting of reused classes. Such results are probably somewhat exceptional; figures of around 50% reuse seem to be more common.

A company must provide resources for any class library project. It is after all an investment – one that will hopefully pay off in reduced development costs of future products. Budgeting the development is difficult – to what project should the initial costs be charged? The development will involve a class library group; initially, this group can acquire expertise in the framework library or any other standard class libraries that are to be used in projects. Each product development team should include a member seconded by the library group; this person would advise on the use of existing components. At the completion of version 1.0 of a product, a member of the development team should be transferred to the library group to explore generalizations of the classes developed for that product.

There are additional managerial problems associated with the generalization of code to form a class library. A company will end up with a class library group that busies itself abstracting generalized reusable classes from the initial versions provided by the different product development projects, while at the same time the teams that created these products are continuing with extensions and other "maintenance" work. Naturally, the result is several diverging sets of "improved" classes. At some stage in the development of an in-house specialized library, management may wish to have existing products retrofitted with the library classes instead of the original classes. Management is likely to find some interesting challenges in accommodating such work into limited "maintenance" budgets, and in dealing with personnel problems such as developers who feel proprietorial about their work.

11.6 BENEFITS?

> Brooks: *There is no single development, in either technology or in*
> *management technique that promises even one order-of-magnitude*
> *improvement in productivity, in reliability, in simplicity.*

On the data reported by Love, Hewlett-Packard once did get an order of magnitude improvement in a project. OK, that result is abnormal and the measure (only 8% of the code for a project being new) doesn't really indicate the true cost so the real productivity gain would have been lower. But, going object-oriented certainly has some benefits, these can include:

• A more uniform representation from analysis through to implementation (this reduces all those burdensome translation steps where a conventional approach requires changes of representation).

• Greater opportunity for the involvement of domain experts and users in analysis and design processes. (Anthropomorphic descriptions of software entities may seem disgraceful to computer scientists, but domain experts can usually relate fairly easily to a ChemStructure that "knows" how to draw itself, add on atoms etc. Because the analyst and domain expert share a language that can be used to describe the desired system, the expert can be more readily integrated with the development team.)

• Less code needs to be written for equivalent functionality (remember all those places where you just use an instance of a standard list class etc.).

• Reduced internal complexity in code (remember how all those switch selection statements get eliminated).

• Simpler extensions (the original specification is not repeated everywhere in the code).

• Easier maintenance (the encapsulation provided by classes means that changes to the implementation of one class will not impact the rest of the system).

It would be nice if such benefits could be quantified, but obtaining the data would be difficult (one would require several systems, each implemented using OO and an alternative technology). Making a few wild guesses, the improvements might be:

• Reduction in the amount of new code required for an application. Code savings must be worth something; but it is probably the saving on design time that one can get from frameworks etc. that counts more.
(say a factor of 1.5 for reused designs, and 1.25 for reuse of code)

• Users can become more integrated into the development process and so
a) are more likely to get what they really want, and
b) will have some understanding of the implications of any new demands that they may make

(maybe that constitutes about an improvement factor of ≈1.25 ?).

- Localization of data access (hence opportunity for errors) through encapsulation and simplification of control flow; both of these help the maintenance programmer
  (worth 10%? i.e. a factor of 1.1).

- Another factor of 1.5 might be accorded to the greater ease of extension consequent on the abstraction away of the details of the original specification.

Compounded together, these should give a factor of almost 4 for improvements. Be cautious, scale it back to "definitely at least a factor of 2 improvement". It is no order of magnitude improvement, but it should be worth something. Managers should be interested in a doubling of productivity – software professionals would be interested in a doubling of pay.

REFERENCES

On first completing an introductory OO course, such as presented in this book, beginners will have to deepen their knowledge of an implementation language. As C++ is the implementation language that they will most likely use in industry, beginners should explore other features of this language (as presented in standard texts such as Stroustrup's or Lippman's) and, then look at some of the more sophisticated uses of these features (as illustrated in Coplien's text).

On a longer term basis, beginners will need to read widely on ideas for OO analysis and design. If the design is correct and contains sufficient detail, implementation is easy. The important skills to acquire from further study are in OO design.

Stroustrup's text contains some suggestions on OO design. Bar-David's book illustrates design suggestions with some examples, but the treatment is not really much deeper than that in this chapter and in Chapter 10. There are also some suggestions on design in Meyer's book on OO software construction.

The books by Wirfs-Brock *et al.*, and by Henderson-Sellers are probably the starting points for further reading. The approach in the Wirfs-Brock book may seem a little naive in places (it is the approach of underlining nouns etc. to identify objects) but the book does provide an easily read overview of OO design. The Henderson-Sellers book is essentially a printed version of a one day industrial tutorial on OO, focussing mainly on analysis and design. It comes complete with printed versions of the overheads from the tutorial. The book offers a cheap way of gaining the benefits from such a course.

The approach taken by Coad and colleagues (three books, OOA, OOD, and OOP) is quite different from that suggested in this chapter. In Coad's approach, classes come first. (Classes appear in Chapter 3 of the analysis book where one finds Classes-&-Objects. Class hierarchies and compositional structures are resolved in Chapter 4. But, it is only in Chapters 6 and 7 that one starts to look for the attributes and services provided by these already identified classes.) A notational system, of some subtlety, is proposed; this notation is supported through a commercially available tool that allows boxes, arrows and labels to be combined to document the results of the analysis and design processes. Programs are expected to involve a "domain component", "a human interaction component", "a task management component", and "a data management component". The OOD volume covers design issues related to these separate components. The OOP book has a series of increasingly complex examples that are each implemented using both Smalltalk and C++.

Booch's original (1991) book *Object-Oriented Design with Applications* covers concepts, "method", and applications. The "method" section presents another notational system; this one is intended for documenting designs (and, again, it is supported through commercially available design tools). The notational system may be suited to use in major systems, but was too complex and subtle for use in an introductory text like this (e.g. there are ten different kinds of arrow with distinct meanings determined by the form of the arrowhead and the style of shaft). The "method" section also covers pragmatics (essentially issues of project management) and the process. The description of the OO design process is a bit brief. The applications section contains detailed examples illustrating design ideas in the context of different applications. Each example application is also used to illustrate a particular implementation language – Smalltalk, C++, Object Pascal, CLOS, and Ada (for an object-based rather than object-oriented example). A second edition expands to cover analysis and provide more detail on the process and pragmatics of design. The example applications in the second edition all use C++.

The book by Goldstein and Alger is worth reading at some stage. (The bit in the book's title about "... *for the Macintosh*" is misleading; the only Macintosh specifics are a few screen dumps and the use of the word Macintosh where others would use computer. This OO design book was published as part of the Inside Macintosh series, so some attachment to the Macintosh was probably a requirement.) This book covers some higher level analysis issues. For example, it goes into the need for the analysts to document the clients' existing problem solving processes and to provide a mapping from these to the proposed computer-based approaches; such things being necessary to plan retraining and to develop documentation. Somewhat less sanguine than the other texts, this book explores some of the problems with OO (like why it is often difficult to find the right classes and why one's "classes" often turn out to be categories that are of no use in design and implementation). A methodology – "Solution Based Modelling" – is proposed. The methodology works in terms of "planes" – the "business plane", the "technology plane", the "execution plane", and the "program plane". The business plane describes the problem in terms of the existing and proposed solutions; the proposed solution is mapped into the technology plane where the user interface and main run-time objects get described. The execution plane presents more detail on relations among objects, calling sequences etc. The book introduces the idea of "calibration" (essentially, a way of tying together concepts and their implementation and providing traceability). Goldstein and Alger make heavy use of scenarios that map out interactions among objects, with these scenarios forming a major part of the design documentation. Another notational system is presented (again, available as a commercial package); this system explicitly supports the sketching of scenarios for object interactions.

Berard has published a set of essays on aspect of OO and software engineering. Some review different methodologies; others attempt to clarify ideas such as "object cohesion" or the differences between "encapsulation" and "information hiding".

The books by Rumbaugh *et al.* and by Jacobson *et al.* are both a bit more advanced. These books present work of industrial developers, and both are based on years of extensive practical experience. Jacobson's "use cases" are related to the scenarios suggested in this chapter, or those of Goldstein and Alger.

References

- Alger, J., *Object-Oriented Software Engineering: Seminar for MacApp Developers Association*, MacApp Developers Association, 1991.

- Bar-David, T., *Object-Oriented Design for C++*, Prentice Hall, 1993.

- Berard, E.V., *Essays on Object-Oriented Software Engineering*, Prentice Hall, 1993.

- Booch, G., *Object-Oriented Design with Applications*, Benjamin-Cummings, 1991.

- Booch, G., *Object-Oriented Analysis and Design with Applications*, Benjamin-Cummings, 1993.

- Brooks, F.P., *No Silver Bullet: Essence and Accidents of Software Engineering*, IEEE Computer, April 1987, pp. 10-18.

- Coad, P., and Nicola, J., *Object-Oriented Programming,* Yourdon Press Computing Services (c/o Prentice-Hall), 1993.

- Coad, P., and Yourdon, E., *Object-Oriented Analysis* (2e), Yourdon Press Computing Services (c/o Prentice-Hall), 1991.

- Coad, P., and Yourdon, E., *Object-Oriented Design*, Prentice-Hall, 1991.

- Coplien, J.O., *Advanced C++*, Addison-Wesley, 1992.

- Goldstein, N., and Alger, J., *Developing Object-Oriented Software for the Macintosh*, Addison-Wesley, 1992.

- Henderson-Sellers, B., *Book of Object-Oriented Knowledge*, Prentice Hall, 1991.

- Jacobson, I., Christerson, M., Jonsson, P., and Övergaard, G., *Object-Oriented Software Engineering*, Addison-Wesley, 1992.

- Lippman, S.B., *C++ Primer*, 2e, Addison-Wesley, 1991.

- Love, T., and Phillips, R., *Choosing Object-Oriented Methods: Seminar for Hewlett Packard*, video "published" via Journal of Object-Oriented Programming, 1990.

- Meyer, B., *Object-Oriented Software Construction,* Prentice-Hall, 1988.

- Rumbaugh, J., Blaha, M., Premerlani, W., Eddy, F., and Lorensen, W., *Object-Oriented Modelling and Design*, Prentice-Hall, 1991.

- Stroustrup, B, *The C++ Programming Language*, 2e, Addison-Wesley, 1991.

- Wirfs-Brock, R., Wilkerson, B., and Wiener, L., *Designing Object-Oriented Software*, Prentice-Hall, 1990.

Intermediate C++

12

This chapter introduces a few of the more sophisticated features in C++. Some of these features, e.g. templates, have only recently become generally available with the release of C++ version 3 compilers. Some, e.g. exceptions, are not yet uniformly available. Most require somewhat greater skill on the part of the programmer and are probably best deferred until beginners have acquired more experience with the straightforward features of C++. Template functions are possibly an exception; they are worth adopting immediately for any procedural style coding. Template functions are relatively simple and their use can considerably improve the intelligibility of "generic functions" (e.g. generic sorting functions) as used in conventional procedural programming.

Template functions are the first extension presented (in section 12.1). Section 12.2 illustrates template classes. The use of template classes is fairly straightforward however their design presents some challenges. Nested classes are illustrated in section 12.3; they offer an alternative to "private classes" (such as the auxiliary Link class that formed a part of the LinkedList cluster introduced in section 5.4). Section 12.4 covers a few minor issues related to inheritance and access controls. A few examples of the use of multiple inheritance are considered in section 12.5. The final section provides an overview of exception mechanisms.

12.1 TEMPLATE FUNCTIONS

It is possible to write "generic" functions in C. These are functions that implement a generalized version of an algorithm in such a way that they can be applied to different types of data. A well known example is the qsort() function in the standard Unix C-library; this function implements a standard sort algorithm in a way that allows it to be used for arrays of any type of data for which the user can define a suitable comparison function.

"Generic functions" in C

However, the coding style for such generic functions in C is obscure, relying on arguments that are pointers to functions etc. The following example code implements a simple shell sort using function pointers and other generic C coding styles (e.g. the treatment of any data type as an array of characters). The code starts with a typedef which introduces the name CompareFn – a CompareFn variable will be a pointer to a function that will return an integer and which will require two pointer arguments of some kind:

```
typedef int (*CompareFn)(void*,void*);
```

The Sort routine takes as arguments a pointer to the start of the array, the number of bytes that each array element requires, the number of elements to be sorted and a CompareFn pointer to the function that is to be used to compare any two elements from the array:

```
void Sort(void* a,int elemsize, int N,CompareFn pf)
{   // a shell sort
    // sorts data in "a"; each data element is "elemsize" ;
    // bytes, there are "N" of them;
    // "pf" points to a function that will compare them.
    // Implementation assumes that can treat data as
    // sequence of characters (should work on all
    // reasonable machines).

    // forward reference to auxiliary routine that will be
    // used to copy array elements:
    void CopyChars(char* to, char* from, int num);

    // rest of code is a more or less standard implementation
    // of a "shell sort".  It "k" sorts the array
    // repeatedly for different k values down to k=1 (when
    // array is completely sorted).  Each step
    // creates a sorted subset within the overall array,
    // with elements k-places apart.

    int         i,j,k;
    char*       temp; // Pointer to space to store an elemen·
    char*       aj;   // Pointers to array elements
    char*       ajmk; // that are being manipulated.

    // Pick starting value of k for comparisons,
    // array element [i] versus [i + k]
    for(k = 1; k<= N/9; k = 3*k + 1);

    temp = new char[elemsize];//space to hold data during swap

    for(; k>0;k/=3)     // repeat until 1-sorting, k=1
        for(i=k;i<N;i++) { // loops to k-sort data
            j=i;
            aj = (char*)a + elemsize*i; // ->a[j]
            CopyChars(temp,aj,elemsize);

            ajmk = (char*)a + elemsize*(j-k); // ->a[j-k
            while((j>=k) && (*pf)(ajmk,temp)) {
                CopyChars(aj,ajmk,elemsize);
                j -= k;
                aj = ajmk;
                ajmk = (char*)a + elemsize*(j-k);
                }
            CopyChars(aj,temp,elemsize);
```

```
            }
        delete [] temp;
    }
```

*Pointer arithmetic
and indirect function
calls*

The body of the function is replete with C idioms, such as getting an address pointer to an array element by performing address arithmetic (with the odd type cast from void* to char* thrown in) e.g. aj = (char*)a + elemsize*i; (i.e. a[i]). The call to the comparison function is a little opaque (*pf)(ajmk,temp) (i.e. use the function pointed to by pf and apply it to arguments ajmk and temp both of which are char* pointers that can be automatically coerced to void* pointers). The implementation needed an auxiliary CopyChars() routine that could copy data elements in the array – treating each data element as an array of characters:

```
void CopyChars(char* to, char* from, int num)
{
    // needed by the shellsort routine to copy elements
    int i;
    for(i=0;i<num;i++)
            *to++ = *from++;
}
```

Use of this sort routine necessitates the definition of some comparison routines:

```
int IntGT(void* a, void* b)
{
    int x = *((int*) a);
    int y = *((int*) b);
    return x > y;
}

int DoubleGT(void* a,void* b)
{
    double x = *((double*) a);
    double y = *((double*) b);
    return x > y;
}

int main()
{
    // Code works with two arrays gData[SIZE] and
    // gRData[SIZE] that contain longs and doubles
    // respectively
    …
    Sort(gData,sizeof(long),SIZE,IntGT);
    …
    Sort(gRData,sizeof(double),SIZE,DoubleGT);
    …
}
```

Such code is usable – as evidenced in the work of tens of thousands of programmers diligently coding C for the past twenty years. However, it does not rate highly for intelligibility. Templates can be used to implement an alternative, more readable routine.

Template functions don't offer anything totally new – they are simply a "better" way of doing the same thing as could be achieved using C-style generic functions. They are "better" in that the programmer can delegate all the grunge work of dealing with pointers to functions etc. to the compiler and can simply encode the general algorithm. A template version of the same sort routine is:

```
template<class Data> void Sort(Data a[],int N)
{
    // a shell sort
    // sorting data in array "a"; it has "N" of elements; "
    // Code is a more or less standard implementation
    // of a "shell sort".  It "k" sorts ...
    // ...
    // elements k-places apart.
    int     i,j,k;
    Data    temp;           // Space to store a data value
    // Pick starting value of k for comparisons,
    // array element [i] versus [i + k]
    for(k = 1; k<= N/9; k = 3*k + 1);

    for(; k>0;k/=3)     // repeat until 1-sorting, k=1
            for(i=k;i<N;i++) { // loops to k-sort data
                    temp = a[i];
                    j=i;
                    while((j>=k) && (a[j-k]>temp)) {
                            a[j] = a[j-k];
                            j -= k;
                            }
                    a[j] = temp;
                    }
}
```

The function declaration, `template<class Data> void Sort(Data a[], int N)`, specifies that it will have a generic implementation that the compiler is to specialize to handle whatever types of "`Data`" are actually used.

In this case, there is only one type parameter in the declaration, `<class Data>`; the implementations of this function will vary only in respect to the type of the array `a[]`. More complex functions can be parameterized with types for several different data elements used in their implementation (e.g. `template <class OutputType, class Displayer> ShowInfo(OutputType x) { … Displayer y; … }`).

The body of the template function does not contain any complexities – it reads just the same as a text book sort routine for a built in type such as `int`. A variable of type `Data` is declared, the code uses a `>` operator to compare two elements of type `Data`, and standard assignment statements are used to copy a value from one array element to another. (It is assumed that `Data` is a relatively simple data type for which only one ordering function `>` can be defined.) It is left to the compiler to bother about how much space will need to be set aside for the variable `temp`, and what code to emit for the `operator>()` and `operator=()` functions.

Such issues are addressed when the C++ compiler finds a call to the `Sort()` function:

```
long        gData[SIZE];
double      gRdata[SIZE];
CString     gSdata[SIZE];
Point       gPdata[SIZE2];

int main()
{
    ...
    Sort(gData,SIZE);
    ...
    Sort(gRdata,SIZE);
    ...
}
```

"Instantiating" specialized versions of a template function

When parsing these calls, the compiler will note that it will have to "instantiate" (make specialized versions of) the `Sort()` function that will handle arrays of integers and doubles. In these cases, it will use built-in operators for '=', '>', etc. The integer version of the sort routine will reserve space in its stack frame for an integer `temp`; the double version of the routine will claim space sufficient for a double (usually, this would require at least twice as much space in the stack).

The same template sort routine can also be instantiated for any programmer defined `struct X` or `class X` – provided that X defines a default constructor (used to create `Data temp`), and the functions `X& X::operator=(const X&)`, and `X::operator>(const X&)`. So, if the program also needs `Point`s:

```
struct Point {
// These get sorted by distance from origin
    Point() : fx(0), fy(0) { }
    Point(int x, int y) : fx(x), fy(y) { }
    int operator>(const Point&);
    int fx;
    int fy;
};

int Point::operator>(const Point& other)
{
    long d1 = fx*fx + fy*fy;
    long d2 = other.fx*other.fx + other.fy*other.fy;
    return (d1 > d2);
}
```

a call to `Sort()` with an array of `Point`s, e.g. `Sort(gPdata,SIZE2)`, will result in another version of the function being instantiated. This version will use the function `Point::operator>(const Point&)` to implement the comparison test (`a[j-k]>temp`). (Struct `Point` does not define `operator=()` function because the version provided automatically by the compiler will have the appropriate behaviour.) The SortTest program, which is among the examples provided, illustrates the `Sort()` function applied to `int`, `double`, `Point`, and `CString` data (class `CString` owns a separately allocated array of characters and so has to define a real `operator=()` function).

Instantiation of versions of `Sort()` for different data types does involve some duplication of code which does not occur with the C-style implementation with its pointers to functions. This cost in additional storage is usually not significant. (That sort routine compiled to about 200 bytes of code on one of the platforms on which it was tested. It really didn't matter that more than one copy was made; besides, if one is quibbling over a few bytes of code, it could be pointed out that it wasn't necessary to define functions like int DoubleGT(…) etc.)

GetData() a generalized input function

A second rather different example is routine GetData() defined below. Often, a program will require code that prompts for an integer (or a real number, or a point, or whatever), and if the input is erroneous has to clear the input stream and repeat the prompt and the read operation. When a program is first written, such input handling code is typically repeated, with minor variations, at every point where input data is needed. Of course it is better to package such code in a subroutine. Templates permit the definition of a single routine that can prompt for and get any kind of data. A simple version is:

```
template <class X>
void GetData(X& val,const char* prompt, const char* errmsg,
        const char* reprompt)
{
    cout << prompt;
    cout.flush();
    for(;;) {
        cin >> val;
        if(!cin.fail()) {
            // Got an OK value
            // Flush anything left in input buffers
            // (? should this be
            // responsibility of this function?)
            if(cin.peek() != EOF)
                cin.ignore(SHRT_MAX,'\n');
            return;
            }
        // OK, something wrong
        if(cin.bad()) {
            cerr << "Input corrupted\n";
            exit(1);  // Can do better - e.g throw a
                    // "bad input" exception,
                    // see section 6 below
            }
        if(cin.eof()) {
            cerr << "Premature termination of input\n";
            …
            }
        // If not end of file, tidy up and reprompt
        if(errmsg != NULL)
            cout << errmsg;
        cin.clear();
        cin.ignore(SHRT_MAX,'\n');
        if(reprompt != NULL)
            cout << reprompt;
        cout.flush();
        }
}
```

Versions of this routine can be instantiated for standard built in types such as `int` or `double` and for any programmer defined `class X` or `struct X` that has an associated global function `istream& operator>>(istream&, X&)`.

GetData() requires an istream& operator>>(...) function
Handling input errors

The arguments are a reference to a variable of the appropriate data type, and various `char*` prompt strings. The first prompt is printed and the output buffer flushed so as to guarantee that the prompt appears before the program waits for input. (The explicit call `cout.flush()` could be omitted if the input and output streams have been "tied" – see details of the iostream library in any C++ compiler's reference manual.) The statement `cin >> val` will compile to a call to the `operator>>(istream&,X&)` function appropriate to the data type for which the routine has been instantiated. If there are no errors (checked with `cin.fail()`), the data value is returned (via the reference argument) after any trailing characters have been removed from the input buffer (the `cin.ignore()` call will clear any characters up to the next newline). If errors did occur then, provided end-of-file has not been signalled, the input stream can be reset (`cin.clear()`), any bad data drained out (`cin.ignore()`) and the user prompted to try again.

Examples of input functions for programmer defined types

The input functions that would have to be defined for classes and structs would be similar to those illustrated in Chapter 5. Thus, for `struct Point` as defined above, an input function that could read point data such as (3, 4) would be:

```
istream& operator>>(istream& ins,Point& x)
{
    // find opening left parens
    ins.ignore(SHRT_MAX,'(');
    // get x coord
    ins >> x.fx;
    // find separating comma
    ins.ignore(SHRT_MAX,',');
    // get y coord
    ins >> x.fy;
    // consume closing parens
    ins.ignore(SHRT_MAX,')');
    return ins;
}
```

while a simple `CString struct` with the declaration:

```
struct CString {
    CString() : fc(0) { }
    CString(char* c) : fc(c) { }
    int operator>(const CString&);
    char* fc;
};
```

might have an input function something like the following:

```
istream& operator>>(istream& ins,CString& x)
{
    const STRMAX = 255;
    char    buff[1+STRMAX];
    cin.get(buff,STRMAX,'\n'); // Read, include whitespace
    int numread = cin.gcount(); // How many characters read?
```

```
                    buff[numread] = '\0';   // make sure its null terminated
                    if(x.fc != NULL)
                            delete [] x.fc;       // avoid memory leaks
                    x.fc = new char[1 + numread];
                    ::strcpy(x.fc,buff);
                    return ins;
                }
```

An example program that instantiates several variants of GetData()

Given these definitions, the compiler will instantiate different versions of the `GetData()` routine for each different data type used. Thus, for the following test program, there would be specialized `GetData()` routines for integers, double precision reals, `CString`s and `Point`s.

```
int main()
{
  for(;;) {
    int i1;
    GetData(i1,"Enter an integer\n","bad data\n",
          "I want an integer\n");
    double r1;
    GetData(r1,"Enter a real number\n","didn't like input\n",
          "I want a real\n");
    CString sp;
    GetData(sp,"Enter a string\n",(char*) NULL,
                      "I want a string\n");
    Point p1;
    GetData(p1,"Enter a point\n",
            "(a point has two integers with formatting,"
          " e.g.(3,4))\n", "try again\n");

    ...
    }
}
```

The coding of a generalized `GetData()` function is made quite simple by C++'s templates, overloaded >> stream operators, and type secure stream input functions.

12.2 TEMPLATE CLASSES

12.2.1 Class buffer: a simple template class

The `<class ...>` parameters for template functions allow a compiler to instantiate different versions of a function that are specialized for different types of data. Similar parameterization with template classes allows a compiler to create (instantiate) a version of a class, e.g. some container class, that is specialized in respect to the types of some of its data members and the return types of some of its member functions.

Circular buffers

The following example illustrates a simple storage structure – a *circular buffer*– that is parameterized to allow versions to be instantiated that can store different data elements (such integers, characters, etc.). Circular buffers are most commonly encountered at relatively low levels in computer systems for their main role is to "buffer" different rates of input and output, as illustrated in Figure 12.1.

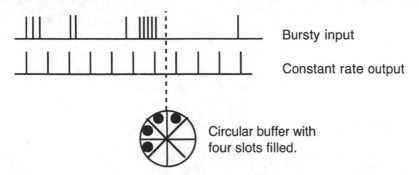

Circular buffer with four slots filled.

Figure 12.1   Circular buffer.

Often, a simple computer system will have input devices that are bursty and output devices that are constant speed. For example, one might have a microprocessor controlling a remote printer attached via modems and telephone lines to a Unix host. The host machine will typically send a burst of characters in a fraction of a second, and then ignore the controller for the next couple of seconds. A simple dot-matrix printer might be capable of printing at a constant rate of 30 characters per second. A circular buffer would constitute a reasonable storage structure in which characters arriving from the host machine could be queued up for later consumption by the printer. Other applications might require similar circular buffers to store integers (e.g. integer values from some analog-to-digital input system attached to a logging device) or other data. So, a generalized, parameterized Buffer class could be useful.

The implementation of a circular buffer is usually based on an array with two *Circular buffer* indices that move cyclically around through the array entries. One index identifies *"design"* the array entry that is to be filled in next; the second index identifies the array position from which data was last taken. After an element is put into the buffer, the input index is incremented modulo the array length – thus making it cycle; removal of an element causes the output index to cycle in a somewhat similar fashion. In addition to functions to put an element into the buffer and to get an element from the buffer, class Buffer should also provide functions that a client can use to determine whether it is full or empty. Handling the buffer correctly is the responsibility of the client. There is little purpose in the buffer code trying to deal with a "put" operation when it is full – about all it can do is terminate execution. The "contract" with the client should require that the client check that the buffer is not full before putting data (or, if the buffer is full, it is the client that takes responsibility for discarding data).

Class Buffer should be parameterized according to the type of data elements *Template class Buffer* that it stores in its array and, possibly, its size. Parameters used to characterize template classes (and functions) don't have to be data-types; expressions that can be evaluated at compile time can also be used. The following declaration defines a version of class Buffer parameterized by the data type of the elements (class Item parameter) it stores and its size (int BuffSize parameter). This version of class

Buffer includes extra data that allow a visual image of the buffer to be displayed by a test program.

```
template <class Item, int BuffSize>
class Buffer {
public:
    Buffer() : f1(0), f2(BuffSize-1), fcount(0)
            { InitImage(BuffSize); }
    ~Buffer();
    Boolean     Empty();
    Boolean     Full();

    void        Put(const Item&);
    Item        Get();
    void        PrintOn(ostream&);
private:
    void        InitImage(int);
    Item        fStore[BuffSize];
    int         f1, f2, fcount;
    char        fImage[BuffSize];
};
```

An instance of an instantiated (specialized) version of class Buffer will own an array that can hold BuffSize data elements of the appropriate type (most commonly characters or integers), the array indices and a counter of the number of elements currently stored, and an extra array fImage of characters. This extra array is only for use in the test program; it contains '*' characters at positions corresponding to filled buffer elements and spaces elsewhere allowing a simple visual display of the state of the buffer (e.g. the buffer in Figure 12.1 would be displayed as |***   *|).

Implementation for class Buffer

The implementation of the class is straightforward. Each member function definition repeats the specification of the template parameters:

```
template <class Item,int BuffSize>
Boolean
Buffer<Item,BuffSize>::Full() {
    return fcount == BuffSize ? true : false;
}

template <class Item,int BuffSize>
void
Buffer<Item,BuffSize>::Put(const Item& i)
{
    fcount++;
    fStore[f1] = i;
    fImage[f1] = '*';
    f1++;
    f1 = f1 % BuffSize;
}
```

Instantiating an instance of a parameterized class

An example program is provided that instantiates a version of class Buffer:. The program models a simple polling loop mechanism for copying data from an input source to an output sink. The functions InputReady(), GetChar() etc. are

provided in an auxiliary file (io.c); they provide a simple simulation of a bursty
input device and constant speed output device.

```
...
#include "Buffer.h"
extern char      GetChar();
extern int       InputReady();
extern int       OutputReady();
extern void      PutChar(char);

int main()
{
    Buffer<char,10>     littleBuf;
    for(int i=0;i<2000;i++) {
        if(InputReady())
            if(littleBuf.Full()) {
                cerr << "Input lost, buffer too small\n";
                exit(1);
                }
            else littleBuf.Put(GetChar());
        if(OutputReady())
            if(!littleBuf.Empty())
                PutChar(littleBuf.Get());
        littleBuf.PrintOn(cout);
        }
    return 0;
}
```

This code requires a version of class Buffer that has an array with ten elements for
character data – Buffer<char,10> littleBuf;. The compiler reading this
declaration will note that it must create this specialized version of the class. When
the file with the routine containing the main program with its declaration Buffer<
char,10> littleBuf; had been processed, an ATT C++ compiler would
proceed to read the file Buffer.c and create the required specialized version of the
code in a subdirectory. The compiled version of this instantiated version of class
Buffer would then be linked automatically with the rest of the program code.

Support for template classes in current compilers and linkers is somewhat
limited and there can be problems. For example, in a larger program composed of
many files one might find references to variables of type Buffer<char,10> in
several files. A compiler might instantiate the class a number of times (most can
avoid this extra work). Copies of the compiled code are quite likely to be included
with each file that uses the instantiated class. Some current systems have special
compiler/linker directives that can be used to arrange for the use of a single global
copy of the code of any particular instantiated version of a template class.

12.2.2 Class LinkedList and friends (again)

Class LinkedList (and its friends class Link and class LinkedListIterator) was
introduced in Chapter 5 to illustrate a simple cluster of classes whose instances
work together to provide some overall service to clients. The implementation

given in Chapter 5 used void* pointers to point to the data elements that were being stored in the list. The use of void* pointers has a number of disadvantages; the compiler cannot in anyway check whether a list is being used sensibly and type casts have to be used whenever data are retrieved. Type casts always increase the opportunity for errors.

A different version of the cluster was illustrated in Chapter 6. In the second version, the list held pointers to instances of class Item or to instances of any class derived from class Item. Class Item was defined with a few member functions (e.g. Compare() and Equal()) that were supposed to be given effective definitions in derived classes. This general style is typical of container classes as defined in most class libraries, including the NIH library, ET++, and MacApp. In all these libraries, there is some base class, class Object, from which all classes are expected to be derived; collection classes hold Object* pointers. Although useful for storing things such as Windows, or Controls, or instances of other framework classes, such collections have limitations. One cannot get a simple list of integers without going through the contortions of defining class IntegerObject : public Object { … }. Apart from the inconvenience, such artificial classes generally involve significant overheads that reduce the value of these library collection classes.

Template classes solve these problems. For example, a template list class (together with template versions of each other class in the cluster) can be specified that is parameterized by the type of the data element stored in the list. Specialized versions of the class(es) can then be instantiated to store integers, characters, instances of class MailItem, instances of class ChemStructure, or whatever other type of data element the programmer might wish to place in a list. Unlike the cluster in Chapter 6, there is no requirement that the things stored in lists be instances of classes derived from some artificial base class. Unlike the version in Chapter 5, the code can be checked by the compiler. If a version of a template list class is instantiated to hold MailItems, the compiler will object to an attempt to store a ChemStructure there. Anything removed from a list of MailItems comes back as a MailItem, and the compiler can check that it is then used appropriately.

Template (parameterized) version of LinkedList cluster

The class cluster illustrated below is a simple reworking of that given in Chapter 5. Apart from the use of templates, the only real change is in the style of the list iterator.

"forward declarations" of auxiliary template classes

The class declarations start with "forward declarations" for classes that will be declared as friends or used as data members; these have to be defined as template classes that are parameterized by a single data type parameter. Specialized versions of this cluster will be created with specific data types used for the parameter <class Datum>.

```
template <class Datum>
class LinkedList;

template <class Datum>
class ListIterator;
```

Template class Link

The first class declared for the cluster is class Link<Datum>. Its member functions are defined in the class declaration (the compiler that was used to develop this example had problems when loading, from a separate #include file, the definitions of in-line member functions of template classes).

```
template <class Datum>
class Link {
    friend class LinkedList<Datum>;
    friend class ListIterator<Datum>;
private:
    Link(Link<Datum>* prev,const Datum& datum,
            Link<Datum>* next)
            : fp(prev), fd(datum), fn(next) { }

    Link*       Prev() { return fp; }
    Link*       Next() { return fn; }

    Datum       GetDatum() { return fd; }

    void        SetPrev(Link *newl) { fp = newl; }
    void        SetNext(Link *newl) { fn = newl; }

    Link*       fp;
    Link*       fn;
    Datum       fd;
};
```

Externally, any uses of Link must be explicitly qualified with the <Datum> parameter. However within the class declaration and definition, it is sufficient to use Link – hence class Link<Datum> can declare its fp and fn (previous and next pointer fields) as being of type Link*. (Actually, some current compilers don't comply with this convention and will require the qualified names – like Link<Datum> – to be used in the declaration and the code of member functions.) Data elements are actually stored in the links themselves, hence the fd data member of type Datum.

The LinkedList class is really the same as that defined in Chapter 5 with the addition of the <class Datum> parameter and consequent change of data types of arguments and return values from void* to Datum. The fhead and ftail data members must be declared as being instances of Link<Datum>*:

Template class LinkedList

```
template <class Datum>
class LinkedList {
public:
    LinkedList();
    ~LinkedList();
    // Asking about list
    int         Length() const;
    Boolean     Find(const Datum& queryitem);
    Datum       FirstItem();
    Datum       LastItem();
    Datum       MarkedItem();

    // Put data in list
    void        Append(const Datum& datum);
    void        Prepend(const Datum& datum);
```

```
        void            PutBeforeMark(const Datum& datum);
        void            PutAfterMark(const Datum& datum);

        // Removing stuff from list
        Datum           RemoveFirstItem();
        Datum           RemoveLastItem();
        Datum           RemoveMarkedItem();

        friend class ListIterator<Datum>;
protected:
        Link<Datum>  *      Head() const;
        Link<Datum>  *      Tail() const;
private:
        Link<Datum>  *      fhead;
        Link<Datum>  *      ftail;
        Link<Datum>  *      fmark;
};
```

Changed design for a
list iterator   The iterator had to be replaced. The earlier design was based on the assumption that the list would store pointers (either `void*` or `Item*` pointers depending on the version). The `Next()` method of class `LinkedListIterator` returned the data item in its current link and moved to the next link; since the data item returned was a pointer, a `NULL` value could be returned when the list had been fully traversed. The style was thus like using a "sentinel" data value to indicate when a loop should terminate. Such a design is inappropriate for the template-based list. The data items stored in the list can be integers, `char*` pointers, instances of class `Point` or whatever else was desired; most of these types would not have any associated "sentinel values" that could be returned to indicate completion of the list-iteration. So, instead of a single method `Next()` that serves the multiple purposes of returning the next stored value, advancing through the list and indicating the end of the list, the substitute `ListIterator` class uses three distinct methods:

```
template <class Datum>
class ListIterator {
// Not robust.  No checks for running off end of list.
// (responsibility of client to honour contract not to
// Advance after More indicates no more items).
public:
    ListIterator(const LinkedList<Datum>& aList) :
            fp(aList.fhead) { }
    Boolean     More() { return fp !=0 ? true : false; }
    Datum       NextDatum() { return fp->GetDatum(); }
    void        Advance() { fp = fp->Next(); }
private:
    Link<Datum>* fp;
};
```

(The member functions are again defined in the class declaration for the convenience of the compiler that had problems with `inline` definitions held in a separate `#include` file.)

Implementation of
class LinkedList   Some examples of member functions from class `LinkedList` are:

```
template <class Datum>
int LinkedList<Datum>::Length() const
{
    Link<Datum>    *     l = fhead;
    int    res = 0;
    for(; l != 0; res++, l = l->Next());
    return res;
}

template <class Datum>
Boolean LinkedList<Datum>::Find(const Datum& query)
{
    // Find item and set mark
    // Don't change mark if item not present
    Link<Datum>    *     l = fhead;
    while(l != 0) {
            if(query == l->GetDatum()) {
                    fmark = l;
                    return true;
                    }
            l = l->Next();
            }
    return false;
}

template <class Datum>
Datum LinkedList<Datum>::MarkedItem()
{
    Datum         res;
    if(fmark != 0) res = fmark->GetDatum();
    return res;
}
```

Implicit requirements for Datum

The implementation code is essentially the same as that of the class defined in Chapter 5, with the substitution of the parameterized types for `void*` etc. The implementation uses functions such as `Datum::operator==(const Datum&)` (for the equality test in the `Find()` routine). Now while these functions are all defined built in types like `int` and `double`, they won't always exist in programmer-defined types. When instantiating a version of `LinkedList`, e.g. `LinkedList<Point>`, the compiler checks whether the type used to instantiate the template defines an `operator==()` function. (The definition for struct Point given in section 12.1 does not define `operator==()`, so a compiler would at some stage report an error associated with a declaration like `LinkedList<Point> ptList`.)

Apart from its use of `Datum::operator==()`, the implementation code also relies on the existence of an assignment operator and a default constructor. These requirements appear in routines such as `MarkedItem()`, shown above, which uses a local variable of type datum (needing a default constructor), an assignment, and also returns an instance of `Datum` as a result. Now these requirements present no difficulties when using `LinkedLists` of `ints`, `doubles` etc. However, there are potential problems if a program is going to have lists of `MailItems`, `Space-Invaders`, or `ChemStructures`.

*Problems when
designing a
parameterized class*
Even if all the required operators etc. are defined for the data type used when instantiating the parameterized class, there can be performance problems. A `ChemStructure` might be a 10,000 byte entity; not the sort of thing that one wants to casually copy around, or initialize using a constructor routine and then change by assignment. For example, the execution of the `MarkedItem()` routine would require initializing a `ChemStructure(Datum res)`, copying a `ChemStructure` during the process of `Link<Datum>::GetDatum()`, and performing an assignment that would again involve copying the structure; all at 10,000 bytes a time. Code written for the member functions of a template class involves assumptions about what operations are reasonable. When a template is instantiated, the compiler verifies that the assumed operations are feasible, but cannot comment on their reasonableness for the instantiating type.

In this specific situation, it would be more appropriate for a programmer using the template list class to instantiate it with `ChemStructure*` – so getting a list of pointers to instances of class `ChemStructure`. The `fd` data member in a `Link` would then be a `ChemStructure*`, as would variables like `res` in `LinkedList <Datum>::MarkedItem()`. There would be no problem in initializing these pointers (initial value `NULL`), assigning pointers, or return pointers as the results of functions. A `LinkedList<ChemStructure*>` list would perform perfectly reasonably. However, the approach of simply using pointers to complex data types is not a general solution.

For example, the list structure might be something a little more sophisticated e.g. a sorted list. Elements inserted in the list would be ordered. Rather than `Append()` and `Prepend()` functions, the sorted list might offer an `Insert(Datum newitem)` function. This function would involve iterating along the list to find the right position for the new item; at each step, the new item would be compared with the next item in the list (e.g. `if(newitem < p->GetDatum()) { /* insert before p */ … }`). When instantiating a version of a template sorted list class, a C++ compiler would verify that comparison operations were supported by the data type used. So, class `ChemStructure` would have to define an `operator<(const ChemStructure&)` member function (the implementation might for example compare the molecular formulae represented as character strings).

A program using a sorted list, `SortedList<ChemStructure> chemList`, would work; but it would run like an arthritic tortoise. All operations involving the function `Link<Datum>::GetDatum()` create a temporary copy of an object in the stack; thus, the comparison test in ordered insert routine would create a temporary copy of a `ChemStructure`, use it to compare formulae with the new item, and then throw it away (at a cost of 10,000 bytes copied as part of each comparison).

If, to improve the speed, the programmer instantiated the sorted list with `ChemStructure*`, the program would still execute, but it wouldn't perform quite as intended. When instantiating `SortedList<ChemStructure*>`, the compiler would find that it had to generate code for `newitem < p->GetDatum()` with the data types of the two elements compared being both `ChemStructure*` pointers. Generating code for that is easy; pointers can always be compared. However, the code generated will keep the `ChemStructures` ordered according to their locations in memory. Now, it is probable that on a small test the structures will, coincidentally, be correctly ordered by composition (structures created in sequence being located at increasing memory locations, first structure ethanol C2H6O, second hexane C6H12, third a decanol C10H20O etc.). Practical use of the program would reveal a problem in the handling of the structure list and maintenance programmers would be given the task of finding "the bug in the code of the function that computes and compares formulae".

It is difficult to design a template class that satisfies all potential users. One approach is to provide different versions of the template classes. This approach is used in the Borland class library (which supported templates somewhat earlier than most other environments). The template-based collection classes associated with the Borland library mostly come in pairs – a "direct" template class, and an "indirect" template class. A direct sorted list would have links that hold actual instances of the data type for which the list was instantiated – this would be fine for `SortedList<int>`, `SortedList< Point>` etc. An indirect sorted list would have links that hold either explicit pointers, or references, to the data elements. So, for example, the instantiated version of `IndSortedList<ChemStructure>` would effectively have links declaring `ChemStructure* fd` (Borland style would dictate a `ChemStructure& fd` reference). The implementations of the "direct" and "indirect" versions of the template class would be slightly different; for example, the argument for an insert function would be a `Datum*` (or `Datum&`) for the indirect class rather than Datum as in the direct implementation. If the indirect class used pointers, the comparison test discussed above would be coded as `if( *newitem < *(p->GetDatum()))  { ... }` (use of references would retain the original form of code).

"Direct" and "indirect" versions of the same template class

"Indirect" classes raise the issue of ownership. A direct sorted list will make copies of the data elements inserted, and can keep these copies long after the originals have been discarded. When a direct sorted list is deleted, it can destroy all its links with the confidence that this cannot leave any dangling pointers (as no other component in the system should hold a pointer to any of its data). An indirect sorted list has to share its data with other components in the system. It relies on the data objects being created correctly (don't put an automatic variable into a sorted list that has a longer lifetime). When an indirect sorted list is deleted, a decision has to be made whether to destroy the data items it holds.

Template classes are well worth using – once programmers have acquired sufficient sophistication to use them correctly. Template classes are a bit more difficult to design, and their design should be left to more experienced team members.

12.2.3 Class Bintree

The final example in this section shows a slightly more realistic use of templates with a binary tree class that is parameterized according to the types of the "keys" and "data values" that are stored in the nodes of the tree. Most binary trees are likely to contain simple data – integer keys, and strings for associated values. But, someone might want `Point` keys (e.g. positions on a map) and complex data describing the corresponding position (e.g. a "bit-image" for a photo or a descriptive text string as the "data value"). So, a general parameterized tree class might well be useful.

The "binary tree class cluster" comprises three classes: User_Rec, Node, and BinTree; all three are parameterized by the types for the "key" (<class Key>) and the data value (<class Value>). A User_Rec has a Key data member and a Value data member; these are set from values passed by reference in the constructor. Overloaded global functions of the form ostream& operator<<(ostream&, const X&) are expected to exist and are used in the implementation of operator<<(,) for the User_Rec class. When inserting a new entry into a binary tree, it is necessary to compare records (in accord with their keys). When searching for an entry (either just to find it or when intending to delete a record with a given key), one usually just has a key. So, class User_Rec defines overloaded Compare() functions that differ in their argument types but have similar implementations that compare the keys. The data type used to instantiate the <class Key> parameter is expected to provide operator==(), operator==()., and operator==() (both Keys and Values will have to have copy constructors needed the User_Rec constructor).

```
enum        boolean { False, True };

enum        Comparison { Less = -1, Equal = 0, Greater = 1 };

template <class Key, class Value>
class User_Rec {
public:
    User_Rec(const Key& k,const Value& v) :
                    fKey(k), fv(v) { }
    Comparison    Compare(const User_Rec<Key,
                        Value>& other) const;
    Comparison    Compare(const Key& query) const;
        ...
private:
    Key           fKey;
    Value         fv;
};
```

The tree will use indirect storage, i.e. its nodes will hold pointers to instances of User_Recs created in the client code. It has been assumed that the tree will "own" the records; consequently, when tree nodes are deleted they expect to delete their associated User_Recs (clients had better not pass addresses of User_Recs created in the stack or in the statics data segment). A Node naturally owns left and right pointers to other nodes in addition to a pointer to an associated User_Rec. Instances of class Node should only be used from within the member functions of class BinTree; so it would have been reasonable to make all members of class Node private and class BinTree a friend as was done with the Links for class LinkedList (this constraint is not included in the code illustrated). The tree-traversal functions in class BinTree will need to ask Nodes for their left and right pointers as well as asking a Node get its User_Rec to perform a comparison with another User_Rec or with a Key. The tree construction and modification routines may need to ask a Node to set a left or a right pointer or to replace an existing

pointer value with a new value. A `Node` can return a pointer to its `User_Rec`.
Member function `DoDelete_ThenDie()` gets invoked from class `BinTree`'s
destructor (illustrated below).

```
template <class Key, class Value>
class Node
{
public:

    Node(Node<Key,Value>* left,User_Rec<Key,Value>* data,
                Node<Key,Value>* right);
    ~Node();
    Node<Key,Value>*    Left() { return fLeft; }
    Node<Key,Value>*    Right() { return fRight; }

    Comparison    Compare(const Node<Key,Value>& other) const;
    Comparison    Compare(const Key& query) const;
    User_Rec<Key,Value>* Rec() { return fdata; }

    void    SetLeft(Node<Key,Value>* newnode)
                { fLeft = newnode; }
    void    SetRight(Node<Key,Value>* newnode)
                { fRight = newnode; }
    void    ReplaceLink(Node<Key,Value>* old,
                Node<Key,Value>* replacement);
    void    DoDelete_ThenDie();

    friend        ostream& operator<<(ostream&,
                const Node<Key,Value>&);
private:
    User_Rec<Key,Value>*        fdata;
    Node<Key,Value>*            fLeft;
    Node<Key,Value>*            fRight;
};
```

Class `BinTree` provides the basic `Insert()`, `Delete()` and `Search()` functions
expected of a binary tree. `Insert()` takes a pointer to a newly created `User_Rec`,
the other functions take `Key` values as arguments. If successful, the `Search()`
operation leaves a marker identifying the `Node` found to have a matching key (this
marker points to the `Node` with the new `User_Rec` operation after an insert
operation, and is `NULL` after a delete operation). Class `BinTree` would be expected
to provide functions allowing access to, or manipulation of the "marked" record;
this version has only a single member function – `PrintSelectedRecord()`. The
class also provides a simple inorder traversal `Print()` function and a crude `Draw()`
function.

```
template <class Key, class Value>
class BinTree {
public:
    BinTree();
    ~BinTree();

    boolean        Search(const Key&);
```

```
    boolean        Insert(User_Rec<Key,Value>*);
    boolean        Delete(const Key&);

    void           PrintSelectedRec();
    void           Print() { DoPrint(fRoot); }
    void           Draw() { DoDraw(fRoot,0); }
private:
    …
    Node<Key,Value>*     fRoot;
    …
};
```

The files of examples associated with this text include a program illustrating the use of `BinTree<Key, Value>` and the `GetData()` template function defined earlier.

12.3 NESTED CLASSES

It is quite common for small auxiliary classes to have to be introduced to serve in support roles for some principal class. Earlier examples have included class `Link` in the `LinkedList` cluster, and class `Node<Key,Value>` in the `BinTree<Key, Value>` example. Clients shouldn't make direct use of instances of these auxiliary classes. It is possible to enforce such a restriction by making all member functions of the auxiliary class private and arranging friend relationships for the other classes in the cluster (as was done for class `Link`). However, even if restrictions on access are imposed, client programmers are aware of the existence of these auxiliary classes – even if its only because their names exist in the global name space. If a program is using the `LinkedList` cluster, it cannot also have a different class `Link`.

```
class Link : public Telecoms {
// Instances of this class represent open communications
// channels in …
    …
};
```

The two `Link`s would clash in the global name space.

"Nested classes" provide one means of getting around such problems. The declaration of an auxiliary class can be nested within that of the major class that it supports.

Classes Graph, Node, and Edge

For example, one might have a system that needs a class `Graph`, with `Graph`s constructed from nodes (or vertices) and edges (or arcs). If operations on `Graph`s are going to involve only the manipulation of individual `Graph`s (e.g. "How many cycles does this `Graph` contain?", "What is the maximal distance across this `Graph`?" etc.) or comparisons of pairs of `Graph`s (e.g. "Is this `Graph` a subgraph of that `Graph`?"), then there is no reason for the nodes and edges to be apparent to the rest of the program. The classes used to represent the internal structure of a `Graph` can be nested:

```
class Graph {

    class Node : public Item{
    public:
            Node();
            Node(int n) : fNum(n) { }
            int         NodeNumber();

            virtual int  Equal(const Item*) const;
            virtual int  Compare(const Item*) const;
            virtual void PrintOn(ostream& os) const;

            ...
            ...
    private:
            int         fNum;
            ...
    };

    class Edge   : public Item{
    public:
            Edge(int nodeId1,int nodeId2);
            Boolean     Connects(int n1,int n2);
            Boolean     IncidentOnNode(int n);
            virtual int  Equal(const Item*) const;
            virtual int  Compare(const Item*) const;
            virtual void PrintOn(ostream& os) const;
    private:
            int         fnode1;
            int         fnode2;
    };

public:
    Graph();

    // Member functions of graph - cycle finding,
    // graph isomorphism, subgraphs, diameter, ...
    void  Show(ostream&);
    ...

private:
    // Support and housekeeping functions of Graph
    // Following are used to interactively construct a
    // graph:
    void        AddNode();
    void        DeleteNode();
    void        Join();
    void        UnJoin();
    void        Chain();
    void        Ring();

    ...

    LinkedList   fNodes;
    LinkedList   fEdges;
};
```

Enclosing class
Graph
Declaration of nested
class Node

Declaration of nested
class Edge

Class Graph itself
starts here

```
ostream& operator<<(ostream& os, Graph& g)
{ g.Show(os); return os; }
```

The code defines a graph of Nodes (that have as data only a distinguishing node number) and undirected Edges (that have as data the numbers of the Nodes that they join). The Graph maintains separate lists of Nodes and of Edges, using the LinkedList class cluster from Chapter 6. Class Node and class Edge have both been derived from class Item so as to make possible the use of the LinkedList class (which requires Item* pointers). The only functionality provided in class Graph is that needed to build a graph structure interactively by adding (and removing) either individual Nodes and Edges or larger components (such as "rings" containing several Nodes and their interconnecting Edges). The real services that a Graph should provide (checking for isomorphism, subgraphs, cycles etc.) are not actually implemented in the example.

Class Node and class Edge are both declared within the scope of class Graph. Consequently, if they are referenced elsewhere, the names Node and Edge must be qualified. For example, part of the code defining the functions of class Node is as follows:

```
Graph::Node::Node()
{
    fNum = ++sNum;
}

int Graph::Node::sNum = 0; // static data member

int Graph::Node::Equal(const Item* i) const
{
    Node* o = (Node*) i;
    return fNum == o->fNum;
}
```

Within the scope of class Graph, the names Node and Edge are "known" and don't require further qualification:

```
void Graph::AddNode()
{
    Node*  aNode = new Node;
    fNodes.Insert(aNode);
}

void Graph::Join()
{
    int nodeId1;
    cout << "Joining nodes, first node:\n"; cout.flush();

    if(!AskForNode(nodeId1)) return;

    int nodeId2;
    cout << "second node:\n"; cout.flush();

    if(!AskForNode(nodeId2)) return;
```

```
        if(nodeId1 == nodeId2) {
                cout << "No loops allowed\n";
                return;
                }

        Edge*  e = new Edge(nodeId1,nodeId2);
        fEdges.Insert(e);
}
```

In a function like `Graph::FindNodeNumbered(int)`, the code is within the scope of class `Graph` and so type `Node` is known, but the return value is specified in global scope and so must use the qualified name `Graph::Node`:

```
Graph::Node* Graph::FindNodeNumbered(int n)
{
    LinkedListIterator itern(&fNodes);
    Node* nd;
    while( nd = (Node*) itern.Next())
            if(nd->NodeNumber() == n) return nd;

    return NULL;
}
```

There are places where templates might have been used in this example. Template list classes (based on the version defined in section 12.2.2) should have been suitable for storing `Nodes` and `Edges`. Use of a template class would have avoided the contrived derivations `class Node : public Item`, and all the type casts associated with the used of the `LinkedList(Item*)` classes from Chapter 6.

Unfortunately, such usage was beyond the capabilities of the compilers used when this example was generated. The compiler could not instantiate a template class with a private class (so, `List<Graph::Node> fNodes` was invalid).

Actually, the template list class of 12.2.2 could be changed to use a nested class for its links (rather than the "private friend" class as introduced in Chapter 5). This would entail something like:

```
template <class Datum>
class List {
    class Link {
            Link(Link* prev,
                        const Datum* datum,
                            Link* next)
                        : fp(prev), fd(datum), fn(next) { }
            ...
    private:
            Datum* fData;
            ...
    };
public:
    IList();
    ...
};
```

Each instantiated version of List, e.g. List<Point>, defines its own nested class Link, e.g. a class whose global name would be List<Point>::Link.

12.4 ACCESS AND INHERITANCE REVISITED

The examples in this section illustrate a few minor aspects of inheritance. Firstly, the following classes serve to illustrate a couple of minor points relating to the names of class members:

```cpp
class BaseClass {
public:
    BaseClass(int a1 = 1) : f1(a1) {   }
    void         DoThis();
    void         Work();
    virtual void DoThat();
    ...
private:
    ...
    int          f1;
};

class DerivedClass : public BaseClass {
public:
    DerivedClass(char a1 = 'a') : f1(a1) { }
    void         DoThis();
    void         DoThat();
    ...
private:
    char         f1;
};
```

Multiple data members with "same" name

A first point to note is that it is perfectly valid for class DerivedClass to have more than one f1 data member. For the example code, a DerivedClass object will have an int f1 data member in its BaseClass part and a char f1 data member in the extra DerivedClass part. It should be obvious that the compiler will not confuse these (though this cannot be said with confidence about maintenance programmers looking after the code). After all, the int f1 defined in BaseClass is private and cannot be seen from within the code of the member functions of DerivedClass. A reference to f1 in any member function defined in BaseClass must mean then int f1; a reference to f1 in any member function defined in DerivedClass will mean the char f1.

What might be less obvious is that the code would still be valid if the protection access status of BaseClass::f1 were changed to protected (or even public). A change to protected access status would obviously make no difference to the code of class BaseClass. However, the int f1 would then be accessible in the code of member functions of class DerivedClass. Many earlier examples have included features like this, e.g. in chapters 3 and 6, the code of Saucer::Move() was able to modify the fh, fv protected data members of its parent Invader class. The difference here is that it seems that there could be an ambiguity – does f1 mean the int or the char variable.

There is actually no ambiguity, because within the code of DerivedClass the correct names for the variables are f1 (the char data member) and BaseClass::f1 (the int data member). If a member function of DerivedClass needs to modify the int f1 data member, it must refer to it as BaseClass::f1. (The code for Saucer::Move() should really have worked with the data members Invader::fh, and Invader::fv; but the compiler could understand the more casual, but still unambiguous use of fh and fv.)

Thus, it is possible for an instance of some class in a hierarchy to have many data members that have the "same" name, provided that the compiler can resolve these names through the use of the class names. Obviously, class DerivedClass wouldn't be able to have a declaration like:

```
class DerivedClass : public BaseClass {
    ...
private:
    double      f1;
    char        f1;
    ...
};
```

because the two f1 names introduced in the same scope would conflict. Similarly, the declaration of a function void BaseClass::ff() would conflict with the declaration of a data member int BaseClass::ff (but one could have void BaseClass::ff() and int DerivedClass::ff because these names are distinguishable).

The public/protected/private access status of the members is really immaterial. A compiler checks whether names can be disambiguated without taking access into account – because, otherwise, a change in access status for a member in one class could have unexpected side effects on some other class.

Although programmers are allowed to use the "same" name for members at different levels in a class hierarchy, such usage should be avoided. The use of similar names is confusing to humans, if not to compilers and their writers.

"Overloading" does of course permit multiple functions with the same identifier. Thus, class Node in section 12.2.3 defined two Compare() functions:

```
template <class Key, class Value>
class Node {
public:
    ...
    Comparison      Compare(const Node<Key,Value>& other) const;
    Comparison      Compare(const Key& query) const;
    ...
};
```

The compiler doesn't use the identifier, e.g. Compare, as the name of a function; instead, the compiler creates names based on a function's signature. A function's signature combines its identifier, details of the arguments, and, in the case of a member function, the class name. So really, the compiler sees these functions being declared with quite distinct names; names along the lines "Compare_

function_of_class_Node_with_reference_to_Node_argument" and "Compare_
function_of_class_Node_with_reference_to_ Key_ argument".

The use of default arguments may result in a class declaration containing
several variants for a single member function declaration:

```
class       Display {
public:
    ...
    // Same function, declared with different default
    // arguments
    void    DrawMessage(char* msg, int h, int v,
                            int size, int fnum = 1);
    void    DrawMessage(char* msg, int h, int v,
                            int size = 12, int fnum = 1);
    ...
protected:
    // overload function name
    void    DrawMessage(char* ErrorMsg,
                            RGBColor r = SCARLET);
    ...
};
```

It is common for the different overloaded functions to have different protection
levels. There could be a function used by clients and another used for some
internal purpose by instances of the class.

In addition to the overloading of function identifiers in a class with several
member functions that possess the same identifier but with different argument lists,
one gets a form of overloading in inheritance hierarchies where classes at different
levels define functions with identical argument lists. This, after all, is the
mechanism that has been used all along to provide polymorphic functions – such a
the DoThat() function in BaseClass and DerivedClass:

```
class BaseClass {
public:
    BaseClass(int a1 = 1) : f1(a1) {   }
    ...
    virtual void DoThat();
    void        DoThis();
    ...
private:
    ...
    int         f1;
};

class DerivedClass : public BaseClass {
public:
    DerivedClass(char a1 = 'a') : f1(a1) { }
    void        DoThat();
    void        DoThis();
    ...
private:
    ...
    char        f1;
};
```

The difference between the `virtual` function `DoThat()` and the non-virtual function `DoThis()` was elaborated on in Section 6.3.

As has been illustrated previously, it is possible for an overriding definition of a virtual function to invoke the implementation defined in its base class:

```
void DerivedClass::DoThat()
{
    // Do as base class does
    BaseClass::DoThat();
    // and now do some more …
    …
}
```

It is also possible for a client to explicitly request the `BaseClass` behaviour:

```
int main()
{
    DerivedClass d1;
    …
    d1.DoThat(); // do that properly as per derivedclass
    …
    // do just the base-class stuff this time.
    d1.BaseClass::DoThat();

    …
    // Similarly --
    d1.DoThis();   // go on, be a derived class
    d1.BaseClass::DoThis(); // but I know you are also a base
}
```

"Access"
declarations

C++ provides some deliberately restricted mechanisms for "adjusting" access to inherited members. A derived class cannot make an inherited member more visible than it was in the class where it was declared. This restriction is necessary because otherwise the concept of `private` members would become meaningless (a programmer wishing to violate privacy controls would simply invent a derived class wherein all the members has their access specifiers changed to `public`). Equally, a derived class cannot make an inherited member less accessible than it was when originally defined. If, for example, `DerivedClass` could make the inherited `BaseClass::Work()` function into a `private` member, then an inconsistency would have been introduced into the program. The compiler could prohibit code with an explicit request that a `DerivedClass` object execute function `Work()`:

```
class DerivedClass : public BaseClass {
    …
private:
    ?? An illegal change of access control!!
    BaseClass::Work;  // Trying to make Work a private member
};

int main()
{
```

```
DerivedClass*D1 = new DerivedClass;
...
D1->Work();    // If that access change were allowed,
               // compiler would have to report that
               // DerivedClass objects don't want to Work
```

but would allow "disguised" use of the function:

```
BaseClass*   ptr = D1;
ptr->Work(); // Force a derived class object to
             //    Work anyway
```

In summary, access declarations can only be used to keep accessibility unchanged from the base class. Of course, this is the normal situation, with *public* inheritance. If one has class DerivedClass : public BaseClass, then all the public members of BaseClass are public members of DerivedClass, and similarly all the protected members of BaseClass are protected members of DerivedClass. (The private members of BaseClass are private!) The only real role for access declarations is in the context of private inheritance. With private inheritance, the default is for all the public and protected members of the base class to become private members in the derived class. However, access declarations can be used to override this default behaviour and preserve accessibility of members.

For example, it is possible that a programmer, with access to the LinkedList code of Chapter 6, might want to make a Queue. The simplest mechanism involves the following definition:

Example "access" declarations"
```
#include "LinkedList.h" // the LinkedList cluster from c. 6
class Queue : private LinkedList {
public:
    LinkedList::Length;
    LinkedList::RemoveFirstItem;
    LinkedList::Append;
};
```

Class Queue now has exactly the functionality that it requires – a Queue can state its length, an extra Item* pointer can be appended at the rear, and an Item* pointer can be removed from its front. All the other methods of LinkedList, that would permit non-standard behaviours (e.g. "queue-jumping" with things taken from the rear etc.) are excluded. Although this arrangement works quite well for this example, it might still be better to have a class Queue that employs a LinkedList rather than the somewhat artificial inheritance structure shown.

12.5 MULTIPLE INHERITANCE

Multiple inheritance is not an essential feature for an OO language, as is evidenced by the fact that languages such as Smalltalk, the Object Pascal dialects, Objective C etc. don't use multiple inheritance. Multiple inheritance was available with the early object extended Lisp dialects ($\approx$1980) and in Eiffel ($\approx$1986) and became part of C++ with version 2 (≈1989). Although around longer than templates, C++'s version of multiple inheritance has probably had less impact on the way that the language is used than have the more recently introduced template facilities.

When designing a class, one is often faced with a trade-off between the use of inheritance and the use of "composition" (or "layering"). In the example at the end of the previous section, a new class Queue was to have some relation to an existing class LinkedList and an inheritance approach was illustrated (class Queue : private LinkedList { … }). A compositional approach would have had:

Composition as an alternative to inheritance

```
class Queue {
public:
    int    Length() { return fList.Length(); }
    void   Put(Item* i) { fList.Append(i); }
    Item*  Get() { return fList.RemoveFirstItem(); }
private:
    LinkedList    fList;
};
```

The inheritance technique used for this example was private inheritance. The implementor was relying on the services of LinkedList but did not want clients of class Queue to know of this relation. Composition is always an alternative to *implementation inheritance* (as in private inheritance). Usually, composition is preferable – it avoids unnecessary dependencies in the class hierarchies and clarifies the role of the classes (class Queue is "buying" the services of a LinkedList subcontractor).

Inheritance implies an *is_a* relationship; an instance of the derived class is, simultaneously, an instance of the base class and can be asked to do anything that a base class object can do. If multiple inheritance is used to create a derived class from several different base classes, then it should be appropriate to view an instance of the derived class as being an instance of any one of its base classes. If such an *is_a* relationship is not present, compositional techniques should be used in preference to inappropriate inheritance relations.

When can an object be both an "A" and a "B"? Some suggestions are presented in the following subsections.

12.5.1 Inheriting interfaces or combining "types"

One possible way for an object be both an "A" and a "B" is for the "A-ness" and "B-ness" to be simply type-descriptions i.e. lists of behaviours that will be exhibited by that object. One can define pure abstract classes that serve simply as type declarations:

```
class Storable {
public:
    virtual ~Storable() { }
    virtual long Size() const = 0;
    virtual void PrintOn(ostream&) const = 0;
    virtual void ReadFrom(istream&) = 0;
};
```

A "storable" type:

A "comparable"
type:
```
class Comparable {
public:
    virtual int  operator==(const Comparable&) const = 0;
    virtual int  operator!=(const Comparable&) const = 0;
    virtual int  operator<(const Comparable&) const = 0;
    virtual int  operator<=(const Comparable&) const = 0;
    virtual int  operator>(const Comparable&) const = 0;
    virtual int  operator>=(const Comparable&) const = 0;
    int Equal(const Comparable& other) const
            { return (*this)==(other); }
    int Eq(const Comparable& other) const
            { return this == &other; }
};
```

An "indexable" type:
```
template <class Thing1>
class Indexable {
public:
    virtual Thing1&  operator[](int) const = 0;
    virtual int MinIndex() const = 0;
    virtual int MaxIndex() const = 0;
};
```

Class `Storable` simply specifies that any "storable" object will be able to respond to requests for the number of bytes it will occupy on disk (`Size()`), to read its data from a stream, and to write its data to a stream. The specification of how these actions are performed is, necessarily, deferred to derived classes; but, all "storable" classes have the same interface (or, approximately, the same "type"). (The declaration includes a virtual destructor; it is plausible that a program might work with `Storable*` variables and want to delete a `Storable`.)

Similarly, a "comparable" object (an instance of any class derived from class Comparable) can respond to all kinds of comparison operator.

Class `Indexable` is meant to capture the notion of an entity that has some structure into which one can index and extract data elements of a particular kind. Examples of such entities are strings (indexing into a string gives a character) and vectors (indexing into a vector returns a specific data element from that vector). An "indexable" object can report the minimum and maximum values for its index (array subscript) and can respond to a `[]` subscripting operation by returning a reference to an indexed element. (Returning a reference allows the subscripting function to go on the left hand side of an assignment operator, as in the example below.) Class `Indexable` is template class – simply because the return type for the `operator[]()` function has to be open for definition in the derived class.

Given these base classes, programmers can define derived classes – specifying things that will be indexable, comparable, and storable, such as a specialized string class `MString` (multiply-inherited string):

```
#include "Comparable.h"
#include "Storable.h"
#include "Indexable.h"
```

```
class MString : public Comparable,
                        public Storable,
                        public Indexable<char> {
public:
    MString(char* s = "");
    MString(const MString&);
    // Not to be used as base class, non-virtual destructor
    ~MString();

    MString&        operator=(const MString&);

    // Implements Indexable:
    virtual char&   operator[](int) const ;
    virtual int     MinIndex() const { return 0; }
    virtual int     MaxIndex()const ;
    // Implements Comparable
    virtual int     operator==(const Comparable&) const;
    virtual int     operator!=(const Comparable&) const;
    virtual int     operator<(const Comparable&) const;
    virtual int     operator<=(const Comparable&) const;
    virtual int     operator>(const Comparable&) const;
    virtual int     operator>=(const Comparable&) const;
    // Implements Storable
    virtual long    Size() const;
    virtual void    PrintOn(ostream&) const;
    virtual void    ReadFrom(istream&);
private:
    char*           fdata;
};
```

A "comparable", "storable", and "indexable" string

Class MString is derived from Comparable, Storable, and Indexable<char> (it is an indexable thing from which one can extract individual chars). As is typical of a C++ string-class, it will own a separately allocated array of characters. Therefore, as a "resource manager" class it has to provide a destructor (defined non-virtual so MString should not be further used as a base class), a copy constructor, and assignment operator. It provides the implementation of all those inherited pure virtual functions. (Note the char& operator[](int) const function which is the effective implementation for Indexable<char>:: operator[](int) const. This function was defined to return a Thing&; here Thing ≡ char).

The implementation of class MString is straightforward; some example member functions are:

```
MString::MString(const MString& other)
{
    fdata = new char[::strlen(other.fdata) + 1];
    ::strcpy(fdata,other.fdata);
}

MString::~MString()
{
    delete [] fdata;
}
```

MString's own unique functions

```
MString& MString::operator=(const MString& other)
{
    if(this != &other) {
            delete [] fdata;
            fdata = new char[::strlen(other.fdata) + 1];
            ::strcpy(fdata,other.fdata);
            }
    return *this;
}
```

MString's realization of class Comparable functions

```
int MString::operator==(const Comparable& o) const
{
    const MString& m = (MString&) o;
    return (::strcmp(fdata,m.fdata)==0) ? 1 : 0;
}
```

Implementing inherited class Storable functions

```
void MString::PrintOn(ostream& out) const
{
    out << setw(7) << ::strlen(fdata)  << fdata << endl;
}
```

Defining pure virtual functions inherited from Indexable<char>

```
int MString::MaxIndex() const
{
    return ::strlen(fdata)-1;
}

char& MString::operator[](int ndx) const
{
    if((ndx < 0) || (ndx>=::strlen(fdata)))
            exit(1);
    return fdata[ndx];
}

// rest of MString functions ...
...
```

The files provided include an example test program exercising some of the features of class `MString`:

```
int main(int,char**)
{
    MString         string1("Hello World");

    // Try some "storable" features
    cout << string1.Size() << endl;
    string1.PrintOn(cout);

    // Try some "indexable" features
    cout << "Characters in string";
    for(int j=0;j<=string1.MaxIndex();j++)
            cout << j << string1[j] << endl;

    // Note use of operator[] on both left and
    //  right hand sides of assignment in this loop
    for(j=0;j<=string1.MaxIndex();j++)
```

```
            string1[j] = toupper(string1[j]);
        string1.PrintOn(cout);

        // Try some of the "comparable" features
        MString       string2("Hi mom");
        if(string1>string2)
            cout << "Greater\n";
        ...
        return 0;
}
```

The assignment `string1[j] = toupper(string1[j]);` (in the second loop) illustrates the flexibility offered by returning a reference to an indexed element (rather than its value). Essentially, the `operator[]()` function returns the address of a data element. When this appears on the right-hand side of assignment, the contents of that address are fetched (in `toupper( string1[j])`). On the left-hand side of the assignment, the address is used as the address where data are to be stored. Standard C++ text books, e.g. Shapiro's *C++ Toolkit,* typically include more complete examples showing how `operator[]()` functions can be used to implement more complete array abstraction.

Classes `Comparable`, `Indexable`, and `Storable` offer *nothing* to the programmer who has to hack out an individual class such as class `MString`. These base classes provide no code sharing, because they don't have any implementation for their functions (apart from the trivial definitions, in class `Comparable`, specifying `Equal()` as being synonymous with `operator==()` and providing a generic specification for `Eq()`). Really, their use involves the programmer in extra work – specifying the inheritance relations and providing implementations for the whole range of pure virtual functions (only a few of which are likely to be needed in any specialized class).

Why use this form of multiple inheritance?

The benefit from this type of inheritance is at a higher level in the software engineering spectrum. Consistent use of such base classes in all the classes developed for a specific project will result in classes that have some similarities in behaviour having identical interfaces. Thus, all "storable" objects will respond to `Size()`, `PrintOn()`, and `ReadFrom()` requests. Such consistency facilitates subsequent work on code maintenance and extension.

The designer of a class library can also exploit this type of multiple inheritance. Most existing libraries have some ultimate base class `Object` that is burdened with the provision of the whole range of capabilities that might be useful in a few of its many subclasses. (Necessarily, most of the member functions defined at the class `Object` level in a hierarchy will either have empty definitions or will be pure virtual functions.) Instead of such an overburdened base class, a designer using multiple inheritance can factor out various facets of "object-life" into separate abstract base classes. Thus, one can have classes `Storable`, `Comparable`, `Indexable`, `Hashable`, and possibly others such as `Displayable`. Actual concrete classes, and partially implemented abstract classes such as `View` or `Document`, can then inherit from selected base classes and so offer just those features that they really require. Such changes can help make the classes in a library easier to understand and more readily reused. (Actually, there is a trade-off. Individual classes are simplified in that they exhibit solely those behaviours that are relevant. However, the class structure becomes a directed graph rather than a simple tree.)

12.5.2 Inheriting from partially implemented and concrete classes

> Stroustrup: *You don't get "re-use" from individual language features, but from design."*

Stroustrup's comment (originally part of an Internet discussion) was quoted earlier in Chapter 6 in the context of an examination of C++ controls on inheritance in single inheritance schemes. Design and planning for inheritance is even more important in the context of multiple inheritance.

Flyingboat = ship + plane?

Don't expect to combine to separately developed concrete classes into some new class through the use of multiple inheritance. It is unlikely that any practical application would ever allow the derivation:

```
class CargoVessel {
public:
    CargoVessel(long displacement, long cargo, long vel, …)
    …
private:
    long    fdisplacement, fcargo;  // units of tons
    long    fmaxspeed; // units of knots
    …
};

class Plane {
public:
    Plane(long wingspan, …);
    …
private:
    long    fwingspan;  // unit of metres
    …
};

class FlyingBoat : public CargoVessel, public Plane {
public:
    FlyingBoat(…)
    …
};
```

Such situations rarely arise in practice. Even if they do, you must remember that the behaviour of such a derived class would be a bit like that of a flyingboat – which is inefficient and expensive when used as a cargo vessel and clumsy when performing as a plane.

Inheritance of "type" abstractions

One possible use of multiple inheritance involves simply an extension of the schemes illustrated in the previous section. When designing a class library, it is possible to factor out "types" (groups of behaviours) into abstract classes. Often, these will be pure abstract classes like those illustrated in the previous section.

However, it is sometimes possible for the library designer to implement sensible
default behaviours in some partially (fully?) implemented abstract class.

Examples are provided by classes Observable and Observer from Interviews
3.1:

```
class Observable {
public:
    Observable();
    virtual ~Observable();
    virtual void attach(Observer*);
    virtual void detach(Observer*);
    virtual void notify();
private:
    ObserverList*observers_;
};

class Observer {
protected:
    Observer();
public:
    virtual ~Observer();
    virtual void update(Observable*);
    virtual void disconnect(Observable*);
};
```

These classes constitute the basic abstractions needed to create a simple
dependency mechanism. A class that wishes to be tracked inherits (publicly) from
Observable in addition to any other base classes that it may have. A class that
needs to track changes in instances of other class will inherit from class Observer.
Class Observer is really a pure abstraction like the classes in the previous section.
Classes derived from class Observer have to provide effective implementations for
its methods:

```
void Observer::update(Observable*) { }
void Observer::disconnect(Observable*) { }
```

Class Observable has default implementations for its member functions that
provide a working, though very simple form of dependency mechanism. Each
Observable maintains its own list of Observers, and changes are handled, through
a notify() function that gets each Observer to perform an update. Of course,
functions like attach() and notify() are virtual so an implementor could
substitute more complex mechanisms if required.

Several of the concrete classes defined in the Interviews library inherit either
from Observer or Observable. Examples are:

```
class Browser : public InputHandler, public Observer { … };
class TellTale : public MonoGlyph, public Observer { … }
class TellTaleState : public Resource, public Observable
    { … }
```

(The class hierarchy in Interviews is still evolving. There are significant differences between the latest version, Interviews 3.1, and the version 3.01 that was described briefly in Chapter 7. The later version seems to be reducing the use of multiple inheritance and avoiding features, such as virtual base classes, that current compilers handle poorly – e.g. by generating inefficient code.)

A somewhat more complex abstraction in the Interviews library is class Adjustable. Class Adjustable captures the concept of an entity whose size can be adjusted in a number of dimensions. (The Interviews library includes a ScrollBox that works in two dimensions; but implementors of client classes can use three, or possibly four dimensions.) Class Adjustable uses several instances of class Observable that are associated with the coordinate data for each dimension. The class provides methods for reading the upper and lower bounds and current value in each dimension, along with functions for "paging" forwards and backwards etc:

```
class Adjustable {
public:
    Adjustable();
    virtual ~Adjustable();

    virtual Coord lower(DimensionName) const;
    virtual Coord upper(DimensionName) const;
    virtual Coord length(DimensionName) const;
    virtual Coord cur_lower(DimensionName) const;
    virtual Coord cur_upper(DimensionName) const;
    virtual Coord cur_length(DimensionName) const;
    ...
    virtual void scroll_forward(DimensionName);
    virtual void scroll_backward(DimensionName);
    virtual void page_forward(DimensionName);
    ...
private:
    AdjustableImpl* impl_;
};
```

All the methods have default definitions; but, again, they are virtual functions and could be altered if necessary. Of course, it is unlikely that any program would instantiate class Adjustable; its intended role is that of a base class. Class Adjustable is one of the base classes of Scrollbox a component in the Interviews library:

```
class ScrollBox : public PolyGlyph, public Adjustable { ... }
```

Class PolyGlyph is a class that helps organize the display of objects that are comprised of multiple separate visual components. From its dual inheritance, a ScrollBox has the abilities needed to display a scrollable list of items.

The examples so far have continued to illustrate inheritance as a kind of type composition device with a set of behaviours, e.g. the ability to adjust dimensions, being added to some instantiable class such as a class representing a composite visual object. In general, this is the most likely use of multiple inheritance; all the examples in Interviews seem to be of this kind.

However, there are cases where different concrete classes do get combined using multiple inheritance. These cases are not opportunistic like FlyingBoat, they are the result of careful design. The most commonly encountered example is in the stream libraries used with C++ programs. Figure 12.2 shows some of the classes from the stream library.

Multiple inheritance in the stream classes

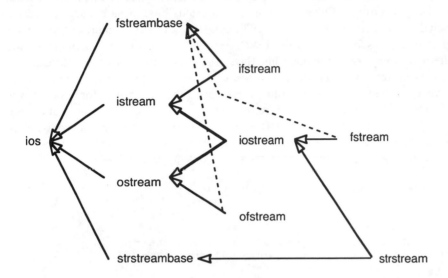

Figure 12.2   The inheritance graph for some of the classes in the stream library.

The library includes streams for terminal i/o, file-based streams, and streams for in-memory formatting (equivalent to C's sprintf etc.). The hierarchy shown is simplified. There are as many different memory-based formatting streams (derived from strstreambase) as there are file-based streams (derived from fstreambase). There are also slightly more specialized versions of istream and ostream (class istream_with-assign etc.). Apart from the streams, the library includes a small hierarchy of specialized subclasses of class streambuf; these provide facilities for buffering data.

The basic ideas of stream i/o were introduced in Chapter 4. There are of course simple input (istream) and output (ostream) streams like cin and cout (strictly these are members of the slightly more specialized subclasses istream_ withassign and ostream_withassign). The concept of a stream can be combined with that of a file – leading to the multiply inherited classes ifstream and ofstream. An ifstream *is_a* kind of input stream and it also *is_a* kind of object that can deal with file-handling requests like open(), close() etc. Sometimes a stream is used for both input and output; such an iostream *is_a* istream and *is_a* ostream. In all these cases, multiple inheritance is a natural way of building up the more complex class.

The stream library illustrates other special features such as repeated inheritance from a virtual base class. All the classes shown in Figure 12.2 are derived (directly

Class ios: base class for ios hierarchy

or indirectly) from class `ios`. Class `ios` defines a few enumerations, some static (class) variables, and a number of private and protected variables; these include a pointer to a buffer, and various parameters and flags used to control formatting (e.g. `x_width` – width of next field, `x_fill` – fill character if width greater than minimum needed, `x_flags` – i/o flags). Some of the member functions of class `ios` permit format controls to be set and read (e.g. `int ios::precision(int)` sets the precision for "scientific" numbers while `int ios::precision() const` reads the current setting). Other functions access status bits; these member functions include `int ios::good()` (is all well?), `int ios::fail()` (failure of last transfer or stream unusable), and `int ios::eof()`. Two operator functions provide alternative ways of testing the same flags; `int ios::operator!()` returns true if there is an error condition, while `ios::operator void*()` returns a 0 pointer if an error is flagged otherwise returning a pointer to the `ios` object for which it was invoked. (These somewhat obscure operators permit concise if somewhat cryptic coding of certain operations on instance of `istreams`, `ostreams`, `fstreams` etc.)

Class istream   Class `istream` adds the "extractors" (i.e. the `istream& istream::operator>>(?)` functions) for all the standard data types of C++:

```
class istream : virtual public ios {
public:
    ...
    istream&      operator>>(char*);
    istream&      operator>>(unsigned char*);
    istream&      operator>>(unsigned char& c);
    ...
};
```

Class `istream` also has the various `get()` functions for reading lines, `gcount()` for returning a count of bytes read by a `get()` function, a `peek()` function and a `putback(char)` function (useful in parsers etc.), and stream positioning functions equivalent to C's `fseek()` and `ftell()`. Since class `istream` is publicly derived from class `ios`, it has the same public functions such as `fail()`, `eof()`, etc. Of course, an instance of `istream` has, in its `ios` part, a pointer to a buffer, some status flags, some format controls etc.

Class ostream   Class `ostream` provides the complementary set of "insertors" (i.e. the `ostream& ostream::operator<<(?)` functions) for all the standard data types of C++:

```
class ostream : virtual public ios {
public:
    ...
    ostream&      operator<<(const char*);
    ostream&      operator<<(int);
    ostream&      operator<<(double);
    ...
};
```

Class `ostream` also has a `flush()` function, and the inherited methods from ios that can be used to set the format, or test the state of the stream. Like class `istream`, class `ostream` has functions for positioning the stream.

By definition, an input/output stream *is_a* input stream and *is_a* output stream:

```
class iostream : public istream, public ostream {
public:
    iostream(streambuf*);
    virtual     ~iostream();
protected:
    iostream();
};
```

Apart from a couple of constructors and a destructor (which is empty), class `iostream` adds nothing to the features that it has acquired by inheritance. Because class `iostream` is publicly derived from `istream`, extractors can be used with an `iostream`; similarly, because an `iostream` *is_a* `ostream`, insertors can be used. Since an `iostream` is derived publicly (albeit indirectly) from class `ios`, all the `ios` functions like `fail()`, `eof()` etc. can be used with an `iostream`.

As is apparent from Figure 12.32, class `iostream` inherits repeatedly (well, twice) from class `ios`. An `ios` has a pointer to a buffer, some integer data members that describe formats, and its flags data member. Should an `iostream` have one or two copies of each of these data members?

This is a design decision. It depends on the behaviour desired in the class. In principle, it would be possible for an input/output stream to have separate input and output buffers. Although possible, it would be inconvenient for this would deviate from common usage on Unix systems and it would be hard for an underlying operating system to maintain consistency of the data in the separate buffers. In this case, it is more natural for an `iostream` to have a single instance of the `ios` data members.

This was designed into the class hierarchy and all the immediate descendants of class `ios` specify `virtual` inheritance:

```
class fstreambase    : virtual public ios { ... };
class istream        : virtual public ios { ... };
class ostream        : virtual public ios { ... };
class strstreambase  : virtual public ios { ... };
```

As explained in the overview in Chapter 3, a derived class has only one copy of the data members for all `virtual` inheritance paths.

In addition to decisions regarding replication of inherited data members, designers of multiple inheritance hierarchies must also plan for name conflicts among member functions inherited from different base classes. In the `ios` library, there aren't many problems. The names of functions tend to be quite distinct. Class `fstreambase` has member functions like `open()`, and `close()`; class `istream` has the `operator>>()` functions and things like `get()` and `gcount()`; while class `ostream` has the `operator<<()` functions and `flush()` etc.

However, both `istream` and `ostream` have *seek* and *tell* operations for positioning in the stream. If both classes had chosen the names `seek()` and `tell()` for these operations, there might be some ambiguity as to which function an `iostream` object was to use when it was asked to perform a *seek* operation. (Actually, it may not make much difference; in some implementations, there is only one pointer into the file.) In the stream library, the potential ambiguity was foreseen and the functions are named distinctly. An istream is used to *get* data so it uses seekg() and tellg() (i.e. *seek* and *tell* operations associated with *get*); while an ostream is used to *put* data so it has seekp() and tellp() functions.

The combinations that appear at the deeper levels in the hierarchy required multiple inheritance:

```
class iostream : public istream, public ostream { … };
class ifstream : public fstreambase, public istream { … };
class strstream : public strstreambase, public iostream { … }
```

In each of these cases, there are multiple *is_a* relations. An input file stream *is_a* input stream and *is_a* file-like thing. Here, composition is not a valid alternative.

It might seem that inheritance from ios was unnecessary. Rather than inheriting from class ios, classes like istream and ostream could each own a data member, fios, that is an instance of class ios. This would work at the top-level. Even there, it might be a bit inconvenient because the classes would have to define member functions that provided indirect access to functions like fail() (so there would have to be functions like int istream::fail() { return fios. fail(); } etc.). But there would be a more serious problem where classes istream and ostream are combined to get class iostream; class iostream would inevitably get two fios data members with separate instances of class ios. (It would be possible to hack around this problem using ios* data members and arranging that both inherited pointers referenced the same object. But this would be a matter of hacking!) The class hierarchy shown in Figure 12.2 is necessary, it can not easily be simplified by use of composition. Of course, this hierarchy was planned as a whole; it did not grow by additions of derived subclasses.

Usually, multiple inheritance is going to lead to complications A few were touched on in Chapter 3. Commentaries by Meyers (*Effective C++*) and Cargill (*C++ Programming Style*) should be read before committing a design to use multiple inheritance. Lippman's text provides many small examples illustrating the fun that can be had with things related to multiple inheritance – things like name dominance, type casting of pointers between base and derived classes etc.

12.6 EXCEPTIONS

Class MString, part of the example of section 12.5.1, included a routine that allowed individual characters to be extracted from the string by use of the operator[]() function. Of course, it is possible that a client might ask an MString for its [-1] element or its [SHRT_MAX] element. What should an MString do in response to such illegal requests? The encoded response:

```
char& MString::operator[](int ndx) const
{
    if((ndx < 0) || (ndx>=::strlen(fdata)))
            exit(1);
    return fdata[ndx];
}
```

i.e. kill the program with an `exit(1)`, was "correct" – but decidedly unhelpful (and potentially dangerous).

Now calling `MString::operator[]()` with an unreasonable argument value is an error. The client has not honoured the implicit contract with the class `MString`; an instance of class `MString`, like any other `Indexable` class, is only required to behave reasonably if the argument to `::operator[]()` is in the range defined by `MinIndex()` to `MaxIndex()`. A correctly coded client program could (should?) always use these functions to verify that it knows the correct range of subscripts and makes only valid requests to an `MString` object. Consequently, it is not totally unreasonable for the implementor of the `MString` class to make the response to an invalid request be one of terminating the contract (with "extreme prejudice" to the client).

Kill clients who don't respect "contracts"?

In other situations, it can be much more difficult for a client to honour a contract with a server class. Consider for example client program that has to import an image file and then get the image displayed with the aid of an instance of a library class `ImageHandler`. An `ImageHandler` is to be given a large array of bytes (via a member function such as `ImageHandler::AcceptImage(char*)`). The bytes encode an image with "operator" byte codes, each followed by some number of data bytes. Such encoding schemes are quite common, one example being the Quickdraw byte protocol for Macintosh PICTs. The image would be displayed via a client call to `ImageHandler::DrawImage()`. The code of `DrawImage()` would involve a loop through the bytes, interpreting operator bytes and consuming data bytes:

Don't kill innocent clients

```
ImageHandler::DrawImage()
{
    for(char* p = fImage; *p;)
            switch(*p) {
case        RECTANGLECODE:
            /* next 8 bytes are coords of rect, */
            /* consume and draw rect */
            p++;
            HandleRectangeBytes(p);
            break;
case        LINECODE:
            /* next 4 bytes define end points of line */
            ...
default:
            /* ??  !!  ?? */
            /* What do we do with an unknown operator byte? */
            /* THROWUP? */
            exit(1);      // ?? Is anything better possible?
    }
}
```

Of course, the implementor of the library class `ImageHandler` has to deal with the possibility of the image data containing an invalid operator byte.

Now such things do turn up. Obviously, if the original image file was corrupt there will be invalid data; but there are other possibilities. Different applications sometimes extend a standard byte operator repertoire with extra operators. For example, a graphics editor might support the display of character strings that follows curves instead of lines – handling this through some extension to the normal byte operator repertoire. Data involving non-standard operators shouldn't be saved to files that are shared with other programs; but there is nothing to stop a user trying to make more general use of a file intended only for use within a specific application. If a user feeds a file with application-specific byte operators into a general program, the `ImageHandler` code will encounter strange operators in the image.

In this situation, killing the client program (e.g. by an `exit(1)` call) seems a bit unfair. The client code cannot really check whether the file selected by the user contains bad data. (If the client did know how to interpret the bytes of an image then there would be no need for the library class `ImageHandler`.)

Of course, the programmer writing the client code would be aware of the possibility of bad data, and would have been able to provide code to deal with the aftermath of a data error (e.g. by displaying an "alert" with a message to the effect that the data in the requested image file could not be displayed).

An unfortunate separation of error detection and handling abilities
A mechanism for communication

Thus, the author of the client code cannot detect errors but knows what to do about them. The author of a library class such as `ImageHandler` can detect errors but doesn't know what to do about them.

"Exceptions" are a language-supported general mechanism whereby information about error conditions can be communicated between different program components.

A component specifies the "exceptions" it may use to describe errors

The author of a component like class `ImageHandler` can publish a list of "exceptions". Essentially, these will be things that will be used to describe error conditions that can be detected by the code of that component but can not be handled locally. Class `ImageHandler` would have a "bad operator" exception and, possibly, some other kinds of exception for other error conditions that it might encounter (e.g. running off the end of the image byte stream while trying to read operand bytes).

At execution time, an exception is thrown

The code of the component can include statement to "throw" an exception at the point where an error is detected.

Client code can "catch" exceptions

The programmer of some client application can review the list of exceptions associated with a component and can plan mechanisms for handling some or all of them. Code to deal with exceptions that might get "thrown" by a component can be packaged with each call to a component function. If, at run-time, bad data makes it necessary for a component to "throw an exception", the client code that called the component may "catch the exception" and resolve the problem. An exception that is "thrown" but not "caught" will cause the program to terminate. There will be situations where the calling code may catch an exception and try to deal with some problem only to find that the difficulty cannot be resolved; in these situations, the catching code can throw another exception so passing responsibility back to some earlier level in the calling sequence.

C's more primitive mechanisms: setjmp and longjmp

Primitive mechanisms for handling similar problems have always been present in C. A calling routine could construct a *closure* using `setjmp()`. A closure contains data identifying the current stack frame and the memory location of the instructions following the `setjmp` call. A call to `setjmp()` would be in an `if` statement:

```
jmp_buf        closure1;
int            problem;
...
if(problem = setjmp(closure1)) {
        /* end up here if called routine hit problems */
        /* integer value "returned" by setjmp encodes */
        /* details of what was wrong. */

        ...
        }
else {
        /* call the routine that must run but which */
        /* may hit problems */
        XXX(&closure1, ...);
        /* everything worked ok */

        ...
        }
```

The call to setjmp() that builds the closure will return 0, causing the else branch of the conditional to be executed. There a call would be made to the routine(s) that might encounter errors that have to be dealt with at the calling level. If no problems arose, processing of the else clause would be completed and the program would then continue with the statements following the conditional.

A called routine (either called directly, or indirectly via other routines) could deal with an error by invoking longjmp(jmp_buf env, int val). The val argument would encode details of the error encountered. Execution of longjmp's code causes the stack to be unwound (with all stack frames being released) and finally fakes a return from the original setjmp() call, with the error code as the value. This leads to execution of code in the first branch of the conditional; there, the error condition could be evaluated and remedial action taken.

Although setjmp and longjmp work, their use is a little confusing. Further, because they involve messing with the stack and registers they require platform specific implementations. There can be inconsistencies between implementations with regard to register variables, leading to some problems with regard to portability. In C++, there are additional complexities with stack unwinding – destructors should be getting called for automatic objects.

The new "exception" features proposed for C++ repackage the kind of control structure that setjmp and longjmp provided, make the scheme reliable and more flexible, and extend the capabilities for returning information about an error. Three new keywords are added to the C language! These are try, throw, and catch.

New keywords: try, throw, catch

The author of a component like class ImageHandler specifies the types of exceptions that may be thrown. They can be simple types, but more commonly they would be classes (probably nested classes to avoid "polluting" the global name space with trivia):

```
class ImageHandler {
public:
    ImageHandler();
    ~ImageHandler();
    ...
    void   AcceptImage(char*);
    void   DrawImage();
```

Defining classes for the exceptions that may get thrown

<table>
<tr><td>

Nested subclass defining a control byte exception

</td><td>

```
...
class UnknownControlByte {
public:
        UnknownControlByte (char b) : fb(b) { }
        char BadByte() { return fb; }
private:
        char   fb;
};
```

</td></tr>
<tr><td>

Nested subclass defining an exception for a data problem

</td><td>

```
class DataByteDeficiency {
public:
        DataByteDeficiency(char control,
            int expected, int got) :
                fc(control), fe(expected), fg(got)  { }
            ...
        };
private:
    /* private parts of ImageHandler */
        ...
    };
```

</td></tr>
</table>

The implementation code for the component will include statements that `throw` exceptions at the points where errors are detected:

```
ImageHandler::DrawImage()
{
    for(char* p = fImage; *p; p++)
        switch(*p) {
case     RECTANGLECODE:
        ...
        break;
        ...
dcfault.
```

<table>
<tr><td>

Throwing an exception

</td><td>

```
        /* Unknown operator byte */
        throw UnknownControlByte(*p);
    }
}
```

</td></tr>
</table>

The code of the client will bracket a call to a method of the component in a `try` clause that will be followed by `catch` clauses that deal with the various possible exceptions:

<table>
<tr><td>

Try – invoke the operation where may hit problems

</td><td>

```
try {
        ...
        myImageHandler->DrawImage();
        ...
}
```

</td></tr>
<tr><td>

Catch — deal with any problems that did arise

</td><td>

```
catch (ImageHandler::UnknownControlByte  unk) {
        // Inform the user that the data file used for
        // input was not suitable
        ...
        }
catch (ImageHandler::DataByteDeficiency bad) {
        // Inform the user that their data file appears
        // to have been truncated or otherwise damaged
```

</td></tr>
</table>

```
                    ...
            }
    ...
```

Where appropriate, the catching code can use information in the data fields of the thrown object for diagnostic purposes.

REFERENCES

Stroustrup illustrates a number of more general ways of using template functions and shows how the default implementation provided by a template function can be overridden for a specific data type. In his design chapter he argues a case for the greater use of nested classes; though in the chapter where these are introduced it is noted that they can complicate and obscure the declaration of the main enclosing class. Stroustrup covers many other aspects of C++ that have been ignored here. One can get pointers to member functions of various classes and use these as arguments to other functions. One can overload operators like ->, and have versions of ++X (pre-increment an X) and X++ (post-increment an X) that have distinct, class specific meanings; one can have "function objects" as with the "iomanip" classes. There are many more variations on the theme of inheritance – protected inheritance, inheritance with templates, nesting and inheritance, "virtual constructors", . etc. Most of these features are illustrated with brief, somewhat formal examples detailing the syntax.

Lippman's text includes rather more illustrative examples of all the varied capabilities of C++. Many of Lippman's examples may seem contrived and somewhat unconvincing. This is inevitable. Several of these language features are somewhat specialized, intended for use in complex situations. Realistic examples would inevitably be lengthy.

Books like that by Murray (*C++ Strategies and Tactics*) can sometimes provide better examples than language texts like Lippman's. These more specialized books don't attempt to cover the entire C++ language, focussing instead on particular topics of general interest. Murray's book should prove very useful to students studying C++ the night before an exam (it provides reasonably clear summaries of a number of topics that examiners are likely to pick as tests of one's understanding of C++). It could be useful as an intermediate text before any attempt is made to use Coplien's book.

The intermediate aspects of C++ should be left until students have acquired skills in more straightforward features such as single inheritance, template functions, and possibly template classes. Then, some of the exercises in the texts of Stroustrup or Lippman might be attempted. Survivors could proceed to sample Coplien's text.

Reiterating: read the advice of Meyers and of Cargill before attempting anything too fancy.

Cargill, T., *C++ Programming Style*, Addison-Wesley, 1992.
Coplien, J.O., *Advanced C++: Programming Styles and Idioms*, Addison-Wesley, 1992.
Lippman, S.B., *C++ Primer*, 2e Addison-Wesley, 1991.

Meyers, S., *Effective C++*, Addison-Wesley, 1992.
Murray, R.B., *C++ Strategies and Tactics*, Addison-Wesley, 1993.
Shapiro, J.S., *A C++ Toolkit*, Prentice-Hall, 1991.
Stroustrup, B.J., *The C++ Programming Language* 2e Addison-Wesley, 1991.

$\mathcal{E}$iffel 13

13.1 INTRODUCTION

Although C++ supports OO programming, those starting on a new OO project can select one of the other OO languages or hybrid OO/procedural languages. Many of these other languages are limited in their scope and their objectives. Thus the various Object Pascal dialects aim only to provide those language features that are essential to permit even a limited form of OO programming; these Pascal dialects are not really intended for serious software development. Then there are the numerous experimental versions of Ada with object-oriented extensions. None is likely to see much use; the role of these experimental languages is more that of identifying those extensions that might most usefully be included in a future Ada-9X. Of course, there is a plethora of purely experimental languages developed by academic and industrial research groups. But, there are a few languages that are sufficiently matured to permit use in serious software development projects.

C++ dominates these mature languages. Objective C does have a small following. Smalltalk has a definite role as a prototyping language; with CLOS filling a rather similar role for the LISP community. But, the only interesting alternative to C++ for the development of large reliable software systems has been Meyer's Eiffel language.

C++ is a compromise, a hybrid. The compromise is reflected in both the form of the language and the style of its use. C++ compromises by maintaining consistency with the earlier C language. In some respects, this has been of enormous benefit. It has allowed C++ to provide an evolutionary escape route for programs and programmers to follow while moving from procedural style to OO style. (The need for a mechanism permitting such evolution was foreseen by Cox, and was one of the motivations leading to his development of Objective C). But, consistency with C, has also been an impediment. For example C++ could not provide a better array abstraction without breaking existing C programs. C++ has been burdened with some of C's more arcane forms for type declarations. The C mindset makes it inappropriate to introduce new keywords – so existing keywords like `static` become further overloaded with additional meanings, and a cryptic construct such as `class X { virtual void f1() = 0; … }` is preferred to a simple keyword scheme like `abstract class X { deferred void f1(); … }`.

C++, a compromise

In terms of usage, C++ offers OO programming as but one style among many. C++ is also to be used for low-level coding – such as the implementation of time-critical code and device drivers. This requirement makes it essential for programmers to be able to obtain explicit control of memory and register allocation. But support for such low-level features makes it less practical for the language to provide a uniform mechanism for automatic memory management. Consequently, high-level programmers must take responsibility for managing memory allocation in their classes and objects. Further, because a programmer can always revert to C-style pointer manipulation etc., it is relatively easy to break the security offered by classes – while such "hacks" may offer short-term advantages they reduce the reliability of code and complicate subsequent attempts at maintenance and extension.

Eiffel: OO is the only way to go

Eiffel brooks no compromise. For Meyer, the object-oriented approach represents the only way in the large scale software systems of the future will be constructed. Eiffel embodies Meyer's ideals for an OO language. It is a high level language that aims to allow a programmer to focus on the application at hand rather than on low-level concerns such as detailed memory management. In this respect, the language is a bit like Lisp or Smalltalk – languages that are intended for programmers exploring the feasibility of some proposed computational process. However, unlike the Lisp/Smalltalk exploratory languages, Eiffel supports practical implementation for real, large scale projects with an appropriate module structure, compile time checking, and reasonably efficient code. (Of course, what is reasonable to me may seem unacceptable to you.)

13.1.1 Origins of Eiffel

Meyer's software engineering ideals

During the early 1980s, Meyer sought ways of improving software. His view was that software needed to be improved with respect to reusability, extendability, and reliability. Reusability could be achieved using generic code – such as Ada's packages, and abstract data types constructed as reusable components. Extendability required some variation on Simula's classes that would allow for extension via inheritance and would permit the use of polymorphism. If classes were to be extendable, they would have to be "open" – the author of an extension might need access to all the data fields and functions of the existing class and might have to change existing functionality. Multiple inheritance would allow new classes to be composed by combining desirable features from separate parent classes. Compile-time type checking would help the construction of reliable code, but both reliability and reusability would require rather more than this.

Contract model

If code is to be reused, the "reuser" has to know what the code claims to do and have confidence that it will perform as advertised. Thus, Meyer was lead to the contractor model – a running system would be composed of many objects that each had subcontracted to handle a specific part of the overall task. Language constructs could provide at least a partial model for this contract model with classes (types of subcontractor) advertising the conditions under which their members would accept contracts (in precondition "require" clauses associated with the member functions of the classes) and detailing the services that they would provide (in similar "ensure" clauses). These constructs could also constrain the implementation of derived classes by ensuring that the semantics of member functions are preserved.

As well as documenting the contractual obligations of class members, these "require" and "ensure" clause could be enforced at run-time – where they would identify those breaching their contractual obligations. Breaches of contract would need to be handled in a prescribed disciplined manner, so an effective OO language would have to have some mechanism for handling run-time exceptions.

Contract violations and exceptions

These ideas led to the Eiffel language, which became available from Interactive Software Engineering (ISE) in 1986. (Most Eiffel development work has been done with Eiffel 2, which was released with a number of revisions in the period 1988 to 1990.) From release 2, Eiffel provided a practical realization of an OO language with multiple inheritance, parameterized (generic) classes, and exception handling.

Eiffel: a realization of ideals

The language is pretty much a pure OO language – everything at run-time is an object (well, almost everything, characters, integers and reals do get special treatment by the compiler). The language uses uniform reference semantics (a variable, or "entity" in Eiffel-speak, is essentially a pointer that can be set to reference an existing object or can be made to point to a newly created object of the appropriate class). Since everything is essentially heap-based, there aren't the complications associated with C++'s static and automatic instances of classes with their assignments, copy constructors etc. The Eiffel language is relatively simple (roughly the same level as Pascal) with a minimum of different constructs.

13.1.2 Inheritance in Eiffel

Inheritance must be a concept that appeals strongly to the aristocratic French for it is used aggressively in Eiffel programming style.

The Eiffel style involves somewhat non-standard uses of inheritance. Inheritance is used extensively as a compositional device allowing new classes to be constructed from existing classes. The facilities for combining existing classes are much more flexible than those in C++. Any inherited function can be redefined – not simply those for which permission was granted, in advance, by the author of the original class (as in C++ with `virtual` functions). The mechanism for renaming of inherited attributes provides much finer control on replication or sharing of data fields.

Inheritance in Eiffel

The entire inheritance mechanism is designed to encourage reuse through serendipitous inheritance; e.g. an existing TREE class might get adopted as an ancestor for a view-like class – so providing a basis for having a tree-structured arrangement of nested views. Really, less is expected of the designer of an Eiffel class than of the designer of a C++ class. The Eiffel designer does not have to plan for reuse

Stroustrup doesn't want programmers just messing with his classes; so programmers developing derived classes only get access to those parts permitted by the designer of a class. Controls imposed on a class by a designer provide security, but limit the scope for reuse. The designer of a C++ class library needs great prescience. All possible ways in which the class might be used have to be foreseen and appropriate access, through `virtual` functions and `protected` access mechanisms, must be provided.

Meyer wants his classes reused and so makes them "open" to subsequent programmers. The Eiffel designer doesn't have to get an Eiffel class exactly right – because subsequent programmers can modify and extend existing code in any way that is compatible with the constraints imposed through the preconditions and "ensure" clauses that specify the semantics of the original class.

Eiffel makes class inheritance the only mechanism for composing software. There are no "header" files etc. This results in some curious hierarchies. If two classes need to share some constants, a new class has to be defined – the only features of this new class being those constants; the class has no data and no functionality. The real classes needing those constants then both inherit from the introduced class. Thus, one may find class `WINDOW` inheriting from class `ASCII` (a class declaring all character constants) – which carries an implication that a `WINDOW` is an `ASCII` character of some kind.

Access and inheritance

Access controls on class attributes and behaviours are handled orthogonally to the inheritance mechanism. This means that it is possible for a class to inherit from an existing base class and yet not exhibit the whole range of behaviours associated with that base class. Inheritance does not necessarily imply a complete *is_a* relation in Eiffel.

The resulting situation is more complex than `private` inheritance in C++ (which can be used as a basis for a compositional inheritance scheme). The use of private inheritance (`class Derived : private Base { }`) precludes subsequent polymorphic assignments in C++ (so, one can't have `Derived* d = new Derived; …; Base *b = d;`) because such assignments imply an *is_a* relation. In Eiffel polymorphic assignments are permitted even where there is no complete *is_a* relation (as when a derived class does not exhibit all the properties of the base class). As illustrated later, this opens "holes" in the type system. Meyer justifies the mechanisms employed in Eiffel on pragmatic grounds.

13.1.3 A complete system – language plus standard class libraries and support tools

Not just a language, more a way of life: standard class libraries

From the beginning, ISE recognized that object-oriented programming is not simply a matter of having a compiler that supports classes. OO programming requires a complete environment – there really do have to be standard class libraries, standard development environments, and standard debuggers. ISE provided these with the Eiffel language implementation.

Eiffel the language came with a quite extensive set of classes – some encapsulate Unix services (such as files and standard i/o), others are the "useful data structures" like linked-lists and binary-trees, there are classes for handling X-Windows displays and classes that provide "windowing" on standard terminals (like the Unix curses library). More specialized clusters in the library include components for the construction of lexical scanners and parsers. The existence of a *standard* class library complementing the language again makes Eiffel programming somewhat like using a high-end Lisp system or Smalltalk system.

Just as Lisp programs can be composed in main from the thousands of built-in functions with only a little basic Lisp coding, so Eiffel programs can be constructed using mainly prebuilt components from its standard libraries.

As well as standard class libraries, ISE's Eiffel came with the "short" and "flat" utility programs, "es" – something akin to Unix "make", an integrated run-time object inspector/debugger, and two browsers – "good" and "eb". The utilities "short" and "flat" can be used separately or in combination to produced printed summary documentation characterising classes. "Flat" flattens a hierarchy by explicitly incorporating all inherited properties in an expanded class declaration. "Short" lists a class declaration showing just the client's interface with signatures of functions and related information.

Short and flat documentation tools

Unix programmers can compose "make" scripts that describe dependencies among program parts – such as dependencies of "....c" source text files on various "....h" header files. Using information on these interdependencies, the Unix "make" utility can determine the minimal set of compilation steps necessary to accommodate any recent changes to files. It is somewhat tiresome to have to keep "make" scripts consistent with the dependencies in the files in a rapidly evolving project – it is the sort of mundane housekeeping task best done by a computer. This is in essence the basis of the "es" system. It is a kind of semi-intelligent "make" utility. The "root" class of a program is identified to "es" along with information identifying the location of class libraries. (All good OO "main" programs consist of just two statements – create the principal object, tell it to run; in Eiffel, the class of this principal object is the "root" class for the program.) The "es" utility will read the source text of the root class, and the text of the classes on which it depends (through inheritance or by client-server relations) and so on recursively searching out dependencies. From the information that it gathers on dependencies, "es" can identify the classes that need to be recompiled and so can then schedule the recompilation processes. This is typical of the more general Eiffel philosophy – give mundane housekeeping work to the computer so that programmers can focus on the real tasks involved in software development. Of course, if a computer is employed to do the work, it has to have the time. Eiffel compilations are not fast.

"es" – a semi-intelligent "make" system

Eiffel's browsing and run-time inspection facilities are similar to those of the environments illustrated in Chapter 7; an example of output from the "good" graphics-based browser is show in Figure 13.4. Eiffel was unusual in providing backward compatible support for ordinary cursor addressable terminals in addition to graphics X-terminals. The "eb" browser requires only simple terminal facilities.

Browsers and inspectors

The rest of this chapter provides a quick survey of Eiffel, illustrated with a few simple examples. The dialect used is ISE's Eiffel 2.3 and the example programs exploit classes from the standard ISE libraries. Eiffel 2.3 is not the language presented in the recent text books of Meyer and of Switzer – these texts focus on Eiffel 3. There have however been problems in getting implementations of Eiffel 3. Sig Computer in Germany has a version for PCs, but this does not use the ISE libraries. When this book was first conceived, ISE was demonstrating their new Eiffel 3 development environment and promising delivery "real soon now". Unfortunately, the "real soon" deadline slipped a bit and Eiffel 3 only started to become available at the end of 1993.

Why Eiffel 2.3?

13.2 CLASSES

At run-time there are only objects present in an Eiffel program, and in its source code there are only classes. There are no include files, there are no globals, there is no main program. (The Eiffel run-time environment provides a kind of `main()` which simply creates an instance of the "root" class for the program and gets it to run.)

Classes as "modules"   An Eiffel class is a syntactic unit, as in Simula or Smalltalk; it is also the unit of modularity. Each class is defined in a separate file. This is generally preferable to the use of separate header files and (multiple) implementation files (as in C++, the Object Pascals, etc.). The use of multiple files tends to complicate the work of the developer; only the best development environments, e.g. "sniff" as described in Chapter 7, can paper over the cracks between files and allow the developer to view a class as a whole. Eiffel's use of separate files for each class also facilitates reuse. If a file contains several different classes, one would typically have the choice of getting them all or having none. The only disadvantage of having individual files for classes relates to the number of files needed in large projects. The Eiffel compiler creates a subdirectory for each source file (to hold the output of intermediate computation steps as well as the final compiled code). In some operating environments, e.g. a large class of students on a shared Unix system, the numbers of files and directories can become a problem.

The general form of a class declaration in Eiffel 2 is:

Class definition:
class DEMO
```
class DEMO
    -- Instance of this class will do … and …
    -- They own …
```
"Services" provided
by an instance of
DEMO
```
export
    -- listing of the services that an instance of this
    -- class provides to clients, e.g.
    data1, data, functionA, functionB, procedureX
```

DEMO's pedigree
```
inherit
    -- details of parent classes together with a statement of
    -- any overrides to inherited functions or renaming of
    -- inherited functions; e.g.:
    PARENT1
        rename
            Work as p1_Work
        redefine
            Work;
```

"Implementation" of
DEMO
```
feature
```
"Data members"
```
    data1        : INTEGER;
    data2        : STRING;
    …
```
"Member functions"
```
    Create is
    do
        …
    end; -- Create
    functionA (s : STRING) : BOOLEAN is
    do
```

```
          ...
    end;  -- functionA

    ...

    ...
    Work is
    do
          ...
    end;  -- Work

    procedureX is
    do
          ...
    end;  -- procedureX

end;  -- class DEMO
```

End of syntactic unit
of class DEMO

Naming conventions require class names to be capitalized and related to file names; class DEMO would be defined in the file named "demo.e". The syntactic unit for the class starts with the keyword class and goes through to the matching end.

A class definition would typically start with one or two lines of comment. In Eiffel, comments start with the -- symbol and terminate at a newline (like C++'s // comments). A class may have an associated indexing clause that lists index keys and values that are then included in an index of all classes in a library. If present, this indexing clause comes immediately before the class keyword.

Comments and "indexing"

A class will normally start with an export clause. Here it specifies the "services" that it is prepared to offer to clients. The export clause will list some of the member functions and procedures defined by the class, together with some or all of the member functions inherited from parent classes. In addition, selected data attributes of the class may be exported – such exports allow clients to have read-access to those data attributes. (One of the changes in Eiffel 3 is the removal of the export clause; details of the export status of class features is included with the rest of their declarations.)

"Export" clause

A service advertised in the normal export list is available to any client. If one has a tightly-knit cluster of classes whose instances cooperate to solve a task, there will often be services that should be provided to other members of the cluster but which should not be generally available to arbitrary clients. Eiffel allows for such selective exports. If the use of an exported feature is to be restricted, that feature will be followed in the export list by a list defining those clients that have privileged access:

Selective exports

```
class ALIST
    export
            length, append, remove_first, -- general access
            first { LISTITERATOR},      -- restricted access
            last { LISTITERATOR}         -- to first and last
            ...
```

This example illustrates the idea with a class ALIST that exports the general services (like length, append, etc.) to all clients and permits a collaborating class LISTITERATOR to have privileged access to first and last (which might be data members, or member functions). These controls on privileged access are more subtle than those that can be achieved through normal use of C++ friend-relationships between classes. If a C++ version of class ALIST were to grant friend access to a C++ LISTITERATOR, then LISTITERATOR would have privileged access to all members of ALIST. Here the controls are precise; privileged access to class features can be granted on an individual basis. One could, for example, have class LISTITERATOR having privileged access only to first, while class RLISTITERATOR (a reverse iterator) having corresponding privileged access only to last.

Inherit clause

The export clause will be followed by an inherit clause where the ancestry of a class is described. With multiple inheritance, the class hierarchy for the classes in an Eiffel program will form a directed (acyclic) graph. This graph has a root in class ANY. All classes are implicitly derived from class ANY; if a class does not provide an inherit clause, it inherits directly from class ANY.

Define and redefine subclauses

The inherit clause will consist of subclauses defining the inheritance relationship with each of a class's parents. These subclauses specify whether the new class is redefining an inherited function (or providing a first definition for a deferred function that was left undefined in an abstract parent class). Any changes to the names of inherited functions will also be listed.

Rename subclause

Name changes may be needed for a variety of reasons. Multiple inheritance can result in name clashes. The parent's implementation of a function may need to be preserved when a child class redefines a function; in this case, the parent's version should be renamed. The renaming may simply be cosmetic; a different name being more appropriate in the context of the child class (e.g. a class WINDOW derived from class ATREE might rename child as subwindow).

Further aspects of these inherit clauses are explored in section 13.7.

Feature

The keyword feature introduces the body of the class with the declarations of its data attributes, functions and procedures. There are no linguistic restrictions on the order of these declarations. The examples given here will, for the most part, declare data attributes and then functions and procedures. The various forms of feature declaration are explored in section 13.4.

Invariant

A class may end with an invariant clause (a simple example is shown later for class SQUARE in section 13.12). The invariant clause will typically list a series of constraints over the values of different data members. While these constraints may be violated during the execution of a member function, they should be satisfied at any time a class instance is supposed to be in a stable state. Normally, a designer would specify the invariants as a guide to the implementor of the class. Some may have to take the form of comments (for example, it would be difficult to express and costly to verify an invariant specifying that all the items placed in an instance of some collection class would have distinct keys). The implementor can arrange to include the executable invariants in the code compiled for a class. They will be checked each time an object of that class is asked to perform a service for a client and an exception will be raised if the constraints are found to have been violated.

There are two variations on the basic form of the class declaration – deferred (abstract) classes, and generic (parameterized) classes.

Eiffel is nice and specific about the declaration of abstract classes. If one needs *Deferred classes*
an abstract `INVADER` class that describes the general behaviour of some space
invaders, the declaration would be something like:

```
deferred class INVADER
export
    Run, Draw, Move,
    ResetPosition, Erase, …

feature
    Move is
    do
            Erase;
            …
            Draw
    end;

    Draw is
    deferred
    end;

    Run is
    deferred
    end;

    …

end; -- class INVADER
```

The keyword `deferred` at the start of the class definition flags it as an abstract
class. The feature list can contain default definitions for functions, e.g. `Move`, while
specifying as `deferred` those functions (like `Draw`) that can only be defined in
subclasses such as `SAUCER`, `BOMB`, and `ROCK`. (The compiler expects at least one of
the functions to be left `deferred` and will generate a gentle warning if it encounters
a `deferred` class wherein all functions have effective default implementations.)

A generic class definition has a list of formal parameters following the class *"Generic" or*
name. For example, the Eiffel "structures" library contains a class `FIXED_LIST` *"parameterized"*
which uses an array to provide an implementation of some standard list-like *classes*
operations. Obviously, the things that get put in `FIXED_LIST`s depend on the
program; one programmer will need `FIXED_LIST`s of `SpaceInvaders`, another
programmer will be dealing with `FIXED_LIST`s of electrical utility bills.
Consequently, class `FIXED_LIST` needs a parameter (just one in this case) for the
type of the object that will be stored. A list of parameters will follow the class
name; so, in this example, the declaration specifies class `FIXED_LIST [T]`:

```
-- Lists with a fixed number of items, implemented as arrays

indexing
    names: sequence, list;
    …

class FIXED_LIST [T] export
    repeat LIST
```

```
        inherit
            LIST [T]
                    redefine
                            i_th, …
                    define
                            count;
            ARRAY [T]
                    rename
                            …


    feature

        Create (n: INTEGER) is
                    -- Allocate list with `n' elements.
                    -- (`n' may be zero for empty list).
                require
                    valid_number_of_elements: n >= 0
                do
                    array_Create (1, n)
                ensure
                    count = n
                end; -- Create

        first: T is
                    -- Item at first position
                require
                    not_empty: not empty
                do
                    Result := array_item (1)
                end; -- first

        last: like first is
                    -- Item at last position
                require
                    not_empty: not empty
                do
                    Result := array_item (count)
                end; -- last


                …
        end -- class FIXED_LIST
```

Naturally, the classes from which a FIXED_LIST is composed, LIST and ARRAY, must also exist as parameterized classes; so, in the inheritance clause they appear as LIST [T] and ARRAY [T].

"Anchored" type declarations
The type of the object returned by function first is naturally type T. The declaration of function last expresses its return type slightly differently using a feature of Eiffel called "anchored types". All that it is saying is that function last will return an object of the same type as is returned by function first (i.e. a T).

Use of anchored types is of little benefit here. Anchored types might be useful for a developer creating a class that is going to create and manipulate collections of other objects and whose functions will return these collections. Initially, it might not be clear whether these collections should be instances of LINKED_LIST, or of SORTED_LIST, or of BINARY_TREE. Anchored types would allow the developer to specify the type of one feature in their class, and then use this feature as the "anchor" in all subsequent declarations. Changing the type of the collection used would then require only a single alteration to the class text.

A program using a generic class will of course specify a particular class type for the parameter:

```
InvasionFleet    : FIXED_LIST[INVADER];
```

or

```
BillsToPrint     : FIXED_LIST[UTILITY_BILL];
```

Naturally, the compiler enforces correct usage of these FIXED_LISTs. The compiler would trap as an error any code that, for example, tried to insert a CUSTOMER into the list BillsToPrint. (This assumes that CUSTOMER is not a subclass of UTILITY_BILL. If this assumption was incorrect, the programmer could store CUSTOMERS in BillsToPrint, and could probably perform many other novel tricks with such an unnatural class hierarchy.)

Eiffel permits the use of constraints on generic parameters. For example, class *Constrained* SORTED_LIST will obviously take a parameter for the type of data object to be *genericity* stored in the list. If the list is to be kept sorted, the contained data elements will have to be compared. This requirement can be expressed by requiring that the classes of the data elements stored be subclasses of class PART_COMPAR (Eiffel's basic definition of a "comparable" object):

```
class  SORTED_LIST [ T -> PART_COMPAR]
export
    ...
end; -- class SORTED_LIST
```

Variables of type SORTED_LIST can only be accepted by the Eiffel compiler if the class used for the parameter type is derived from class PART_COMPAR.

C++ parameterized classes don't have quite the same concept of constrained genericity. The C++ compiler sorts things out in a different way. When a C++ parameterized class is instantiated, the compiler really has to generate additional code. During this process it can verify whether the type used to instantitate the parameterized class actually supports all the operations used in the functions of that class (actually, just those functions that get used in the program). Eiffel does not have a similar "instantiation" process, because the different versions of an Eiffel SORTED_LIST class can share the same code. Since Eiffel variables are all really pointers, they all have the same size. In a C++ parameterized class, one is likely to have automatic (stack-based) variables that are (value-based) instances of the parametric type. Obviously, different types will require quite different sizes for stack frames. Typically, there is less scope for code sharing among different variants of a C++ parameterized class than there is for the corresponding Eiffel generic classes.

Class ANY   As noted earlier, all classes in an Eiffel system are derived directly or indirectly from class ANY and thus all share a minimum set of common functionality. Part of the summary produced by the "short" documenting tool applied to class ANY is:

```
class interface ANY exported features
      io, out, … print, … copy, equal, deep_clone,
      deep_equal, …

feature specification

    io: STD_FILES
                -- Standard input and output

    out: STRING
                -- Terse printable representation of current ,
                -- instance field by field.

    print
                -- Write terse external representation of
                -- current instance on standard output.

    copy (other: like Current)
                -- Copy `other' into current instance, field
                -- by field.
            require
                ...

    equal (other: like Current): BOOLEAN
                -- Is current object field by field
                -- identical to `other'?

    deep_clone: like Current
                -- A copy of the current instance,
                recursively duplicating the entire
                -- dependent object structure.

        deep_equal (other: like Current): BOOLEAN
                    -- Are the object structures starting at
                    -- Current and `other' recursively identical?
        ...

    end interface -- class ANY
```

All objects share an instance of STD_FILES – this provides access to stdin and stdout on Unix. (How objects get to share a common i/o path is explained in section 13.11.) Code for a function in any Eiffel class can contain statements that read-from or write-to standard i/o. Objects of all classes can print themselves (the compiler arranges this automatically).

Different instances of a class can be checked for equality. Naturally, as explained more in the next section, there are different kinds of equality. The = operator tests for identity (do two variables refer to the same object), the equal function (which checks whether the data attributes in the two objects identical), and deep_equal that tests whether two structures are isomorphic. Obviously, the deep_equal function involves a form of recursive search — but the Eiffel compiler handles this. The Eiffel compiler also handles the similar problem of providing a "deep clone" — i.e. making a distinct but equivalent copy of an object.

13.3 OBJECTS

For the most part, Eiffel uses reference semantics. Variables (or "entities" in Eiffel's preferred terminology) are declared as being of particular class types but really they are pointers (references). These pointers are initially null or "void". Pointers get to reference objects by assignment, or through the process of creating a new object. Naturally, polymorphic assignments are permitted. A variable (pointer) of type INVADER can reference an instance of class SAUCER, class BOMB, or any other of class INVADER's subclasses.

The following fragment of a class declaration is from the example developed in Section 13.8. This example is a reworking of the ASSOC (association list) example from Chapter 5. The program is to maintain a list of country names and their international dialling codes, with the list being searchable by name or code number. Class PHONES is the "root" class for the system. At run-time, there will be a single instance of class PHONES present. It needs a list to store data in memory, and it needs to access a file where the list can be stored permanently. Consequently, it declares entities liz for the list in memory and book (phonebook) for the file that holds the persistent version:

```
class PHONES

feature

    liz    : STOREDLIST[ENTRY];
    book   : FILE;

    Create is
        do
                book.Create("PHONEBOOK");
                liz.Create;
                ...
        end; -- Create
```

As class PHONES is the "root" class for the program, an instance of this class will be created in the heap by the Eiffel run-time system. This PHONES object will be a data structure with fields for the two pointers.

Eiffel 2 has a Create "instruction". This can be applied to an entity (variable) *Create "instruction"* and will cause the creation of a new object in the heap and set the entity to reference the new object. The Create instruction looks like a call to a procedure and in fact classes generally define Create procedures. There are subtle differences between a class's Create procedure and its normal procedures; for example, a class doesn't list its Create procedure in its export list.

A class's Create procedure gets called immediately after a new object has been created on the heap and its data fields have received their default initial values —

reference entities are set to void (null), integers are zero, booleans are false etc. In the example, because class PHONES is the root class for a program, its Create procedure would be invoked by the Eiffel run-time system.

The Create procedure in class PHONES starts by creating the list object, liz.Create, and the file, book.Create("PHONEBOOK"). A Create procedure can take arguments. Class FILE (one of the standard Eiffel classes) provides an access path to a Unix file; its argument is a filename.

The objects stored in a PHONES's list are instance of class ENTRY. An ENTRY object has an INTEGER data field for the dialling code and a STRING for the country name (class STRING is from the standard Eiffel libraries):

```
class ENTRY export
    countryname,
    code


feature
    code         : INTEGER; -- The international code
    countryname  : STRING; -- Country name
```

INTEGER – an "expanded" type

An ENTRY is an object in the heap with an integer field and a pointer field. Integers, booleans, reals and a few other types are "expanded" – an object needing an integer doesn't have a pointer to some object that is an instance of class INTEGER ; it has a data field for the actual integer value. The entity countryname is initially a null pointer that will reference a separate STRING object. Entity countryname is set by assignment in one of the procedures of class ENTRY:

```
get_country is
    do
            io.next_line;
            io.putstring ("Country name: ");
            io.readline;
            if io.laststring.Void then
                    countryname := "Unknown"
            else
                    countryname := io.laststring.duplicate
            end;

    end -- get_country
```

Setting an entity by assignment

This code requests that the shared io object prompt the user for input and then read a line (using something much like C's gets(…) function). If the io object can find a characters forming a valid string, it creates a STRING object which is accessible via the laststring accessor. A duplicate of this STRING object is created and its address is assigned to entity countryname.

Figure 13.1 uses instances of class ENTRY to illustrate the different types of equality. (As suggested in Figure 13.1, a STRING object is itself a composite with a string descriptor part and a separate character string.)

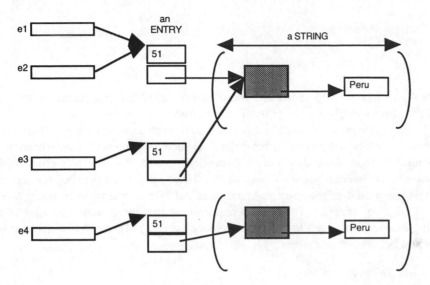

Figure 13.1 Example objects illustrating variations on the theme "equals", e1 = e2, e1 ≠ e3 but e1.equal(e3) is true, e1.equal(e4) is false but e1.deep_equal(e4) is true.

Entities e1 and e2 reference the same ENTRY object and therefore e1 = e2. Entities e1 and e3 don't refer to the same object so they are not "equal" in the sense of the = operator; they do however have identical values in their data fields (they are sharing the same STRING) and so satisfy the "equal()" function that tests for surface similarity. Entities e1 and e4 reference totally separate structures, but the structures they reference are isomorphic so e1 and e4 satisfy the deep_equal() predicate.

Usually, only entities of INTEGER, BOOLEAN, and similar basic classes are "expanded". So, if one had the classes:

```
class POINT export
    x, y, …
feature
    x, y   : REAL;   -- coords of a point
    …
end;   -- class PT

class RECT export
    top, left, bottom, right, …
feature
    top, left, bottom, right : REAL;
    …
end; -- class RECT
```

and

```
class INVADER export
    …
feature
```

```
        Position      : POINT;
        Area          : RECT;
        ...
    end; -- class INVADER
```

then an `INVADER` object would have some pointer fields. The actual `POINT` and `RECT` objects would need to be explicitly created.

Entities that are instances of expanded types

The data `Position` and `Area` are unique to each `INVADER` object. These data should never be used by any other object. In such a situation, it is somewhat more natural to have these data fields "expanded" so that an `INVADER` object really possess six "internal" `REAL` number fields. This would avoid the need for separate create steps and might result in some marginal improvements in memory layout and speed of access. (There is no requirement that the memory be allocated a single block, that is left to the Eiffel compiler writer.) The modified declarations would be:

```
    class INVADER export
        ...
    feature
        Position      : expanded POINT;
        Area          : expanded RECT;
        ...
    end; -- class INVADER
```

There are a few restrictions. The class that is "expanded" should have a `Create` procedure, possibly just the default one, that takes no arguments (this is somewhat analogous to the situation in C++ where if you want an array of objects their class must provide a default constructor that doesn't require arguments). Fitting expanded types into a predominantly reference based system necessitates lots of additional rules concerning assignments, parameter passing etc. In Eiffel 2.3, it is possible to write code that results in an entity of an expanded type in one object being shared, via a reference entity, in some other object; this kind of sharing seems undesirable.

Void

A "Void test expression" is used to check when a reference entity is unassigned; for example:

```
    class POINT
    export
        SetX, SetY, SetName, ...
        Describe, ...
    feature
        x, y          : REAL;
        ptname        : STRING;
        Describe is
        do
                io.putstring("Point (");
                if(ptname.Void) then
                        io.putstring("un-named")
                else
                        io.putstring(ptname)
                end;
                io.putstring(",");
```

Using a Void test

```
            io.putreal(x);
        …
    end;  -- Describe
```

The code fragment assumes that this class POINT has no Create procedure and that
the only way that a POINT can be named is via the SetName procedure. In this
situation, it is likely that some POINTs will be unnamed, with their ptname fields
being null. Obviously, this must be checked in the Describe routine because
otherwise an error would be generated when attempting to print a non-existent
STRING, so the if statement uses a "Void test expression".

The link from an entity to an object can be cut using the Forget "instruction" –   *Forget*
e.g. x.Forget – which makes the entity Void.

The Eiffel system provides a garbage collector. The "system description file"   *Automatic garbage*
that defines the libraries and root class for a program also specifies whether the   *collection*
garbage collection facilities are required. If garbage collection is enabled, the
collector runs as a coroutine with the main program. The garbage collector will
reclaim heap space from any objects for which there are no remaining links from
live entities.

The Eiffel system does maintain some run-time data describing the class of an   *Run-time*
object. The functions defined in class ANY include conforms_to(other) which   *identification of an*
can be used to check whether the current object is of the same type (or a descendant *object's class*
type) of the object used as an argument.

Checks on the dynamic type of an entity are relatively rare. But, some such   *Reverse assignment*
checks are often necessary when dealing with persistent objects – objects created
by one program, stored on file, and recovered later by the same program or by
some other program. A general purpose "restore persistent object" function will
have no choice other than to return a reference to an object of type ANY. Of course,
a program can't do much with an ANY. A program will be expecting something
specific, like a UTILITY_BILL or a CUSTOMER, rather than an ANY. If one is lucky,
the file will only contain objects of the expected type (and they would probably be
grouped by type so one would know what to expect next in the file). But it is still
necessary to have some way of checking whether a retrieved object is of the
expected type. Eiffel handles such checks through the reverse assignment operator
?=. For example:

```
    retrieved_item       : ANY;
    next_bill            : UTILITY_BILL;
    …
    -- get something(!) out of the file
    retrieved_item := filer.retrieve;

    next_bill ?= retrived_item;                          Reverse assignment
                                                         attempt
    if(next_bill.Void) then
            -- the item retrieved was not a utility bill,
            -- so, … ?
            …
    else
            -- can process bill, append it to list etc
            …
```

```
        end;
        ...
```

The assignment to next_bill would only be made if retrieved_item turned out to be a UTILITY_BILL (or an instance of some subclass of UTILITY_BILL); otherwise, next_bill would be Void. Essentially, the ?= operator involves a conforms_to check at run-time.

13.4 FEATURES

Entity declarations   Features include entities (variables), procedures, functions, and constants. Entity declarations look a bit like Pascal:

```
    <entity_name1>, <entity_name2>: <type>;
```

e.g.

```
    p1, p2      : POINT:
    ptname      : STRING;
```

Anchored types provide a minor variation on normal type declarations; for example in a generic class with THING as a parameter, one might have declarations like:

```
    p           : THING;
    q           : like p;
```

Arrays aren't built in, as in languages like Pascal. Array like facilities are provided through class ARRAY (a generic class). So one gets declarations like:

```
    invasion_fleet   : ARRAY[INVADER];
```

(The declaration specifies a pointer; the actual array would be created using a Create instruction e.g. invasion_fleet.Create(1,100). The Create procedure for class ARRAY takes arguments specifying lower and upper bounds.)

Constant declarations   Constants can also be declared. These include "manifest" constants:

```
    min_width : INTEGER is 100;
    max_width : INTEGER is 400;

    wname     : STRING is "Untitled";

    sqrt_2    : REAL is 1.4142135621
```

and "unique" values, such as a set of integers that a particular class might need to represent distinct error conditions:

```
    No-error        : INTEGER is unique;
    Locked_floppy   : INTEGER is unique;
    Write_fail      : INTEGER is unique;
```

(These are a bit like enumerated types; the actual values used don't matter, provided that each is distinct.)

Procedure and function declarations have a name, an argument list (if any arguments are required), and in the case of functions a return type. For example, the library class STRING includes the following procedures and functions:

```
to_lower is                                                    Procedures
-- Convert string to lower case
do
    ...
end; -- to_lower

append(s : STRING) is
-- append a copy of s at end of string
do
    ...
end;

item(i: INTEGER) : CHARACTER is                                Functions
-- Character at position i
do
    ...
end; -- item

duplicate  : STRING is
-- make a copy
do
    ...
end; -- duplicate

empty      : BOOLEAN is
do
    ...
end;
```

The library class STRING exports a feature count; count gives the actual length of the string. Now it could be that count is an integer feature of class STRING that is exported in read-only form:

Equivalence of readable attributes and functions without arguments

```
count  : INTEGER;
```

or it could be that count is a function (in a Unix environment, it might map onto strlen() from the C libraries):

```
count  : INTEGER is
    ...
do
    ...
end; -- count
```

A client using class STRING does not to be concerned with these issues; the two different implementations of count present the same interface. Access to the length is as msg.count, as in the following code, – msg : STRING; ... if msg.count > MAX_WIDTH then ...; – irrespective of whether count is a read-only attribute or a function.

Operator functions   It is possible to declare "operator functions". For example, if one wanted a class of (rectangular) AREAS that could be sorted, then a possible declaration would be along the following lines:

```
class AREA
export ..., ...,
    infix "<", ...
inherit
    PART_COMPAR
            define infix "<="
feature
    width, height: REAL;
    ...
    infix "<" (other: like Current) : BOOLEAN is
    do
            Result := width*height < other.width*other.height;
    end;

    ...
end; -- class AREA
```

13.5 FUNCTIONS AND PROCEDURES

The structure of a function or procedure is:

```
name [(arguments)] [ : <return type>] is
    [Precondition]
    [Externals]
    [declaration of local variables]
    Body
    [Postcondition]
    [Rescue]
    end;
```

Precondition and postcondition clauses for individual routines form part of Eiffel's mechanism for contract specification and enforcement. They are illustrated briefly in Section 13.9.

External functions   An external clause identifies routines written in other languages that are called from a function. (Actually, the only other language supported is C.) Many low-level operations use routines from Unix's standard C libraries. For example, class STRING implements its to_lower procedure using a C function str_lower (which on a Unix system would probably be implemented using tolower(...) from <ctype.h>); this auxiliary function is declared as an external:

```
to_lower is
    external
            str_lower language "C";
    ...
    do
```

```
            str_lower( ...);
    end;   -- STRING's to_lower
```

Similarly, a routine needing random numbers would most likely use `rand()` from
`<stdlib.h>`:

```
reposition is
    external
            random: INTEGER language "C";
            abs(x : INTEGER) : INTEGER language "C"
    local
            h,v    : INTEGER;
    do
            h := random;
            h := abs(h);
            ...
    end; -- reposition
```

Naturally, procedures and functions will generally require local variables. If *Local variables*
needed, these are declared in a local clause, like `h` and `v` in the code fragment
above. Local variables are always initialized (integers to 0, reals to 0.0, booleans
to `false`, while reference types are void). Apart from simple types like INTEGER
(and any expanded types), locals will be references (pointers) that must be set by a
`Create` instruction or by assignment. Any objects that are created in a routine and
are only referenced by locals will be "forgotten" on exit from that routine.

Functions have an implicit local variable named `Result` whose type is identical *"Result"*
to the return type of the function. Naturally, like all other local variables, `Result` is
initialized. The value returned by a function is the value in this local variable. The
following example function, used earlier to illustrate an infix operator function,
also illustrates the use of `Result`:

```
infix "<" (other: like Current) : BOOLEAN is
    do
            Result := width*height < other.width*other.height;
    end;
```

The statements forming the body of a routine follow the `do` keyword, as shown *Body of routine*
in the many small examples already presented. If the routine is abstract, it will
have no body – simply having the keyword deferred as illustrated earlier in Section
13.2 (with function `Draw` for deferred class `Invader`).

A routine may have a `rescue` clause. This forms part of the mechanism for *Rescue clause*
dealing with exceptions, and is described briefly in section 13.10.

13.6 STATEMENTS AND CONTROL STRUCTURES

Like all other imperative languages, Eiffel has assignment statements, loop
constructs, and conditionals. There are no procedure calls or function calls –
because there are no global procedures or functions. Of course, there are numerous

"feature calls" where a request is made to some other object asking it to perform an operation or return a value.

Assignments

Most assignments are going to be reference assignments – the entity on the left hand side of the expression gets to point to the object identified by the expression on the right hand side. Value assignments take place for simple types like INTEGER, and also when the entity on the left is an expanded instance of a class. Assignments look much like Pascal – `a := b`; `x := y.first`; etc.

Loop structures

A single loop construct suffices:

```
from
    -- whatever initialization steps are needed
    ...
until
    -- definition of termination test
    ...
loop
    -- do whatever must be done
    ...
end
```

A construct `from <1> until <2> loop <3> end` is equivalent to `for( <1>; <2>; ){ <3> }` in C. This basic loop construct can be used to create counting loops (like Pascal's `for i:= 1 until n do`), `while` loops, and `repeat` loops. As a specific example, the following code fragment is taken from the program presented in section 13.8 (the program manipulating a list of countries and their international dialling codes):

```
from
    liz.start
until
    liz.offright
loop
    e := liz.item;
    if e.code = number then
            io.putstring("Country is ");
            io.putstring(e.countryname);
            io.new_line;
            found := true;
    end;
    liz.forth;
end; -- loop along list
```

This code iterates along a list (`liz`) extracting individual `entry` objects (referenced via entity `e`). The dialling code for each is compared with the number entered in a query, and if it matches the country name is printed. (International dialling codes are not unique; for example, all countries that were part of the former Soviet Union currently share code 7). The loop setup, `from liz.start`, requests the list object position a mark at its first entry. The termination test, `until liz.offright`, will be satisfied if the list object reports that its mark has passed the last entry. The loop body has the code to retrieve the item at the current mark, `liz.item`, and use this item. The final statement of the loop gets the list object to advance its mark, `liz.forth`.

Eiffel has a typical `if` ... conditional construct with the variations: *Conditionals*

```
if <condition> then <action> end

if <condition> then
    <action 1>
else
    <action 2>
end
```

and

```
if<condition 1> then
    <action 1>
elsif <condition 2> then
    <action 2>
elsif
...

...
else
    <default action>
end
```

Examples of some of the simpler forms have been shown in earlier code fragments.

The slogan "*Case statement considered harmful*" has been attributed to Meyer.   *Selection statement*
Multiway selection statements like Pascal's `case` ... or C's `switch` should be relatively rare in OO programs. After all, most such statements express the idea "case 1 of original specification, case 2 of original specification, ... etc." and a good OO design should have factored out all these details of the original specification, relying instead on the polymorphic behaviour of related types of objects. However, sometimes selection statements are useful. For example, an input routine in some lexical scanner might need to read a character and, depending on the character, call routines that handle input of numeric data, identifiers or reserved symbols, comments, operators etc. A conditional construct like `if` ... `then` ... `elsif` ... `elseif` ... `else` could be used, but it would be unwieldy. Eiffel's `inspect` multi-branch selection statement can be used in such situations:

```
io.readchar;
ch := io.lastchar;
inspect
    ch
when 'a' .. 'z' then
    read_identifier
when 'A'..'Z' then
    read_reservedword
when '0'..'9' then
    read_number
...
else
    -- illegal character in input, abandon processing
```

```
        ...
   end
```

Current   The calls `read_identifier` etc. in the above fragment invoke other functions of the parser object that inspects the input character. This could have been made more explicit by writing the calls as `Current.read_identifier`, `Current.read_number` etc. Within the body of any member function of an Eiffel class, references to data members and other member functions are implicitly qualified by "`Current`" (just as in Object Pascal they are implicitly qualified by "`Self.`" and in C++ they are implicitly qualified by `this->`). It is legal to make the `Current .` qualifier explicit, but this is not the normal style of most Eiffel programmers. (`Current is` technically an expression; it is not an entity as it cannot appear on the left hand side of an assignment.) Usually, `Current` (i.e. the implicit pointer to the object executing a function) only appears in code that sets up networks of cooperating objects, where a reference to the current object is being passed to some newly created object (there are examples in the code illustrated in section 13.13).

Feature calls   As noted earlier, there are no calls to globals functions or procedures because there are no such things. All calls are feature calls – "hey object, do this". There are numerous examples in code fragments illustrated earlier. A coding style using concatenated feature calls (or member function calls) is possible in most OO languages, but seems particularly common in Eiffel; a simple example is

```
   io.laststring.duplicate
```

The feature call `io.laststring` invokes the `laststring` function of object `io`; this function returns a `STRING` object. This `STRING` object is then requested to perform its `duplicate` function. Concatenated sequences of as many as five calls sometimes appear in Eiffel code.

These few simple constructs are about all there is to Eiffel as a coding language. There are obviously arithmetic and logical expressions; these use the normal operators and normal precedence relations. But there are no address-of operators, or pointer manipulations, no short form increments of counters etc. The coding aspects of the language are limited, simple, and straightforward.

13.7 INHERITANCE

As noted in Section 13.2, a class's inheritance structure is specified in an `inheritance` clause; if a class does not have an explicit `inheritance` clause, it is implicit that it is derived immediately from Eiffel's class `ANY`. The `inheritance` clause may contain `define` subclauses (when a class defines functions left as deferred by its ancestors), `redefine` subclauses (where a class provides a new implementation that overrides an inherited routine), and `rename` subclauses. The `rename` subclauses have a number of uses; in particular, in the context of multiple inheritance, `rename` clauses are used to resolve name conflicts among functions and control the replication/sharing of data fields.

Eiffel's design presumes that inheritance will be used extensively to as a mechanism for software reuse. This necessitates that classes be "open" so that the programmer adopting an existing class can have untrammelled access to existing features and can (almost) freely override inherited behaviours. The controls provide by Eiffel are flexible as will be illustrated in the context of examples based on those in Chapters 3, 6, and 11.

A simple case of inheritance would be one involving INVADER and its SAUCER, ROCK and other subclasses as in the Space Invaders example. Class INVADER is a deferred class:

Simple inheritance

```
deferred class INVADER
export
     Run, Draw, Move,
     ResetPosition, Erase, …

feature
     hpos, vpos    : INTEGER;

     Move is
     do
             Erase;
             …
             Draw
     end;

     Draw is
     deferred
     end;

     Run is
     deferred
     end;

     …

end; -- class INVADER
```

An abstract base class

The subclasses would each specify their inheritance with an inheritance clause that would contain define subclauses and, possibly, redefine subclauses. (In this case, it would not be necessary to rename any inherited features.) Thus, one would get:

```
class ROCK
export
     repeat INVADER
inherit
     INVADER
             define Draw, Run, …
feature
     Draw is
     do
     …
     end; -- Draw
```

A derived concrete class providing effective implementation of deferred procedures

```
    …
    end; -- class ROCK
```

The export clause is discussed in more detail later in this section. In this case, the client interface for class ROCK is identical to that for class INVADER – nothing added, nothing removed. Class ROCK simply provides definitions for those functions left deferred in the abstract INVADER class.

Class SAUCER might be slightly more complex in that it might need to change the default Move behaviour provided by the INVADER base class, in addition to providing effective definitions for deferred procedures:

<table><tr><td>

A derived class overriding inherited procedures

</td><td>

```
class SAUCER
export
    repeat INVADER
inherit
    INVADER
            define Draw, Run, …
            redefine Move
feature
    Move is
    do
    …
    end;  -- Move

    Draw is
    do
    …
    end; -- Draw

    …
    end; -- class ROCK
```

</td></tr></table>

The intention to override an inherited routine must be explicitly stated in a redefine subclause.

Changing the signature of a redefined routine

The signature of a redefined routine should be generally consistent with that of the original, but the consistency requirement does allow some changes in the types of arguments (and of the returned value from a function). Arguments and returned value of a redefined routine must be compatible with the originals. As a simple example, class INVADER might have a procedure scavenge_parts_from(wrecked : INVADER); in class SAUCER, it would be quite reasonable to redefine this as scavenge_parts_from(wrecked : SAUCER). The whole business of redefining signatures along with "covariant" and "contravariant" policies for the allowed changes in argument types is a somewhat complex area of Eiffel and related languages like Sather. It is touched on further in section 13.12.

Preserving the original implementation of a routine

Of course it is very common for a new class to need to redefine an existing routine so that some additional actions are taken either before or after the normal processing. For example, class SAUCER's Move procedure might want to do everything that a normal INVADER does and then provide an option for the SAUCER to fire at its target. All languages with inheritance must provide a mechanism allowing for such constructs. In Eiffel, it is achieved through a combination of redefinition and renaming. The renaming preserves the original implementation of a function, providing it with a distinct name valid in the context of the redefining class. The replacement implementation can involve a call to the renamed routine. So, in the case of the example, one would get something along the lines:

```
class INVADER
export
    …
feature
    Move is
    do
    …
    end;


    …
end; -- INVADER

class SAUCER
export
    …
inherit
    INVADER
    rename                                              Rename to preserve
            Move as I_Move                              original
    redefine                                            Redefine to change it
            Move
    …
feature
    …
    Move is                                             Use original in
    do                                                  redefined version
            I_Move;
            Fire_at_will
    end; -- Move

    Fire_at_will is
    do
            …
    end;

    …

end; -- class SAUCER
```

C++ programmers should not confuse Create procedures with constructors. In *Create procedures* C++, class INVADER's constructor gets invoked as part of the process of constructing an instance of a derived class such as a SAUCER. This is not the case in Eiffel. An Eiffel programmer must handle initialization of base class components explicitly. A Create procedure for class SAUCER would need to invoke the Create procedure in class INVADER. This necessitates renaming; and it is very common to see things like:

```
class INVADER
…
feature
```

```
                    Create is
                    do
                    …
                    end;


                …

        end; -- INVADER

        class SAUCER
        …
        inherit
            INVADER
            rename
                    Create as I_Create

            …
        feature
            Create is
            do
                    -- create standard invader bits
                    I_Create;
                    -- now specialized initializations, paint saucer,
                    -- prime guns etc
                    …
            end;

        end; -- SAUCER
```

Inheritance: the only mechanism for sharing

As noted in section 13.2, inheritance is used as a way of providing shared declarations of constants etc. There is no other "include" mechanism so if two classes, e.g. some display WINDOW class and some interactive SKETCHER class, need to know the minimum and maximum dimensions of a drawing area one would need something like:

```
    class DIMENSIONS
    feature
        min_width    : INTEGER is 100;
        min_height   : INTEGER is 160;
        max_width    : INTEGER is 400;
        max_height   : INTEGER is 640;
    end; -- DIMENSIONS

    class WINDOW
    …
    inherit
        DIMENSIONS
        …
    …
    end; -- WINDOW

    class SKETCHER
    …
    inherit
        DIMENSIONS
        …
    …
```

```
end; -- SKETCHER
```

A client's view of a class is defined by its export list. A class will name those *Exports*
of its features that it wishes its clients to access. It can also choose to export either
all, or just some of the features that it has inherited from its ancestors. If a class
does not rename any inherited feature, then it can simply repeat the export list of its
parent:

```
class INVADER
export
    Draw, Move, Run, …
…
end; -- INVADER

class SAUCER
export
    repeat INVADER,
    color
feature
    color  : INTEGER;
…
end; -- SAUCER
```

*Repeating the export
list of a parent*

As noted earlier, exports can be selective; thus, the export specification export
special_feature { PRIV1, PRIV2} would restrict use of special_feature to
the named classes PRIV1 and PRIV2 (together with all their descendants). (This
selective export mechanism is used to guarantee that the basic features defined for
class ANY are always available. Class ANY exports features like out and equal to
ANY and its descendants, i.e. to everything.)

A class is not obliged to export all of the features that were exported by its *Exporting inherited*
parent, and it can chose to export inherited features that were not exported by its *features*
parent. The example program in the next section includes the (generic) class
STOREDLIST[T]. This is built from existing classes STORABLE and
LINKED_LIST[T]; it exports all the features defined in STORABLE, but only a subset
of the features defined in LINKED_LIST:

```
class STOREDLIST[T] export
    repeat STORABLE,
    empty, first, item, put_right, forth, start, finish,
     search_equal, offright, remove, count

inherit
    LINKED_LIST[T];
    STORABLE

…
end; -- STOREDLIST
```

Class LINKED_LIST has many other features, e.g. wipe_out and split, that were
not required in class STOREDLIST of the example program and so were not included
in its export list. Such usage is perfectly reasonable for a class developed for a
specific, quite limited purpose. However, a combination of certain uses of

Multiple inheritance

polymorphic assignments and limited export lists causes some problems; these are examined briefly in Section 13.12.

In contrast to C++, multiple inheritance is probably more common than single inheritance for classes in an Eiffel program. Some uses of multiple inheritance are essentially trivial. In the earlier example with class DIMENSIONS and its heirs WINDOW and SKETCHER, multiple inheritance would almost certainly be used with DIMENSIONS being only one of a number of parent classes for the other classes. The real inheritance relationships would be between SKETCHER and some "command" class or between WINDOW and some "screen manager" class. In similar fashion, classes that want to use mathematical functions like sine have to "inherit" from class SINGLE_MATH. Again, this inheritance relationship is really an artefact, a contrivance to avoid any "global" constants or functions.

The standard ISE libraries for Eiffel have several classes that capture general concepts such as PART_COMPAR, COMPARABLE, or STORABLE. Many of these are partially implemented abstract classes. For example, PART_COMPAR exports the infix functions "<", "<=", ">", and ">="; only "<" is deferred, the others being defined in terms of equal and "<". It is very common for application specific classes to inherit from several of these standard classes, providing their own effective implementations for any deferred functions.

In the Eiffel libraries and examples, one does find cases of inheritance overkill. Some of the use of inheritance as a composition device don't simplify programs, they make things more complex to understand. As an example, the Eiffel library contains class ITERATOR:

```
deferred class ITERATOR export
    iterate, start, over, action, next
feature
    iterate is
                -- Loop
        do
            from
                    start
            until
                    over
            loop
                    action;
                    next
            end
        end; -- iterate

        start is
                    -- Initialize loop.
            deferred
            end; -- start

        -- functions over, action next all deferred
        ...

end -- class ITERATOR
```

This class is simply a repackaging of the standard `loop` construct. It is not a class definition in the sense of a description of a type of entity that owns resources and provides services based on those resources. It describes a single capability (the ability to iterate). So, it is not like a class like PART_COMPAR which provides a consistent way of describing the many facets of behaviour that make up the concept of "something that can be placed in a defined order". But it exists and can be reused. In fact it can be multiply re-used. For example, one of the standard Eiffel examples includes a class ANIMATED that inherits twice from class ITERATOR:

```
class ANIMATED export
    insert_image, …
inherit
    ITERATOR
            rename
                    action as snapshot
            redefine
                    snapshot;
    ITERATOR
            rename
                    iterate as wait,
                    action as snapshot,
                    next as increment,
                    start as start_waiting,
                    over as stop_waiting
            redefine
                    start_waiting, stop_waiting,
                    increment, snapshot
    feature

    …

    end -- class ANIMATED
```

The doubly inherited ITERATOR essentially provides two functions (one still named `iterate`, the other called `wait`) that are really just normal loops. The renaming of the auxiliary functions results in distinct behaviours; one loop is `from  start until  over  loop  snapshot;  next  end` while the other is really `from start_waiting until stop_waiting loop snapshot; increment end`. Class ANIMATED provides the definitions of `start`, `start_waiting`, `over`, and other functions. This use of repeated compositional inheritance seems an overly complex way to define two iterative methods in a class.

Repeated inheritance obviously engenders name conflicts. But, fortuitous name collisions can of course arise with any general use of multiple inheritance. These are resolved by renaming as in the following example where it is necessary to resolve name conflicts for the procedure `travel` and the data attribute FuelWeight:

Renaming to solve name conflicts

```
class INTERSTELLAR_TRANSPORT
…
feature
    FuelWeight   : INTEGER;
    travel is -- leap into hyperspace
```

```
        do
        …
        end;

    end; -- INTERSTELLAR_TRANSPORT

    class INVADER
    …
    feature
        FuelWeight   : INTEGER;
        travel is -- move in to attack
        do
        …
        end;
        …
    end; -- INVADER

    class  SAUCER
    …
    inherit
        INTERSTELLAR_TRANSPORT
                rename travel as t_travel,
                        FuelWeight as t_FuelWeight;
                redefine travel;
        INVADER
                rename travel as i_travel;
                redefine travel;

    feature
        travel is
        do
                -- move in
                t_travel;
                -- and attack
                i_travel
        end;
        …
    end; -- SAUCER
```

In class SAUCER, routines inherited from INVADER would manipulate the FuelWeight field introduced in that class, while routines inherited from INTERSTELLAR_TRANSPORT would use the second FuelWeight field. Any code defined in class SAUCER would need refer to these fields as FuelWeight and t_FuelWeight.

Replication of data attributes   With repeated inheritance, one can get either sharing or replication of data attributes; the process is controlled by renaming. A feature repeatedly inherited from a common ancestor is shared unless it has been renamed along one of the inheritance paths. Thus in the standard situation of an amphibian DUKW repeatedly derived from a DIESEL_VEHICLE base class (which involves a repeated inheritance "diamond"):

```
class DIESEL_VEHICLE
…
feature
```

```
        e              : ENGINE;
        steerwheel     : PART;
    ...
    end; -- DIESEL_VEHICLE

    class TRUCK
    ...
    inherit
        DIESEL_VEHICLE
    ...
    end; -- TRUCK

    class BARGE
    ...
    inherit
        DIESEL_VEHICLE
    ...
    end; -- BARGE

    class DUKW
    ... -- describe an amphibious vehicle
    inherit
        BARGE
                ...;
        TRUCK
                ...
    ...
    end;
```

The designer of the DUKW class can select whether a DUKW has one or two engines and one or two steering wheels. It is the designer of the DUKW class who makes these decisions, not the designers of the intermediate classes. Further the selections of replicated versus shared status can be made independently for each individual attribute (and you can't do that in C++). Thus, if two engines are needed, class DUKW would use a rename subclause in its inheritance clause (e.g. inherit BARGE rename engine as b_engine); by default, it will get just one engine and one steering wheel.

In principle, Eiffel's approach here is very flexible and gives the designer of the derived class the kind of control that would be required. In practice, this supposedly flexible inheritance scheme is not that much use in Eiffel 2. There are restrictions on polymorphism (e.g. a DUKW object can be referenced by an entity of type DUKW or DIESEL_VEHICLE, but not by an entity of type BARGE, or TRUCK). More importantly, if class DIESEL has a function, e.g. service, that uses its engine field then the compiler cannot sort out what to do and terminates compilation. After all, in the context of a DUKW, it is not clear which engine should be serviced. These issues have been explored in more depth in Eiffel 3 which has provided additional controls that make practical this type of inheritance structure. In any case, although it may be a pretty feature, it is rare for practical projects to need this ability to combine concrete classes like TRUCK and BARGE.

13.8   A SIMPLE EXAMPLE

Sufficient of the Eiffel language has now been presented to allow a first small
example. It is a variation and extension of the "association list" example from
Chapter 5. The program is to maintain a list of countries and their international
dialling codes that can be searched by country name or by number. The program
provides more facilities than that in Chapter 5; these lists of names and numbers are
persistent.

The program will require a "root" class, PHONES, that will handle interaction
with the user. There will have to be some class, ENTRY, that holds a country-name
and number pair, and there will be a list structure to hold the entries in memory and
organize transfer to disk.

Reuse, reuse!   It is at the point that a difference in design style becomes obvious. If one were
working with just an OO language, one would proceed to design some suitable list
structure. With a proper OO environment, one's first response is to find
components that can be reused. Here, a persistent list (i.e. a list that can easily be
saved to and restored from file) is required. Obviously, the Eiffel library includes
several types of list in its structures cluster and there are facilities for persistence in
its support cluster. A persistent list can be created by simply combining these
existing classes:

```
class STOREDLIST[T] export
    repeat STORABLE,
    empty, first, item, put_right, forth, start, finish,
            search_equal, offright, remove, count

inherit
    LINKED_LIST[T];
    STORABLE

feature

end; -- class storedlist
```

As defined, a STOREDLIST can do all the important things a LINKED_LIST can do,
and also everything a STORABLE can do.(e.g. store_by_name and retrieve_
by_name for transfers to/from a named file). The STORABLE features inherited by
STOREDLIST handle all aspects of file transfers. There is no need for the author of a
class like ENTRY (the item stored in the list) to write any specialized functions or
include any macros in order to exploit system support for object i/o (so, object i/o is
much simpler to use than it is in the systems described in Chapter 9). The Eiffel
compiler provides a specialized version of function out (a function inherited from
class ANY) in every class; out provides a printable representation of an object.

An ENTRY owns a string and an integer data field; clients are allowed read
access to these fields:

```
class ENTRY export
    countryname, code

feature
```

```
code          : INTEGER; -- The international code
countryname   : STRING; -- Country name
```

Entries will normally be created interactively when a user has to add an extra country to the list. Here, the Create procedure has been written to call routines that will prompt for the individual data elements (Create is not like a C++ constructor, it doesn't get called automatically and, in particular, it is not invoked when a STOREDLIST is executing is retrieve_by_name procedure):

```
Create  is
      do
            get_code;
            get_country;
      end; -- Create

get_code is
      do
            io.putstring ("Code: ");
            io.readint;
            code :- io.lastint;
      end; -- get_code

get_country is
      do
            ...
      end -- get_country
end; -- class ENTRY
```

The i/o code is typical of Eiffel. Functions shouldn't really have side effects in Eiffel, while procedures can. Consequently, there isn't a readint function returning an integer (which would have a side effect in changing the state of an i/o stream). Instead, readint is a procedure; if it executes successfully, it will change the state of the stream and also the state of the io object (where it records the value of the integer read). This value can be subsequently accessed using the function lastint.

The root class PHONES owns a STOREDLIST and a FILE; since STOREDLIST is a generic class, it is explicitly instantiated as STOREDLIST[ENTRY].

```
class PHONES
feature
    liz          : STOREDLIST[ENTRY];
    book         : FILE;
```

As it is the root class, the Eiffel run-time system will create an instance of class PHONES and invoke its Create procedure. The Create procedure sets everything running by restoring the list from a named file (provided the file exists), printing various prompts and then calling the interact routine. When interact returns, the program will terminate; but, before terminating, the possibly modified list is again stored on file.

```
Create is
      do
```

```
                                      book.Create("PHONEBOOK");
                                      liz.Create;
Restore list structure                if book.exists then
          from file                          liz.retrieve_by_name("PHONEBOOK");
                                              liz := liz.retrieved;

                                      end;

Prompt user                           io.putstring("Phone codes demo");
                                      io.new_line;
                                      io.putstring("Commands:");
                                      io.new_line;
                                      io.putstring("a -- add code/country,");
                                      ...

                                      ...
                                      io.putstring("q -- quit");
                                      io.new_line;
                                      Interact;

Save list to file                     liz.store_by_name("PHONEBOOK");
                              end; -- Create
```

The `Interact` routine is simply a loop that repeatedly iterates until a "quit flag" has been set (by the user entering command letter q). The routine needs a local variable of type BOOLEAN, and a `from ... until ... loop ... end` construct. The user interface is trivial; single letter commands are entered to select processing options. The options are handled by separate routines that prompt for an additional data that may be required.

```
Interact is
local
        quitflag : BOOLEAN
do
        from
        until
                quitflag = true
        loop
                io.putstring("Enter command:");
                io.new_line;
                io.readline;
                io.laststring.to_lower;
                if io.laststring.equal ("a") then
                        add_entry
                ...

                ...
                elsif io.laststring.equal ("c") then
                        lookup_country
                elsif io.laststring.equal ("q") then
                        quitflag := true
                else
                        io.putstring("I don't understand.")
                end
        end -- loop
end; -- Interact
```

The `add_entry` routine creates a new ENTRY (with ENTRY's Create procedure arranging for prompts to the user for name and number data) and then inserts this at the end of the list. The standard list classes in the Eiffel library work by setting a mark and then performing manipulations at the mark position; so, to append an entry at the tail of the list, one moves the mark to the end, `liz.finish`, and then inserts to the "right" of the mark.

```
add_entry is
local
        e : ENTRY;
do
        e.Create;
        liz.finish;
        liz.put_right(e);
end; -- add_entry
```

The routine to identify all countries sharing a particular international code requires a number of local variables, a conditional construct and a loop. If the list is empty, nothing can be processed; this is tested using function empty from the LINKED_LIST. If there are entries in the list, the user is prompted for the number of interest and then a loop is used to iterate along the list. Movement in a list is effected using procedures like `liz.start` (move mark to start), `liz.finish` (move mark to finish), and `liz.forth` (move mark to next entry). The iteration will terminate if the mark is moved off the end of the list, `liz.offright`. The data item, of type ENTRY, at the mark position is retrieved using `liz.item`.

```
lookup_number is
local
        number : INTEGER;
        e      : ENTRY;
        found  : BOOLEAN;
do
        if liz.empty then
                io.putstring("The list is empty, ");
                io.putstring("you can't look up a number.");
                io.new_line
        else
                io.putstring("Enter number : ");
                io.readint;
                number := io.lastint;
                io.next_line;
                from
                        liz.start
                until
                        liz.offright
                loop
                        e := liz.item;
                        if e.code = number then
                                io.putstring("Country is ");
                                io.putstring(e.countryname);
                                io.new_line;
                                found := true;
                        end;
```

```
                                    liz.forth;
                        end; -- loop along list
                        if not found then
                                io.putstring("Sorry don't know.");
                                io.new_line;
                                end;
                end;

        end; -- lookup_number
```

The other routines like lookup_country and remove_entry are similar. Naturally all these routines that are part of class PHONES are bracketed in the features list of the class:

```
class PHONES
feature
        liz     : STOREDLIST[ENTRY];
        book    : FILE;

        Create is
        ...
        end; -- Create

        Interact is
        ...
        end; -- Interact

        add_entry is
        ...
        end; -- add_entry

        remove_entry is
        ...
        end; -- remove_entry

        lookup_number is
        ...
        end; -- lookup_number

        lookup_country is
        ...
        end; -- lookup_country
end; -- class PHONES
```

The "system description file" (SDF) serves as the input for Eiffel's intelligent "make" mechanism. The SDF identifies the root class, in this case class PHONES, and the location of necessary libraries. Other options allow enabling/disabling of things like assertion checks at run-time, garbage collection (in this case it's turned off), along with more specialized options like C package generation (generation of a version of the complete program in C for cross-compilation to some other platform). Eiffel's make system, es, does a graph walk of the classes starting at PHONES. This identifies all the classes that are used either via inheritance relations or by client server relations. This information is then used to organize the compilation process.

```
ROOT: phones                                                A "System
UNIVERSE: $EIFFEL/library/support $EIFFEL/library/structures Description File"
EXTERNAL:
NO_ASSERTION_CHECK (N):
PRECONDITIONS (Y): ALL
ALL_ASSERTIONS (N):
...
GARBAGE_COLLECTION (N)
C_PACKAGE (N):
...
```

13.9 ASSERTIONS

Meyer has suggested viewing programming as being a contractual process. Authors of components, individual classes or clusters of classes, should provide a contractual specification of what they will do for clients. Of course, the client has some responsibilities in any contractual arrangement (e.g. if you contract with a builder to put up a house on a plot of land, it is probably not the builder's responsibility to determine whether you own the land and whether you have planning permission). The contract style proposed for software components entails specification of what the contractor can assume will be done by the client (fair dealing arrangements), what the contractor will deliver, and for possibly, for the contractors own internal management, some specification of how the work will be accomplished reliably.

The contract notion is possibly most useful in the design stage of system development, but Eiffel does provide `require`, `ensure`, and `invariant` constructs that allow expression of these contractual arrangements in implemented code.

Each routine in an Eiffel class can have a `require` clause prior to any declarations of external functions and local variables. For example, in class `STORABLE`, the function `store_by_name` starts

```
store_by_name (filename : STRING) is
    require
            filename_not_void:
                    not filename.Void
    ...
```

It is the client's responsibility to pass a valid string as a filename. If the client cannot manage to deal with this part of the work, then class `STORABLE` really cannot be expected to do much about saving data to a file. Obviously, this started as part of `store_by _name`'s initial design specification ("this routine will be passed a string with the filename") which here has been converted into a formal precondition. A `require` clause can have several subparts. Typically each subpart has a label (name) and then a description of the constraints. Sometimes, the description will take the form of a comment, but often the description will be in the form of an executable Eiffel statement as in the example where a test is made to determine whether a void reference has been passed.

These `require` clauses form part of the public interface for a class. In an Eiffel environment, a client programmer will use the class summary produced by the "short" tool to obtain a specification of the services provided by a class of interest. "Short" lists the exported routines with their arguments, initial comments, and any `require` clauses. So, in this case, a client programmer will realize that it is necessary to choose and set the filename before a `store_by_name` operation is attempted.

It is possible to have executable preconditions compiled into the code generated. (The system description file, shown at the end of Section 13.8, uses most of the default settings and so does include precondition checks in the executable code). Any precondition associated with a routine will be evaluated whenever that routine is called. If a condition is violated, an exception will be raised. If the exception is not caught in the invoking environment (see Section 13.10), it will result in termination of the program. The label associated with the violated constraint will be used to identify the condition that has failed. So, if a programmer carelessly encoded a call to `store_by_name` using an unset string pointer as argument, the program would be terminated with a `filename_not_void` exception being reported.

The `require` statements specify a client's obligations to the contractor; `ensure` statements specify the contractor's promises to their clients. Again, each routine in an Eiffel class can have an `ensure` clause. This clause will specify the changes made to a data structure, or constraints on a result being returned by a routine. Some examples from deferred class `LIST` (a parent of `LINKED_LIST[T]`) are:

```
start is
    -- Move cursor to first position;
    do
            if not empty then
                    go (1)
            end
    ensure
            empty or isfirst
    end; -- start

back is
            -- Move cursor backward one position.
    require
            not_offleft: position > 0
    do
            move (-1)
    ensure
            position = old position - 1;
    end; -- back

wipe_out is
-- Make the list empty.
deferred
ensure
        empty
end; -- wipe_out
```

An example like wipe_out illustrates an ensure condition being used primarily as a design tool. Class LIST is deferred; programmers have to create concrete list classes by derivation. Any programmer creating a list class must provide an implementation for wipe_out. It doesn't matter quite what they do in their version of wipe_out provided that they guarantee that when they've finished the list is empty. (In a list built on an array, wipe_out might simply zero out an index while a version of wipe_out for a list built from links might have rather more work to do.) Here the ensure condition summarizes what the designer of class LIST requires from implementors of specialized lists.

The other ensure clauses in the example functions really act to provide a guarantee of consistent implementation of the separate routines of class LIST. If procedure go has been correctly implemented then, unless the list is empty, go(1) will mean that the mark is at the first position in the list.

The ensure clause for procedure back illustrates how Eiffel allows checks on changes in the values of data fields. The clause

```
ensure
        position = old position - 1;
```

requires that the final value in data field position is one less than the value this field had on entry to procedure back (i.e. the "old" value of position).

Many of the ensure clauses that appear in implementation code are really pretty trivialized. Thus, in one of the Eiffel example programs, there are routines like:

```
start is
    do
            times := 0
    ensure
            times = 0
    end; -- start

start_waiting is
    do
            waiting_step := 0
    ensure
            waiting_step = 0
    end; -- start_waiting

increment is
    do
            waiting_step := waiting_step + 1
    ensure
            waiting_step = old waiting_step + 1
    end; -- increment

stop is
    do
            stopped := true
    ensure
            stopped
    end; -- stop
```

These are really improper usages of the ensure construct.

The real value of ensure clauses is in design – either design of a whole class or of a deferred routine in a class. Thus one might have a class SET with a design specification something like:

```
deferred class SET[T]
...
feature
    ...
    put(v: T) is
    deferred
    ensure
            -- set contains a single instance of v
            -- no change if v already present
    end; -- put

end; -- set
```

The ensure clause, in the form of a comment, provides a concise summary of what the implementor must achieve – a set that now has the element v, or an unchanged set if v was already present.

Class invariants again serve mainly as a design tool, but once again they can be included in executable code as a run-time check. If used in design, they would specify conditions that must be satisfied whenever a class instance is in a stable state. Class LIST has the invariants:

```
invariant
offright_is_empty_or_beyond:
    offright = (empty or (position = count + 1));

offleft_is_pos_zero:
    offleft = (position = 0);

empty_is_offleft_and_offright:
    empty = (offleft and offright);
```

The various kinds of assertions are also important in the context of class reuse. If a programmer extends a class and changes some of the existing functionality, they are still bound to honour the original contract. A derived class can relax preconditions – offering to do similar work under more general conditions, and can strengthen postconditions – offering to do more or better work than originally done; but such extensions still satisfy the original contract. This is important for it allows a designer to create library classes that use precondition and postcondition constraints to express the semantics of a class confident that the same semantics will be honoured by all derived classes.

13.10 EXCEPTION HANDLING

Where programs deal with the real world, they must allow for failure. Failures like bad storage media and unreliable communication lines. Consequently, all

programming systems need to make some provision for detecting and handling
such problems.

Meyer chose to integrate into the language the handling of failures, and other
"exceptional conditions". Exceptional conditions include things trapped by the
hardware (e.g. division by zero) and the operating system (e.g. write fail to disk),
violated assertions in routines, and exceptions generated explicitly by the code of
routines. (Eiffel classes can inherit from class EXCEPTIONS which includes a
method raise that can be used to generate a programmer-defined run-time
exception). Each routine in an Eiffel class can make provision for dealing with
exceptions raised during the execution of its body. This takes the form of a rescue
clause after the body of the routine. (It is also possible to have a rescue clause
associated with a class as a whole.)

A rescue clause can effect some processing intended to remediate a problem   *Rescue clause*
and then cause processing to be retried. If it is not appropriate to retry the
operation, a routine is considered to have failed and an exception will be
propagated to the calling environment.

The following provides a simple example:

```
batter_up is
local
    strikes                         : INTEGER;
    three_strikes_and_you_are_out   : INTEGER is 3;
do
    -- bat up at baseball
    ...
rescue
    strikes := strikes + 1;
    if strikes < three_strikes_and_you_are_out then retry;
end; -- batter_up
```

Normal processing causes an attempt to execute the body of the routine; all being
well, this will complete and return before the rescue clause. If something goes
wrong, an exception would be raised causing the rescue clause to be executed.
The count of attempts is incremented (all local variables are initialized so the initial
value of strikes is known to be zero). If appropriate, another attempt is made to
execute the main body of the routine. Otherwise, processing is abandoned with an
exception being raised in the calling environment.

13.11 "ONCE"

In addition to deferred, there is another alternative form for the body of a routine.
Instead of deferred, or the normal do, this form uses the keyword once. For
example in class ANY there is the function:

```
io : STD_FILES is
    once
            Result.create;
    end; -- io
```

Both functions and procedures can be declared as being once routines.

The body of a once routine is only executed on its first call. If it is a function, the result that is to be returned is remembered. On subsequent calls to a once function, the saved result is again returned. Subsequent calls to a once procedure have no effect.

These once routines provide a means of avoiding "global shared variables" and cutting down on administrative overhead for communication among classes. Thus, although all Eiffel classes effectively share a single instance of STD_FILES (that is used for things like io.putstring(…) etc.), the language does not have to support the concept of a shared global variable. Each time an object that is executing a routine performs a feature call like io.putstring it will execute function io, inherited from class ANY, to obtain a reference to the STD_FILES object and then request that this STD_FILES object perform operation putstring. Since function io is a once function, the very first call causes a STD_FILES object to be created; subsequent calls return references to the same STD_FILES object. If one didn't have the once mechanism, some form of global variable would be required.

As another example of where such a feature might be useful, consider a program like one of the integrated office packages available on PCs (e.g. Microsoft Works). Typically, one can have several different windows open onto different "spreadsheet" and "database" documents, together with a single "data entry bar". The data entry bar is used to change data in the "currently selected" document. Now each "document" object needs to have a reference to the "data entry bar" object. (A document responds to selection of one of its data elements by copying information to the data entry bar, and handles "enter" commands by copying data from the data entry bar to a data element.) In most programming styles, it would be necessary either to have a global variable referencing the data entry bar, or an arrangement by which the data entry bar object was created by the overall application and a reference passed to each document that was created. Here a once function would provide the basis of a cleaner alternative. The document class could have a once function that creates the data entry bar. The first document to use the data entry bar would bring it into existence; subsequent documents would all obtain references to the existing instance.

13.12 "HOLES" IN THE TYPE SYSTEM?

Some of the features of Eiffel that are intended to encourage aggressive reuse actually introduce minor anomalies or "holes in the type system". One way that problems may arise is with derived classes that choose to export only a subset of the features exported by a parent class.

For example, one might have a class RECTANGLE whose features include a point origin (represented as an instance of class PT), horizontal and vertical dimensions (another PT), maybe a function computing an area, and, possibly, functions that allow rectangles to be independently scaled along the x and y axes.

```
class  RECTANGLE
export
    origin, dimension, area, scalex, scaley, …
```

```
feature
        origin : PT;
        dimension    : PT;

    Create (…) is do … end;
    …

    scaley(scale : real) is
    do
            dimension.sety(scale*dimension.y);
    end; -- scaley

end; -- class RECTANGLE
```

Given that class RECTANGLE was available, a programmer might decide to implement SQUARE as a special case; deriving class SQUARE from class RECTANGLE . Of course, the horizontal and vertical dimensions of a square should always be the same; this requirement might be specified as an invariant. Since one cannot independently scale the sides of a square, the class should not export the scale routines:

```
class SQUARE
export
    origin, dimension, side, …

inherit
    RECTANGLE
            rename
                    Create as Rect_Create
            redefine
                …

feature
    …

    side : REAL is
    do
            Result := dimension.x;
    end; -- side

…
invariant
    is_square: dimension.x = dimension.y;
end; -- class SQUARE
```

Under Eiffel's rules, although a SQUARE cannot be asked to do everything a RECTANGLE can do (e.g. you can't ask a SQUARE to perform operation scaley), it is still a RECTANGLE, and so can be included in a list of RECTANGLES:

```
class TEST
feature
    liz    : LINKED_LIST[RECTANGLE];
```

```
            Create is
            do
                    liz.Create;
                    liz.start;
                    Run;
            end;


            Run is
            local
                    r       : RECTANGLE;
                    sq      : SQUARE;
            do
                    r.Create(1.00,2.0,5.0,10.0,"First rect");
                    liz.put_right(r);
                    sq.Create(0.0,0.0,10.5,"First square");
                    liz.put_right(sq);
                    ...
```

and, of course, one can walk a list of RECTANGLES asking them to scale themselves:

```
            from
                    liz.start
            until
                    liz.offright
            loop
                    liz.item.scalex(3.5);
                    liz.forth;
            end;
```

This is legal, but has the effect of causing a SQUARE to rescale itself along its x axis, so violating the class invariant that requires the horizontal and vertical dimensions be identical. The next time that a feature of the SQUARE object is invoked, the invariant will be rechecked, and a run-time exception will be generated.

Really, the fault here is in the program design (and it is possibly somewhat unfair to ask a compiler to check a design). A program could have been designed on the premise that all shapes like rectangles were essentially similar. Given this as a design premise, it would be reasonable to store them in a single collection and have iterative constructs that run through the collection asking each item to perform an operation such as changing its scale in one axis. The addition of a non-scalable shape like the squares, necessitates a revision of the design for the program. The program should have been redesigned to reflect the fact that it now must work with objects that have dissimilar characteristics. There is an opportunity to introduce a new specialized class, SHAPECOLLECTION, that manages shapes. Its implementation could use separate lists of scalable and non-scalable shapes along with associated functions for adding shapes. The SHAPECOLLECTION class would provide functions that deal with scaling etc., – allowing the explicit loops and knowledge about shapes to be factored out of the main program.

Meantime, there is a problem. The compiler isn't catching these errors. There is a simple solution – restrict polymorphic assignments. If a derived class exports only a subset of the export list of a parent (like SQUARE only re-exporting a few of RECTANGLES exports), the compiler could prohibit polymorphic assignments i.e.

SQUARE s; RECTANGLE r; s.Create;...; r := s; would be illegal. Such a policy would result in something very much like C++'s private inheritance with access declarations for those functions that were being re-exported.

Meyer has rejected this approach, arguing that it imposes excessive restrictions. The argument is that it is not inherently wrong to include SQUARE in a collection of RECTANGLES; it is only a problem if the program uses features like scalex on items in such a collection. It is proposed that future versions of an Eiffel compiler might perform system-wide checks that would identify the possibility of such misuse, allowing the compiler to warn of situations where polymorphic assignments could potentially cause problems.

Problems to be solved by type checking complete system

The other "hole" in the type system results from Eiffel's allowing changes in the signatures of routines. The example given earlier had class INVADER with a routine scavenge_parts_from(wrecked : INVADER); the idea being that an INVADER object might refurbish itself by recovering parts from some wreck. Quite reasonably, subclass SAUCER wants to redefine this as scavenge_parts_from(wrecked : SAUCER). After all, a SAUCER should be able to take advantage of any SAUCER specific components of the wreck as well as any bits common to all INVADERS.

Although eminently sensible in terms of the problem domain, the fact that the procedure's signature is changed does raise problems. Generally, a derived class should honour the contract of its parent. It is implicit in the definition of INVADER's version of routine scavenge_parts_from(wrecked : INVADER) that one will get reasonable behaviour whenever it is called with another INVADER as the argument. But class SAUCER is really requiring a stronger precondition (that is illegal, derived classes can weaken preconditions but shouldn't strengthen them); its version of the routine will only work reliably when given a SAUCER to work on. Again, if one prohibited polymorphic assignments, there wouldn't be any real problems. The problems arise through the combination of a change in signature and polymorphism.

A SAUCER can of course be placed in a collection of INVADERS. Then, this collection can be processed in a loop that gets each individual INVADERS to perform its scavenging actions:

```
scrap(x : INVADER) is
do
from
    invaderlist.start
until
    invaderlist.offright
do
    invaderlist.item.scavenge_parts_from(x);
    invaderlist.forth
end; -- scrap
```

Of course, because of dynamic binding, each invader does its scavenging in its own unique way. But, as should be obvious, it is now going to be possible for a SAUCER to try to scavenge saucer parts from an invader of a different type. An error, probably leading to a run-time exception, will occur when the code results in an attempt to access a saucer-specific data field.

Covariance and
contravariance   Eiffel allows "covariant" redefinitions of routine signatures. As one moves
through the class hierarchy to more specific classes, arguments to routines can
similarly become more specific. "Theoretically" this is incorrect; after all it does
result in a routine in a specialized class having an effective precondition that is
stronger than that of the equivalent routine in the parent class. The "correct"
theoretical model allows only "contravariant" redefinitions of routine signatures; as
one moves to more specialized classes, the types allowed for arguments change in
the opposite direction becoming less specialized. It is perfectly correct to allow a
subclass to redefine a signature so that it generalizes the type of argument. Thus,
from a language perspective, it would be correct for class SAUCER to require a more
general type of argument for its scavenge_parts_ from routine. The
generalization of INVADER would be, presumably, ANY; so, a correct contravariant
redefinition rule would permit scavenge_parts_from(wrecked : ANY). While
this might be satisfactory linguistically, it obviously has no useful semantics with
respect to the problem domain.

Again, Meyer argues that here pragmatics should outweigh linguistic purity. A
covariant policy for redefinition does introduce a possible source of error in a badly
designed program. (The design is bad for the same reason as the earlier example;
things with distinct behaviours are being grouped into a collection and then
presumed to have similar behaviours.) But, in general, it allows a form of
specialization that is useful in practice.

13.13 EXAMPLE

This second example Eiffel program utilizes the WINPACK libraries from Eiffel
2.3 and illustrates some conventions for interactive, menu-based Eiffel programs.
WINPACK provides "windowing" facilities on a cursor addressable terminal, in a
manner somewhat analogous to the Unix "curses" library routines. All the
interactive elements in a program will be represented as "windows"; there is
typically a background window, windows for the various individual data elements
that must be displayed, and pop-up windows for menus etc. A "front-to-back"
ordering relation determines the appearance of the final display when windows are
overlaid.

Like most interactive programs, the example makes some use of "command"
objects. There is a deferred class COMMAND in the Eiffel libraries with deferred
methods like execute, undo, and redo; however, as the example does not require
reversible commands it uses its own somewhat simpler structure for command
objects. The approach used to handle menu selection is based on standard Eiffel
examples and illustrates a useful variation on a familiar theme. The menu
structures are arrays of entries combining a selection title and an instance of the
specialized command class that will deal with that menu selection. When an item
is selected, the prototype command object associated with the menu is cloned, and
the newly created command object is given responsibility for processing the
request.

13.13.1 The "assignment" – a chase game

The example program implements a simple version of the game "Dr. Who and the Daleks". This is a form of chase game where the user controls the movements of the quarry ("Dr. Who") and the behaviour of the hunters (the "daleks") is programmed.

The player controls Dr.Who's movements by using the keys [q,a,z,w,s,x,e,d,c] ('q' -> north west, 'a' -> west, 'z' -> south west, 'w' -> north, 's' -> stay put, etc.). There are also options for "teleporting" Dr. Who and of using a "sonic screwdriver" (this incapacitates all daleks within a limited range). The player can also elect to abandon the current game and start a new game, or quit. These special options are provided through a pop-up menu. Teleporting provides no guarantee of safety – Dr. Who may land beside (or even on top of) a dalek. The sonic screwdriver may only be used once in a game. In the first game, Dr. Who need only contend with three daleks; in each successive game, one more dalek joins in (however, there is a maximum of 20 daleks).

After each move by Dr. Who, all the daleks move. They move in a straight line towards Dr. Who's current position. If they collide with him the game is over. If two daleks collide, they are both destroyed leaving wreckage. Anything (Dr. Who or a dalek) that subsequently collides with wreckage is destroyed.

The player's strategy should be to move Dr Who so that the daleks will collide. The doctor cannot "win"; if he survives the attacks of the initial group of daleks, the game continues with an invasion of a larger number of daleks.

The required game display is illustrated in Figure 13.2. Dr. Who's position is represented by the "W"; the daleks are "D"s; wreckage (from collided or sonically-bust daleks) is represented by "#"s. Normally, the pop-up menu is not shown; it should only be displayed when the user presses the space bar (or types any character < space, e.g. return). The program processes the characters [q,a,z,w,s, x,e,d,c] – interpreting these and moving the doctor appropriately (all other character input will simply be ignored).

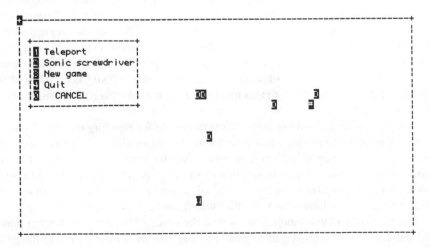

Figure 13.2   Screen display from the example program.

A game ends if the daleks catch the doctor (or if he teleports onto a dalek). The user must then elect to play a new game or quit. If all the daleks are destroyed, Dr. Who can wander around and teleport to his heart's content; when he gets bored, a new game should be selected or the program should be exited.

13.13.2 Finding the objects and their classes

As always, a few of the objects are obvious although relations amongst their classes are still to be clarified. The playing pieces, Dr. Who and the daleks, are all obvious objects, presumably from distinct though closely related classes. Because of the way the WINPACK class cluster works to display information, the classes used for the playing pieces will need to inherit from class WINDOW. (WINPACK WINDOWS show text in a frame of defined size; the frame can itself be shown as a border, and the textual data may be displayed in inverse mode, blinking, with underlines etc.)

The WINPACK class cluster includes class POPUP_MENU[T]; this generic class is a specialized subclass of class WINDOW. The program will use some POPUP_MENU[COMMAND] where COMMAND is the general class for all the different kinds of command object that are needed. Four specialized kinds of command object will be required – one for each of the menu options.

Some kind of "screen manager" object has to initialize the terminal before windows can be created and displayed. Screen management capabilities are defined in class SCREENMAN, another member of the WINPACK cluster. A reasonable design for the overall program would be to have the root class as an instance of a specialized variant of SCREENMAN. This object could i) set up the screen, ii) create some kind of "game manager" object to handle the rules of the game and the maintenance of the playing pieces, and iii) look after low-level interaction with the user (e.g. causing the pop-up menu of options to appear when needed).

A game manager object is required. A single game object can be used; starting a new game will require that the game object re-initialize some of its internal data records etc. The game object will need to own a list of the dalek playing pieces. Obviously, a list class from the library would be used; linked-lists seem pretty well ubiquitous but in this context a FIXED_LIST might be suitable (the maximum length of the collection being known). The game object should keep track of the number of dalek pieces used, so that this can be increased each round until the limit of 20 is reached.

The game object should probably be responsible for building the actual pop-up menu. The command objects associated with the menu will need references to the game object that they affect, so it is reasonable for them to be initialized by the game object in some setup phase prior to the user playing the first round. The game object might also be made a kind of WINDOW; it would then provide the background frame within which all other visual elements are displayed.

The game object will handle the rules of the game. Moves by the human player and daleks should alternate; after every move, checks should be made for game termination and events such as dalek-collisions. The game object will need various local data, e.g. a boolean field to flag whether the "sonic screwdriver" has already been used in the current round. The game object would initialize each round. As the daleks are going to have to be placed randomly, somewhere in the code there will be calls to external C routines like the rand() random number generator declared in Unix's <stdlib.h>.

Figure 13.3 shows the classes that appear to be required, together with some details of their ancestry. Of course, all classes in an Eiffel program form part of a directed acyclic graph that has class ANY at its root; links back to class ANY are not shown in Figure 13.3.

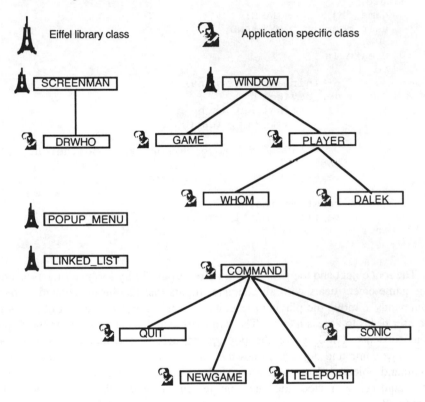

Figure 13.3 Classes used in example program; only the more important inheritance relationships are shown.

13.13.3 Implementation

The root class, DRWHO, inherits from Eiffel's screen manager class SCREENMAN, and also from class VIEWABLE. Class VIEWABLE provides features needed if one wants to use Eiffel's run-time object inspector. When created, the root DRWHO object has first to get the screen into cursor addressable mode; this is handled by one of the inherited procedures from SCREENMAN. Occasionally, there could be a problem in setting terminal modes that would result in an exception; a rescue clause associated with the Create procedure catches the exception (it tries to get the terminal set back to normal and prints a message in routine report_ terminal_problem). If the terminal is correctly set up, the game object is created and then the Run procedure is invoked:

```
class DRWHO
inherit
    SCREENMAN;
    VIEWABLE
feature
    game_obj     : GAME;
    user_command : COMMAND;

    Create is
    do
            -- Initialize terminal in raw mode.
            set_device;
            if terminal_set then
                    game_obj.Create;;
                    Run;
                    game_obj.move_cursor (game_obj.height, 1)
            end;
            reset_device   -- Reset terminal.
    rescue
            reset_device;
            report_terminal_problem
    end; -- Create
```

Cloning a command object from a menu   The root object and the game object share responsibility for handling user input. The game object deals with all character inputs that can be interpreted as being commands to move the playing piece, but returns control to the root object when the user needs to access menus. The Run routine in this class organizes the display (and subsequent hiding) of the pop-up menu; gets the menu selection (as a prototype command object); clones the selection and lets the cloned copy of the command object and the user-requested action. A check is made for a QUIT command being created; the Run routine can terminate when it detects a QUIT command.

```
    Run is
    require
            not game_obj.Void
    local
            last_choice  : COMMAND;
            game_over    : BOOLEAN;
            quit_command : QUIT
    do
            from
                    quit_command.Create(game_obj);
                    game_obj.display;
                    game_obj.move_cursor (1, 1)
            until
                    game_over
            loop
                    -- Ordinary moves
```

```
                        game_obj.move_around;
                        -- deal with a menu command
                        game_obj.prepare_choice_menu;
                        game_obj.display;
                        game_obj.choice_menu.hide;
                        -- get user's choice from menu object
                        user_command :=
                                game_obj.choice_menu.user_answer;
                        if not user_command.Void then
                                -- Command creation and execution
                                last_choice.Clone (user_command);
                                last_choice.execute;
                                if user_command.equal(quit_command) then
                                        game_over := true end
                        end; -- handle command
                        -- get game_obj to update screen
                        game_obj.display
                end
        end; -- Run
```

Obviously COMMAND objects get created in that loop (at last_choice.Clone(
user_command)) but they are never explicitly destroyed. If the garbage collector is
running, any discarded COMMAND objects will eventually get tidied away.

Figure 13.4 presents a collage of views from Eiffel's "good" graphics-based
browser. Graphic views include the obvious displays of class hierarchies (e.g. pane
1 in Figure 13.4) and various displays of client-server relations among classes. As
shown in pane 2, the "universe" of classes in a system can be displayed (i.e. all the
classes used in a program); selecting a class from this list allows a quick change of
focus of the main graphical display. After selecting a class in the main graphics
display, the user can open text windows showing either the full class text or various
more restricted selections of information. The "good" browser" is a convenient
mechanism for exploring the organization of a program, although it is a little slow
in an X-terminal/host environment.

Using Eiffel tools to explore the structure of the program

Class GAME repeats the export list of its parent class WINDOW, adds those of its
features that are needed by the DRWHO root object or the various command objects,
and actually re-exports a feature bell (a feature provided in one of WINDOW's
parents that is not normally exported).

Repeating and modifying parent's export lists

```
    class GAME export
        repeat WINDOW,
        bell, prepare_choice_menu, move_around, choice_menu,
        new_game, move_the_doctor, game_over, sonic_blast,
        sonic_used
    inherit
        WINDOW
    feature
        initial_army : INTEGER is 3;
        game_over    : BOOLEAN;
        sonic_used   : BOOLEAN;
        over_msg     : WINDOW;
        theDoctor    : WHOM;
        army         : LINKED_LIST[DALEK];
```

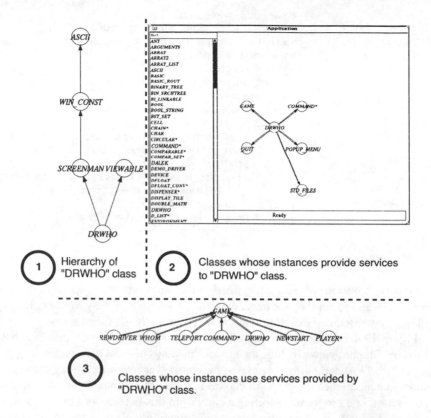

Figure 13.4   Collage of views from Eiffel's "good" graphics-based browser.

The routines in class GAME group into categories: routines used for set up, for handling normal play, for responding to menu commands, and for creating the pop-up menu. The set up routines arrange display dimensions then create and randomly position the playing pieces. Some of these routines need to use C library functions like time() and srand() and so declare external function interfaces.

```
-- SET UP GAME etc
Create is
do
        set_up;
        create_army;
        land_tardis
end; -- Crate

set_up is
        -- Organize the window hierarchy with menus.
do
        set_width (on_device.width - 6);
        set_height (on_device.height -3);
        put_at (2, 2);
        set_frame;
        over_msg.Create();
        over_msg.associate_text("Game Over Man.");
```

```
            over_msg.set_height(3);
            …
            …
            general_menu
    end; -- set_up

    create_army is                                      Calls to external C
            external                                    functions from Unix
                    time(x : INTEGER) : INTEGER language "C";   libraries
                    srandom(seed : INTEGER) language "C";
            local
                    count  : INTEGER;
                    adalek : DALEK;
                    temp   : INTEGER
            do
                    army.Create;
                    temp := time(0);
                    srandom(temp);
                    from
                            count := 0
                    until
                            count = initial_army
                    loop
                            adalek.Create();
                            adalek.initialize(Current,"D");
                            army.put_right(adalek);
                            count := count + 1
                    end; -- loop
            end; -- create_army
```

Routine `move_around` is the driving routine for the actual game code. It reads characters entered by the user; a character like a space or a return is interpreted as a request for the display of the menu (so `move_around` returns). Other characters are passed to the object representing the doctor playing piece for interpretation as movement instructions. The dalek pieces have a chance to move each time the user moves. After every move, it is necessary to check for game termination:

```
    move_around is
            -- deal with standard input, characters that move
            -- doctor,
    local
            key_stroke: INTEGER
    do
            from
                    key_stroke := charcode (getchar);
            until
                    key_stroke <= sp
            loop
                    if not game_over then
                            theDoctor.run(key_stroke);
                            check_the_doctor;
                            end;
                    if not game_over then
                            army_advance;
                            check_the_doctor
```

```
                    end;
                    key_stroke := charcode (getchar)
             end;
     end; -- move_around
```

A check to determine the doctor's safety involves a simple iteration along the list comparing the doctor's position with that of each dalek in turn. Checks for dalek collisions (that are necessary as part of the `army_advance`) code are a bit more complex. Obviously, a double loop is needed to take pairs of entries from the same list. Since Eiffel's list-iterators work with a notional mark position that is part of the list, multiple iterators are inconvenient but can be done.

Handling commands   Command handling code is simple. Thus, procedure `move_the_doctor` (invoked by the execution a `TELEPORT` command object) simply repositions the playing piece (and checks for game terminating conditions):

```
move_the_doctor is
do
        theDoctor.reposition;
        check_the_doctor
end; -- move_the_doctor
```

while a `SONIC` command object invokes another routine that needs simply to iterate along the list of daleks checking their distance from the player's piece:

```
sonic_blast is
    external
            abs(x : INTEGER) : INTEGER language "C"
    local
            adalek : DALEK;
            dh,dv  : INTEGER;
    do
            sonic_used := true;
            from
                    army.start
            until
                    army.offright
            loop
                    adalek := army.item;
                    dh := adalek.xpos - theDoctor.xpos;
                    dh := abs(dh);
                    dv := adalek.ypos - theDoctor.ypos;
                    dv := abs(dv);
                    if (not adalek.dead) and
                            (dh <= 3) and (dv <= 3)
                    then
                            adalek.is_exterminated;
                    end;
                    army.forth;
            end;
    end; -- sonic_blast
```

A pop-up menu is filled in by creating the various command objects that will handle menu selections, associating them with appropriate text strings, and adding

them to the menu structure. The POPUP_MENU class automatically provides an extra
"cancel" entry in the final menu.

```
choice_menu: POPUP_MENU [COMMAND];
```

*Building a menu
containing prototype
command objects*

```
general_menu is
        -- Set up menu.
local
        quit_state    : QUIT;
        aretry        : NEWSTART;
        jump          : TELEPORT;
        screw  : SCREWDRIVER
do
        quit_state.Create(Current);
        aretry.Create(Current);
        …
        choice_menu.Create (Current);
        choice_menu.associate ("Teleport", jump);
        …
        choice_menu.associate ("Quit", quit_state)
end; -- general_menu
```

The deferred class COMMAND exists as an abstract generalization of the specific
command types only so that it will be possible to create the POPUP_MENU [
COMMAND]. All kinds of COMMAND will need an execute routine and a link to the
GAME object, so at least these features can be specified. Of course, there is no need
for reversibility in these commands so there is no suggestion of undo, redo etc.

Command classes

```
deferred class COMMAND export
    execute
feature
    target: GAME;

    execute is
            deferred
            end; -- execute

end -- class COMMAND
```

The different specialized commands provide their specific implementations of
execute (and Create procedures, although the Create procedures are all
essentially identical it isn't possible to define Create in class COMMAND as the
Create instruction does not use "inherited" Create procedures).

```
class NEWSTART export
    execute
inherit
    COMMAND
feature
    Create(g : GAME) is
            do
                    target := g
    end; -- Create
```

```
        execute is
                -- Start a new game
                do
                        target.new_game
                end; -- execute

end -- class NEWSTART
```

A deferred class   A "deferred" class PLAYER can capture commonalities in the behaviour of the two kinds of playing pieces. All playing pieces are little "windows" that display a character string, they all respond to requests to randomly reposition themselves, they can be "exterminated" etc. The initialization routine sets up the window parts of a PLAYER object; in addition to setting the dimensions, a "priority" is set – this really defines the front to back ordering of windows.

```
deferred class PLAYER
export
    repeat WINDOW,
    initialize, reposition, is_exterminated, dead
inherit
    WINDOW
feature
    target : GAME;
    dead           : BOOLEAN;
    rname   : STRING;

    is_exterminated is
            do
                    dead := true;
                    associate_text("#")
            end; -- is exterminated

    initialize(mytarget : GAME;ident : STRING) is
            do
                    target := mytarget;
                    attach(target);   -- "make subwindow of …"
                    set_height(1);
                    …
                    set_priority(2);
                    …
            end; -- initialize

    reposition is
            external
                    random: INTEGER language "C";
                    abs(x : INTEGER) : INTEGER language "C"
            local
                    h,v   : INTEGER;
            do
                    h := random;
                    h := abs(h);
                    h := h mod ( target.width - 1 );
                    …
                    show;
```

```
                        dead := false
              end; -- reposition
end -- class PLAYER
```

Since class PLAYER does not specify any routines as being deferred, it is not strictly a deferred class.

Classes DALEK and WHOM specialize class PLAYER to provide the two kinds of playing piece. Class DALEK includes an advance routine (that takes as input the coordinates of the "doctor" target); its execution causes the DALEK object to move directly toward the target. Class WHOM has a movement routine that selects the direction in accord with the character data entered by the user. This is one place where Eiffel's inspect multi-way branch construct can reasonably be employed:

```
class WHOM
export
    repeat PLAYER,
    un
inherit
    PLAYER
feature

    run(intcode : INTEGER) is
          local
                  dh,dv : INTEGER
          do
                  if dead then bell
                  else
                  inspect
                          intcode
                  when lower_q then
                          dh := -1; dv := -1
                  when lower_a then
                  ...
                  ...
                  else
                          dh := 0; dv := 0
                  end;

                  if dh < 0 then
                          if xpos = 1 then dh := 0 end
                  else
                          ...
                  end;
                  ...
                  end;

                  end -- if dead tests
          end;
end -- class WHOM
```

"Inspect" multi-way branch

The code provided in the files has full definitions for all classes.

13.14 POSTSCRIPT

The future of Eiffel is uncertain. On the positive side, there is an international committee promoting the development of the language. This committee (the Non-profit International Consortium for Eiffel) has a subcommittee focussing on the libraries as well as one for the language itself. This will be a benefit because standard libraries are really as important as the actual OO language if software reuse is to become a reality. There are a number of large scale projects under the EEC ESPIRIT research umbrella that use Eiffel. Several research groups are exploring language extensions such as concurrency (see for example, Communications of the ACM, September 1993). Some additional compilers are in beta test, coming from companies such as Tower Technology; while the Eiffel/S system for PCs is being improved. (Suppliers like Tower and SIG are actually cooperating over the definition of library classes that will be supplied with their systems.) ISE has begun deliveries of its Eiffel 3 development environment. If this proves to be as far ahead of its rivals as was the original Eiffel 2 of ≈1988, then Eiffel's fortunes might revive.

Meantime, there is a continuing loss of interest in industry. (Eiffel does get used on real applications – there is even a recent book to prove this.) Currently, there seems to be a possibility that either Eiffel will die, or will suffer a fate worse than death by becoming the plaything of academic computer science departments.

References   B. Meyer, *Object-Oriented Software Construction*, Prentice Hall, 1988.
B. Meyer, *Eiffel: The Language* Prentice Hall, 1992.
B. Meyer, and J.M. Nerson (Eds), *Object-Oriented Applications*, Prentice Hall, 1993.
R. Switzer, *Eiffel: An Introduction*, Prentice Hall, 1993.

Index